Boston Studies in the Philosophy and History of Science

Volume 325

More information about this series at http://www.springer.com/series/5710

Nimrod Bar-Am • Stefano Gattei
Editors

Encouraging Openness

Essays for Joseph Agassi on the Occasion of His 90th Birthday

Springer

Editors
Nimrod Bar-Am
Communication and Mass Media Department
Sapir College
Shaar Hanegev, Israel

Stefano Gattei
California Institute of Technology
Pasadena, California, USA

ISSN 0068-0346 ISSN 2214-7942 (electronic)
Boston Studies in the Philosophy and History of Science
ISBN 978-3-319-57668-8 ISBN 978-3-319-57669-5 (eBook)
DOI 10.1007/978-3-319-57669-5

Library of Congress Control Number: 2017943335

Printed on acid-free paper

This Springer imprint is published by Springer Nature
The registered company is Springer International Publishing AG
The registered company address is: Gewerbestrasse 11, 6330 Cham, Switzerland

Joseph Agassi (Photo by Charles Sawyer)

Preface

Karl Popper, the father of critical rationalism, noted that all schools come to defend dogmas. The dogma shared by critical rationalists is that intellectual progress comes from undermining dogmas, wherever they are found. Rationality is in criticism not in proof. A school, Popper emphasized, is rational to the extent that it allows transcending its own dogmas. Accordingly, we critical rationalists are neither a family nor a tribe. And the extent to which we should form a cohesive community is continuously under debate: it is our enduring *lebensfrage*, one that we must constantly nurture and face. For communities are formed, cradled, and sustained by networks of dogmas and commitments: by blood ties turned allegiances (and vice versa) and by psychobiological propensities turned culture (and vice versa). Dogmatic commitments of this kind are rarely, if ever, easily reconcilable with the cultivation of rationality as our individual and institutional ideals. We discover and progress by critically reflecting upon our commitments. Yet we often commune and cooperate by virtue of our precritical internalization of our culture and our shared habits and rituals and sometimes by sheer suspension of our disbeliefs (perhaps even of our ideals). Both tacit commitments and their public criticism, then, are required for our progress. The pursuit of an open and yet cohesive community is thus a fragile and sometimes paradoxical existential and institutional quest.

Many of the merits and shortcomings of the papers that have been assembled here are easily traceable to the fragility of this quest of ours. Among the merits, it is a pleasure to observe the playful freedom with which some of the writers of this volume pay tribute to Agassi by openly criticizing his work. This is unprecedented in festschrifts honoring a highly esteemed and dearly beloved teacher and colleague. Their bold assuredness (yes, this combination, too, is paradoxical) that in doing so they show him the highest honor strikes us as among the greatest compliments that a critical rationalist could ever hope to receive. Agassi has earned it fair and square: he is easily the most devoted critical rationalist around. It is by virtue of that resourceful and compassionate steadfastness that he has so significantly contributed to the intellectual and emotional development of so many of us assembled here. Among the shortcomings of this volume, traceable to the fragility of our quest for a community of autonomous individuals, let us mention the academic loneliness that some of the bibliographical lists included here may reveal to the keen-eyed observer. As a community of learners, we have clearly not yet found the right balance between individual autonomy and cohesiveness. Let us hope that our next festschrift will exhibit improvement in this respect.

This brings us to say a word about our policy as editors of this volume: it was intentionally minimal. As the Zen saying goes, we tried neither to shorten the leg of the stork nor to extend the leg of the squirrel. We invited esteemed colleagues and precious friends to an almost spontaneous intellectual jam session with Agassi's work as a suggested (but by no means required) theme. The concept we had in mind stems from an experience that many of us who have been fortunate enough to work with him share. You wrestle with a problem. It can be intellectual, personal, or somewhere in between. You wish to discuss it with the most brilliant (and most available) person around. You call Agassi. He instantly invites you to come over. You arrive, and the most engaging and instructive conversation that you could hope for ensues. Then the phone rings. Someone else is grappling with a difficulty. It too is burning. Agassi invites her over. Before you know it, you are surrounded by a group of people of markedly different interests and talents, all of whom eagerly engrossed in the stimulating conversation. Some offer you good advice. Some seem to be in need of your help. Judith and Joske serve coffee and cake and join in the conversation with their ever-abundant wisdom and grace, encouraging openness. Your frustration at not working out the problem through and through wanes: you can always stay up last or come again tomorrow. For the most burning problem always takes precedence, and luckily it is not always yours. In any case, no meaningful discussion is ever truly complete. In the meantime you've made good progress and had the most exciting of experiences. Perhaps you've even made some friends along the way and were able to make a small difference. That's all there is to it, really. That is critical rationalism at work.

February 2017

Nimrod Bar-Am
Stefano Gattei

Joseph Agassi with the editors, May 2008
Nimrod Bar-Am (*left*), Stefano Gattei (*right*)
(Photo by Gali Drucker Bar-Am)

Contents

Part I
Reflections on Rationalism as a Tradition and a Choice

Chapter 1
Why Don't Scientists Respect Philosophers?

Mario Bunge

1.1 The Hume Cult

An opinion poll published in 2013 showed that David Hume is nowadays the philosophers' favorite philosopher of all times. Hume himself would be surprised, since only one of his many books, namely his *Enquiry* (1748), was properly philosophical. Moreover, that work was neither successful nor very original, since it only elaborated in great detail the scholastic principle *Nihil est in intellectu quod prius non fuerit in sensu.* Science from physics to biology to historiography has bypassed this principle, and Hume realized this, as he had the nerve to attack Newtonian mechanics despite lacking the mathematics required to understand it.

Hume disliked Newtonian mechanics for containing such nonempirical concepts as those of mass and action at a distance, and for formalizing such counter-intuitive models, as the heliocentric view of our solar system.

As a cynic noted long ago, philosophy is where science goes to die. Indeed, Hume's empiricism, which played no role in the making of modern science, exerted an enormous influence on philosophy. Suffice it to recall Berkeley, Kant, and their two main offsprings: positivism from Comte and Mill to Mach and the logical positivists – phenomenalists all of them and therefore allergic, at last initially, to the very idea of matter, in particular imperceptible matter, be it corpuscular or field-like.

Around 1900 positivism became so ingrained in the teaching profession, that even atomic physicists, evolutionary biologists, psychologists and historians parroted a scientific methodology which they did not practice, namely the one according to which one always starts by making observations, and proceeds to condensing their results into inductive generalizations, that are subjected to tests in order to attain certainty.

M. Bunge (✉)
McGill University, Montreal, QC, Canada
e-mail: marucho.bunge@gmail.com

N. Bar-Am, S. Gattei (eds.), *Encouraging Openness*, Boston Studies in the Philosophy and History of Science 325, DOI 10.1007/978-3-319-57669-5_1

True, Poincaré and Popper had criticized inductivism using historical counterexamples, but they did not explain why it is wrong, did not say where hypotheses come from, and did not analyze any experiments except for some imaginary ones. Moreover, just like his rivals, Popper regarded experiment as the umpire, albeit not an infallible one.

So, which is a truer account of science? Let us peek at only a few traits of interest to philosophy: the genesis of hypotheses, the role of the prevailing worldview, the indicators involved in every measurement, and the battery of indicators of factual truth.

1.2 Genesis of Scientific Hypotheses

Nearly everyone agrees that we should distinguish several kinds of scientific hypotheses: ordinary empirical generalizations, scientific empirical generalizations, low-level theoretical statement, high-level ones, philosophical principles, and wild speculations. In this paper we shall confine ourselves to exemplifying these kinds and making brief comments on them. Here are some examples:

Ordinary empirical generalization: "All adult domestic dogs can bark."
Scientific empirical generalization: Galileo's law of free fall.
Low-level theoretical statement: Huygens' pendulum law.
High-level theoretical statement: Newton's laws of motion.
Philosophical principle: All material things are changeable.
Wild speculation: There are ideas outside human brains, and some of them – surely mine – will outlast humankind because they are incorporated into books, disks, pictures, and other lasting inmates of the *Weltgeist*, or World Three.

Theoretical physics includes extremely general statements, such as the Newton-Euler laws. However, even these proved to be limited, and were eventually generalized for generalized coordinates, which can be interpreted in non-mechanical terms, and thus utilizable in electrodynamics and even in economic vocabulary. The resulting equations are the Euler-Lagrange ones, which in turn are derivable from the variational principle about the system's lagrangian or action L, which states that the integral of L between any two instants is either a maximum or a minimum.

Hamilton's and similar variational principles are of particular philosophical interest for the following reasons. First, they are too far removed from measurements to be directly testable. In other words, they are empirically *untestable*, yet at the same time the pinnacles of theoretical physics, for they entail all the other general law statements.

Second, the earliest variational principle, namely the Maupertuis-Euler principle of least action (1744), seems to have been the formalization of the metaphysico-theological principle that "nature is thrifty," as Maupertuis himself put it, or of the teleological assumption that every change is goal-directed. When Hamilton rewrote it in 1834, the principle lost all traces of its birth.

The point I wish to make is that some powerful scientific ideas are rooted not only in the extant body of scientific knowledge, but also in some of the rather obscure ideas of some worldview or other.

Another, much more important, instance of the role of worldviews in the generation of scientific ideas and research projects is the materialist principle that everything mental is cerebral. This metaphysical hypothesis, first stated by Alcmaeon (500-450 BCE), and which Hippocrates and Galen adopted, is no less than the spinal column of cognitive neuroscience, the contemporary phase of psychology.

An even more important case is that of Lucretius' principle of universal conservation: *Ex nihilo nihil fit* – nothing comes out of nothingness. Although Lucretius and the other ancient materialists were banned by all the theocracies since Justinian, that broad metaphysical principle resurfaced in the French Enlightenment.

Historians of science seem to agree that Lucretius' materialist poem *On the nature of things* inspired the many parents of the principle of conservation of energy, in particular the medic Julius Robert Mayer (1845) and the brewer James Prescott Joule (1843). This principle is not just one more hypothesis, but the first axiom of thermodynamics and a charter member of the modern *Weltanschauung*. One of its consequences, the impossibility of perpetual motion, is so important to technology that the employees of patent offices spend much of their time examining the designs of the fake perpetual motion machines proposed by ingenious inventors who share Popper's belief that, since energy conservation is hypothetical, it may eventually be confuted.

1.3 External Consistency

Everyone except the Hegelians and some postmodernists admit that internal consistency is a must for any theory. How about external consistency, or compatibility with the bulk of received opinions, or even with the main postulates of the prevailing worldview? Let us recall three famous cases in twentieth century physics: those of beta decay, steady-state cosmology, and multiverse cosmology.

Careful radioactivity measurements performed in 1911 seemed to show that beta radioactivity violated energy conservation: the energy of the products of a radioactive decay involving the emission of an electron seemed to be less than the energy of the input. In 1933, the great physicist Enrico Fermi suggested that an unknown particle carried away the missing energy. Because the hypothetical particle lacked both mass and electric charge, he called it a *neutrino*. But, because it was massless and electrically neutral, the neutrino defied all detection attempts until 1956.

Because collisions involving neutrinos are rare due to their small energy, most neutrino detectors are gigantic. So far, the largest is the IceCube, that occupies one cubic kilometer and is situated near the South Pole, and started to work in 2004. So much for a useless particle invented with the sole goal of saving energy conservation.

Second example: in 1948 the cosmologists Hermann Bondi, Thomas Gold and Fred Hoyle, parents of the steady-state cosmological theory, attempted to save its "perfect cosmological principle" from the standard interpretation of observed red-shifts, namely the expansion of the universe – or, more exactly, the increase in inter-galactic distances, since the predicate "expansion" makes sense only for finite things, and, so far as we know, the universe may be spatially infinite. The theory attracted the attention of those who disliked the Big Bang hypothesis, which suggested that the universe had a beginning – anathema for the prevailing secular and materialist worldview.

The cosmologists in question attempted to save their theory by adding to it the ad hoc hypothesis that matter was gushing out of nothing at the rate needed to balance the decrease in overall mass density accompanying the expansion. Of course, there is nothing wrong about ad hocness as long as it is independently testable. In my paper "Cosmology and magic," published in *The Monist*, in1962, I criticized harshly this conjecture because it contradicts all the conservation laws accepted in physics, starting with the conservation of energy.

This exercise led me to add external consistency to my list of scientificity criteria. Karl Popper reacted angrily, arguing that the conservation principles are just hypotheses, hence subject to confutation: he did not admit the idea that firmly entrenched hypotheses deserve special treatment. Besides, Bondi was his friend. When I met Bondi in 1965 at Bedford College, we chatted cordially. By then the continuous creation hypothesis had been buried.

My third and last argument for the external consistency criterion is the battle over the mind, which has been going on over three millennia. Though published in 1977, the Popper & Eccles volume *The Self and its Brain* does not contain a single reference to papers in cognitive neuroscience. This was quite a feat, since the earliest victories of this discipline go back to Philippe Pinel, the founder of scientific psychiatry (1793), and Paul Broca's seminal paper of 1861 on the localization of speech production in the left cerebral cortex (Broca's area).

Philippe Pinel's philosophical idea that mental diseases are brain diseases, and consequently mental patients should be treated just as humanely as any other sick people, was put into practice nearly half a century before Broca's paper. Besides, Pinel's work inspired the 1838 French law enforcing the "moral" (humane) treatment of mental patients. In earlier times these had been chained, beaten, and sprayed with cold water. (This brutal tradition remains in the name *loco*, Spanish for 'insane', given in Chile and Perú to certain prized marine molluscs that have to be softened by beatings before being cooked.)

In any event, the above-mentioned book by Popper and Eccles appeared more than a century too late, and only to support an unscientific and moth-eaten view of the mind. Moral: Tell me which scientific advances your philosophy favors or hinders, and I will tell you its worth.

1.4 Indicators

The main problem that is currently being tackled by astrobiologists is to identify which if any of the thousands of recently discovered planets are homes to living beings. Given that those exoplanets are trillions of kilometers away, that is a daunting task involving biosignatures, or life indicators, in otherwise unknown planetary atmospheres. Prima facie, water would be a good candidate, but it would be hard to find because sunlight can decompose water, thus freeing its hydrogen, which could easily escape into space.

Therefore, only more stable gases would do as reliable biosignatures. Stay tuned.

Francis Bacon, usually presented as a crude empiricist, intuited that the oscillation period of a pendulum, or of a pendulum clock, depends on the gravitational pull, and that in turn the latter is the weaker, the greater the height. To test this hypothesis, he proposed to compare the time shown by a pendulum clock raised to the highest church spire in town with that shown by a similar clock bound to the ground.

This experimental design was ingenious, but hard to implement with the coarse instruments available at the time. The same experiment was tried again two centuries later, and it confirmed Bacon's hunch: the gravitational pull decreases with the distance from the planet, a fact that settled the age-long controversy over whether weight is an intrinsic or a relational property of bodies.

Bacon's ingenious albeit failed experiment was only an episode in the centuries-long saga of indicators – also called signs, markers, or proxies. This story that started with the search for reliable indicators of time intervals, weight, specific gravity, time, viscosity, acidity, sweetness, and other properties. We now have precision anemometers, chronometers, scales, galvanometers, spectrographs, pH meters, Geiger counters, and other physical and chemical contraptions, far more precise than the ancient sand-clock, aqua regia, and litmus paper. All science students become familiar with indicators in their lab practices.

Physiologists and psychologists too have invented a number of quantitative indicators. For example, Pavlov used the amount of saliva secreted by his dogs at the sight of a meat morsel as an indicator of their hunger. And the pupil's contractions and dilations have been used routinely as hate and love indicators, and a commercial eye tracker is routinely used to find out the point of gaze, and thus the subject's movement intentions.

Archaeologists and anthropologists have been using indicators, such as tools, earthworks and drawings, ever since our species was baptized *Homo faber* (working man). The alternatives *Homo sapiens* (knowing), *adorans* (worshiping), *ludens* (playing) and *loquens* (speaking) have also been proposed, possibly under the influence of idealist philosophers. But all the indicators of learning, worship, play and speech are material artifacts such as flint quarries, harpoons, totem poles, and knuckle bones used as dice.

Even the believers in that speech is what makes us unique, should grant that the only evidence for this conjecture is what is also evidence for cooperative actions, such as remains of defense earthworks, communal buildings, whaling, and big-game hunting, that involve social organization, discussion and planning, all which in turn require speech more sophisticated than grunting. But glossocentrists dismiss sociolinguistics: they focus on syntax, forgetting that sentences indicate or represent ideas, and that the main function of speech is communication.

Since about 1960, a number of new social indicators have been devised for studying or assessing contemporary societies. The best known of them is the UN human development indicator (1990), far more reliable than the GDP introduced in the 1930s. Since 1974 there has been a whole journal devoted to social indicators research. Yet, the very notion of an indicator or marker does not occur in the most popular philosophies of science, or even in theoretical physics textbooks. Take, for instance, Dirac's *Principles of Quantum Mechanics* (1930), which in my time was regarded as the bible of quantum mechanics, but was seldom used for teaching because it reflects its author's proverbial laconism.

That work postulates that the eigenvalues of "observables" (dynamical variables) are the values that their measurement may yield. This assertion comes from the operationist philosophy tacitly accepted uncritically by the faithful of the Copenhagen school, but is at variance with experimental physics, as will be seen anon.

Consider, for instance, the hamiltonian or energy operator H of a quantum-mechanical object such as an atom. H and the state function Ψ occur jointly in the basic formula (axiom) of quantum mechanics, namely $i\hbar\partial\Psi/\partial t = H\Psi$, where i is the imaginary unit and $\hbar$ the Planck constant divided by 2π. Of the three variables, only t, interpreted as time, is measurable – though of course not directly but through some indicator such as the angle swept by the needle of a chronometer.

As for the eigenvalues E_i of H, they are not measurable either; only the difference between the energy levels E_i and E_j is measurable, for it equals the energy of the radiation emitted or absorbed during the $i \rightarrow j$ quantum jump. But, of course, this measurement too is indirect: one uses a comparator to measure the corresponding wavelengths. As for Ψ, every eigenvalue measurement is also a test of the corresponding state function, likely to be the most elusive of physical magnitudes.

The following diagram supplements and summarizes the preceding discussion.

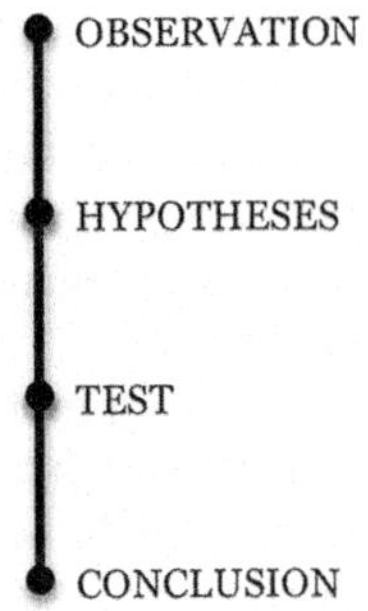

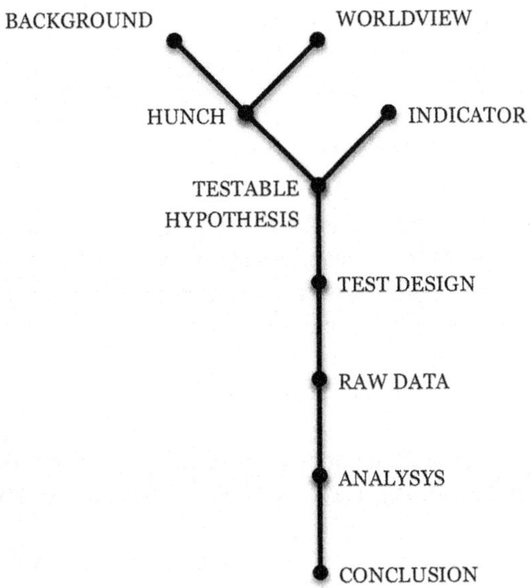

How does the external consistency criterion work in microphysics? It works through what Bohr called the *correspondence principle*. According to it, the quantum-theoretical results must approach the corresponding classical formulas for large quantum numbers. The problem with this condition is that most quantum-theoretical formulas have no classical counterparts. Indeed, recall that QM was invented because classical physics does not even allow for the existence of atoms.

This hitch may be overcome by rewording the said principle, stating that the quantum-theoretical formulas must not contradict the classical ones – should they exist. This would apply to such macrophysical objects as blackbody objects (e.g., ovens), liquid bodies at "normal" temperatures, and even grandmothers.

Bohr had stated, tongue in cheek, that you could write a quantum-theoretical Hamiltonian for your grandmother (no mean feat!) by first writing its classical hamiltonian and then replacing every classical dynamical variable in it with its quantum-mechanical counterpart. However, some classical variables, such as viscosity, have no quantum partners, and some quantum variables, like spin, lack faithful classical partners.

So, let us grant that the said problem persists and that, given its importance, it warrants devoting it some research projects. Incidentally, this is one of the functions for a philosophy in touch with the science of its time: to detect holes in the body of extant scientific knowledge.

Unfortunately, the converse is usually found, namely big philosophical holes, such as a social philosophy capable of helping economists going forward instead of assuring them that they have been on the right track since about 1870, while in fact most of them have been complicit with those who generated some of the worst economic bubbles and their subsequent bursts.

1.5 What About Induction?

Where do the previous consideration leave induction, a component of the standard or empiricist view of scientific research? The short answer is that induction is still being used in the so-called soft sciences, like anthropology, but is useless in the hard sciences. Induction has a limited use even in biomedical research, where receptors, genetic editing and immunity factors must be imagined because they do not occur among the empirical data, or rather peta (or sought), as I prefer to call them, since they are not given.

I have just suggested that induction has a role to play in finding and testing low-level scientific generalizations, such as Ohm's law and the fever-inflammation connection.

But to understand and perfect such inductions we need higher-level generalizations, that is, law statements that do not contain any of the predicates occurring in the descriptions of empirical data or peta. For example, we explain Ohm's law, along with the other laws concerning electric and electronic circuits, in terms of electrons dragged by an electric field, and bent by a magnetic one.

Let us take a quick look at the equations for the electromagnetic field that James Clerk Maxwell guessed around 1860. The primitive (undefined) concepts are those of electric current density and electric and magnetic field intensities. They occur in Mawell's six differential equations for the field, and one for the conservation of electric charge. None of the seven basic magnitudes is directly measurable. In particular, we measure the force $F = eE$ that an electric field of intensity E exerts on an electric charge e. The force exerted by a magnetic field on a charge is $e\ (v \times B)$, where v represents the velocity of the particle divided by the speed of light in a vacuum, and B for the magnetic induction. None of these magnitudes is directly measurable. The measurement of each of these magnitudes calls for its own experimental device. For example, the electric charge on an electron was measured by Robert Millikan only as recently as in 1909.

Maxwell's equations are very beautiful but mathematically hard to handle. To solve special cases one transforms them into second-order equations for the potentials ϕ and A defined tacitly by $E = -\nabla\phi - (1/c)\partial A/\partial t$, $B = \nabla \times A$. The resulting equations are $\nabla^2 A_\mu - (1/c^2)\partial^2 A_\mu = (4\pi/c)j_\mu$, where μ varies between 0 and 3. These four equations are testable only if enriched with the four formulas for the components of the four-density current j_μ, or field source. The reader may wish to draw the corresponding deductive tree, starting with the preceding equations, then conjoining them with formulas for a specific current density, conjoining these with indicators, and ending with the testable consequences.

The theory of antennas, which belongs in the intersection of science with communications engineering, is centered in these formulas. Warning: although the formulas look straightforward, they are part of the solution to a problem that can be made as hard as desired. Indeed, the electrical engineer's task is to figure out the form of the current density corresponding to the desired shape, intensity and direction of the electromagnetic waves represented by a solution of those field equations. In short, the engineer is asked to solve an inverse problem of the form: Given the desired output, find the required input. For example, given the desired magnetic induction B, find the potential A such that $B = \nabla \times A$. Tip: this inverse problem has no general solution. One must use trial and error for every particular field configuration. Regrettably, the very notion of an inverse problem is alien to the vast majority of philosophers.

Finally, because of the operationalist confusion between testability and meaning, sanctified by Rudolf Carnap in his famous, if somewhat pathetic, 1936 paper with the same title, it is usually said that the electrodynamic potentials are just mathematical auxiliaries devoid of physical meanings. But the equations they obey are equivalent to Maxwell's, which are central to classical electrodynamics.

Besides, A occurs in the so-called kinetic momentum $\pi = p - (e/c)A$ of a charged particle in an electromagnetic field. That is, such a particle is accelerated when immersed in a field A. This occurs even when B vanishes while A does not. In other words, the presence of A has a measurable effect even when its curl is zero. This fact, puzzling at first sight, is called the Aharonov-Bohm effect.

1.6 Conclusion

The standard philosophies of science are hardly of interest to the practicing scientist or technologist because they do not detect, let alone analyze, most of the philosophical problems raised by scientific or technological research. In particular, logical positivism failed because it kept phenomenalism and focused on low-level generalizations. And logical negativism, that is, Popper's refutationism, ignores that Nobel prizes reward discoveries or inventions, not confutations of myths.

More importantly, neither of the best-known philosophies of science has inspired any cognitive breakthroughs, partly because neither of them contains an ontology or metaphysics, and so neither of them belongs in the worldview that has inspired scientists and technologists since the Scientific Revolution.

The good news is that, if we grant that there is a big hole in our cognitive net, we may do something to mend it, such as hanging a realistic epistemology from a materialist and systemic ontological hook, as suggested in my book *Doing Science* (2017). After all, all the factual sciences study the concrete things, us included, found in the real world, provided we succeed in piercing the thick fog of phenomena that blankets it.

Chapter 2
Evaluative Selection in Multimerit Situations

Nicholas Rescher

Throughout his long career, Joseph Agassi has stressed the importance of rationality in matters of cognition and the guidance of action. This topic has accordingly seemed to me to be well suited for a Festschrift in appreciation of his contribution to our subject.

Rationality is a paramount value: there is no better advice than to urge people to do the rational thing. But to delineate this and determine just exactly where rational optimality lies in a particular case can prove to be a task of considerable theoretical and practical difficulty. How this is so is vividly illustrated in matters of rational selection and choice.

Decision theory as it has evolved in recent decades standardly deals with issues of rational decision and choice in conditions of uncertainty. However, quite interesting and challenging theoretical problems can already arise with selections made under conditions of certainty, where definitive objects of choice are at issue. It is such choices under circumstances of all-in information regarding the objects available for selection that will be are issue here.[1] We shall accordingly be concerned with the choice between definite objects as such, and not with such abstractions as "a 50% chance of getting" an object. After all, when shopping for a car, people set out to buy one definite sort of vehicle or another: they are simply not prepared to conjure with probabilistically varying choices of obtaining different sorts of vehicles. One is not going to persuade them to fill in the blanks in a statements like "I am going to pay $X to get a 50% chance of car *A* and a 50% chance of car *B*;

[1] One of the few studies dealing with the sort of cases examined paper is Ralph Keeney and Howard Raiffa, *Decisions with Multiple Objectives* (New York, etc.: John Wiley, 1976). However its close gearing to classical utility theory soon enmeshes this book in a focus on risks and probabilities that leads it outside the domain of evaluation *per se*.

N. Rescher (✉)
Distinguished University Professor of Philosophy, University of Pittsburgh, Pittsburgh, USA
e-mail: rescher@pitt.edu

N. Bar-Am, S. Gattei (eds.), *Encouraging Openness*, Boston Studies in the Philosophy and History of Science 325, DOI 10.1007/978-3-319-57669-5_2

whereas I am going to pay only \$Y for a 40% chance of *A* and a 60% chance of *B*." In the ordinary course of things, people just don't think in these terms.

Suppose, then, that you face the choice among several alternative items, and let it be that the items at issue exhibit various different modes of merit, as with domiciles in point of desirable location, physical condition, and layout convenience. And to simplify the problem—and perhaps even to oversimplify it—let us take a quantitative approach to evaluation. In assessing the status of a factor one might accordingly use a zero-to-ten scale along the lines of Display 1.

Display 1
Evaluative gradations

10	altogether satisfactory
9	highly meritorious: strongly positive
8	superior: rather positive
5	average: middlingly meritorious
2	inferior: rather negative
1	strongly negative: very low in merit
0	totally unacceptable

This coordination of quality with quantity implements the instructive and advantageous verbal shift made by contrasting *evaluation* with *valuation*, subject to the idea that only when numbers are introduced do we have the latter—i.e., valuation.

Suppose, then, that we are looking to rent an apartment, and let it be that within the range of affordable alternatives three factors are of paramount concern: size, lay-out, and convenience of location to be represented by m_1, m_2, m_3 respectively. And let us further suppose that three alternatives (*A, B, C*) stand before us, as follows:

	A	*B*	*C*
m_1	3	4	7
m_2	8	5	4
m_3	4	8	9

In situations of this sort, comparative evaluation is automatic *within* rows simply in virtue of the way in which those numbers are assigned. However, column comparisons are another matter altogether. Thus in renting a domicile, proximity to one's workplace (near, rather distant, quite far) will be one key consideration, as will the modernity of the kitchen ("state of the art," modern, dated). So while comparison, *within* these dimensions is straightforward, preferential comparisons *among* them can be rather a puzzle. Just this issue of determining the role of various factors in effecting an overall comparison constitutes a paramount challenge to rational evaluation.

The question of the way in which different factors play off against each other in evaluation renders this project a complex and challenging issue.[2] Consider an automobile. Here the parameters of merit clearly includes such factors as speed, reliability, repair infrequency, safety, operating economy, aesthetic appearance, road-handling ability. But in actual practice such features are interrelated and often actually conflict with one another.

What might be termed *merit complementarity* arises whenever different sorts of merit stand in a teeter-totter relationship it is clearly inevitable that they cannot both achieve a maximal degree at one and the same time. This sort of situation is a clear indication that the idea of absolute perfection is simply inapplicable and inappropriate in many evaluative situations. Thus with a car one cannot maximize both *safety* and *economy of operation* (which demands lightness of weight). Accordingly, an automobile—however splendid—is not, and cannot possibly be, absolutely perfect—perfect in *every* respect—because this sort of absolute perfection at issue with the concurrent maximization in every relevant positivity is simply unavailable. All that one can ever reasonably ask for is an auspicious *combination* of values—an overall optimal *profile*.

But how is such an optimal combination to be determined?

It might seem on casual thought that a process of *weighted averaging* is all that is ever needed. But this idea is not workable. For there is sometimes just no way of assigning weights that will give the right answers to problems of multi-factor evaluation. This circumstance is manifested by so simple a case as that of the following:

Display 2
Assessment methods
Dominance. The value of the whole is the value of that one of its aspects that has priority.
Maximum. The value of the whole is the value of its highest-value aspect.
Simple Average. The value of the whole is the average of the values of its aspects.
Weighted Average. The value of the whole is the appropriately weighted average of the values of its aspects.
Product Maximization. The value of the whole is the product of the values of its aspects.
Maximin: The value of the whole is that of its weakest link: the lowest of its aspectival values.
Average Excess: The value of the whole is the sum of all those aspectival values that exceed the average.

[2] It is not that economists have not recognized the problem, but just that they have no workable handle on its solution. As Kelvin J. Lancaster put it in one of the few attempts "we shall make some assumptions which are, on balance, neither more nor less heroic than those made elsewhere in our present economic theories." (See his "A new Approach to Consumer Theory," *The Journal of Political Economy*, vol. 74 (1966), pp. 132–57; see p. 135.) As an admission of defeat, this ranks close to the Emperor Hirohito's statement that "the war situation has developed not necessarily to Japan's advantage."

	A	B	C
m_1	8	2	3
m_2	2	8	3

Let us suppose here that for an alternative to be acceptable both the m_1-rating and the m_2-rating must exceed 2. But while C is now the only acceptable option nevertheless there is no possible assignment of weights that can yield this result. For this would require having weights x and y for m_1 and m_2 respectively such that both

$$8x + 2y < 3x + 3y \left(\text{or equivalently } 5x < y\right)$$

and

$$2x + 8y < 3x + 3y \left(\text{or equivalently } 5y < x\right)$$

Display 3
Alternative evaluations

Procedure	**Possible Optima**
Dominance	Any of A, B, C
Maximum	C
Simple Average	C
Weighted Average	Any of A, B, C
Product Maximization	C
Maximin	Either of B, C
Average Excess	C

There just do not exist weights x and y which allow these conditions to be realized concurrently. Clearly, weighted averaging is not going to be sufficient by itself to meet the needs of rational evaluation.

The reality of it is that there are various different alternative ways of assessing the overall comparative value of an item on the basis of its aspectival merits. The prime prospects here are registered in Display 2. And when these alternative modes of value compilation are employed in the preceding illustrative example the possible outcomes for our various assessment procedures will stand as per the data of Display 3. The upshot is a prospective-optimum score of 2 for A, 3 for B, and 7 for C. So here C enjoys a clear advantage and on this basis may plausibly be regarded as the top choice in our selection problem, with B and A following in that order. (By contrast, if the m_1-value of alternative A were 10, then A would clearly prevail.)

The governing idea here is to prioritize that choice alternative which has the greatest potential for realizing optimality among the selection-methods that qualify as plausible in the circumstances. In using this procedure to access the respective claims of those three alternatives we have, in effect, adopted a methodological democracy of procedure: one method, one vote. (And clearly if one were to introduce further evaluative procedures, this method for choice optimization would automatically extend its reach.)

All the same, it is, of course, possible that this selection procedure would prove unsatisfactory in some particular domain of operation. The course of experience in some range of preferability-determination might show that some one particular method would yield optimally satisfactory results in the circumstances. Thus in a certain context of application it might be imperative to avoid the insufficiency in any one mode of merit, irrespective of advantages elsewhere. In this event, Maximin would be the sole relevant standard and it would make perfectly sound sense to abandon methodological democracy. But in default of special considerations of situational expediency the Principle of Procedural Democracy provides a plausible method for choice resolution among alternatives of multicritical merit.

Chapter 3
Joseph Agassi from Metaphysics to Politics

Christopher R. Donohue

Joseph Agassi's defense of speculation and of metaphysics is the result his overall conception of philosophy and his epistemological stance. Philosophy for Agassi is the "merciless" search for principles which invariably ends in the central tenant of critical rationalism: "we are all in error." Critical rationalism is the relentless search for truth through the testing of ideas. Philosophy thus conceptualized can abide neither Karl Popper's fideism nor Wittgenstein's mysticism as both delimit rationality, resulting in a kind of pseudo-rationalism. Because rationality must be relentless, it seeks ideas from all sources and tests them, including the metaphysical. Pluralism is to be advocated. Because error is ubiquitous, a scientific idea and a metaphysical idea are both to be equally tolerated. The toleration of metaphysics is not only a consequence of extremism in thinking but the result of Agassi's commitment to pluralism in science and in politics. Because we are most likely in error and because our ideas are most likely wrong we must be tolerant of both metaphysics and the metaphysician. Thus, philosophy must be relentless but open to all ideas, the philosopher constantly testing and extreme in his thinking but gentle in his manner. It is this way that Agassi goes from metaphysics to politics.

3.1 Introduction

In this paper, I will articulate what I consider to be a consistent line of inquiry in the large philosophical and historical corpus of the philosopher and historian of science Joseph Agassi: a robust and provocative defense of metaphysics as essential to

---

C.R. Donohue (✉)
National Human Genome Research Institute, National Institutes of Health, Bethesda, MD, USA
e-mail: christopher.donohue@nih.gov

N. Bar-Am, S. Gattei (eds.), *Encouraging Openness*, Boston Studies in the Philosophy and History of Science 325, DOI 10.1007/978-3-319-57669-5_3

science and to philosophy. Agassi is keen throughout his many writings to defend speculation and metaphysics as such, even if they are not strictly scientific and even if metaphysics is in danger of degenerating into a pseudoscience. Agassi's defense of metaphysics, moreover, is important to his articulation and critique of the critical rationalism of Karl Popper and can even be extended into Agassi's political theory. Agassi's defense of metaphysics may be tied to his advocacy of compromise, tolerance, and pluralism.

Agassi emphasizes that any idea can make a lasting contribution to science and society. Agassi's emphasis on outlandish ideas and on metaphysical speculation is central to his conception of critical rationalism as being distilled in one principle that the result of philosophical speculation is the admittance of one principle, "The supreme principle is the principle of error; we are all liable to err." And because we are liable to error we should test everything out.

Moreover, the duty of the philosopher is to test every idea "relentlessly" because philosophy is nothing but "merciless logical extremism" (Agassi 2016) Indeed, Agassi underscores that most [nearly all] ideas will fail tests in this manner. Thus, we must be open to any idea, no matter how strange, and conversely, the failure of any principle, no matter how cherished.

This has significant implications for political philosophy as when individuals present to us an idea [even if it is outlandish] we must be tolerant of them. We should encourage a plurality of viewpoints whenever possible. In his philosophical anthropology, Agassi underscores that a "Darwinian" (Agassi 1977) approach to ideas where multiple lines of inquiry are offered in the hope of achieving a beneficial result. We must be tolerant of these ideas and the people who hold them because it may very well be the case that we are in error and that they are closer to the truth. Agassi's defense of metaphysics and of speculation generally is the consequence of these stances. We must be tolerant of metaphysics and of the metaphysician.

Agassi's openness to any idea, his defense of metaphysics and speculation, and his account of the unlimited critical nature of philosophy is the result of his engagement with the philosophy of Karl Popper as well as that of Ludwig Wittgenstein. Both reject metaphysics and both delimit rationality. My argument will begin then with Agassi's engagement with both authors. I will then return to the issue of how Agassi's defense of metaphysics is critical to both his account of philosophy as "merciless" and to his account of politics as the promotion of mildness due to the ubiquity of error.

In this way, I hope to highlight a consistent strand in Agassi's thought which seeks like Agassi's work generally, to interrogate and ultimately resolve dichotomies, between philosophy and politics and between science and metaphysics. (Agassi 1977). It is the resolution of these dichotomies which have so plagued philosophy which makes Agassi's philosophy important today.

3.2 "In Praise of Idle Speculation"[1]

Agassi notes in his philosophical anthropology (Agassi 1977) that Karl Popper would have objected to the entire enterprise of the book. Agassi's defense of metaphysics against Popper's never-fully articulated but still ever-present hostility towards metaphysics was a clear development away from Popperian notions of the status of metaphysics. While Popper granted that metaphysical and philosophical theories, such as the idealism of Berkeley or the skeptical challenge of Hume, could be the subject of critical discussion, he treated metaphysics as differing in kind than empirical knowledge or the insights gained from the mathematical sciences. Agassi notes "Popper decidedly gives metaphysics its right of existence, and he also admits that an individual can scarcely avoid some metaphysical view or other."

According to Agassi, Popper confesses himself to be "a believer in realism, the simplicity of nature, etc.", Agassi continues, "he makes this confession" and makes it clear "that he will not allow metaphysics into science and keeps his own metaphysical views strictly private." More explicitly, Popper's views of metaphysics in the *Logik der Forschung*, published in 1935, "is clearly tinged with old-fashioned positivism" while the *Open Society and Its Enemies* is "inconsistent, seemly ranging in admiration for Plato's theory of ideas to a violent positivist attack on Whitehead's endorsement of metaphysics." Popper in the *Open Society* in his attack on Whitehead concludes that metaphysics cannot progress. This is more ironic considering that Popper "introduces a new philosophical anthropology" which describes man as "suffering under the strain of civilization" due to the conflict between his "natural tendencies" which is towards collectivism and the tendencies of civil life "which are individualistic in their moral character" (Agassi 1977). Popper then regards metaphysics and all sorts of other "what is…?" questions as "taboos."

Instead, Agassi suggests a kind of "Darwinian approach" "which suggests studies along diverse lines, including diverse sorts of idle speculations in the hopes that at times something worthwhile may come of it." Agassi does this even though this position places him in opposition with the clear majority of "rationalist" philosophers, retorting that metaphysics is a bit like democracy, one should not reject either out of hand because "of the many follies" committed under both banners (Agassi 1977).

Just because we do not like Kantian determinism or Pyrrhonean skepticism, or Leibniz's monadology, does not mean we can or should reject metaphysics out of hand. Agassi rightly points out that many critics of democracy signal out exemplars of its idiocy in their effort to repudiate democracy entirely. On both counts, Agassi contends, this is an inappropriate way of excluding both metaphysics and democracy. Indeed, he points out that the positivistic rejection of metaphysics is not only metaphysical, it is embedded in a very specific understanding of the nature of man as lazy and as dogmatic. In contrast, the positivistic view of man and nature consuls that "we should not waste ourselves in idle and conceited speculation." Accordingly,

[1] This section and the following borrow Joseph Agassi's chapter titles.

its view of science is that of "strict demonstration" the consequence of strenuous labor of the hard work of verification (Agassi 1977).

3.3 "Was Wittgenstein Really Necessary?"

The philosophy of Ludwig Wittgenstein is in many ways antithetical from that of Karl Popper, but both present existential threats to the work of philosophy Wittgenstein and Popper both inveighed against philosophers. Wittgenstein went further saying that philosophy was in no way a doctrine, simply an activity.

This Agassi points out perfectly fits the definition of a theology. The entire purpose of a theology is to keep the faithful amassed. Theology has a "huge and ever expanding corpus of theological theory and argument, and the argument is, indeed, to and fro, and so within its limits perfectly rational; but staying within its limits is not in the least radical, it is dogmatic" and as such, theology is concerned only keeping the faithful in faith and not with the development of any ideas. Theology has as its object only the faith of the faithful and as such is a kind of "pseudo-rationalism."

The theology of Wittgenstein and his followers – that there is no philosophy – is much more difficult to conceptualize. The faith of the follower of Wittgenstein is manifold, not only are their limits to science, but that beyond that there is a great deal that is mystical. Moreover, Wittgenstein contended that "language is limited just like science." This was, Agassi pointed out, the central tenant of Wittgenstein's "picture theory of language." This picture theory of language allows Wittgenstein to argue that there is no way for us to understand the mystical. In this way positivism is very much a kind of "competitor" to religion, having many of the same rituals and functions as religion, as well as many of the same views towards reason and its limits. Agassi argues that Wittgenstein's picture theory of language engages in the same delimiting of rationality as religious dogma, both are pseudo-rational (Agassi 1981).

For Agassi, the claims that (1) all disputes must be capable of being settled (2) all claims which are not capable of being settled are somehow outside the bounds of rational inquiry and therefore "anything goes" are objectionable because the exclusion of an area of criticism is not rational, or at the very least concedes too much to the irrationalist. This is a kind of fideism, a leaving of things to faith and of a delimiting of rationality. In this instance, Popper's fideism and Wittgenstein's mysticism engage in the same kind of pseudo-rationalism.

Agassi points out Popper's grave mistake in Chapter Twenty-Four the *Open Society*. Agassi writes "one point, he proposes is better placed beyond criticism: one better stay dogmatic about one's readiness to change one's mind and keep one's hope to improve because of the criticism." This for Agassi is an unfortunately concession to the fideist that the "skeptical critique of classical rationalism needs to be overcome by some prerational choice…" (Agassi 1987).

The accusation of fideism has a long history. It was a live issue for Popper's critical rationalism. W.W. Bartley in *The Retreat to Commitment* (Bartley 1962; Agassi 2012) pointed out the dilemma of accepting criticism in every aspect of one's life but the commitment to criticism itself. He concluded that critical rationalism was only truly thus if such criticism extended to the very capacity of rationality itself. Bartley accused Popper of fideism later (Radnitzky et al. 1987) and it lead to much discussion, ending when Popper finally convinced that their positions were identical. Agassi notes, correctly, that this is untrue (Agassi 1987).

For Agassi, Popper's protection of rationality from itself underscores a painful truth, belief and faith has nothing to do with rationality, since "When one sees a refutation of a seemingly unshakable view one feels something known by the expression, I am unable to believe my eyes; yet one does not believe one's eyes at times." Belief in the virtues or irrefutability of a statement or of a conception, such as the goodness of Newtonian mechanics or Einstein's physics, has nothing to do with the refutability of said ideas. Strongly held beliefs fall as readily as any other (Agassi 1987).

Thus, believing in the power of critical rationalism does not in some way insulate it from falsity. Indeed, although Agassi does not point this out, it may be the case that strongly held belief in the unassailability of rationalism may be extraordinarily harmful to it. In this scenario, the belief in rationality has become dogmatic and in its dogmatism, functions in the opposite way of a truly critical rationalism. This too, as Agassi points out, constructs a hypocrisy at the very root of rationality. It allows opponents of rationality to easily and justly point out, I think correctly, that rationality is not truly such if it shields itself from its own criticism.

The irrationalist or foe of rationality who is always on the lookout for arguments against reasoned dicussion sees in this protection an easy and glib point. The subjection of critical rationalism to itself, furthermore, is a kind of bravery. It opens the rationalist to criticism of his most cherished principles. This in turn sets a good standard for debate. The rationalist can justly declare that all beliefs are to be subject to interrogation, including the belief in the necessity of interrogation. This gives the foe of critical rationalism no quarter; it gives him no cheap argument. It promotes some good rules for open discussion.

The critical rationalist can argue first off that in this discussion, everything is open to discussion, including 'himself.' This would have the effect of convincing his audience and his opponent that he is indeed serious and that he as a philosopher is relentless.

Such is the philosophy of critical rationalism that it must be intellectual unforgiving. But because of its intellectual unforgiving nature, it must be tolerant and merciful towards individuals. Such is the work of the philosopher that he or she must entertain any idea, including bizarre speculations and metaphysics, because as Agassi declares "I extol some metaphysics as the physics of the future" (Agassi 1975).

3.4 From Metaphysics to Politics by Way of "Logical Extremism"

Philosophy for Agassi is the "the readiness to develop an idea and its logical consequences." Philosophy too means that "anyone interested in logic and in philosophy even to a small extent, has to know that intellectual mercy means cruelty." Philosophy is the relentless search for truth and improvement through the correction of error through testing or critique. It is only with the weak of mind and the weak of will that we relent in our criticism. Agassi writes in this regard, "The classic example is the story of Don Quixote who tested his helmet by hitting it with a sword and who thus broke it, then mended it, then intended to test it again, but finally preferred to forego second the test. Here it is clear that common-sense conflicts with the Don's conduct, but in other cases commonsense sides with the Don as against logic, as I shall describe in detail on another occasion. This is how philosophers have learned to consider compromisers and moderates as people with weak intellects, as irredeemable muddle-heads." Agassi relates that therefore Lenin preferred speaking with reactionaries to moderates, as only reactionaries were clear in their logic and clear in their thinking (Agassi 2016).

Philosophy is about getting out of "muddles", designed towards the development of austere, even extreme principles. He understands that this extremism drives many people away from philosophy and makes philosophy unsuited to the trivialities of everyday life. However, philosophy is compatible with politics according to Agassi if the principle arrived at by "merciless logical extremism" is "the principle of error." By this Agassi means, "we are all liable to error." As noted earlier, this is the principle of critical rationalism (Agassi 2016).

Such a conception of philosophy is consistent with Agassi's embrace of metaphysics as well as this understanding of Popper's great error of fideism and Wittgenstein's delimiting of rationality. Both are another example of a Don Quixote-like reasoning. If one is a critical rationalist, one cannot maintain that any realm is beyond the scope of critical rationalism, even the idea of rationalism itself. We may think that not subjecting the belief in rationalism to the same kind of scrutiny is a mercy, but the truer cruelty may be allowing individuals to persist in their falsehoods. Popper's position in this conception resembles the helmet which once broken and mended is not tested. This for Agassi is not an enviable position.

And because we are all liable to be in error, all have to be tolerant of all ideas, within limits. Agassi underscores that entertaining a variety of theories leads to the increase of knowledge because of the differing qualities that all theories have. Some theories in physics or in biology may be intriguing but ultimately untestable, abstract, more or less general and so on. Thus, "we may admire one theory for its boldness, another for its explanatory power, another for its elegance; and yet another…for the light that it throws on some topical metaphysical issues" (Agassi 1975). This, for Agassi, is pluralism. But we also must be very careful and we should ruthlessly test since it was very possible for metaphysics to become pseudo-scientific when metaphysics becomes a framework for ad-hock explanations.

In this way, metaphysical research programs can become dogmatic (Agassi 1987), but so long as metaphysical programs lead to testable ideas than Agassi is all for them. It is possible in this regard to distinguish good metaphysics and bad metaphysics. Metaphysics at its best can support "crucially important regulative ideas." Here science to can add metaphysics in the debate over its importance. Agassi underscores that "some of the greatest single experiments in the history of modern physics are experiments related to metaphysics" and while metaphysics is not part of science proper there "is a requirement that metaphysics be inspiring and lead to the development of scientific theories." Agassi makes it very clear that metaphysics, while not changing the facts of the matter can very well lead to novel interpretations. This in turn will lead to an increase in the usefulness of a theory (Agassi 1975).

Agassi then believes metaphysics can be useful. They can provoke inspire and baffle. This does not mean that we should tolerate pseudoscientific nonsense or the work of liars. Agassi opened his review of Feyerabend's *Against Method* with: "How do you read a book which extols lies?" (Agassi 1976) Agassi answered simply that we do not. Nor should we.

The idea that we are extremists in philosophical reasoning leads to mildness and pluralism in politics is perhaps a bit strange. But Agassi here as throughout his writings insists on resolving false dichotomies. If philosophy is the relentless search and testing of principles, philosophy finally settles on the principle of universal error. This leads in politics to the principle of compromise. Agassi explains how this works in politics: "The principle of compromise says, we should remember that we are liable to err and that our sticking to our opinions may incur high costs—to us or to our neighbors. The principle of compromise should reduce—if and when it is possible to reduce—the likelihood that my neighbor will pay the high price for my mistake." Such prices are reduced through recourse to pluralism, as a plurality of views and opinions will decrease the chances that an individual or group is incorrect, thus lowering the price.

Error, toleration and pluralism connect with politics in the following way. If we are all liable to error in our day to day lives, in municipal planning, in the development of public policy, we should try to have as many voices as possible which may still all be wrong, but the chances are greater that they could indeed be correct. This does not mean that we should tolerate all opinions or encourage those which extols lies, but it does mean that more opinions about the course of our daily lives are better than fewer. What we should pursue, according to Agassi, with logical extremism is the principle that because all we know is error, we should as much as possible promote a diversity of opinions, while tolerating as many as possible. Toleration moreover is not only mental, but moral. Toleration too is part of the rules of debate, and in this way, debates become part of ethics.

In this way, Agassi links rationality and politics, bridging the impasse between the utopia of philosophical speculation and the compromise of decision-making. If we are extremists in our pursuit of error and in the pursuit of compromise, we become moderates in politics and in our everyday lives, while being philosophers all the same.

References

Agassi, Joseph. 1975. *Science in flux, Boston studies in the philosophy of science*. Dordrecht/Boston: D. Reidel Pub. Co.

———. 1976. Against method: Outline of an anarchistic theory of knowledge. *Philosophia* 6 (1): 165–177.

———. 1977. *Towards a rational philosophical anthropology*. The Hague: Martinus Nijhoff.

———. 1981. Was Wittgenstein really necessary? In *Science and society: Studies in the sociology of science, Boston studies in the philosophy of science*, ed. Joseph Agassi, vol. 65, 33–43. Dordrecht/Boston: D. Reidel Pub/Kluwer Boston Inc.

———. 1987. Theories of rationality. In *Rationality: The critical view*, Nijhoff international philosophy series, ed. Joseph Agassi and Ian C. Jarvie, 249–263. Dordrecht/Boston: Martinus Nijhoff.

———. 2012. Critical rationalism, comprehensive or qualified: The Popper-Bartley dispute. In *Der kritische Rationalismus als Denkmethode und Lebensweise : Festschrift zum 90. Geburtstag von Hans Albert*, ed. Giuseppe Franco and Hans Albert, 34–55. Klagenfurt: Kitab.

———. 2016. Letters to my sister concerning contemporary philosophy, with special reference to the philosophy of education, the philosophy of science, and the Yom Kippur War. Unpublished manuscript in the author's possession.

Bartley, William Warren. 1962. *The retreat to commitment*. 1st ed. New York: Knopf.

Radnitzky, Gerard, William Warren Bartley, and Karl R. Popper. 1987. *Evolutionary epistemology, rationality, and the sociology of knowledge*. La Salle: Open Court.

Chapter 4
Horror Dogmatis

Zuzana Parusniková

Horror dogmatis was a strong (and perhaps the strongest) driving force behind Popper's effort to formulate a new conception of reason identified with criticism. Dogmatism is inherent in all forms of justificationism; strategies seeking the confirmation of a theory are not only logically flawed and thus irrational, but tend to immunize theories against criticism. Dogmatism is also present in our expectations of (and strong belief in) regularities; Popper set out to solve Hume's problem by treating the need for regularities as biologically inborn and by reformulating belief-sentences in terms of logic (according to the principle of transference). Dogmatism is dangerous in the social sphere, too. Holistic philosophies when practically implemented lead to authoritarianism and oppression. In sum, dogmatism – apart from being irrational – presents an obstacle to the growth of knowledge and to the emergence of the open society. Should critical rationalists adopt a mercilessly hostile attitude to dogmatism? Further, could dogmatism gradually diminish in the course of the evolution of mankind?

4.1 The Role of Dogmatism

In his autobiography Popper mentions that "*the problem of dogmatic thinking and its relation to critical thinking*" occupied him even during his early studies (Popper 1982a, 41); the first impulse to turn a attention to this relation was at that time the Marxist movement in Vienna and Popper's short-lived sympathy with it. This experience, however, triggered his interest in social issues, most importantly in the philosophical sources of totalitarian ideologies; his two famous works, *The Open Society and its Enemies* and *The Poverty of Historicism* were the main results of this

Z. Parusniková (✉)
Institute of Philosophy, The Czech Academy of Sciences, Prague, Czech Republic
e-mail: parusnikova@volny.cz

N. Bar-Am, S. Gattei (eds.), *Encouraging Openness*, Boston Studies in the Philosophy and History of Science 325, DOI 10.1007/978-3-319-57669-5_4

wartime research. However, in the pre-war era Popper had become equally interested in the philosophy of science due to his involvement in the positivist movement of the Vienna Circle. Though not a member he took part in the seminars led by Moritz Schlick, and in this environment he formed his critical views concerning induction, verification and demarcation which constituted the core of the positivist doctrine. This paper deals only with this second area comprising of Popper's conception of critical rationalism in the field of knowledge.

For Popper, then, criticism and reason are linked. On a general level, this is a thesis that no philosopher, methodologist or scientist would dispute. However, the originality of critical rationalism lies in the claim that merciless criticism is the *only* rational way to evaluate hypotheses and theories. Popper's orientation towards the negative powers of reason, consisting in error-elimination instead of finding positive evidence in support of a theory, marked an epistemological revolution that redefined the methods and goals of science as well as the conception of rationality. Rational method proceeds by testing the consequences of a conjecture *in order to refute the conjecture* and – if we are lucky – to discover new, deeper problems and further challenges that stimulate further growth of knowledge. In Miller's words, "criticism is the life-blood of reason" and its task is not to "consolidate our knowledge, but to liquidate it" (Miller 1994, 67, 111). Popper identifies the scientific attitude with the critical attitude; criticism is the essence not only of a scientific method but of a rational attitude to problems in both in life and knowledge; "[the critical attitude] does not look for verifications but for crucial tests; tests which could *refute* the theory tested, though they could never establish it" (Popper 1982a, 38). Popper's conception of reason is deliberately destructive as reason can only undermine our beliefs and remind us of our acute fallibility. Yet Popper does not view this predicament as shattering or paralyzing – quite the reverse; he invites us to embrace the imperative of criticism, not fear it. It is not just rational but exciting, too. It encourages bold guesses, conflicts and disagreements – in sum, it makes the process of cognition dynamic and adventurous.

This approach stands in a complete opposition to all dogmatic strategies that, by contrast, cling uncritically to established "truths", hold on to beliefs (for belief in ancient Greek is "*dogma*") using all kinds of the justificationist trickery, such as ignoring counter-evidence and marginalizing the warning signals when our expectations, based on a certain hypothesis, are not fulfilled, when crucial tests uncover inconsistencies in the proposed hypothesis. For Popper, "a dogmatic attitude is clearly related to the tendency to *verify* our laws and schemata by seeking to […] confirm them, even to the point of neglecting refutations" (Popper 1985, 50). Dogmatism thus serves like a deterrent, illustrating bad method both in science and in cognitive attitudes generally In this light it seems very surprising that Popper advocates a positive – though limited – role for dogmatism in his methodology.

Throughout his major writings Popper repeats the claim that a certain amount of dogmatism is the initial and necessary phase in handling a new hypothesis. According to Popper, a theory should show its mettle before it is submitted to the destructive effect of criticism; and complementarily, the hypotheses will become strong enough in this phase to be a worthy target for criticism. An emerging theory must therefore "gain weight" even at the price of dogmatism: "a critical attitude needs for its raw material, as it were, theories or beliefs which are held more or less dogmatically" (1985, 50). Criticism should thus be put on hold in the initial phases of scientific discovery and the theory in question should be granted something like a closed hunting season. At times, as Popper suggests, the theory should not only be spared criticism but also allowed an active defence: "somebody has to defend a theory against criticism, or it would succumb too easily, and before it had been able to make its contribution to science" (Popper 1979, 30).

So on the one hand, "[e]very attempted solution, every theory, is tested as rigorously as it is possible for us to test it. But a rigorous examination is always an attempt to discover the *weakness* in what is being examined" (Popper 2003, 10); but on the other hand, "(o)nly when a dogma is established as a kind of background can one begin to criticize, and only later can the dogma – the background of critical discussion – itself be included in the criticism" (Popper 2003, 41). Therefore "the critical attitude is not so much opposed to the dogmatic attitude as super-imposed on it" (Popper 1985, 50). Popper's message is that there can be no critical phase without a preceding dogmatic phase, and that though we must try to refute a theory we should not do so too soon. These are strong statements which seem to go against the spirit of critical rationalism.

4.2 Is Dogmatism Necessary?

What precisely is Popper's point? He may be saying something completely innocent, with no bearing on his philosophical project of negative reason. For instance, the positive role of dogmatism may only imply that most newly-formed hypotheses are rather imprecisely formulated and it may well be unclear what they logically imply. If so, the investigator does well not to be over-impressed by criticism that takes advantage of the sloppiness of what is under investigation. In such situations some imperfections in the presentation of a new hypothesis *should not serve as a reason to attack the theory as such*. The imperfections are criticized not in order to falsify the theory but to improve its form, clarity and strength. Yet this attitude need not entail any dogmatism (as Popper assumes) – it is driven by *an intense interest in the problems that the theory attempts to solve and by the envisaged potential of the theory to solve them*. The investigators, led by their curiosity, are themselves their first critics and as the new theory attracts more and more attention the imperfections are gradually corrected; when the theory becomes consistent it is then easier to design crucial tests which will determine its fate. In other words, scientists take care of this evaluating process, in which they combine specific criticism and partial defence of the theory, and – given the challenging nature of the theory – there is no danger that, on the grounds of minor inconsistencies, a front-line falsificationist attack against the main content of the theory will occur.

Popper sharply distinguishes between subjective thought processes, which are the subject of psychology, and the objective logic of discovery which is the sole object of his investigations. For Popper, the rational reconstruction of the psychological factors involved in a discovery is only partially possible: "in so far as the scientist critically judges, alters or rejects his own inspiration we may, if we like, the methodological analysis undertaken here as a kind of 'rational reconstruction' [...] but [it] would not describe the processes as they actually happen [...] there is no such thing as a logical method of having new ideas" (Popper 1965, 31–32). Scientists use criticism (and dogmatism) continuously in all stages of their research regardless on the completeness of a hypothesis; at the stage of exploring a problem they make, just as Popper says, bold guesses, they speculate, they free their imagination and indulge in fantasy – yet they simultaneously scrutinize these guesses, check them, test them and in so doing come by many new ideas and discover new problems which may not even be associated with the initial focus of enquiry. These processes contain both dogmatism and creative intuition; they depend upon individual talents, determination, personal feelings and ambitions, or outside influences such as chance and the influence of the scientific community, and in this field no methodological rules can be proclaimed. However, as Popper emphasizes, the psychological (and social) aspects of the cognitive process must be distinguished from the *objective pattern of the growth of knowledge*. This pattern complies with logic (the principle of retransmission of falsity, in Popper's case) and is thus a guarantee of the rationality of science. Popper's attitude is reminiscent of the positivist approach in which logic and science are the essence of rationality.

Therefore, Popper's endorsement of dogmatism which brings to the forefront the subjective defensive strategies seems both unnecessary and dangerous. It is unnecessary in such case as it appeals to scientists' subjective attitude regarding how to handle a newly-emerging hypothesis. It is dangerous when applied as a methodological norm but this is exactly what Popper does in his recommendation to apply dogmatism (though in limited dose) in the first stage of evaluating hypotheses. As a result, for progress to be made in science suspension of critical activity (dogmatism) is not just *permissible* but *obligatory*; Popper hereby lets dogmatism – and justificationism – into his method. Criticism thus ceases to be the non-negotiable *normative methodological principle* and the inviolable standard of rationality.

Defensive dogmatic methods should then be incorporated in the objective pattern of the growth of knowledge in the same way as negative testing methods. Popper's famous model of the evolution of knowledge divides into four chronologically distinct phases (Popper 1985, 287):

$$\begin{array}{c} TT_a \rightarrow EE_a \rightarrow P_{2a} \\ P_1 \rightarrow TT_b \rightarrow EE_b \rightarrow P_{2b} \\ TT_n \rightarrow EE_n \rightarrow P_{2n} \end{array}$$

where P stands for problems, TT for tentative theories, and EE for error elimination.

Knowledge, then, starts with an objective problem, and proceeds to many conjectures which aim to solve the problem; these are submitted to criticism and error detection in the proposed conjectures stimulates the emergence of new problems and the discovery of new areas of knowledge or, rather, new areas of ignorance. The successive phases stand in a causal relation and are thus interlinked but each has a distinctive task in place and time. The dogmatic phase, as formulated by Popper, should be also added to the scheme. It could be something like this:

$$P_1 \rightarrow TT \rightarrow (dd) \rightarrow EE \rightarrow P_2$$

where the '(*dd*)' indicates the dogmatic defence stage, short-lived and limited in scope. However, this concession to dogmatism would compromise Popper's imperative of falsification and thus also the logic of discovery:

> [...] empirical science may be defined by means of its methodological rules. [...] First and supreme rule is laid down which serves as a kind of norm for deciding upon remaining rules, and which is thus a rule of a higher type. It is a rule which says that the other rules of scientific procedure must be designed in such a way that they do not protect any statement in science against falsification (Popper 1965, 54).

Critical rationalism is based on a strict opposition to dogmatism for logical reasons (justification is impossible), for evolutionary reasons (justification blocks the growth of knowledge) and for ideological reasons (dogmatism encourages totalitarianism). Popper's recommendation of dogmatism is incompatible with the supreme rule of falsification.

Let us consider that Popper's concern regarding the dogmatic phase in evaluating a theory might be rather pragmatic. Yet in this case, Popper moves towards an analysis of science which takes into account various psychological and sociological factors – in other words those that Popper himself excludes from critical rationalism. If Popper ventured into this territory any clear dividing line between dogmatism and criticism would be blurred. The use of dogmatism in real research is determined by many factors; it depends upon the potential of the theory, the interest it raises among scientists, how their critical abilities are developed and the critical environment in the scientific community. Popper does not provide any objective criterion of when dogmatism should be abandoned and the critical phase – responsible for the growth of knowledge – should begin. His instructions are vague: dogmatism is necessary *to some extent*, it should not be given up *too soon*, theories should be *more or less* held dogmatically in the *initial phase* of evaluation. These are just appeals that can easily be misused by dogmatists to stretch the initial phase indefinitely. By introducing dogmatism as a legitimate (and even obligatory) method of defending hypotheses Popper violates the supreme rule of a total ban on dogmatism, quoted above. Moreover, he comes perilously close to Kuhnian normal science.

4.3 Dogmatism and Normal Science

It is well known that Popper rejects Kuhn's picture of 'normal' science as puzzle solving within a given paradigm and presents instead the development of science in terms of permanent revolution.[1] "In my view", Popper says, "the 'normal' scientist, as Kuhn describes him, is a person one ought to be sorry for" (Popper 1982b, 52); in contrast to this uninspired labour work he posits scrupulously critical work consisting of bold conjectures and refutations. Yet even in this polemic with Kuhn Popper does not forget to add that "the dogmatic scientist has an important role to play. If we give in to criticism too easily, we shall never find out where the real power of our theories lies" (Popper 1982b, 55); his main objection against Kuhn consists in the fact that Kuhn believes in the domination of a ruling dogma over *considerable periods*. Yet how are we to determine and control the *appropriate period* for defending our belief in a theory against criticism if it cannot be formulated as a universal methodological rule? Popper often emphasized that even mild (appropriate?) dogmatism has a tendency to spread and degenerate into excessive dogmatism – such as is typical of normal science – under the pretext of exploiting the strength of a theory.

For instance, there is a stage in research which Popper calls "working on a problem", characterized by our fully acquainting ourselves with the problem; at this stage "we begin to see the ramifications of the problem, its sub-problems, and its connection with other problems" (Popper 1979, 260). This sounds exactly like the phase in which dogmatism should be practiced. It seems close to Kuhn's description of the first reactions to Newton; Kuhn shows that Newton's work in his time suffered from limited agreement with experiment and observation yet this never served as a reason to question its validity; Newton's *Principia* "retained some of the clumsiness inevitable in a first venture" (Kuhn 1970, 33). Scientists in many fields then worked on reformulations of some aspects of Newton's theory, on devising new experiments, on the possibilities of application: "their work produced not simply new information but a more precise paradigm, obtained by the elimination of ambiguities that the original from which they worked had retained. In many sciences, most normal work is of this sort…" (Kuhn 1970, 34). Of course a work like Newton's *Principia* is a special (revolutionary) case, triggering a new stage of physical science, but Kuhn uses the notorious Popperian example in the form of a simple general statement "all swans are white" to the same effect. Even here, Kuhn claims, a scientist might be a loser if (s)he abandoned the statement should a black swan be observed.

The merit of normal science resembles Popper's description of the *initial* dogmatic phase in research in which the immunization of a theory against criticism enables the *full articulation* of a theory; in Kuhn's words, "[c]onfronted with the

[1] Only this specific aspect of Kuhn's philosophy of science is considered in this paper, leaving aside all other major differences between him and Popper, such as the long-term reign of normal science, the dominance of one single paradigm, incompatibility between paradigms, the relativity of truth and the reduction of scientific norms to the practice of the community.

unexpected, he [the scientist] must always do more research in order further to articulate his theory in the area that has just become problematic" (Kuhn 1982, 19). Kuhn can very well say that "[t]he scientist who pauses to examine [critically-Z.P.] every anomaly he notes will seldom get significant work done" (Kuhn 1970, 82). However, as Kuhn remarks and as Popper should realize, there are no exclusively logical criteria for defining this dogmatic 'pause' and for determining when an anomaly (a contradiction) should be critically scrutinized and when it should be considered just a minor imperfection. For Kuhn, this decision is not methodologically dictated but is *internal*; above all it depends on a general situation in science, specifically on whether a paradigm is still strong and can resolve anomalies from its own resources – whether it still has, in Lakatos' term, strong heuristic power.

Darrell Rowbottom discusses the role of dogmatism in Popper's methodology and compares it to Kuhn's concept of normal science; although he admits that "it is unclear how long we should stick with a theory in the face of trouble spots" he rather emphasizes the obvious differences between the two philosophers of science (Rowbottom 2010, 112). Kuhn, of course, was much more extreme in his claims about the value of shielding theories from criticism; Popper represents the other extreme of permanent criticism in science. According to Rowbottom, Popper's small concession to dogmatism does not present any danger that dogmatism would usurp too a great role. Rowbottom's goal in developing critical rationalism is to find a possibility to bring the two approaches, Kuhn's and Popper's, together. The first step is to look at science through its various functions (dogmatic, imaginative, and critical) and also to consider these functions at the level of a group rather than an individual scientist (as Popper does). In this light we can appreciate these functions as working together, not in opposition. As Rowbottom says, "it is possible for science to perform a critical function with wide scope even when none of its participants have (completely) critical attitudes. One simple way to see this is to imagine a scenario in which each scientist holds different assumptions dogmatically, but in which no peculiar assumption […] is held dogmatically by all scientists. So at the group level no statement is beyond criticism" (Rowbottom 2010, 119).

Agassi and Jarvie develop an even stronger argument concerning dogmatism; they defend the *rationality of dogmatism*. They agree with Popper's thesis that criticism is logically binding whilst dogmatism (justificationism) is not but claim that dogmatism may be binding by some other cannons. For them, it is rational to adopt dogmatism on a *specific* level and apply it in open-mined group debates, represented by a *scientific workshop*. Agassi and Jarvie outline a complex picture of two opposite epistemologies, dogmatism and scepticism. On the general level "justificationist dogmatism is the *general* view that we may and ought to close certain debates once and for all" and "general scepticism keeps debates endlessly prolonged by demanding in general impossibly high standards for closure" (Agassi and Jarvie 1987, 436, 438). They see neither of these positions satisfactory; the former lacks any challenging and dynamic factors and the latter ultimately leads to nihilism or fideism.

Their definitions of general dogmatism and scepticism seem, however, somewhat dubious. Viewing dogmatism as a strict closure of inquiry, forbidding any further questioning of the problem under investigation, fits rather the ancient sceptics (to which the authors refer several times); it also fits the dogmatic implications entailed in Fries trilemma. However, for Popper it is essentially *justification* that characterizes dogmatism; in other words, the desire to find empirical support for a theory and *the belief that positive evidence should count* – that it proves or at least strengthens the theory. The closure of inquiry may be a part of the justificationist strategy but not the sole defining feature of dogmatism. For Popper, scepticism (the critical stance) is logically binding, as Agassi and Jarvie say, but is not a *game* in which permanent undermining of a statement is the end in itself (as in the ancient sceptics' case). *Criticism defines rationality and enables progress* – it is the engine of the evolution of knowledge and gives us hope of getting nearer to the truth; Agassi and Jarvie do not fully appreciate these aspects of Popper's strictly anti-dogmatic methodology. Instead they propose a version of critical rationalism which should, like Rowbottom's thesis, combine stability, continuity and challenge; *specific* dogmatism allows *specific* criticism and both attitudes thus retain balance. A scientific workshop provides the ideal environment in which "critical debate about open questions, employing tentative answers which are accepted either as grounds for action or as agenda for further debate or both" (Agassi and Jarvie 1987, 440).

Rowbottom's proposals overlap Agassi's and Jarvie's in the emphasis on the productive, challenging interaction between dogmatism and criticism, and justification and falsification. They thus shift the focus of the philosophy of science onto the scientific *enterprise*; that is, of course, an interesting field for inquiry but it was not the Popper's theme. As previously stressed, his focus was the logic of investigation embodied in the deductive falsificationist model. The model captures the evolution of knowledge as a process leading to ever deeper problems; the engine of this evolution is the phase of criticism in which we discover errors through the falsification of empirically testable predictions of the theory. Dogmatism *cum* justification contributes nothing to this process when seen from this objectivist perspective. The recommendations to reconcile dogmatism and criticism on the specific or individual levels, involving psychological or social factors such as the *willingness* of the investigators to be critical, their critical (or dogmatic) *temperament* and an overall open-minded *atmosphere* in the workshop seem rather a departure from critical rationalism than an enrichment of it.

Popper develops the modern ideal that philosophy should provide universal foundations for rationality; despite many specific differences he shares this ideal with logical positivism especially with respect to linking logic, science and reason. Steve Fuller emphasizes this point and notes that "Popper invested science with symbolic import as the standard-bearer for critical rationality" (Fuller 2003, 15); he contrasts this approach to Kuhn's who "was satisfied with contingent foundations drawn from the actual history of science" (Fuller 2003, 35). For Kuhn, scientists set their own standards of rational behaviour, dependent on the variable internal factors that determine scientific research. In practice, science suffers from too many undesirable constraints that suppress its critical spirit. They involve dubious methodological standards (demands for good reasons), professional concerns (where errors are interpreted as a failure on the part of scientists) and social concerns (including jobs, grants, prestige, power, etc.). Yet that is the precise reason why Kuhn's historical and practical descriptions must not be elevated to a methodological norm. Popper's own legitimation of a positive role for a time-limited dogmatism seems an unintended mistake, and a completely needless one. Had Popper stuck to the ideas developed in the logically purer *Logic of Scientific Discovery* he would never have to face the problem of a premature, rushed refutation of a theory. In this early text he makes it quite clear that falsification depends not on the occurrence of a single counterexample (that could make us reject a theory too soon) but on the *successful corroboration of a falsifying hypothesis*. There is no need to introduce dogmatism; by doing so Popper opened his methodology to the threat of arbitrary and uncontrolled practice of dogmatism and thus denied the constitutive feature of critical rationalism.

4.3.1 Horror Dogmatis

As the main argument of this paper contends, Popper's philosophy had been driven by *horror dogmatis*; he ascribed to dogmatism the cunning to sneak into a rational discourse in various disguises and undermine it from within. Agassi, for instance, emphasizes that "[i]n an attempt to block all attempts at dogmatism Popper has entirely isolated the open-minded [critical-Z.P.] from the possibly dogmatic and dogmatic" (Agassi 1975, 30). Though Agassi does not agree with such an extreme rejection of dogmatism (as discussed above) he rightly notes that Popper feared "hosts of shifty techniques and clever twists [… by which] dogmatism may creep into science through the back door" (Agassi 1975, 30). Once one makes an exception (or two, or three) the 'amount' of dogmatism cannot be controlled and the doors to its influence are wide open. Therefore extreme methodological measures must be taken to fend off this danger, measures that for Popper consist in adopting the imperative of falsification.

Popper's *horror dogmatis* was motivated by logical, epistemological and ideological considerations. The epistemological reasons had been gaining prominence throughout his work. In his *Logic of Scientific Discovery* Popper considered the positivist conception of probability that is linked to justification – the ideal of high probability of our theories leads to epistemological stagnation. For Popper, probability and content are inversely related and therefore the aim to achieve the highest possible probability of our statements inhibits our willingness to risk, and stifles the boldness and creativity of our thinking. Instead, unchallenging, banal statements with low empirical content that can pass in tests are favoured. Popper emphasizes that the growth of knowledge requires an adventurous spirit, welcoming the high-risk nature of the cognitive business. Justification is a defensive attitude that prevents us from daring (improbable) guesses, suppresses intellectual conflicts, and fails to exploit the creative potential of erring. Later, in his *Objective Knowledge* Popper explored the epistemological themes concerning the dynamic aspects of cognition in the wider context of Darwinism and applied his tetradic scheme of the growth of knowledge to the evolution of life.

But there is yet another – less frequently discussed – reason behind Popper's fear of dogmatism. Popper claims that on the biological level we have some inborn expectations, most importantly the expectation of regularities. However, the belief in regularities is dogmatic: "[o]ur propensity to look out for regularities, and to impose laws upon nature, leads to the psychological phenomenon of *dogmatic thinking* or, more generally, dogmatic behaviour: we expect regularities everywhere and attempt to find them where there are none; [...] and we stick to our expectations even when they are inadequate and we ought to accept defeat" (Popper 1985, 49). Dogmatism is thus imprinted in our genes and as such is the *primordial force* affecting our beliefs and behaviour.

This position of Popper unfolds from his critique of Hume's associationist psychology in which the belief in regularities is a result of repeated experience. According to Hume, after observing two instances conjoined together – say, sun rising each morning or fire always producing heat – we get into the habit of expecting that this will always occur in the future; we believe that these instances are in causal relation and the one implies the necessity of the other. This conclusion is the effect of our psychological constitution: "nothing leads us to this inference but custom or a certain instinct in our nature" (Hume 2005, 159). Popper objects that "induction – the formation of a belief by repetition – is a myth" (1979, 23). Our knowledge and our life experience develop in a deductive framework; "expectations may arise without, or before, any repetitions" (Popper 1979. 24). Yet is there not a danger of an infinite regress lurking, or is this not the kind of an insoluble question of what was first, the chicken or the egg? Popper rejects this possibility and claims that the chicken and egg question is soluble in his epistemology: "[g]oing back to more and more primitive theories and myths we shall in the end find unconscious, *inborn* expectations" of which the most dominant is the expectation of regularities (1985, p. 47). The original frame of reference determining both our cognitive focus and instinctive behaviour consists in *priori dogmatic* dispositions that are part of our genetic hardware. The danger that dogmatism can rule our nature is felt by Popper as very acute; our need for regularities expresses our need for stability that is satisfied when our expectations are confirmed. Our nature thus steers us towards dogmatic behaviour which can be concealed beneath critical proclamation.

Not only, then, is there a dogmatic streak in our nature, but it is a primal instinctive force. As Popper says, "this dogmatic attitude, which makes us stick to our first impressions, is indicative of a strong belief; while a critical attitude … is indicative of a weaker belief" (Popper 1985, 49). Furthermore, since the world around us is so diverse that it is easy to find a confirmation for almost any statement, if that is what we really seek, dogmatism has perfect conditions to flourish. *We are natural-born justificationists and dogmatists*, just like amoeba and other species. We instinctively yearn for our expectations or hypotheses to be confirmed and we believe that empirical support does the job. This disposition contradicts the critical disposition and thus *sabotages* the growth of knowledge. It is thus criticism that is the weak link and needs a boost. That is also why Popper felt the urge to counterbalance inborn dogmatism by the imperative of falsification *defined as a non-negotiable rule*. A methodological exclusion of dogmatism must be total, allowing *no bargains and no compromises*. Any compromise would devalue the very purpose of Popper's methodology in which criticism is the essence of reason. From the evolutionary point of view, criticism is a hard-earned ability that has been achieved on three levels – through the evolution of life on Earth leading to the emergence of the human species, through the advancement from primitive cultures (similar to the Lévi-Strauss's concept of 'cold cultures', relying on the oral transmission of traditions) to 'civilized' societies in which argumentative language and the written word have developed, and through the individual maturing of a child into an adult. Popper views all three developments as transitions from dogmatism to criticism; as he observes, "dogmatic thinking, an uncontrolled wish to impose regularities, a manifest pleasure in rites and in repetition as such, are characteristic of primitives and children" (Popper 1985, 49).

On the animal level Popper illustrates this process through the difference between an amoeba and Einstein; "the difference between an amoeba and Einstein is that, although both make use of the method of trial and error-elimination, the amoeba dislikes erring while Einstein is intrigued by it: he consciously searches for his errors in the hope of learning by their discovery and elimination" (Popper 1979, p. 70). The difference between an amoeba and Einstein, a shaman and Einstein or the baby-Einstein and the adult Einstein "lies in the constructive attitude towards error" (Popper 1985, p. 52). Human adults, especially those with a strong critical temperament, develop imaginative, reflexive and critical forms of intelligence. Popper further illustrates the deepening critical behaviour by the principle of transference. He claims that it is possible to weaken dogmatic beliefs by reformulating belief-sentences in objective (criticizable) terms: for example, "instead of speaking of a 'belief', I speak, say, of a 'statement' or of an 'explanatory theory'" (Popper 1979, 6). Popper applies this procedure to Hume's psychological problem of induction. He argues that "on the basis of the following *principle of transference*, what is true in logic is true in psychology" (Popper 1979, 6).

Popper hopes that this principle can solve Hume's psychological problem of induction; as discussed above, the belief in regularities is instinctive for Hume – yet it is fact of life. Popper agrees with Hume's conclusion that this belief is not rational but denies it is unshakeable. He borrows Hume's example concerning the necessity of sunrise: "[o]ne wou'd appear ridiculous, who wou'd say, that 'tis only probable the sun will rise to-morrow, or that all men must dye; tho' 'tis plain we have no further assurance of these facts, than what experience affords us" (Hume 1981, 124). Popper then restates the problem as follows: "if we look at a theory critically, from the point of view of sufficient evidence rather than from any pragmatic point of view, do we always have the feeling of complete assurance or certainty of its truth, even with respect to the best-tested theories, such as that the sun rises every day? I think that the answer here is: No." (Popper 1979, 26). Popper's conclusion, however, is experimentally unsupported, as I argue elsewhere (Parusniková 2004, 2011), and rather than being convincing it shows his belief in criticism eventually prevailing in our nature.

Popper's attitude can be explained in a wider context. Popper continues the Kantian and the positivist tradition endorsing the universal liberating mission of reason; in the (post)war period the need to defend this role of reason against various forms of irrationalism was felt with an increased urgency. Therefore, Popper called himself the" last laggard of the Enlightenment" (Popper 1988, 177) in matters both of epistemology and social theory. As Hacohen remarks (2002, 69), Popper's position grew from the progressive intellectual atmosphere in Vienna – termed *Spätaufklärung* – that developed before the First World War in opposition to various conservative clerical and nationalist tendencies. The horrors of the war, followed by the disintegration of the Austrian Empire, during which lives were destabilized and all hitherto valid values undermined, made the task of reaffirming the Enlightenment ideal and its optimistic ethos even more pressing.

This task grew even more urgent during and after the Second World War. It was necessary to do more than merely dust off the good traditional ideal. Having experienced the failure of reason, more precisely the failure of societies in the 'enlightened' age to act in a rational, civilized, humane way, rationalism needed to be reformulated to regain its credibility in the new historical circumstances. For Popper, reason had to be identified with criticism, embodied in the falsificationist methodology and, in the social context, with democratic societies legally guaranteeing the possibility of public critical discourse. The progress of history is ultimately a process of *maturation of mankind* in which critical reason gradually takes over dogmatism; this explains why he considers the defence of dogmatism – as in Kuhn's concept of normal science – "a danger to science and, indeed, to our civilization" (Popper 1982b, 53).

To sum up, Popper's *horror dogmatis* is linked to his acknowledgment that *dogmatism is an objective biological determinant of our lives.* Thus it is criticism which needs to be protected from the dogmatic disposition in our nature. This can be done, in Popper's view, only via the methodological imperative that defines criticism as a *purely destructive force*. Thus Popper turns traditional rationalism (termed by him 'comprehensive or uncritical') upside down; so far, as Miller says, traditional rationalism has been "hooked on justification" and now it is time that we "kick the habit"; cold turkey from the justificationist addiction is recommended (Miller 1994, 49). Instead of defining reason as the agent of justification of knowledge, falsely considered a guarantee of the rational (scientific) status of knowledge, Popper's critical rationalism implies a negative definition of reason – *ratio negativa* (Parusniková 2009). However, the negative power of reason is embodied only in the testing procedures; rationality consists not in the rules forced upon discovery but in the critical investigation of all hypotheses and beliefs. In epistemology, Popper changed the focus of the rational attitude *from proving to seeking, from grounding to exploring*. Any declaration of the legitimacy or even rationality of dogmatism – however limited it may be – tames criticism. However, in critical rationalism a methodological norm must not tame criticism or it consequently loses its regulative appeal.

Acknowledgement I would like to thank David Miller for extensive critical comments.

References

Agassi, Joseph. 1975. *Science in flux*. Dordrecht/Boston: Reidel.
Agassi, Joseph, and Ian C. Jarvie, eds. 1987. *Rationality: The critical view*. Dordrecht: Martinus Nijhoff.
Fuller, Steve. 2003. *Kuhn vs Popper*. Cambridge: Icon Books.
Hacohen, Malachi H. 2002. *Karl Popper. The formative years*. Cambridge: Cambridge University Press.
Hume, David. 1981 (1739). *A treatise of human nature*. ed. L.A. Selby-Bigge, Revised P.H. Nidditch, 2nd ed. Oxford: Clarendon Press.
——— 2005 (1748). *An enquiry concerning human understanding*. ed. L.A. Selby-Bigge, Revised P.H. Nidditch, 3rd ed. Oxford: Clarendon Press.
Kuhn, Thomas S. 1970. *The structure of scientific revolutions*. Chicago: Chicago University Press.
———. 1982. Logic of discovery or psychology of research? In *Criticism and the growth of knowledge*, ed. Imre Lakatos and Alan Musgrave, 1–25. Cambridge: Cambridge University Press.
Miller, David. 1994. *Critical rationalism. A restatement and defence*. Chicago/La Salle: Open Court.
Parusniková, Zuzana. 2004. Two cheers for Karl Popper. In *Karl Popper: Critical assessments of leading philosophers*, ed. Anthony O'Hear, 79–99. London: Routledge.
———. 2009. *Ratio negativa* – The Popperian challenge. In *Rethinking Popper*, Boston studies in the philosophy of science, ed. Zuzana Parusniková and Robert S. Cohen, 31–47. Dordrecht: Springer.
———. 2011. Criticism and dogmatism in Popper's evolutionary epistemology. In *Knowledge, value, evolution*, ed. Tomáš Hříbek and Juraj Hvorecký, 109–125. London: College Publications.
Popper, Karl R. 1965 (1959). *The logic of scientific discovery*. New York: Harper and Row.
———. 1979 (1972). *Objective knowledge. An evolutionary approach.* Oxford: Clarendon Press.
———. 1982a (1974). *Unended quest. An intellectual autobiography*. Glasgow: Fontana.
———. 1982b. *Normal science and its dangers*. In Imre Lakatos and Alan Musgrave, eds, 51–59. Cambridge: Cambridge University Press.
———. 1985 (1963). *Conjectures and refutations. The growth of scientific knowledge*. London/Henley: Routledge & Kegan Paul.
———. 1988 (1983). *Realism and the aim of science*, ed. W.W. Bartley III. London: Hutchinson.
———. 2003 (1999). *All life is problem solving*. London/New York: Routledge.
Rowbottom, Darrell. 2010. *Popper's critical rationalism: A philosophical investigation*. London: Routledge.

Chapter 5
Comprehensively Critical Metapolitics

Michael Chiariello

5.1 Introduction: The Search for a Radical Metapolitics

As a graduate student in the late 1960s, my politics and philosophy were a variant of the radicalism propounded by Herbert Marcuse, which held that a revolutionary consciousness, or a radical critique, may only be advanced within a new alternative intellectual framework, which is incommensurable with the established conservative framework. Within the established framework it would remain impossible to ever justify the resort to a radical analysis or revolutionary political strategy. The case for revolutionary change, or radicalism, thus introduces a clash of frameworks making critical discussion and rational choice impossible in the absence of a common understanding of rationality (Chiariello 1983).

At the time, I was Agassi's graduate assistant, but not yet his student. But when we discussed my dissertation topic, the problem of justifying a revolution, Agassi advised me to look into the nature of justification itself and recommended I read Bartley and Popper, the former for the distinction between justificationist and non-justificationist criticism (Bartley 1984), the latter for the proposal of a nonjustificationary and critical view of rationality (Popper 1966). Needless to say, I was thus awakened from my justificationist slumbers.

The context for Popper's proposal of critical rationalism is his defense of the open society. This defense is developed in opposition to comprehensive rationalism that claims it possible to hold all views to the standards of reason, including the belief in reason itself. Comprehensive rationalism supports the politics of radicalism, wherein all social relations are subjected to such a standard. Popper rejects radicalism as well as its concomitant comprehensive rationalism. As an alternative, Popper offers critical rationalism, supported by an "irrational commitment to rationality," which he characterizes as a "minimal concession to irrationalism" (Popper

M. Chiariello (✉)
St. Bonaventure University, St Bonaventure, NY 14778, USA
e-mail: mchiarie@sbu.edu

N. Bar-Am, S. Gattei (eds.), *Encouraging Openness*, Boston Studies in the Philosophy and History of Science 325, DOI 10.1007/978-3-319-57669-5_5

1966, 230). Rejecting the need for such a concession, Bartley argued that Popper unnecessarily opens the door to irrationalism, while a thoroughly critical rationalism can meet its own standard of rationality and be held comprehensively without problem.

In my view, the debate prompted by Bartley's claim is unduly abstracted from the metapolitical question of the best defense of a rationalism supportive of an open society. This essay will survey the debate over the limits of critical rationalism in Popper's original metapolitical project.

5.2 Metapolitics

First, I want to clarify my use of the term "metapolitics." I am not the first to use the term "metapolitics;" claims to the meaning and use of the term range from right to left. The term was first used in English by American poet and political writer, Peter Vierek to describe a nineteenth century version of German nationalism which, he claimed, later became the cornerstone of Nazi ideology (Viereck 2003). Vierek misattributes such illiberal sentiments to Constantin Frantz who coined the term in1878 (Wagner 2013). Today European proponents of right-wing anti-immigrant politics use the term in a Gramscian sense of the cultural conditions for political change. On the left we find Alain Badiou who writes: "By 'metapolitics' I mean whatever consequences a philosophy is capable of drawing, both in and for itself, from real instances of politics as thought" (Badiou 2005). But none of these interpretations have much connection with my interest here, as far as I can tell. I employ the term metapolitics in two senses: (1) the effort to advance political aims through philosophical arguments; and (2) an inquiry into the conditions for the rationality of political action or *praxis*. Indeed, sometimes the debates regarding (2) are instances of (1).

Examples of metapolitical issues of the second type would include theories regarding the choice between revolution and reform, and the limits of political possibility. These themes are richly developed in Popper's metapolitical writings, which encompass both forms of metapolitics. Popper himself claims that *The Poverty of Historicism* and *The Open Society and Its Enemies*, were part of his "war effort" thus advancing a political aim through philosophical argument (Popper 1976).

Another example of metapolitics, in the sense of philosophy as politics, is the exchange known as the *Positivismusstreit* or positivism dispute, matching Popper and proponents of critical rationalism against Adorno representing Critical Theory. Students of this dispute note with some frustration the disputants' "apparent process of talking past one another" (Adorno et al. 1976, xv) David Frisby hypothesized that the debate involves competing and incommensurable theories, and thus lacking the conditions for a rational, critical debate. Frisby writes, "Yet this might in turn have its origin not in conceptual confusion as such but in the situation which Feyerabend refers to as the incommensurability of theories" (Adorno et al. 1976, ix–x; see

Feyerabend 1975). To extend Frisby's hypothesis, we may see the competing theories as reflecting rival metapolitical frameworks, raising second order questions about the rationality of controversies of this sort. So we must ask whether reason is comprehensive enough to frame a meta-level discussion such as the positivism dispute, or whether we have opposing and incommensurable frameworks underwritten by irrational commitments about which further discussion is pointless.

Popper's proposed critical rationalism, expounded in the famous chapter 24 of the *Open Society* (1945), is part of a general critique of radicalism, and is offered as both a critique and alternative. As I read it, the most fundamental argument Popper frames against radicalism is the charge that it rests on a overly ambitious conception of rationality, comprehensive rationalism. And he eschews comprehensive claims for the scope of reason arguing that the adoption of rationality is itself irrational. According to Popper, the aim of the critical rationalist is to minimize the irrationality of the commitment to rationality. Bartley's response is that critical rationalism can be reflexively, and thus comprehensively, maintained and the concession to irrationalism be eliminated totally.

Assuming the success of Bartley's proposal, my new question was whether the reintroduction of a comprehensive form of rationality removes a central barrier to the acceptance of radicalism – and if the answer is yes then what sort of radicalism are we talking about. The answer to this question I refer to as "comprehensively critical metapolitics".

5.3 The Problem of Rationality

What is the problem of rationality? According to Agassi, we may see the problem as one of abstract philosophy, i.e., "what is foundation of my intellectual system?" or as a more-or-less concrete problem of life, "on what shall I base my life?" (Agassi 1981a, b, c, 1991) It is thus commonly understood as the quest for a foundation or a starting point, a so-called "Archimedean point" as Descartes expressed it (1960, 23).

However, if every decision process has a starting point then how does one decide on a starting point? Or, how does one avoid circularity or infinite regress? Or, is there a non-arbitrary way to choose our beliefs? That is, is there a rational choice of a criterion for rational choice? Thus the question is whether we can be completely rational, that is, whether we choose our starting point rationally. And if we consider rationalism, as a philosophy, a point of view, or a matter of choice we see that the question of whether to be rational cannot be answered affirmatively without circularity.

Popper concludes that rationalism is a matter of faith, an irrational or non-rational commitment. But Bartley asks, do we not undermine our choice, specifically the choice to be rational, by inviting equivalence with those who reject rationality? We are already at a high level of sophistication concerning the problem and the difficulties of available approaches. Do these difficulties contribute to a rational case against rationality, or for irrationalism? This is the famous *tu quoque* problem, the rationalist excuse for irrationalism, which has been discussed by Bartley (1984, 71–78). Ironically, Popper's intent in stipulating the irrationality of rationalism was completely opposite, to prevent the defeat of rationality.

The abstract philosophical problem of rationality is to show that the decision for rationality is not just another arbitrary choice, but rather that it is one which is superior to other choices and particularly the choice to avoid rationality in belief and life-style. For Bartley "the dilemma of ultimate commitment or of the limits of rationality" is expressed as the question "How is it possible any longer for a man to remain a consistent rationalist?" (1984, xxv–xxvi).

The failure of classical, or comprehensive, rationalism to meet its own standard of rationality undermines the authority of rationalism in general, and enhances the appeal of irrationalism. The choice, however, is not simply between irrationalism and comprehensive rationalism; there remains Popper's own alternative, critical rationalism.

5.4 Popper's Critical Rationalism

Popper offers his theory of critical rationalism as a modest alternative to "comprehensive rationalism." The latter is characterized by the claim: "I am not prepared to accept anything which cannot be defended by means of argument or experience." But this claim "cannot in its turn be supported by argument or experience" and is thereby inconsistent (1966, 224). Since a rationalist attitude must already be adopted before argument and experience can be effective, rationalism cannot be based upon argument and experience. Thus the adoption of rationalism must rest on some ultimate assent to a belief or proposal. Popper describes this, in what I consider an unfortunate phrase, as "an irrational faith in reason" (1966, 230).

According to Popper the distinctive feature of critical rationalism is that unlike comprehensive rationalism, it admits the irrationality of its basic commitment. As Popper says the adoption of the "rationalist attitude [...] may be called irrational. [...] an irrational *faith in reason*" (1966, 230). This spares critical rationalism from self-defeat by exempting its basic commitment from the conditions of rationality and by keeping this irrational element to a minimum.

Having offered a more tenable version of rationalism, we may contrast it with irrationalism, and more importantly we may better understand what is at stake in the choice between them. We may consider arguments favoring rationality such as the moral reasoning offered by Popper (Gattei 2002). Rationalism, he asserts, best supports such attitudes as tolerance, impartiality, and even nonviolence. "Critical reason is the only alternative to violence so far discovered." The use of "method of rational critical discussion [...] would make the use of violence obsolete", and therefore it is "the obvious duty of all intellectuals" to promote this outcome by supporting rationality. (Popper 1994a, 69). Popper's concession to irrationalism makes the dispute between rationalism and irrationalism more rational.

But what, we might ask, is this rationality to which we are thus irrationally committed? Popper answers, it is an attitude characterized by the statement: "I may be wrong and you may be right, and by an effort we may get nearer to the truth." (Popper 1966, 225) It is an attitude of openness to, or perhaps eagerness for, criticism. Popper proposes a philosophical program which generalizes the ethos of the open society. (Bartley 1984, 140) The Popperian program is intellectually revolutionary: we should give up our traditional search for justification and certainty, and replace it with the search for criticism and the elimination of errors. This rationalism urges us to express our views clearly enough to expose them to criticism and to be willing to surrender views when criticism becomes refutation, rather than shielding them from criticism.

5.5 Bartley's Comprehensively Critical Rationalism

It is important that we distinguish Popper's proposal of a nonjustificationist theory of rationality – as outlined above – from critical, or limited rationalism which may be justificatory or not. The former concerns the nature of rationality, whereas the latter concerns its limits. And Bartley maintains that the consistent application of the nonjustificationist theory of rationality allows us to resolve the question of the limits of rationality in a way which is both more comprehensive than and more consistent with Popper's other views.

According to David Miller, Popper has failed to note that the split between comprehensive and critical rationalism does not parallel the division between justificationism and nonjustificationism. Therefore the question that is not clearly enough addressed is whether the failure of comprehensive justificatory rationalism is a consequence of the requirement that its conditions be reflexive (comprehensive rationalism) or whether it fails because rationality is identified with justification (justificationism) (Miller 1994).

In *The Open Society*, Popper concludes that rationalism must acknowledge its foundation is an unjustified, and hence irrational, decision. Consequently, *critical* or noncomprehensive rationalism could be either justificatory or nonjustificatory. Here Miller identifies Bartley's immense contribution to the discussion of rationality with the discovery that critical justificatory rationalism undermines rationalism unnecessarily. Thus critical rationalism, as originally formulated in *The Open Society*, is not an adequate solution to the problem of rationality. Popper's concession that even rationalists hold their basic beliefs irrationally offers a rational excuse for irrationalism; this is the *tu quoque* argument. However an alternative, and more consistent formulation of a Popperian theory of rationality obviates this difficulty. This alternative was originally named "comprehensively critical rationalism" (CCR) by Bartley who later renamed it "pancritical rationalism." (Bartley III 1990) CCR combines the idea of rationality as criticism, which Bartley accepts from Popper, with a comprehensiveness which was rejected by Popper. His proposed modification is that we maintain this nonjustificationist view comprehensively. That is, we should hold our openness to criticism open to criticism. And this eliminates the claim that the adoption of rationalism is irrational.

According to Bartley, critical rationalism avoids one problem but invites another: the *tu quoque* challenge from irrationalism. That is to say that irrationalists can claim that their initial commitment, to whatever form of irrationalism, is the same as the adherents to critical rationalism, because we must all hold our initial commitments irrationally regardless of the substance of those commitments, and this implies that all commitments are equally irrational. The irrationalist is entitled to hold beliefs in the same way as the critical rationalist.

As Bartley interprets the problem of rationality, the difficulty is not the expectation of reason's comprehensive scope, but the combination of such expectations with justificationism. Bartley asserts that expectations of comprehensiveness can and should be met but by a nonjustificational critical approach. What should be abandoned is the idea that rationality is identified with justification (Bartley 1984).

Bartley maintains that unlike comprehensive justificatory rationalism, CCR can be practiced without inconsistency, i.e., it can be held open to criticism. Therefore CCR can meet its own standard of rationality. Moreover since it maintains the rationality of its own theory of rationality, CCR gives no quarter to irrationalism, thus cutting off the *tu quoque*. (Bartley 1964, 7) Comprehensive rationalism could not meet its own standard of rationality, justification of all our beliefs, so it was in a sense a logical impossibility. But we can criticize CCR, Bartley claims, or at least hold the theory open to criticism. This last point is the target of a debate concerning CCR to which we turn.

5.6 The Debate Over Comprehensively Critical Rationalism: Watkins's Critique

Bartley claims that by abandoning justificationism we can maintain a consistent and comprehensive rationalism because although we cannot justify rationalism without logical difficulty we can hold our acceptance of rationalism open to critical examination. Thus rationality meets its own standard (comprehensively applied to itself) in a way that deprives irrationalism of recourse to a *tu quoque* accusation. Rationalism is as rational, by its very own lights, as any other view which may be held open to criticism. However, a necessary condition of the rationality of rationalism is its criticizability, since as Bartley claims "[…] someone could devastatingly refute this kind of rationalism if he were to produce an argument showing that at least some of the unjustified and unjustifiable critical standards necessarily used by a comprehensively critical rationalist were uncriticizable to boot". (Bartley 1984, 149). Is CCR criticizable?

Among the first to take up the challenge, J. W. N. Watkins charges that CCR cannot be criticized because it is so structured that "a defender of CCR will always be assured of victory over his critics however good their criticisms may be." (Watkins 1969, 60) Watkins formulates Bartley's thesis thus:

> CCRw: "That a rationalist can and should hold all his positions open to criticism is itself a position that he can and should hold open to criticism" (Watkins 1969, 58).

Paradoxically any attempt to falsify that thesis will have the opposite effect by showing it to be criticizable and hence true. Says Watkins, "If the critic comes nowhere near to meeting the challenge, Bartley wins; and if the critic does come near to meeting the challenge, the critic loses, since his nearly successful criticism establishes criticisability" (Watkins 1969, 60). Either way Bartley wins and critics lose: "Bartley's claim that CCR is criticisable will be reinforced by every attempt to refute it" (Watkins 1969, 89). Not only does Watkins show the uncriticizable nature of CCR but he exposes it to be a form of reinforced dogmatism (Watkins 1971).

To expand: If we are to hold openness to criticism as a standard of rationality there must be an interesting distinction between criticizable and uncriticizable statements (Watkins 1971, 58). The notion of criticism must contain the possibility of refutation. That is, some criticisms must be decisive. A decisive criticism, a refutation, would demonstrate that CCR is uncriticizable (Watkins 1971, 59). But paradoxically such a refutation of CCR would backfire for it would also establish the criticizability of CCR. If uncriticizability is a criticism then CCR can never be found uncriticizable, therefore CCR can never be refuted, and therefore CCR cannot be held open to criticism. However, the critique of CCR invariably will strengthen it by establishing criticizability.

No doubt it is paradoxical to offer as a decisive criticism of a theory the claim that it is immune to decisive criticism. Watkins sees in the self-characterization of CCR something of a ruse: Bartley invites criticism no doubt, but his invitation is something of a "heads-I-win-tails-you lose proposition (Watkins 1971, 60). Similarly, it is ironic to find that "a position presented as the *ne plus ultra* of intellectual anti-authoritarianism should turn out to have this 'dictatorial' or 'dogmatic' character contrary to the author's intentions." (Watkins 1971, 60).

Watkins's comments do not establish the uncriticizability of CCR unless that were the only possible criticism. It remains possible to criticize CCR. We could, for example, claim that CCR had immoral implications, that it deterred research, that it had dangerous political implications and so on. These are certainly not trivial criticisms. Moreover, they establish good and rational grounds to abandon CCR. Having thus established CCR's criticizability we have done little more that invite further critical efforts.

The failure to demonstrate uncriticizability through efforts at criticism does not establish the dogmatic character of CCR. The latter would be dogmatic if it turned criticism into positive support – this is what is meant by the 'reinforced dogmatism' with which Bartley is charged (Bartley 1964, 7). CCR neither seeks nor needs such positive support which would make it a justificationist theory. Watkins seems to think that demonstrating the criticizability of CCR would be a justification – seeing criticizability as a sufficient rather than necessary condition of the success of CCR (Agassi et al. 1971). But what do we make of Watkins' charge? Clearly Watkins has pointed us to the paradoxical nature of CCR in making uncriticizability into a criticism. Is such a paradox a decisive criticism? Or, does the paradox establish criticizability which in turn establishes validity? This reads criticizability as both a necessary and sufficient condition for accepting CCR. That is a misinterpretation of Bartley. Moreover I think since we can simply stipulate that semantic paradoxes cannot be used to immunize CCR it is rather unfair to charge dogmatism (Agassi 2012).

In response to Watkins's critique, Agassi et al. argue that CCRw is based on a misplaced maxim that in a critical debate we should endorse the winning, or "victorious view" (Agassi et al. 1971, 44–5).

Moreover, contrary to Watkins, any demonstration of criticizability of CCR need not be seen as a "heightened inducement to hold CCR" but rather "might force one to reject CCR" (Agassi et al. 1971, 44). Likewise John Kekes argues that it is true that "no criticisms will show the uncriticisability of CCR, but some may show that [CCR's] description of rationality is inaccurate" and therefore we may be able to "specify the conditions under which CCR would have to be abandoned" (Kekes 1971, 51).

Stefano Gattei argues that Bartley's comprehensive version of critical rationalism is more concerned with refuting various forms of irrationalism to which he thinks Popper's concession gives unnecessary comfort, and thereby misses the ethical element in critical rationalism. Gattei is no doubt correct in reporting the profound ethical dimension of Popper's defense of reason. Recall that Popper had spoken of rationalism as "the obvious duty of all intellectuals" (1994a).

According to Gattei, Bartley's analysis of Popper's position "seems entirely drawn from logic, thus disregarding the profound ethical nature of Popper's choice" (Gattei 2002, 247). But I ask why are logic and ethics proposed as exclusive disjuncts? Certainly the act of choosing, or deciding, and the object of one's decision may be distinguished but I cannot see how they can be separated, or why attention to one may exclude attending to the other, or further, why one needs to embrace Popper's critical dualism to support this split? Gattei says, "an attitude is neither true nor false […]" and therefore "cannot be replaced by a theory of rationality" (2002, 248). I do not know whether this means that an attitude cannot be expressed by a theory of some sort. Nonetheless, Gattei does say that this attitude or decision "can be supported by argument" (249) leaving me wondering what kind of an "irrational faith" we have here?

For Popper ethics is a matter of decision, rather than truth-functional statements. Yet the tenets of critical rationalism can be formulated and evaluated as we have seen. I believe this is what Gattei means when he complains that the debate is more logical than ethical. He has a point. A theory of rationality may be examined logically and it is no surprise that questions of comprehensiveness lead to paradox. So I think there may be two questions here: What is the theory to which I am committing and how does it respond to the skeptic's challenge? As we learn from Agassi, the famous *tu quoque* argument is nothing more than the generalization of the skeptical charge of inescapable circularity of any comprehensive theory. If so, Bartley's criticism of Popper's proposed critical rationalism is nothing more than a restatement of the skeptical critique.

It appears that there are two options here: (1) Popper's critical rationalism, with its irrational commitment, is not rational enough by its own critical lights, and therefore we must replace it with a theory that allows a more thoroughgoing and reflexive defense, i.e., Bartley's comprehensively critical, or pancritical, rationalism. Or (2) Popper's critical rationalism, if properly understood, is already open to criticism, and thus sufficiently rational. And I credit Bartley with moving us toward a better understanding of the strength of critical rationalism.

Does Popper need to characterize his embrace of rationalism as irrational? I think not. It would be irrational if we were looking for a non-question-begging justification, but this is no longer necessary to maintain rationality. Is Bartley correct in seeing Popper's concession as a license for a more thoroughgoing irrationalism or dogmatism? To the extent that Popper's concession is unnecessary as Bartley argues, I think not. But more to the point, Popper's own commitment is tentative.

It is clear that Popper intends to hold his so-called irrational commitment, tentatively and with the attitude of "I may be wrong [...]". According to Agassi this is anything but fideism, where beliefs are held with stubborn tenacity even in the face of critical challenges (Agassi 2012, 25). So before an irrationalist is able to proclaim *tu quoque*, he would need to maintain his commitment following Popper's tentative and critical manner. These observations imply that Popper does hold his rationalism comprehensively if by that is meant his commitment is thoroughgoing. Indeed it is possible to hold to a form of critical rationalism stubbornly, if by critical rationalism we mean a discursive theory, but it makes no sense to hold the critical attitude any way other than critically, or as Popper says, "critical rationalism never ceases to criticize itself" (1994b, 191).

Bartley has shown it unnecessary to ground rationality in a decision not subject to rational discussion regardless of whether he allows this to Popper. In other words, I maintain that the success of Bartley's efforts is not to be judged by the tenability of CCR as a theory but rather by his success in eliminating some dogmatic or allegedly fideistic elements from critical rationalism (here defined as nonjustificationism) although more may remain. The weakness of Popper's irrational act of faith in rationality is removed if we can correctly characterize such an act of faith as rational and that may be accomplished by a willingness to entertain criticisms of the faith.

I think this sounds too simple, but Agassi offers a more subtle account of the matter. (Agassi 2012) Both Popper and Bartley agree that the traditional skeptical critique renders the program of classical rationalism question begging, and both reject the idea of rationality as proof or justification. However whereas Popper insists that rationalism cannot be held comprehensively, Bartley says that it can and should be, and that Popper's great contribution, the move from justification to criticism, renders this possible. Critical rationalism thus construed makes unnecessary Popper's "concession to irrationalism" because it is possible to hold the commitment to rationality open to criticism. The real issue involves the degree to which Bartley's alternative, comprehensively critical, or pancritical rationalism constitutes an alternative theory of rationality. Of course, this is a different matter. To Bartley's claim that Popper's critical rationalism presents a rational excuse for irrationalism, or the *tu quoque* argument. Agassi responds that the *tu quoque* is a variant of the skeptical argument against classical rationalism as question begging, which both already accept (Agassi 2012, 25).

According to Agassi, Bartley wrongly presupposes that the Popper's concession to irrationalism is a form of fideism in which a chosen fundamental axiom is exempted from critical discussion. But this is clearly not the case. Perhaps it is not necessary to choose between Bartley and Popper. It would be enough to accept Bartley's criticism and the conclusion that we may achieve a theory of rationality which can be subjected to critical discussion. As far as tu quoque is concerned the question remains whether it is a serious argument indicating a weakness in critical rationalism or perhaps mostly a piece of rhetoric, prompted by an unfortunate choice of words by Popper, rather than a valid argument.

5.7 The Political Significance of a New Theory of Rationality

What needs to be examined is the following: (1) Assuming that Bartley has improved the Popperian view of rationality, does this set the stage for improvements in Popperian social philosophy? (2) Does Bartley eliminate the limits of rationality in such a way that allows the rationality of radicalism in social philosophy and metapolitics? Has he eliminated the basis for Popper's critique? Or (3) does this expansion of rationality strengthen the case against irrationalism or pseudo-rationalistic social theories or enable us to distinguish among more or less rational expressions of radical theory?

Bartley's contribution is to show how unnecessary is Popper's "concession to irrationalism" in support of rationality. Popper's concession is also quite harmless if properly understood. Disregarding this concession helps us to see how nonjustificationism represents a new departure in rationality. What need be seen is how the removal of this concession affects the critique of radicalism and other metapolitical issues.

Moving from Popper's limited critical rationalism to a comprehensive critical rationalism may better support the rationality of radicalism. It should be noted that neither Popper nor Bartley himself saw such implications: indeed successive republication of the *Open Society*, even after his encounter with CCR, leave Popper's critique of radicalism untouched (See also Popper 1948 [1963]). And of course this also questions what advance in social theory is achieved by Bartley's version of Popper's views. For example, Agassi notes the absence in social theory of an analogy to the criteria by which scientist's judge the initial worth of hypotheses. That is, the *Open Society* lacks a discussion of "a criterion for attempting to implement social reform" (Agassi 1993, 218). I would add that also lacking are criteria for the selection and prioritization of problems and needs for reform, and this is even logically prior to the choice of solutions.

As Watkins writes, the discussion of the problem of rationality since Bartley has distinguished two dichotomies and thus four possible theories: comprehensive or limited, justificationist or nonjustificationist (Bartley and Radnitzky 1987, 270). Thus the critique of comprehensive justificationism is resolved by eliminating justificationism rather than by eliminating comprehensiveness. On which of these two does radicalism depend for its rationality?

Whether or not it is held comprehensively, nonjustificationism involves both new doctrines and new attitudes. CCR combines the belief, that we can (or should) be rational about all of our beliefs and the attitude of openness to criticism, including that which is self-limiting (Agassi et al. 1971, 45). Offering CCR as an alternative to comprehensive justificationism provides an alternative to irrationalism for a rationalist seeking a consistent conception of rationality and faced with the self-defeat of comprehensive justificationism.

Resolving the problem of the rationality of radicalism requires the resolution of the problem of rationality, then we must submit the terms of our controversy to an examination similar to the one given above. The collective pursuit of a new theory of rationality which we have outlined above will be our guide in eliminating the attitudes and doctrines which defeat efforts to rationalize radical controversies.

Popper's proposal for a nonjustificational rationalism must be seen as the initiation of a new philosophical program, which includes the effort to clarify and develop the new theory of rationality, such as we have surveyed above, and tasks directed at establishing new areas of application for the theory. The latter involves testing the adequacy of Popper's philosophy in specific areas. Most importantly, we consider how critical nonjustificatory rationalism makes it easier to resolve the problems of rationalist social philosophy.

At one time it seemed to me that Bartley's critique of the tu quoque was correct: Popper's critical rationalism opened the door too widely for social theories indisposed to defend first principles rationally. I am grateful to find the discussion had clarified the nature of initial commitments appropriate to critical rationalism. Rather than reserving the field to forms of political thought not open to dialogue across ideologies, a deeper understanding of Popper's solution to the problem of rationality signals a new comprehensively critical metapolitics, or more simply a self-critical radicalism where "critical" is understood to mean nonjustificatory and "comprehensive" to mean self-critical.

References

Adorno, Theodor W., Hans Albert, Ralf Darendorf, Jurgen Habermas, Harold Pilot, and Karl R. Popper. 1976. *The positivist dispute in German sociology*. Trans. Glyn Adey and David Frisby. New York: Harper Torchbooks.

Agassi, Joseph. 1981a. *Science and Society*. Dordrecht: Reidel.

———. 1981b [1968]. Science in flux: Footnotes to Popper. In *Boston studies on the philosophy of science*, Vol. 3, ed. Robert S. Cohen and Marx W. Wartofsky. Dordrecht: Reidel.

———. 1981c. Rationality and the Tu Quoque argument. In *Science and society: Studies in the sociology of science*, ed. J. Agassi. Dordrecht: Reidel.

———. 1987. Theories of rationality. In *Rationality: The critical view*, ed. J. Agassi and I. Jarvie. Dordrecht: Martinus Nijhoff.

———. 1991. *The siblinghood of humanity: An introduction to philosophy*. Delmar: Caravan Books.

———. 1993. *A philosopher's apprentice: In Karl Popper's workshop*. Amsterdam: Rodopi.

———. 2012. Critical rationalism, comprehensive or qualified: The Popper-Bartley dispute. In *Der Kritische Rationalismus als Denkmethode und Lebensweisse: Festschrift zum 90. Geburtstag von Hans Albert*, ed. Giuseppe Franco. Klagenfurt: Kitab.

Agassi, Joseph, and Ian C. Jarvie, eds. 1987. *Rationality: The critical view*. Dordrecht: Martinus Nijhoff.

Agassi, Joseph, Ian C. Jarvie, and Tom Settle. 1971. The grounds of reason. *Philosophy* 46: 43–50.

Albert, Hans. 1985. *Treatise on critical reason*. Princeton: Princeton University Press.

Badiou, Alain. 2005. *Metapolitics*. London/New York: Verso.

Bartley, W.W. III. 1964. Rationality vs. theories of rationality. In *The critical approach to science and philosophy: Essays in honor of K. R. Popper*, ed. M. Bunge. London: Macmillan.

———. 1984. *The retreat to commitment*. 2nd ed. La Salle/London: Open Court.

———. 1990. *Unfathomed knowledge, unmeasured wealth*. La Salle: Open Court.

Bartley, W.W. III, and Gerard Radnitzky. 1987. *Evolutionary epistemology, rationality, and the sociology of knowledge*. La Salle: Open Court.

Chiariello, Michael. 1983. The philosophical critique of radicalism and its limits. *Philosophy Research Archives* 9: 193–216.

Descartes. 1960. *Meditations*. Trans. L.J. Lafleur. Indianapolis/New York: Bobbs-Merrill.

Feyerabend, Paul. 1975. *Against method*. London: New Left Books.

Gattei, Stefano. 2002. The ethical nature of Karl Popper's solution to the problem of rationality. *Philosophy of the Social Sciences* 32 (2): 240–266.

Kekes, John. 1971. Watkins on rationalism. *Philosophy* 46 (175): 51–53.

Miller, David. 1994. *Critical rationalism: A restatement and defense*. Chicago/La Salle: Open Court.

Popper, Karl. 1945 [1966]. *The open society and its enemies*. 5th rev ed. London: Routledge & Kegan Paul.

———. 1948 [1963]. Utopia and violence. In *Conjectures and refutations: The growth of scientific knowledge*, 355–363. London: Routledge & Kegan Paul.

———. 1976. *The unended quest*. London: Open Court.

———. 1994a. The myth of the framework. In *Defence of science and rationality*, ed. Mark A. Notturno. London: Routledge.

———. 1994b. Reason or revolution? In *The myth of the framework*, ed. Mark A. Notturno. London/New York: Routledge.

Viereck, Peter. 2003. *Metapolitics: From Wagner and the German romantics to Hitler*. Piscataway: Transaction Publishers.

Wagner, Richard. 2013. Open letter to Richard Wagner. In *Bayreuther blätter-primary source edition* (German Edition). Charleston: Nabu Press.

Watkins, John. 1969. Comprehensively critical rationalism. *Philosophy* 44: 45–51.

———. 1971. CCR: A refutation. *Philosophy* 46: 56–61.

———. 1987. Comprehensively critical rationalism: A retrospect. In *Evolutionary epistemology, rationality, and the sociology of knowledge*, ed. G. Radnitzky and W.W. Bartley III, 270–271. La Salle: Open Court.

Chapter 6
Feng Shui: An Overlooked Topic for Critical Rationalists

Michael R. Matthews

Joseph Agassi has published extensively on history, philosophy, sociology and politics of science. Less known, but also extensive, he has over a long span of time published on education (Agassi 2014), including a long review of my own book on 'History, Philosophy and Science Teaching' (Agassi 2014, Chap. 16). The research field he has carved out is properly labeled 'Critical Rationalism'. At many points, Agassi discusses pseudoscience, superstition, 'intellectual rubbish', and other such common commitments. He endeavours to tease out the identifying philosophical features of such positions; and their social psychology, specifically why are they believable?

One subject not mentioned in Agassi's extensive *oeuvre* is Feng Shui. This is no special fault, as it is equally not mentioned by any modern philosopher or educator who has discussed pseudoscience, the demarcation question, or the responsibility of schools for engaging in appraisal of worldviews. There is a peculiar critical blind-spot about the subject. This *Festschrift* contribution is meant as an entrée to the history, philosophy, and cultural studies of Feng Shui; and to show that it is worthy of, indeed demanding of, philosophical and educational attention by critical rationalists.

Feng Shui is an ancient Chinese worldview. It is a system of beliefs and practices originating some three to four thousand years ago that is concerned with identifying, charting, and utilizing the supposed all-encompassing flow of chi or *qi*, the putative universal life force, so that people's lives and their habitat can be brought into harmony with it, made more natural, and so improved.[1] It is a significant feature of Chinese and Southeast Asian cultures. And it has long migrated from Asia: millions

[1] Feng Shui's origins, history, philosophy and applications have been exhaustively written upon. See, among others, Bruun (2008), Bruun and Kalland (1995), Eitel (1873/1987) and Smith (1991, Chap. 4). There are countless sources on the web including an informative Wikipedia entry.

M.R. Matthews (✉)
School of Education, University of New South Wales, Sydney, NSW 2052, Australia
e-mail: m.matthews@unsw.edu.au

N. Bar-Am, S. Gattei (eds.), *Encouraging Openness*, Boston Studies in the Philosophy and History of Science 325, DOI 10.1007/978-3-319-57669-5_6

of believers throughout the world regulate their life by Feng Shui theory, worldview or ideology. Yet it is a neglected topic in science education and is ignored in most philosophical discussions of pseudoscience and the demarcation dispute; discussions where it might be expected to be mentioned and used as a case study.

For both educational and philosophical purposes, it is informative to examine two historical occasions where Feng Shui was engaged by Western science and philosophy. It is noteworthy that it was Christian missionaries that prompted the engagements. First, the Jesuit priest Matthew Ricci in the late sixteenth century; second, the Lutheran pastor Ernst Eitel in the mid-nineteenth century. Both episodes have historical, philosophical, political, cultural, theological, scientific and educational dimensions.

6.1 Matthew Ricci (1552–1610)

Matthew Ricci was one of the first Europeans to give an informed and detailed appraisal of Feng Shui belief and practice in China, and certainly one of the first whose account gained wide readership.[2] He was among the earliest, and foremost, of the Jesuit missionaries who were sent to Ming-dynasty China in the late sixteenth century following the thwarted efforts of Francis Xavier.[3] As well as philosophical and religious training at Jesuit universities in Rome, Ricci had serious mathematical, scientific and technical competence. At the *Collegio Romano* he studied mathematics and astronomy with the famed German Jesuit Christopher Clavius, who was known as 'the Euclid of the sixteenth century'. He was director of the Gregorian calendar reforms and a confidant of Galileo to whom he sent his mathematical and astronomical papers.

With fourteen fellow Jesuit missionaries, he left Lisbon in 1578 sailing to Goa where he spent three years completing theology studies; then sailed on to and arrived in Portuguese Macao in 1582. There he immersed himself in the study of Chinese language, literature, customs and history. His quick mastery of so much completely novel material was abetted by his astonishing, and much commented-upon memory. After one reading, he could repeat random lists of hundreds of characters. He had written an essay on memory during his studies in Rome.[4] Ricci's life in China is a justly celebrated 'first contact' between Western philosophy and theology and Eastern religion, metaphysics and culture, and consequently something from which a better understanding of both traditions can be obtained.[5]

---

[2] For studies of Ricci's life and influence see: Cronin (1955), Hsia (2012), Spence (1984), Rule (1972, 1986), Tang (2015a) and Wright (2010).

[3] On the early history of the sixteenth-century Jesuit missions, and their precursors, see Rule (1986).

[4] On Ricci's memory feats, see Spence (1984).

[5] Famously, Marco Polo (1254–1324) travelled through China but his travel *Journal* does not deal with Chinese beliefs - Taoism, Buddhism and Confucianism - in a historical and scholarly way.

In 1583 Ricci began his 27-year sojourn in China. It was a country for whose culture and achievements he had the greatest admiration, incorporating many customs within the practice of the embryonic Chinese Catholicism that he and fellow Jesuits were establishing. He moved first in the south – Nan-chang, Shao-chou and Chao-ching – then finally, in 1600, gained admission to Beijing where he spent the final decade of his life. He had the highest regard for Confucius and the social, political and personal commitments of early Confucianism; he defended, to a point, Confucian beliefs, which he and the Jesuits regarded as more compatible with Christian belief, against imported Buddhism and neo-Confucianism; he adopted the dress and demeanor of Confucian teachers. Ricci died in 1610 in Beijing where his grave is still maintained and respected, and was the site of the Communist Party's 1960 celebration of the 350th anniversary of his death – one of the few places in the world where the anniversary was celebrated. Ricci was a pioneer of European-Chinese cultural engagements. The *Encyclopedia Britannica* claims that 'Probably no European name of past centuries is so well known in China as that of Li-ma-teu (Ricci Matteo)'.

In 1615 Ricci's travel journal, *On the Christian Mission among the Chinese*, was published posthumously in Rome in Latin with the oversight of his fellow Jesuit, Nicolas Trigault. Within a decade four Latin editions appeared along with three French, and one each of German, Italian and Spanish editions. Three centuries passed before the first English edition which was the 1953 translation by Luis Gallagher SJ of Trigault's 1615 Latin text. Gallagher's translation, 620 pages, was given the title: *China in the Sixteenth Century: The Journals of Matthew Ricci, 1583–1610* (Ricci 1615/1953).[6]

Ricci's admiration of China, its culture and its Confucian traditions is evident throughout his *Journals*. Gallagher wrote of Ricci that:

> The author of numerous works on science and religion, written in Chinese, Ricci was well known to the educated classes of China as a prominent professor of physics, mathematics and geography, as a learned philosopher of Chinese and of extraneous doctrine, as a prominent commentator on Confucius, and particularly as an eminent teacher of the Christian religion. Some of his Chinese compositions [...] are included in the official index of the best Chinese writings of all time. [...] His first Chinese *mappamondi* stood as a model for European cartographers for a century after its publication; and earned him the title, The Ptolemy of China. His correction of the Chinese calendars was a preface to a century and a half of scientific advancement in China.
>
> The nobility of character of the Chinese people, their love of liberty, of order, and of learning, their devout tendency toward religion, and their keen sense of justice and of ethical interpretation, were never more clearly revealed than when set forth in what Ricci calls

There had been earlier Franciscan missions and Macao-based Portuguese traders and missionaries in China, but none left such a large, well-circulated, and influential trove of letters, reports and manuscripts as did the Jesuits. These informed much Enlightenment discussion of China and non-European religion and worldviews.

[6] There has been some debate about how much Trigault inserted himself into the 1615 Latin edition. But it seems that with some exceptions it is a faithful translation of Ricci's original Italian text (Rule 1972, pp. 122–24).

> his summary study of the customs, laws, institutes and government of the Chinese people. (Ricci 1615/1953, p. xxi).

As a Jesuit missionary, Ricci's overwhelming preoccupation was with effective preaching of the Christian Gospel to the Chinese; especially with laying out its wisdom in comparison to esteemed Confucian belief, which was the belief of the *Literati* or the Chinese scholarly sect. To further this mission Ricci wrote many studies, in Chinese, Latin and Italian, on Confucian philosophy, history and literature. In one of his Chinese studies, *T'ien-chu shih-i* he counselled the *Literati* on how best to support Confucius and oppose Buddhism and Taoism. This text displays something of Ricci's character and why he enjoyed contemporary and subsequent esteem:

> It is better to argue against them than to hate them, and better to convince them by reasoning than to argue against them. The followers of the two sects are all made by the Father of the Lord of Heaven, and so are our brothers. [...] In my extensive reading of Confucian books I have frequently come across expressions of hatred for the two sects, ranking them with barbarians, and accusing them of heresy, but I have not seen the application of first principles in order to refute them. [...] But if we both use reasoning in conducting our argument, then without harsh words the right and the wrong will appear, and the three schools will return to unity. (Rule 1972, p. 149).

Ricci was also occupied with the internal task of formulating a view about the appropriate 'style' or 'organisation' of the Catholic Church in China – the 'China-Rites' issue as it has been called.[7] But his journal provided, along with much else of historical and cultural value, a much commented upon account of the Chinese Taoist metaphysical and geomancy traditions. It is a 'first contact' document of Western religion and science with Chinese religion and worldview.

Ricci's mathematical, astronomical and chronological knowledge was taken up in a purely utilitarian way by the Chinese court and mandarins; it was seen to improve astronomy, astrology and calendar calculations upon which so much of the functioning of the Emperor's state apparatus depended. Ricci's 'natural philosophy', or his Western science, had no impact on Chinese culture; this was seen by the court, mandarins and *literati* as a positive outcome.

Chapter Nine of Book One of the *Journal* is titled 'Concerning Certain Rites, Superstitious and Otherwise'. Ricci's opening admonition is: 'I would request of the reader that he recognize in the two following chapters a reason for sympathizing with this people and for praying God for their salvation' (Ricci 1615/1953, p. 82). He then proceeds by noting:

> No superstition is so common to the entire kingdom as that which pertains to the observance of certain days and hours as being good or bad, lucky or unlucky, in which to act or to refrain from acting, because the result of everything they do is supposed to depend upon a

[7] Ricci argued with other missionaries and with the Vatican over how much of Confucian ritual concerning ancestors, burial, etc. and metaphysical beliefs, could be maintained by converts to Catholicism. He wanted to maximise retention. Against Ricci's advice, the Vatican ultimately said that converts had to leave their cultural rituals and language about the 'Lord of Heaven' at the church door.

> measurement of time. This imposture has assumed such a semblance of truth among them that two calendars are edited every year, written by the astrologers of the crown and published by public authority. These almanacs are sold in such great quantities that every house has a supply of them [...] and in them one finds directions as to what should be done and what should be left undone for each particular day, and what precise time each and everything should be done. In this manner the entire year is carefully mapped out in exact detail. (Ricci 1615/1953, pp. 82–83).

Ricci's observations have been confirmed by all commentators on Chinese history. Richard Smith in his work on *Divination in Traditional Chinese Society* writes: "Although its particular manifestations and social significance may have varied from time to time, place to place, and group to group, divination touched every sector of Chinese society, from emperor to peasant" (Smith 1991, p. 9). And later that 'geomancy served remarkably well as shared lingua franca for Chinese at every level of society (Smith 1991, p. 170). Ricci elaborated on divination:

> These people worry a great deal about judging their whole lives and fortunes as dependent upon the exact moment of birth, and so everyone makes an inquiry as to that precise moment and takes an accurate note of it. Masters of this kind of fortune-telling are numerous everywhere [...]. (Ricci 1615/1953, p. 83).

And he observes, not unexpectedly, that:

> Fraud is so common and new methods of deceiving are of such daily occurrence that a simple and credulous people are easily led into error. These soothsayers frequently have confederates in a gathering who declare to a crowd that everything that was told to them by the performer came to pass just as he had predicted it. Sometimes, too, when strangers are brought in as confederates and relate marvels of the past, the followers of the local imposter respond with loud applause. The result is that many deceived by this trickery, have their own fortunes told and accept what is predicted for them as the certain truth. (Ricci 1615/1953, pp. 83–84).

Geomancy was thoroughly embedded in Ming dynasty (1368–1644) life, the practice being codified in the 1445 *Daoist Canon* that contained entire sections of geomantic charts for divination purpose. Few civil and military decisions were made without geomantic input; by law, no official building construction could commence without geomantic certification. Ricci regarded all of this as a *superstitio absurdissima* observing that:

> What could be more absurd than their imagining that the safety of a family, honors, and their entire existence must depend upon such trifles as a door being opened from one side or another, as rain falling into a courtyard from the right or from the left, a window opened here or there, or one roof being higher than another? (Ricci 1615/1953).

These dimensions of Chinese culture preceded Ricci by thousands of years and despite twentieth-century modernisation and Maoist communism, have remained ingrained in Chinese life for the subsequent four hundred years. The sale of auspicious Chinese astrological calendars is a contemporary multi-million-dollar business.

Ricci had no hesitation in pronouncing Feng Shui beliefs as absurd, but the equally widespread and related belief in the power of devils, or more generally of the interference of supernatural entities in earthly affairs presented a more complex problem for him. All three of the Abrahamic religions were committed to the reality and power of devils so Ricci could have no 'in principle' objection to invoking devils as explanations of daily events.[8] This is well illustrated later in the *Journal* where Ricci recounts an instructive episode when a major Nankin city official offered him, and his fellow missionaries a palace as a residence. Ricci writes of the official:

> "A short time ago", he said, "I built a palace, at public expense, for my colleagues; the staff of my tribunal. As soon as it was finished and given over to them, ghosts and devils took possession of it, and no one could live there without being harmed. [...] if you are not afraid of ghosts, it is yours to purchase and there will be no wrangling about the price, because you may determine that yourself, as you see fit". (Ricci 1615/1953, p. 345).

Ricci did fully realise that agreement with European natural philosophy, had consequences.

> And in truth it thus happened that many, having learnt our mathematical sciences [astronomy], laughed at the law and doctrine of the idols, saying that if they taught so much error in natural matters and those of this life, there is no reason to give them credit in supernatural matters and those of the other life. (Rule 1972, p. 166).

This comment was directed specifically at Chinese Buddhists, but it has become a perennial issue in all cases of contact between traditional belief systems and modern science.

Ricci's journal comments might appear like a textbook example of insensitive missionary denigration of foreign cultures. But his story is more complex. He had a renowned appreciation of Chinese culture and custom and he sought an 'accommodation' between Catholicism and Confucianism (Rule 1972, pp. 128–140). He thought his own historical studies of Confucian texts established that divination belief and practice was a corruption of original Confucian belief; it was unnecessary and unwanted baggage.[9] He fought the Vatican to allow the carry-over of many Chinese rites, specifically ancestor worship, by converts in the emerging Chinese Catholic Church. When finally, in 1773 the Roman Catholic Church banned these Chinese rites, the Emperor cast the Jesuits out of China, but this did not affect the high regard in which Ricci was held.

Upon Ricci's death in Beijing on May 10, 1610, his friend Li Chih-tsao wrote to the Court requesting a fitting burial place:

> Your servant, Li Ma-tou [Ricci] from the time he entered your court, began to absorb your brilliant culture, to read (Chinese) books and penetrate their meaning. Morning and evening, reverently and respectfully he burnt incense and prayed to Heaven, reciting your praises in poor return for your kindness. The loyalty of his heart is known to everybody, high or low, in the city, and we would not dare to embellish it. While he was yet alive he was reputed to be a lover of scholarship and a writer of no mean ability. In his earlier days across the seas he was known as a famous scholar, and when he came to this country he was

[8] For elaboration and literature, see Matthews (2015, pp. 361–63).

[9] The issue of what is 'corrupt' and what is not, what is authentic development and what is misguided, is something that all belief systems and ideologies face. The more so, of course, for any based upon putative divine revelation. The raptures within Christianity and Islam arise precisely from this issue.

> praised by high officials who did not fear to liken him to the hermits who withdraw to lonely places. (Rule 1972, p. 119).

In 2010, among the very small number worldwide of those who celebrated the 400th anniversary of his death was the Chinese Communist Party who approved a long Beijing newspaper account of his life, complete with photos of his tomb, portraits, and listing of places with which he was associated.

6.2 Ernst Johann Eitel (1838–1908)

Two hundred and fifty years after Matteo Ricci's landmark account of the early engagement of European natural philosophy and religion with Chinese culture, Ernst Johann Eitel, a German Lutheran missionary with a doctoral degree in Classics from the University of Tübingen, published a longer and more substantial appraisal of the history, metaphysics and functioning of Feng Shui in his book *Feng Shui: The Rudiments of Natural Science in China* (Eitel 1873/1987). The book provides opportunity for elaboration of wider issues concerning Feng Shui, science and society, science and experiment, science in China, and science and worldviews. These are all issues that can easily and naturally be discussed in science classes.

6.3 Feng Shui as a 'Black Art'

Eitel notes how the European construction in the Treaty Ports that followed the mid-nineteenth century Opium Wars, was constantly embroiled in Feng Shui disputes about the work's expected positive and negative impacts on the flow and distribution of chi in the port area. He observes how much of the foreigners' building work in Hong Kong inadvertently turned out to be in 'good' Feng Shui sites (access to water, protection from high wind, away from malaria swamps, and so on) which led many Chinese to impute advanced Feng Shui knowledge to foreigners.

This is the relatively harmless, near trivial face of Feng Shui – do what common sense and accumulated experience suggests and then call it 'Feng Shui guided'. For instance, all Feng Shui site-selection manuals point out that invisible chi is blown away by wind and is accumulated and borne by water (Han 2001). Thus, attention to wind and water (the original meaning of Feng Shui) as indicators of chi is of paramount importance and consequently ideal Feng Shui sites should not be wind-swept and should be located near water. This is hardly rocket-science.

Water is beneficial in all ways: for drinking, plant growth, temperature control, transport, and so on. Good ecological and commonsense reasons suggest proximity to water and avoidance of wind tunnels make for good living places. People claiming this conclusion as warrant for Feng Shui (FS) belief are committing the elementary logical mistake of affirming the consequent:

FS → site A will be healthy and beneficial
Site A is healthy and beneficial
∴ FS is proved/warranted/confirmed

But, of course, any number of other ‘theories’, including scientifically informed ones, can equally suggest site A. Further, if site A turns out not to be healthy and beneficial, then other theories, especially scientifically-based ones, can suggest local reasons why this might be so and consequently can give guidance on rectification or remediation of the local problem. It is not clear how Feng Shui theory can, on its own, do this. The theory is just a distraction.

Feng Shui cannot make considered adjustments in the way that scientifically informed predictors of site A can make adjustments: ecological (acid rain), climatological (rising sea levels), geological (seismic activity) or economic (property values) theories can. With water, sun and chi plants grow. A proper science education should lead students to inquire whether plants will grow with just water and sun. A challenge will be to see how chi can be removed from the experimental set-up. But if it cannot be, then equally gravitation or space could be identified as necessary for plant growth. It becomes apparent that chi just an idle passenger; along for the ride.

But Eitel recognises that Feng Shui site selection and medical guidance goes beyond harmless and benign commonsensical practices:

> Well, if Feng Shui were no more than what our common sense and natural instincts teach us, Chinese Feng Shui would be no such puzzle to us. But the fact is, the Chinese have made Feng Shui a black art, and those that are proficient in this art and derive their livelihood from it, find it to their advantage to make the same mystery of it, with which European alchemists and astrologers used to surround their vagaries. (Eitel 1873/1987, p. 1).

Eitel here articulates a standard criticism: what is good in Feng Shui practice is merely dressed-up commonsense – build and plant in proximity to water and not in a wind tunnel or on a mountain top; have living areas orientated to the sun, and sleeping areas away from the sun; avoid having front and rear doors opening in an uninterrupted line; do not have toilets opening into living areas, etc. A Feng Shui advocate recently wrote in an American Institute of Architects publication:

> Feng Shui is all about what nurtures a building’s occupants and makes them feel comfortable in a space. As architects and designers, we instinctively do a lot of these things. Feng Shui provides the framework and the philosophy to support our instincts. (Knoop 2001).

The argument of this chapter is that all the associated cosmology is just hand-waving. As the saying is: “With the cosmology and two dollars you can ride the subway”. But it is not idle or harmless hand waving, as it opens the cultural door to mystification and worse still, to manipulation by charlatans. If people get accustomed to believing the fantastic, evidence-free, or evidence-neutral, chi narrative, then what other evidence-free or neutral beliefs might they be prepared to accept?

6.4 Chinese Proto-science and Avoidance of Experiment

Eitel regards Feng Shui as simply Chinese natural science, moreover a science that did not 'grow up' (Eitel 1873/1987, p. 3). As a proto-science it was hampered by cultural factors that impeded and limited the progress of Chinese natural science. These limiting or constraining factors have been long identified: a stratified society where there was a clear separation of manual work and technology from intellectual or philosophical speculation; the primacy given in the Confucian tradition to 'inner' knowledge and personal wellbeing; the rigid court control of the pursuit and dissemination of higher knowledge; an in-grained preference for utility rather than understanding, etc. For these reasons, as Needham pointed out, there would be no 'Scientific Revolution' in China (Needham 1963).

Importantly Eitel points to the absence of a Chinese experimental tradition that "shrank from the idea of dissecting animal bodies and never took to hunting beetles and stuffing birds or chemically analyzing inorganic substances" as the central cause of the failure of science to develop in China. Towards the end of his book, he writes:

> There is one great defect in Feng Shui, which our Western physicists have happily long ago discarded. This is the neglect of an experimental but at the same time critical survey of nature in all its details. (Eitel 1873/1987, p. 69).

One century later Joseph Needham made the same point:

> Controlled experimentation is surely the greatest methodological discovery of the scientific revolution of the Renaissance, and it has never been convincingly shown that any earlier group of Westerners fully understood it. I do not propose to claim this honour for the medieval Chinese either, but they came just as near it theoretically, and in practice often went beyond European achievements. (Needham 1963, pp. 147–48).

6.5 Influence of the Heavens

Given that the heavens were seen to so definitely impact on *inanimate* earthly circumstances (the sun's movement on seasons, the moon's movement on tides, etc.) it was a small step to see them impacting on *animate* human and personal circumstances. Indeed, the unity of heaven, earth and man was at the center of the Chinese tradition whether Confucian, Taoist or Buddhist; and the Christian claim that man was made in the image of God resonated with Chinese worldview. Feng Shui accepts and reinforces this belief in a cosmic harmony that unites the heavens, earth and man.

Ancient belief in the powers of the moon to affect human affairs was long-standing and widespread; it was not just a Chinese belief. Aristotle attributed different kinds of madness to the effect of full moons. European astrologers right through to the eighteenth century and beyond identified lunar causes for mental disorders, hence the once common term 'lunatic' (Harrison 2000). But by Eitel's time, such lunar beliefs had been banished from informed European thinking and medical practice.

6.6 Astronomical Problems for Feng Shui

Leaving aside any scientific or 'metaphysical' problem in translating from motions of the sun and moon to earthly and human affairs, Eitel identifies two major physical problems for Feng Shui theory: the precession of the equinoxes and missing planets.

The Chinese astronomers/astrologers did not know of the precession of the equinoxes from east to west, the effect of which has been to disconnect the signs of the zodiac from their respective constellations. Eitel comments that:

> The Chinese, not knowing of the precession of the equinoxes, are rather perplexed by the discrepancy, but caring less for accuracy and more for ancient tradition, ignore the actual discrepancy, and still represent the twelve signs, not as they appear now, but as they appeared to their ancestors two thousand years ago. (Eitel 1873/1987, p. 11).

This is a matter that at least some contemporary Feng Shui consultants deal with by simply severing the connection of auspiciousness from the position of star groups in the galaxy. The above mentioned Grand Master Skinner says of his Flying Star system that:

> This is the first book in English on Flying Star Feng Shui which explains how it compatibly relates to Eight Mansion Feng Shui, and provides 216 lower kua Flying Star charts with 24 Mountain direction indicator, as well as all of the variant Substitution Star (ti kua) charts. (Skinner 2015).

But then concedes that:

> Although the 'Flying Stars' were originally related to the stars of the astronomical Big Dipper asterism, in their Feng Shui usage they are simply terms for changing types of subtle *ch'i* (*qi*) energy present in our living environment, our homes, and our workplaces. (Skinner 2015).

Needless to say, this is a major concession: either the galactic position of certain stars has influence, but the tables of the original diviners are now out of synch with star movement; or there is no such influence, and Feng Shui can get by with just attending to the configuration of local chi. The latter option simplifies the task of consultants, but probably does not lessen their fees.

Eitel identifies another major conceptual and evidential problem: the Chinese recognised only five planets – Jupiter, Venus, Saturn, Mars and Mercury. Given the centrality of these in the whole Chinese cosmological picture, it is easy to understand why Chinese astronomers did not seek to discover new planets. Chinese astronomers may or may not have seen additional planets, but if so, they did not identify them as planets; they would have been regarded as comets, stars or some aberration, just as Uranus was by those who happened upon it from antiquity till William Herschel's identification of it in 1781 (Miner 1998). Similarly, the spectacular 1846 discovery of Neptune by Johann Galle following its mathematically precise prediction by Urbain Le Verrier (Grosser 1962) did not and could not have happened within the Chinese astronomical tradition. These are merely examples of the commonplace observation that culture impacts on the conduct of science, and that recognising and being alert to such a nexus is an important outcome of a good science education.

A ‘research problem’ for the Feng Shui tradition was to determine whether the newly discovered Uranus and Neptune had or did not have auspicious influences, and with what numbered constellations it should be linked. And whatever problems Western astronomy has had with the planetary status of Pluto, these are magnified in the Feng Shui tradition which additionally has to divine the planet’s impact, if any, on earthly affairs.

6.7 The Five Elements

Eitel relates how the principal agents through which heaven, and especially the five planets, act upon all living creatures are the five elements of nature. The Chinese ‘theory’ of elements is fundamental to Feng Shui.[10] He correctly cautions that:

> By these however we must not understand five material substances, chemically indissoluble, but rather spiritual essences, each characteristically different from the other and forming the generative causes of all material substances. These five elements are wood, fire, earth, metal and water, the first of them being the agent of Jupiter, the second that of Mars, the third that of Saturn, the fourth that of Venus, the fifth that of Mercury. (Eitel 1873/1987, p. 14).

Belief in the defined powers of the five planets, acting in and through the five elements, flows over to a fivefold division of the body (muscles, veins, flesh, bones and hair); the five primary colours (white, black, red, blue and yellow), the five fortunes (money, honour, longevity, children and peaceful death). There is also the fivefold division or ‘aspects’ of chi. The fact of five fingers and five toes could also be called upon to bolster the primacy or foundational place of five.

6.8 Eitel’s History of Feng Shui

In chapter six of his *Feng Shui* book (Eitel 1873/1987) Eitel provides an outline history of Feng Shui in which he maintains that while the leading principles of Feng Shui have their roots in antiquity, and it did not rise as a distinct branch of study or professional practice till the Sung dynasty (AD 960–1126), nevertheless the “history of the leading ideas and practices of Feng-shui is the history of Chinese philosophy” (Eitel 1873/1987, p. 51). Eitel marks down Confucius and his influential early disciples for not engaging with and correcting the proto Feng Shui systems of his age:

---

[10] The title of a recent expository book is *Feng Shui: The Chinese System of Elements* (Rolnick 2004).

> It was in the power of Confucius and his disciples, Mencius and Sun-tze, who exercised a strong influence on the minds of their countrymen during this period [BC 500–200], to repress and rectify the superstitious notions already floating about among the people and tending towards a regular system of geomancy, by assuming a definite attitude, denouncing superstition and substituting an enlightened theory on the subject. But he and his disciples, though personally free from superstition, contented themselves with urging a reform of morality according to the pattern of the ancient sages, without venturing to grapple with the superstitions that were gathering around the ancient form of ancestral worship. In one word, they remained neutral, and the consequence was that superstition spread farther and farther. (Eitel 1873/1987, p. 54).

Many historians and commentators on Confucianism make the same point: Confucius was not engaged by or interested in how the world worked; he had little if any concern with 'natural philosophy'. For example, in the present day, Yijie Tang, a leading Chinese philosopher and cultural historian, writes that:

> Confucianism is a humanistic mode of thinking. Confucianism does not proceed along the path of science nor does it subscribe to principles of science. (Tang 2015b, p. 76).

Other contemporary Chinese philosophers have lamented that:

> [...] those experiments which had no practical purpose but were rather intended to verify scientific theory were relegated to an inessential position in ancient Chinese science and technology. This was even more true after the appearance of neo-Confucianism. (Jin et al. 1996, p. 147).

The primacy afforded in the Confucian tradition to leading a useful moral life, and to living responsibly or dutifully in the State, meant that spending time on scientific experiments, especially ones intended to confirm or disprove hypotheses about the basic workings of nature rather than, at best, utilitarian hypotheses about better production, transport, weapons or manufacturing – was seen as frivolous, as a distraction from the main social game. Eitel notes that:

> With the rise of the Tang dynasty (AD 618–905), which is famous for its revival of literature generally and of poetical literature especially, which had hundreds of Buddhistic works translated from Sanskrit into Chinese, a new era opened, particularly favourable to the propagation of mystic and fanciful doctrines assuming, as geomancy had learned to do, the garb of national as well as Tauistic and Buddhistic philosophy. (Eitel 1873/1987, p. 61).

And:

> The notion of five planets (Venus, Jupiter, Mercury, Mars, and Saturn) influencing the earth and every living being, made its first appearance about this time, and was eagerly taken hold of by the professors of Feng Shui. (Eitel 1873/1987, p. 61).

Eitel observes that in his day the adherents of Feng Shui are divided into two schools, the Tsung-miau (ancestral temple) school, which took its rise in Foh-kien, and the Kwang-si school. He says that the former school of geomancers attributes the greatest importance to the doctrines of the order of nature (Li) and of the numerical proportions of nature (Su). They are therefore specially attached to the use of the compass. The second school placed the greatest stress on the doctrines of the breath (K'e) and outlines (Ying) of nature. He says that they use the compass too, but only as a subordinate help in prospecting the country, for 'their principle is, first to look for the visible symptoms of dragon and tiger and of a good breath, and then to judge of the surrounding influences by consulting the compass' (Eitel 1873/1987, p. 63).[11]

6.9 Eitel's Appraisal of Feng Shui and the Educational Task in China

Eitel concludes his exposition by writing that while two thousand years ago the embrace of some of the fundamental ideas of Feng Shui was understandable and 'rational', but such embrace no longer is. He writes:

> Feng Shui is at present a power in China. It is an essential part of ancestral worship, which national religion, neither Tauism nor Buddhism managed to deprive of its all-pervading presence. [...] Feng Shui is indeed the refined quintessence of Tauistic mysticism, Buddhistic fatalism and Choo-he's materialism, and as such it commands if not the distinct approval yet the secret sympathy of every Chinaman, high or low. (Eitel 1873/1987, pp. 65–66).

He notes that Feng Shui exponents:

> had studied nature, in a pious and reverential yet in a very superficial and grossly superstitious manner, but which, trusting in the force of a few logical formulae and mystic diagrams, endeavoured to solve all the problems of nature and to explain everything in heaven above and on the earth below with some mathematical categories. The result, of course, is a farrago of nonsense and childish absurdities. (Eitel 1873/1987, p. 69).

Eitel had no hesitation in pointing to the educational project, specifically the science-education project, that was needed in his day:

> The only powerful agent likely to overthrow the almost universal reign of Feng Shui in China I conceive to be the spread of sound views of natural science, the distribution of useful knowledge in China. [...] Let correct views be spread regarding those continually interchanging forces of nature, heat, electricity, magnetism, chemical affinity and motion; let these views be set forth in as forcible and attractive but popular a form as Choo-he employed, and the issue of the whole cannot be doubtful. (Eitel 1873/1987, p. 69).

Although a Lutheran clergyman, he spoke with a voice informed by science and the Enlightenment:

> The fires of science will purge away the geomantic dross, but only that the truth may shine forth in its golden glory. (Eitel 1873/1987, p. 69).

Importantly, he did not underestimate the educational task; it could not be confined just to classrooms:

> Feng Shui is, moreover, so engrafted upon Chinese social life, it has become so firmly intertwined with every possible event of domestic life (birth, marriage, housebuilding,

[11] Modern Feng Shui commentators and expositors refer to these schools as the 'compass school' and the 'form school' of Feng Shui.

> funerals, etc.) that it cannot be uprooted without a complete overthrow and consequent re-organisation of all social forms and habits. (Eitel 1873/1987, p. 65).

Not surprisingly, this contention has drawn down on Eitel the ire of some who have read his work. It is the kind of statement that is put up as 'Exhibit A' to demonstrate cultural insensitivity, Western chauvinism and objectionable scientism and positivism. Notwithstanding all of this, Eitel's recognition of the material embeddedness of Feng Shui beliefs, and the consequent social and cultural adjustments needed if they were to be abandoned or significantly changed, is both realistic and commendable. Further, Eitel's view has been independently adopted by many contemporary Chinese scientists, philosophers, educators and cultural critics concerned with the modernization of Chinese culture. For example, by Ch'en Tu-hsiu (1879–1942), Yu-Lan Fung (1895–1990) and Yijie Tang (1927–2014).

Eitel was not a naïve idealist who thought that important core beliefs of a person or culture can change without any collateral social or ideological dislocation. This idealist outlook is, unfortunately, common in multicultural and indigenous science education discussion. It does need be recognized that the changing of some deep and substantial beliefs entail wholesale changes in personal and cultural epistemology; the status and power of traditional authority figures and sources; relations between men and women; relations between parents and children; between teachers and students; in many cases the legal system of countries; and so on. There are some cultural beliefs and commitments that cannot be changed as easily as shirts; Feng Shui is one of them, and Eitel recognized this.

Both Matthew Ricci and Johann Eitel were aware of the cultural embeddedness of science and the consequent magnitude of the task facing any orthodox science education that challenged extant 'science-like' worldviews in sixteenth- and nineteenth-century Chinese society. Contemporary critical rationalists can with profit retrace the philosophical, cultural and educational steps they took in their engagement with Feng Shui beliefs that have now spread far beyond China and draw appropriate conclusions for the present time.

References

Agassi, J. 2014. In *The hazard called education*, ed. R. Swartz and S. Richmond. Rotterdam: Sense Publishers.

Bruun, O. 2008. *An introduction to Feng Shui*. Cambridge: Cambridge University Press.

Bruun, O., and A. Kalland, eds. 1995. *Asian perceptions of nature – A critical approach*. London: Curzon Press.

Cronin, V. 1955. *The wise man from the west: Matteo Ricci and his mission to China*. New York: Harper Collins.

Eitel, Ernest J. 1873/1987. *Feng Shui: The rudiments of natural science in China*. Hong Kong: Lane, Crawford & Co.

Grosser, M. 1962. *The discovery of Neptune*. New York: Dover.

Han, Ke-Tsung. 2001. Traditional Chinese site selection-Feng Shui: An evolutionary/ecological perspective. *Journal of Cultural Geography* 19(1): 75–96.

Harrison, M. 2000. From medical astrology to medical astronomy: Sol-lunar and planetary theories of disease in British medicine, c. 1700–1850. *The British Journal for the History of Science* 33 (1): 25–48.

Hsia, Po-Chia R. 2012. *A Jesuit in the Forbidden City: Matteo Ricci 1552–1610*. Oxford: Oxford University Press.

Jin, Guantao, Hongye Fan, and Qingfeng Liu. 1996. The structure of science and technology in history: On the factors delaying the development of science and technology in China in comparison with the west since the 17th century (part one). In *Chinese studies in the history and philosophy of science and technology*, ed. F. Dainian and R.S. Cohen, 137–164. Dordrecht: Kluwer Academic Publishers.

Knoop, A. 2001. Feng Shui for architects. *American Institute of Architects, Newsletter*. http://www.lezley.info/feng-shui.htm.

Matthews, M.R. 2015. *Science teaching: The contribution of history and philosophy of science*. 20th Anniversary Revised and Enlarged Edition. New York: Routledge.

Miner, E.D. 1998. *Uranus: The planet, rings and satellites*. New York: Wiley.

Needham, J. 1963. Poverties and triumphs of the Chinese scientific tradition. In *Scientific change: Historical studies in the intellectual, social and technical conditions for scientific discovery and technical invention, from antiquity to the present*, ed. A.C. Crombie, 117–153. New York: Basic Books.

Ricci, M. 1617/1953. On the Christian mission among the Chinese. In *China in the sixteenth century: The journals of Matthew Ricci, 1583–1610*, ed. L. L. Gallagher. New York: Random House.

Rolnick, H. 2004. *Feng Shui: The Chinese system of elements*. Hong Kong: FormAsia Books.

Rule, P.A. 1972. *K'ung-tzu or Confucius? The Jesuit interpretation of Confucianism*. PhD thesis, Australian National University.

———. 1986. *K'ung-tzu or Confucius? The Jesuit interpretation of Confucianism*. Sydney: Allen & Unwin.

Skinner, S. 2015. *Advanced flying star Feng Shui*. Singapore: Golden Hoard Press.

Smith, R.J. 1991. *Fortune-tellers and philosopher: Divination in Chinese society*. Boulder: Westview Press.

Spence, J.D. 1984. *The memory palace of Matteo Ricci*. New York: Penguin.

Tang, Y. 2015a. The attempt of Matteo Ricci to link Chinese and Western Cultures. In *Confucianism, Buddhism, Daoism, Christianity and Chinese culture*, ed. Y. Tang, 179–189. Dordrecht: Springer.

———. 2015b. Some reflections on new Confucianism in Chinese Mainland culture of the 1990s. In *Confucianism, Buddhism, Daoism, Christianity and Chinese culture*, ed. Y. Tang, 67–78. Dordrecht: Springer.

Wright, J. 2010. *Mission to China: Matteo Ricci and the Jesuit encounter with the East*. London: Faber.

Part II
The Sceptic Quest for a Sensible World

Chapter 7
Verisimilitude: Why and How

Ilkka Niiniluoto

When Karl Popper's attempt to define verisimilitude or truthlikeness failed, some of his followers suggested that critical rationalists do not really need this notion. Some others took up the challenge of rescuing Popper's definition or finding a better one. I first met Joseph Agassi at this intellectual crossroad in 1974. We share a sense of the importance of the problem of verisimilitude, but we approach it from different backgrounds: Agassi from the LSE anti-inductivism, I from my education with Jaakko Hintikka's inductive logic, so that we disagree on some issues about the *why* and the *how* of this problem. This paper gives me an opportunity to respond to some interesting comments and critical reviews by Agassi on my books *Truthlikeness* (1987) and *Critical Scientific Realism* (1999).

7.1 The Breakdown of Popper's Definition of Verisimilitude

I met Joseph Agassi for the first time in the Conference for Formal Methods in the Methodology of Empirical Sciences in Warsaw, June 17–21, 1974. Personally this conference was a turning point in my career. In December 1973 I had defended my doctoral thesis at the University of Helsinki, where I applied Jaakko Hintikka's system of inductive logic to illuminate the role of theories and theoretical concepts in scientific inference. In such an "anti-inductivist" inductive logic, I argued, one can incorporate some methodological insights of Karl Popper, especially the significance of information content as a goal of science (see Niiniluoto and Tuomela 1973). In Warsaw, I gave with Hintikka a joint paper which improved the Carnap – Kemeny axiomatization of inductive probability measures. I also met there for the

I. Niiniluoto (✉)
Department of Philosophy, University of Helsinki, Helsinki, Finland
e-mail: ilkka.niiniluoto@helsinki.fi

N. Bar-Am, S. Gattei (eds.), *Encouraging Openness*, Boston Studies in the Philosophy and History of Science 325, DOI 10.1007/978-3-319-57669-5_7

first time Theo Kuipers, who was working on induction and has later shared my interests in many topics in formal philosophy of science.

Agassi, who is almost 20 years older than me, was already a well-known scholar from the tradition of critical rationalism, with education at the London School of Economics, and professor at Boston and Tel-Aviv. He did not present a paper in the official program, but a special session was arranged where he sharply stated his criticism of induction. After a heated discussion, Agassi concluded that he can attack his own position better than anyone else in the room. As a shy Finn I was a little embarrassed, but an older colleague whispered to me that such a style of debate is common at the LSE. Something of the nature of such exchanges can be seen in the special issue of *Synthese* in 1975, which includes colorful papers from a 1973 conference at South Carolina (with Richard Jeffrey and I. J. Good on the side of the Bayesians, Agassi and David Miller among the Popperians).

The Warsaw conference was a turning point also in the study of the notion of truthlikeness as proposed by Popper (1963). The news about Miller's refutation of Popper's comparative definition of verisimilitude had reached Helsinki in the fall of 1973 via my colleague Raimo Tuomela. I had a vague idea that Hintikka's measure of corroboration could be related to the epistemic problem of verisimilitude – better than Popper's own measure of corroboration. But the real revelation came when in Warsaw I listened to Risto Hilpinen's lecture, where he explicated the notions of approximate truth and truthlikeness by using the nested spheres of possible worlds of David Lewis. I immediately realized that, instead of taking the similarity of possible worlds as an undefined primitive notion, one could replace such worlds with Hintikka's constituents and then there are natural ways of giving explicit definitions of the distance between these constituents in monadic predicate logic – the framework which was familiar to me from inductive logic. In the fall of 1974 the papers of Miller and Pavel Tichý were published in the *BJPS* (see Miller 1974), and besides refuting Popper's definition Tichý gave a positive proposal by applying the method of constituents in propositional logic. I developed my account of verisimilitude for monadic generalizations, including the proposal that degrees of truthlikeness can be estimated by expected values using inductive probabilities from Hintikka's systems, in my invited paper for the fifth international conference of Logic, Methodology and Philosophy of Science in the summer of 1975 in London, Ontario (see Niiniluoto 1977).

In the same congress in London, Ontario, Jagdish N. Hattiangadi informed that Carl G. Hempel had proved the negative Miller-Tichý theorem for finitely axiomatizable theories during his oral doctoral examination at Princeton in 1970 (see Hattiangadi 1975). Popper was furious about these news and broke his relations with Hattiangadi for several years. But when his favorite student David Miller, who was a faithful ally in the campaign against Carnapian induction, proved that no two false theories are comparable for their truthlikeness by Popper's definition, Popper was compelled to admit his mistake. Agassi's (1975) attempt to avoid Miller's negative results by appealing to the amplification of quantitative errors was quickly dismissed by Miller (1975). The Hempel-Hattiangadi refutation of Popper's definition was then reproduced in Agassi (1976), which appeared as "a paper presented by title" in the Proceedings of the 1974 Warsaw conference.

7.2 Reactions to the Breakdown: Dismissing or Saving Verisimilitude?

The refutation of Popper's attempted definitions of comparative and quantitative verisimilitude elicited various responses among philosophers of science (cf. Niiniluoto 1987). Many leading philosophers in different camps remained indifferent and never directly contributed to the debates on truthlikeness (e.g. Carnap, Hintikka, Jeffrey, Alfred Tarski, Wilfrid Sellars, Hilary Putnam, Wesley Salmon, Adolf Grünbaum, Richard Boyd, Nicholas Rescher, Peter Gärdenfors, Bas van Fraassen). Some philosophers thought that the whole concept is absurd or illegitimate (e.g. Kurt Hübner, Wolfgang Stegmüller, Larry Laudan), concluding that Thomas Kuhn was right in his defense of the problem-solving account of scientific progress against Popper's account of progress as increasing verisimilitude. Isaac Levi argued that truth and information define the aims of science without the addition of distances from the truth. Hattiangadi (1983) proposed to save fallibilism by analyzing scientific progress as the generation of new problems. Some former critical rationalists made scornful remarks about the "monsters" of verisimilitude (e.g. Paul Feyerabend, Gerhard Radnitzky), and some scientific realists were frustrated by their attempts to come to grips with this technical notion (e.g. Philip Kitcher). Miller has continued his endorsement of Popperian critical rationalism in spite of his skepticism about the possibility of giving a language-independent definition of truthlikeness. Some critical rationalists argued that one may continue to use the concept of verisimilitude in an intuitive sense in spite of the collapse of Popper's definition (Hans Albert, Gunnar Andersson, Noretta Koertge), and this view is shared by some contemporary scientific realists (e.g. Stathis Psillos).

Popper's first public reaction to Miller's result in 1976 was to restrict comparisons of theories to their truth content, but this leads to the fatal "child's play objection" by Tichý: any false theory can be improved by adding arbitrary falsities to it. In the first 1972 edition of *Objective Knowledge* Popper insisted that "a major desideratum of the methodology of the natural sciences" is "to be able to say with good conscience" that "we have reason to conjecture that Einstein's theory of gravity is not true, but that it is a better approximation to truth than Newton's" (p. 335), and in the second 1979 edition he was still "optimistic concerning verisimilitude" (in agreement with Tichý) (see Popper 1979, p. 371).

Jeffrey (1975) argued that Popper's attempts to explicate verisimilitude and corroboration are "misguided", not only because he is "a clumsy player of Carnap's game of formalization", but because "Carnap-style explication is irrelevant and distracting a medium for Popper's message". This turned out to be incorrect as a prediction, since the similarity approach to truthlikeness by myself and by Tichý and his student Graham Oddie from New Zealand used logical methods from the Carnap-Hintikka toolbox. In December 1977 I had the chance to invite John Watkins and Oddie to Helsinki to a conference on "The Logic and Epistemology of Scientific Change". Oddie had started a doctoral dissertation at the LSE, with the hope that he would be the person who saves verisimilitude. In 1980 I participated with Miller and Tichý in a session on verisimilitude at the LSE. Popper was present, but did not seem very happy about my presentation, which probably was too technical for his taste. According to his view on meaning-analysis, "we should never try to be more exact than is necessary for the solution of the problem in hand" (Popper 1982, p. 277). When I was ready to make a slight revision in my own definition in the light of Oddie's criticism, Watkins friendly advised me to keep on supporting my own program.

In his Introduction to the volume of *Postscript to the Logic of Scientific Discovery*, Popper claimed that the admitted failure of his definition of verisimilitude has only negligible impact on his theory of science (see Popper 1982, pp. xxxv–xxxvii). A formal definition of verisimilitude is "not needed for talking sensibly about it". A new definition is needed only if it strengthens a theory, and it is "completely baseless" to claim that this unfortunate mistaken definition weakens his theory. Popper added that no one has ever shown "why the idea of verisimilitude (which is not an essential part of my theory) should not be used further within my theory as an undefined concept".

But could Sir Karl really pronounce this with good conscience? Some critical rationalists like Joseph Agassi (1976), who largely accepted Popper's anti-inductivism, still thought that any theory of verisimilitude should answer the question about scientific progress. Moreover, many scientific realists like myself, Theo Kuipers, and Roberto Festa, who had also worked on inductive logic, joined the efforts of the similarity explication of truthlikeness. I know that Popper wrote a friendly letter to Kuipers about his first study of nomic truthlikeness in 1982, but I never received any feedback from Popper. The first book-length expositions, Oddie's *Likeness to Truth* (1986) and my *Truthlikeness* (1987), both employed the machinery of Hintikka's constituents and distributive normal forms extended to full first-order logic. The main difference is that the average measure of Tichý and Oddie does not satisfy Popper's crucial adequacy condition that among true theories logically stronger ones are more truthlike, while my min-sum measure respects this principle. But verisimilitude is a hard nut: no consensus among rival approaches has been reached so far (for surveys see Niiniluoto 1998 and Oddie 2014).

7.3 Agassi on Verisimilitude

Agassi (1975) noted that Popper's comparative notion of verisimilitude has no reference to experience or evidence of any sort, suggesting that a definition of verisimilitude should be relative empirical data. In his reply, Miller (1975) wanted "to discourage the view that the metalogical problem of 'defining' verisimilitude is basically a problem about the relation of theories to evidence". He added that "if progress to truth on the back of theories is a phantasm, so is it when steered by observations". Agassi (1976) further remarked that Popper's theory of verisimilitude applies to all content, asking whether it could be qualified to empirical content only. This idea was developed in Agassi (1981) with the proposal "to save verisimilitude" by applying Popper's comparative condition to empirical content: verisimilitude increase means "the combination of an increase of true empirical content and a decrease of false empirical content". Restricted to crucial evidence, this means that "a theory is more verisimilar than its predecessor if and only if all crucial evidence concerning the two goes its way". Hattiangadi (1983) replied, agreeing with Miller, that "there is no sense in adopting corroboration as a criterion of progress towards the truth unless there is such a thing as progress towards the truth". Verisimilitude like truth is an objective matter and "quite independent of what we happen to believe", so that Agassi's clauses about empirical content and crucial evidence defeat the purpose of looking for a formal definition of verisimilitude.

In my comments on Agassi, I agreed with Miller and Hattiangadi that one has to separate the logical and epistemological problems of truthlikeness, and the former one is prior to the latter (see Niiniluoto 1986, pp. 609–610; Niiniluoto 1987, p. 483). The logical concept of verisimilitude cannot be defined in terms of empirical content, since two theories may have the same potential falsifiers but still one of them may be closer to the truth in its theoretical part. If the language in Popper's definition would be restricted to empirical statements only, Miller's theorem would still be valid for purely empirical theories. And if empirical truth content and falsity content would consist merely of singular basic statements, then their comparison would be a relatively simple special case of verisimilitude measures for Carnapian state descriptions.

Agassi's proposal, even though it is not formally developed in any detail, still merits reconsideration. Some logicians have argued that the failure of Popper's definition indeed is in his treatment of theories as deductive closed sets of sentences. If the truth and falsity content of a theory is restricted to a suitably characterized class of "relevant consequences", then Popper's original definition is saved from Miller's trivialization (Gerhard Schurz & Paul Weingartner, Ken Gemes). If truth and falsity contents are defined as classes of content elements, i.e. negations of state-descriptions, as Carnap proposed in his treatment of semantic information, then Popper's definition is again saved from Miller's refutation – even though not from other problems (see Cevolani 2016). While these new developments are definitions of logical verisimilitude, some philosophers have given accounts of evidence-relative or epistemic verisimilitude in a Popperian spirit. So Theo Kuipers (2000) defines "empirical progress" by the condition that the new theory has more empirically established true consequences and less established incorrect consequences. But for Kuipers such empirical success is not the whole story, as it is for Agassi's definition in terms of empirical content, since continued empirical success of theory A over B justifies the "success hypothesis" that A is objectively more truthlike than B.

7.4 Debates with Agassi

In my view, Agassi deserves credit for the fact that among critical rationalists he has consistently defended the indispensability of the notion of verisimilitude in defining scientific progress. His keen interest in this notion was evident in our communication in the 1980s. Agassi came to listen to my invited paper "The Significance of Verisimilitude" in the PSA 1984, where I for the first time formulated my min-sum measure (see Niiniluoto 1986), and he actively participated in discussion. After the publication of my *Truthlikeness*, he sent me a manuscript "Verisimilitude" on December 20, 1987. In my letter of July 22, 1988, I replied with comments, and on July 28, 1988, Agassi sent me a polite answer. Suggestions for my further work were included in Agassi's (2001) review of my *Critical Scientific Realism*. Finally, a somewhat revised version of the old manuscript "Verisimilitude" appeared in 2011, giving me here an opportunity to continue my debate with Agassi by elaborating some arguments given in already in my letter in 1988.

One of our disagreements is rooted in our different philosophical backgrounds. Agassi thinks that my acceptance of the program to study verisimilitude means discarding the program to study induction, and that I would agree with him (and Imre Lakatos) that Popper "clearly won" his debate with Carnap (Agassi 2011, p. 65). I do think that Popper made good points about information content, but he was wrong in claiming that inductive logic is impossible. But neither was Carnap right. They both made the mistake in claiming that the probability of laws is zero (see also Agassi 2001, p. 79). For me the real winner of these debates was Jaakko Hintikka, who showed in his system in the mid-1960s how inductive generalization is possible as a fallible method of justification (see Niiniluoto 2011). All methods of comparing the verisimilitude of theories by empirical evidence – including my expected values and Agassi's empirical content – involve some kind of inductive reasoning, so that the epistemological problem of verisimilitude is as difficult as the problem of induction (see Niiniluoto 1987, p. 263).

It is correct that the logical problem of truthlikeness can and should be studied without reference to evidence, but this notion is not primarily needed as a rival to induction but rather as a tool for making sense of the idea of scientific progress. Without it Popper's falsificationism would collapse to a form of skepticism, and his attack against Kuhn's anti-realist view of progress would be powerless. Hence, like Tarski's semantic concept of truth, verisimilitude is needed in a philosophical position which combines realism and fallibilism.

Agassi finds the idea of verisimilitude to be implicit in the writings of Albert Einstein who treated old theories as special cases and approximations of new ones (see Agassi 1975, p. 336; Agassi 2011, p. 62). This important form of theory-change is known also as Niels Bohr's Correspondence Principle, which shows its compatibility with instrumentalism. But in my view it is important to distinguish the general problem of approximation, involving the idea of distance between two theories (e.g. Newton and Einstein), and the problem of verisimilitude or truth approximation, involving the distance of a theory from the *true* theory. The logical concept of verisimilitude should allow us to say with good conscience that Einstein's theory is more truthlike than Newton's, but both of them may approximations to a future true theory.

With all respect to Einstein, the first hints about approximations to the truth were given by Cicero with his concept *veri simile* in discussion about Academic skepticism. In the late nineteenth century the most powerful advocate of the self-corrective nature of science was Charles S. Peirce.

Agassi's formalism morver(A,B,e) and ver(A,e) explicitly relativize comparative and quantitative judgments of verisimilitude to evidence e (*ibid*. p. 80). I agree with his comparisons of Galileo G, Newton N, and Einstein E, which I can obtain retrospectively with my measure of expected verisimilitude ver(A,e) by including the best theory E in the evidence e (see Niiniluoto 1986, p. 605). In other words, if E is hypothetically assumed to be true, then G, N, E is a progressive sequence of truth approximation. But to know that E is true or at least truthlike (in some fallible sense) would involve a solution to the problem of induction. I contend that we need more illustrations and applications of the verisimilitude formalism to case studies from the history of science.

Agassi's principle "If e contradicts A and is explained by B, then morver(B,A,e)" is sound as a sufficient (but not necessary) condition of comparative verisimilitude, but raises interesting questions. A false theory A may be close to the truth, so that ver(A,e) may be high even when e refutes A. This is the main difference between my estimated ver-measure and Popper's degree of corroboration, which drops to zero when e refutes A. This also means that Popper's degrees of corroboration are not adequate indicators of verisimilitude. If B explains e, then e gives abductive confirmation to B, which is reflected in my ver-measure.

Agassi characterizes my scientific realism as "halfway pro- metaphysical" while his view is "fully pro-metaphysical" (Agassi 2011, pp. 63–64). Now I do not share the horror of metaphysics of the logical positivists, and I defend ontological realism by arguments in Niiniluoto (1999). But I am not a "metaphysical realist" in Putnam's sense, since I wish to combine the correspondence theory of truth with conceptual pluralism. Moreover, I fail to see that I "smuggle" a specific metaphysics with my logical tools. For most of the applications, ordinary first-order logic with Tarskian model theory is sufficient. I use modalities only occasionally for physical necessity statements and quantitative laws, but I do not rely on Saul Kripke. Besides, Stig Kanger and Hintikka invented the possible worlds semantics before Kripke already in 1957.

Agassi gives a misleading description of my logical tools in stating that they are "state-descriptions (generalized)" (*ibid*. p. 79). Carnap's state-descriptions are singular sentences, but my account in Niiniluoto (1987) includes existential and universal generalizations, quantitative laws, probabilistic laws, and arbitrary first-order theories. Oddie (1986) has considered higher-order languages as well. But in my conceptual pluralism I do not assume there is some unique language which could express the "whole truth" about the universe (cf. Agassi 2011, p. 78). To say that the world is in this sense inexhaustible is not bad metaphysics by my standards.

It is this attempt to be as general as possible which gave me reason to write 500 pages about the explication of verisimilitude in *Truthlikeness* (1987). I realize that this is against Popper's warnings against "excess rigor". Every reader of this book has to decide, as Agassi tells about himself, whether the study of the required formalism is a waste of time or not. My conviction is that verisimilitude deserves a rigorous examination. When the theory is made precise, it can be critically discussed and hopefully improved. I also advice the potential readers to go to my next book *Critical Scientific Realism* (1999) to judge for themselves whether this notion is useful in assessing arguments for and against realism. Further, scientific realism and critical rationalism give important support to the ideals of Enlightenment.

In his review, Agassi (2001) mentions many interesting topics which could be discussed in connection with scientific realism (e.g. Russell, Wittgenstein, Ryle, Bunge). This is fair enough. But I do not agree that Kuhn is ignored, as I repeatedly return to the assessment of his brand of anti-realism – even suggesting a realist interpretation in terms of conceptual pluralism. I am also a little puzzled by the remark that Agassi's "chief dissent" with me concerns belief (*ibid.* p. 83). A fallibilist, who acknowledges that most scientific hypotheses and theories are idealizations and even known to be false, cannot really literally consider science as a "belief system". Indeed, there is need for a conception of conjectural and truthlike "knowledge", which does not involve belief in the same way as the classical notion of *episteme* (see Niiniluoto 1999, p. 84). This position is compatible with the view the dynamically evolving and progressive body of scientific "knowledge" gives the best basis for the scientific worldview. This worldview is a rival and alternative to religious belief systems, but it is not based upon faith but rather upon rational arguments. In this sense my sympathies are on the side of what W. W. Bartley III calls comprehensive or "pancritical rationalism".

I learned to know Joseph Agassi as a sharp critical rationalist who loves disputes with his philosophical opponents. He also appreciates well-informed intellectual conversation on important issues. Concerning my work, he has offered both compliments and challenges, and I am grateful for both. The chance of discussing with him has strengthened my trust in progress both in science and philosophy.

References

Agassi, Joseph. 1975. Verisimilitude: Comments on David Miller. *Synthese* 30: 197–204.

———. 1976. Verisimilitude: Popper, Miller, and Hattiangadi. In *Formal methods in the methodology of empirical sciences*, ed. M. Przelecki, K. Szaniawski, and R. Wojcicki, 335–352. Wroclaw/Dordrecht: Ossolineum/D. Reidel.

———. 1981. To save Verisimilitude. *Mind* 90: 576–579.

———. 2001. The complete critical realist – Review of Ilkka Niiniluoto. *Iyyun: The Jerusalem Philosophical Quarterly* 50: 77–84.

———. 2011. Verisimilitude. *Discusiones Filosóficas* 12: 61–86.

Cevolani, Gustavo. 2016. Carnapian truthlikeness. *Logic Journal of the IGPL* 24: 542–556. (advance access).

Hattiangadi, Jagdish. 1975. After verisimilitude. In *Contributed papers, 5th international congress of logic, methodology and philosophy of science*, 49–50. London, Ontario.

———. 1983. To save fallibilism. *Mind* 92: 407–409.

Jeffrey, Richard. 1975. Probability and falsification: Critique of the Popper program. *Synthese* 30: 95–117.

Kuipers, Theo. 2000. *From instrumentalism to constructive realism: On some relations between confirmation, empirical progress, and truth approximation*. Dordrecht: Kluwer.

Miller, David. 1974. Popper's qualitative theory of verisimilitude. *The British Journal for the Philosophy of Science* 25: 166–177.

———. 1975. The accuracy of predictions: A reply. *Synthese* 30: 207–219.

Niiniluoto, Ilkka. 1977. On the truthlikeness of generalizations. In *Basic problems in methodology and linguistics*, ed. R.E. Butts and K.J. Hintikka, 121–147. Dordrecht: D. Reidel.

———. 1986. The significance of verisimilitude. In *PSA 1984*, ed. P.D. Asquith and P. Kitcher, vol. 2, 591–613. East Lansing: Philosophy of Science Association.

———. 1987. *Truthlikeness*. Dordrecht: D. Reidel.

———. 1998. Verisimilitude: The third period. *The British Journal for the Philosophy of Science* 49: 1–29.

———. 1999. *Critical scientific realism*. Oxford: Oxford University Press.

———. 2011. The development of the Hintikka program. In *Handbook of the history of logic, vol. 10: Inductive logic*, ed. D.M. Gabbay, S. Hartmann, and J. Woods, 311–356. Amsterdam: Elsevier.

Niiniluoto, Ilkka, and Raimo Tuomela. 1973. *Theoretical concepts and hypothetico-inductive inference*. Dordrecht: D. Reidel.

Oddie, Graham. 1986. *Likeness to truth*. Dordrecht: D. Reidel.

———. 2014. Truthlikeness. In *The Stanford encyclopedia of philosophy*, ed. E. Zalta. Stanford: Metaphysics Research Lab, Center for the Study of Language and Information, Stanford University.

Popper, Karl R. 1963. *Conjectures and refutations: The growth of scientific knowledge*. London: Routledge and Kegan Paul.

———. 1979. *Objective knowledge*. 2nd ed. Oxford: Oxford University Press.

———. 1982. *Realism and the aim of science*. Totowa: Rowman and Littlefield.

Chapter 8
The Story of the German Translation of Tarski's Book on Truth

Jan Woleński

> *His [Popper's] admissions of changes of opinions seems to me sparse and forced, as if they were admissions of some guilt. [...] The clearest example is his attitude to Tarski's theory of truth.*
>
> (Joseph Agassi 2008, p. 216)

This paper describes the story of the German translation of Tarski's seminal work *The Concept of Truth in Fomalized Languages*, originally published in Polish in 1933; the German translation was published in 1935 and became a sensation in logic and philosophy. The story is based on the letters exchanged mostly between Tarski and Kazimierz Twardowski (other persons were involved as well); this collection was discovered 3 years ago. The letters clarify some terminological points, changes in Tarski's views, participation of other people (Kazimierz Ajdukiewicz, Rudolf Carnap, Maria Kokoszyńska and Karl Popper) in doing translation as well as the ways of the early influence of Tarski's work.

The German translation of Alfred Tarski's famous book on truth was one of the most important publication-events in the history of logic and analytic philosophy.[1] Due to that translation, the semantic definition of truth became widely accessible to communities of philosophers of the West. I use the adjective 'widely', as it would be incorrect to say that Tarski's ideas were completely unknown before 1935. Tarski

[1] Polish original: Pojęcie prawdy w językach nauk dedukcyjnych (The Concept of Truth in Languages of Deductive Sciences), Warsaw: Towarzystwo Naukowe Warszawskie, 1933 (Tarski 1933 hereafter). Translated into German as "Der Wahrheitsbegriff in den formalisierten Sprachen", *Studia Philosophica*, 1, 1935, pp. 261–405 (Tarski 1936 hereafter). English translation: "The Concept of Truth in Formalized Languages", in A. Tarski, Logic, Semantics, Metamathematics: Papers from 1923–1939, translated (from German) by J. H. Woodger, Oxford: Clarendon Press, 1956, pp. 152–278. The original edition had appeared as a separate book. German and English translations are formally papers, because the former appeared in a journal, the second in a collection. However, it is a common practice to refer to this work as a book. There are a few reprints of Polish, German and English editions and numerous translations into other languages (I skip bibliographical details). The German translation has different title than the Polish original; the English edition preserves that change. I return to this issue in the next sections.

J. Woleński (✉)
University of Information, Technology and Management, Rzeszów, Poland
e-mail: jan.wolenski@uj.edu.pl

N. Bar-Am, S. Gattei (eds.), *Encouraging Openness*, Boston Studies in the Philosophy and History of Science 325, DOI 10.1007/978-3-319-57669-5_8

had been visiting Vienna several times in the first half of 1930s and, as we can learn from Carnap 1963, pp. 60–61 or Popper 1972, pp. 319–320, all of them (but never together) discussed with each other the concept of truth. Popper reminds his readers that he had had a discussion with Tarski on the occasion of the Prague conference (before the 8th International Congress of Philosophy) in 1934. On January 21, 1932, Tarski delivered a talk in Vienna on the concept of truth in the languages of deductive sciences; the summary of that lecture was published in Tarski 1932. Finally, Tarski's and Kokoszyńska's papers (Tarski 1936a, 1936b; Kokoszyńska 1936) inspired very heated discussions (see Neurath 1935; Carnap 1963, p. 61) at the Paris Congress for the Unity of Science in 1935. Though semantics had enemies (including Neurath), the discussions in Paris essentially contributed to increasing interest in logical semantics and the concept of truth based on it (see Mancosu 2008 for an extensive and detailed report).

Carnap, who suggested the idea that Tarski should deliver lectures about semantics at the Paris Congress, was acquainted with the galley proofs (Popper also read that material) of Tarski's book. Thus preparations for the publication of Tarski 1935 played a very important role in the reception of Tarski's ideas among logicians and philosophers. This concerns also Poland, even though the Polish original had appeared in 1933. In fact, Tarski's conception of truth had been known only to a limited number of Polish scholars in the early 1930s.[2]

The above reported data suggest that it would be interesting to reconstruct the road to Tarski 1935. Apart from the appointment of the translator of Tarski's study, nothing had been known about the work on German text. However, Dr. Katarzyna Gan-Krzywoszyńska (University of Poznań) discovered at the end of 2013 a collection of letters shedding light at how Tarski 1935 was prepared.[3] The collection contains 60 items (letters, postcards and telegrams), 55 between Twardowski and Tarski, 2 letters of Twardowski to Popper and 3 of Popper to Twardowski. The first letter (Tarski to Twardowski) is from September 30, 1933, and the last one (from Twardowski to Tarski) is from March 15, 1937. Most of the letters were written in 1935, when Tarski visited Vienna (January–August) after receiving a Rockefeller grant. Twardowski's letters are typed, but those of Tarski are hand-written. The way they addressed each other shows the very ceremonial customs (patterned on the

[2] Only one review of Tarski 1933, namely Kotarbiński 1934, appeared in Poland.

[3] The collection is located in the Archive of Polish Philosophical Society, Poznań Branch. It was probably taken from Lvov to Poznań by Kazimierz Ajdukiewicz (he became a Professor of the Poznań University) in 1945 or 1946. It constitutes a part of materials concerning the journal Studia Philosophica. That journal (its subtitle was Commentarii Societas Philosophicae Polonorum) was established in the 1933 and its first volume appeared in 1935. Ajdukiewicz, Roman Ingarden and Kazimierz Twardowski served as the editors. The main task of the journal consisted in publishing works of Polish philosophers in foreign languages. The second volume was printed in 1937. After the World War II, the journal was renewed and two volumes appeared, namely III (dated 1939–1946; it was assumed to stress the continuity of Polish philosophy) and IV (1950–1951; published in 1951). Since Ajdukiewicz became the editor-in-chief, the journal was produced in Poznań. This fact explains why the collection of letters was located in Poznań. However, the collection is not ordered by typical bibliographical standards. In particular, there are no signatures of particular items. I quote by dates of letters.

German and Austrian style) that were employed in Polish academic circles at that time. Tarski used the form *Hochgeehrter Herr Professor* (there is no good translation of this phrase into English) and Twardowski replied to Tarski as *Geehrter Herr Kollege* (only Twardowski's first letter begins with *Sehr geehrter Herr Doktor*). This form of addressing confirms that Tarski was fully recognized as a member of the Polish academic community. The letters deal with many problems, particularly financial matters, details of translation, proofs, offprints, bibliographical issues, consultations with other scholars, their participation in preparing of the translation and other issues. Most of the topics touched upon in the correspondence are of a technical-editorial character. However, the letters also contribute to a better understanding of Polish and German terminology used in the 1930s (and earlier) and shed new light on some long-debated interpretative problems concerning Tarski's semantic ideas or allow more precise dates to be fixed for the history of semantics and philosophy. Although the correspondence does not force us to change the standard interpretation of the semantic theory of truth, it clarifies some historical points.

The letter that initiated the entire exchanged of letters is lost. From Tarski's answer we may know that the Editorial Board informed Tarski about the new journal and proposed that he submitted something to be published in *Studia Philosophica*. Tarski replied to the Editorial Board (no-one of name is mentioned) that he is very pleased by the new initiative and grateful for invitation to co-operate, but that he is afraid that his works, due to their technical character, do not fit philosophical standards. Twardowski answered on November 11, 1933, and recommended the translation of Tarski 1933 as suitable for *Studia Philosophica.* He added that that contribution could be translated into German in its full version or even in part. Twardowski's proposal is not surprising. He knew Tarski very well. First of all, Twardowski had supported Tarski as a candidate, against Leon Chwistek, for a professorship of mathematical logic at the Lvov University in 1928 (Chwistek won this competition). Secondly, Twardowski had heard Tarski's lectures in Lvov in 1930. He noted (Twardowski 1997, p. 189): "today [December 15, 1930] the last Tarski's lecture in P. T. F. [Polish Philosophical Society] – very interesting as well as nicely constructed" (that lecture concerned the concept of truth in formalized deductive sciences; see Tarski 1931).

Tarski replied on November 25, 1933. He said that Twardowski's proposal conformed to his wishes, but he expressed two reservations. Firstly, Tarski suggested that his work should be translated in full, because a partial translation would be not accessible for people not familiar with deductive inferences. Secondly, he raised the problem of payment for a translator. Since Twardowski himself admitted (see above) that the entire text of Tarski 1933 might be translated, Tarski's first reservation was immediately and positively resolved. The financial question required some efforts, but it was successfully arranged very soon (I skip details). Thus, translating could start. Leopold Blaustein, a philosopher and a Twardowski's student was appointed. He was not a logician, but his German was perfect.[4] Twardowski wrote (April 10,

[4] Twardowski noted in Twardowski 1997, p. 331 that Zygmunt Schmierer, a logician, was ready to translate Tarski's book, but he was rejected; Twardowski did not explain why Schmierer was not appointed as the translator.

1934) that Blaustein submitted a translation sample, and it was approved. Tarski replied on April 21, 1934. He expressed his pleasure that Blaustein served as the translator, but he proposed some changes concerning terminology. The correspondence shows that Blaustein had translated the text literally from Polish. Twardowski checked the result and sent it to Tarski for approval; also Ajdukiewicz served as a consultant for making corrections of Blaustein's translation. Moreover, Ajdukiewicz also translated the *Nachwort* (Afterword) the significant addition to the German edition, in which Tarski stated his famous undefinability theorem (the set of true sentences of Peano arithmetic is undefinable in this theory) and explained some historical details; the *Nachwort* was also read by Popper. The whole book passed three proof procedures before publication. Tarski himself had been working on proofs in Vienna in Spring and Summer of 1935. The final version was checked by Kokoszyńska during her stay in Paris in 1935, also on the occasion of the aforementioned Paris Congress.[5] The year 1935 was historically significant for three reasons. Firstly, Tarski could present the proof-pages to Popper (see below) and Carnap in Vienna. Secondly, it is very probable that he used the final version at the Paris Congress. And finally, Tarski decided to add the *Nachwort* after his visit in Vienna. It is rather certain that doing so, he was influenced by the Viennese logicians, perhaps Carnap and Gödel.

Most of letters are devoted to the details of translation. Two principal kinds of questions are discussed in the correspondence.[6] First, there were several purely linguistic problems, namely those concerning style, syntax, and spelling of some words or using the ordinary terminology. These types of questions naturally arise in the case of translating from one language to another. Twardowski was very helpful in recommending some changes in Blaustein's translation. The same concerns were had by other consultants, including Popper, but only Twardowski's suggestions are noted in the correspondence. Those problems are fairly unessential for philosophical and logical reflections on Tarski 1935 and I skip them here. Second, several translation issues were connected with a specialized lexicon. I begin with the new title of the Tarski's monograph. The literal wording of the original Polish title is *Der Wahrheitsbegriff in den Sprachen der deduktive Wissenschaften* (the title of Tarski 1931 is almost identical, except *Disziplinen* replaces *Wissenschaften*). In his letter on March 26, 1935 Tarski informs Twardowski about two points, namely that

(1) He wants to change the title of his study into *Der Wahrheitsbegriff in den formalisierten Sprachen*,
(2) Since he changed of some of his views since 1933, he would like to give notice of this fact. Tarski did not, however, announce that he would like to write *Nachwort*.

[5] Tarski also mentioned (in the letter to Twardowski on May 2, 1934) Jan Łukasiewicz and Stanisław Schayer as consultants.

[6] See Gruber 2016 for a comparison of the Polish original to the German translation. The author had no access to the Twardowski-Tarski correspondence while working on the book. Fortunately, I provided her with a copy before the first draft of the book was completed.

Now, it is unclear whether (1) and (2) are somehow connected. The new title was possibly chosen for aesthetic reason. In particular, the new title is shorter and, at least, more elegant.[7] Now I see that Tarski's motivation was deeper and more substantial. Probably after discussions with Carnap and reading Carnap 1934, Tarski came to the conclusion that exact (it was identified with scientific) semantic investigations must be carried on in syntactically well-defined languages. However, Tarski decided to use not the adjective *formale,* but *formalisierte* in order to characterize languages for which the concept of truth is to be defined.[8]

One could claim that ultimately it does not matter very much, which of two adjectives is employed. To justify that the opposite view is correct, however, I quote the following passage from Tarski 1956 (pp. 166–167; it is the exact translation of Polish original and its German translation):

> It remains perhaps to add that we are not interested here in 'formal' languages in one special sense of the word 'formal', namely sciences of the things and expressions to which no material sense is attached. For such sciences the problem here discussed [the problem of truth – JW] has no relevance, it is not even meaningful. We shall always ascribe quite concrete and, for us intelligible meanings to the signs we shall consider. The expressions which we call sentences still remain sentences after the things which occur in them have been translated into colloquial language. The sentences which are distinguished as axioms seem to us to be materially true, and in choosing rules of inference we are always guided by the principles that when such rules are applied to true sentences the sentences obtained by their use should also be true.

The expressions 'material sense' (in Polish, *intuicyjny sens*; in German, *inhaltliche Sinn*) and 'intelligible meaning' (in Polish *zrozumiałe znaczenie;* in German *verständliche Bedeutung*) are of the utmost importance in the above quotation. Roughly speaking, we can assume that the adjectives 'material', 'intelligible' and 'intuitive' are close in meaning in Tarski's usage.

Partially implicitly, Tarski introduced three contrasts:

(a) formal languages vs. formalized languages;
(b) formal languages vs. informal languages;
(c) formalized languages vs. informal languages.

The intended meaning of 'formalized' is that formalized languages are those languages which arise as the result of formalization. Thus, formalized languages are always interpreted. So-called intuitionistic (better to say – intuitive) formalism, a view of Stanisław Leśniewski, Tarski's doctoral thesis supervisor, constitutes the philosophical background of Tarski's attitude to languages. Tarski rejected formal languages as objects of semantic investigations. Due to the antinomies, he was also skeptical about the consistency of the semantic treatment of ordinary (colloquial)

---

[7] Note that cases of nouns in Polish are indicated by endings of words, but not by such qualifications as *der, die, dass*, etc. Hence, the expression *Pojęcie prawdy w językach nauk dedukcyjnych* is not so complicated as its German counterpart. Perhaps it was Tarski's reason to skip the word *Wissenschaften* in the final version of the title.

[8] Note that Tarski 1931 has *sformalizowany* (formalized) in its title. Thus, it seems that Tarski used the concept of formalized language from the inception of his works in semantics.

languages, at least in their entirety. Consequently, the homological relation between formalized languages and their informal counterparts, belonging to colloquial parlance, was essentially limited so as to exclude the antinomies. At that point, the problem of intuitive elements in semantics and logic enters the stage.[9]

Tarski frequently refers to intuition (see Tarski 1956, pp. 153, 155, 211, 215, 224, 228, 229, 234, 235, 244, 265). This caused some problems for the translation. Twardowski informed (April 9, 1935) Tarski about Ajdukiewicz's proposal to translate the phrases *z intuicyjnego punktu widzenia* (from the intuitive viewpoint), *intuicyjnie rzecz biorąc* (taking the issue intuitively), etc. not literally, that is as *von intuitiven Standpunkt, intuitive genommen*, etc. but by using phrases *vom inhaltlichen Standpunkt* (referring to content), *inhaltlich genommen* (taking content into account), etc., eventually using both translations. Tarski replied the same day (incidentally, it shows how efficient post was at that time). He said[10]: "In Polish text I constantly use, and perhaps overuse, those terms ['intuition', 'intuitive', etc.] following the practice, which became common among Polish mathematicians and logicians. On the other hand, logicians in Vienna say that these terms occurring in such contexts as functioning in my work, are almost unintelligible for German readers".

Tarski followed Ajdukiewicz recommendation in some cases, but not in all. For instance, we have *intuitive Kenntnis* (Tarski 1935, p. 265), not *inhaltliche Kenntnis* which has no precise meaning and could be misleading. However, English version returns to 'intuition', 'intuitive', etc., because adjectival derivatives for 'content' (English counterpart of *Inhalt*) are not applicable in contexts pointed out by Ajdukiewicz.

The difference between Polish (as employed in the Lvov-Warsaw School) and Viennese ways of taking account of the sense of 'intuition' and 'intuitive' (more precisely, the related Polish and German words) is remarkable from the philosophical point of view. Tarski refers to the parlance of Polish mathematicians and logicians. However, the issue had deeper aspects. Leśniewski once said that logic is a formal exposition of intuitions.[11] It was an essential component of his intuitive intuitionism (see above) as a view concerning the foundations of logic and mathematics. On this view, we have an understanding of language—ordinary or specialized—for example, of the mathematical one, and we can rely on it as basically correct. It does not mean that this understanding is sufficient for all our cognitive tasks. On the contrary, for instance semantic antinomies show that language should be corrected in some cases. Logic enters the scenario in such situations. Yet logic should be

---

[9] I do not claim that Tarski maintained those views until the end of his career. In Tarski 1956, p. 62, we find the remark that he abandoned the intuitive formalism. Several passages in Tarski 1944 suggest that Tarski became more liberal in respect to the possibility of consistent semantics for ordinary language. However, it seems that he always claimed that non-interpreted languages are not subjects of semantic analysis. This suggests that to do semantics, one must explicate an interpretation of a language in question. Strictly speaking, the interpretation can be arbitrary and artificial, but it must be specified.

[10] In the same letter, Tarski informs Twardowski that he has finished (in Polish) the text of *Nachwort*.

[11] Personal communication of Henryk Hiż.

applied even if no troubles, antinomies or other defects, particularly, ambiguities would appear. Thus, logic, according to Leśniewski and Tarski as his follower, is inherently connected with our ways of employing language and its resources. This is why Tarski ascribed the fundamental role for formalized languages and not to formal ones. Consequently, linguistic intuition and what is generated by it was considered by Tarski as providing fundamental insights which cannot be entirely eliminated. This issue is also important for comparisons between the standpoint of the Vienna Circle and that of the Lvov-Warsaw School, because logical empiricism entirely rejected intuition as a relevant cognitive device. Due to this fact, that even if the object language becomes fully formalized, its metalanguage must be informal to some extent, even if contains a part of standard mathematics, for instance, set theory. Thus appeal to intuition in the Tarskian sense cannot be avoided.

The first serious translation problem in preparing Tarski 1935 concerns quotation-mark-names (see Tarski 1956, p. 156; *nazwy cudzysłowowe*, for instance 'logic').[12] Tarski's earliest choice (the letter on May 2, 1934) was *Anführungszeichennamen*; he had recognized that problem as difficult but after consultations with Łukasiewicz and Schayer he changed his mind because of the ambiguity of that term (it can be understood either as referring to quotation-names or to names of quotes). Tarski asked Carnap but the latter did not help very much (Tarski's letter on May 10, 1934); Carnap only observed that the word *Name* (or *Eigenname*, a proper name) should be not used. He suggested to follow Carnap 1934 and to distinguish descriptions (*Kennzeichungen*), for example, 'the word consisting of the letters 'a' and 'm'' (as a description of the word 'am') and proper names, for example 'es' as the proper name of the letter 's'; moreover, he reserved quotation-(mark)-names only for the contexts consisting of the left quote ', the single word or the group of letters and the right quote'; according to him complex phrases in quotes should be considered as hidden descriptions. In the end, the term *Anführungsnamen* was adopted for the German translation ('quotation-mark-names' is not the literal translation of this German word and is closer to Polish original) of *nazwy cudzysłowowe*. Troubles with translating *nazwy cudzysłowowe* into German reveal some peculiarities of the Polish logical vocabulary. Due to the lack of articles in Polish, the Polish word *nazwa* (name) covers common nouns, proper names as well as descriptions. Carnap did not agree with such a wide usage of the phrase *Name*. Perhaps Twardowski recommended Carnap's solution but Tarski opposed. According to him (his letter of May 10, 1934), quotation-(mark)-names should be considered as proper names of expressions standing between quotes. In fact, Tarski preferred so-called structural-descriptive names (see Tarski 1956, p. 156), for instance, the context 'the expression consisting of the letters: es, en, o, double-u' as the proper name for 'snow'. Note that this convention had been already used in Tarski (1933), that is, before Tarski asked Carnap about the translation of *nazwy cudzysłowowe*.

---

[12] It seem that some letters concerning these problems are lost. Clearly Tarski's letters on May 10, 1934 and May 14, 1934, formulate answers concerning the translation of *nazwy cudzysłowowe* but we do not exactly know what Twardowski had said.

Tarski (the letter on June 7, 1936) had serious reservations to such phrases as *Bedeutung der Begriffe* (meaning of a concept) or *in der Sprache gefundene Inhalt der Begriffe* (content of a concept presented in a language). He observed that if *Begriff* were replaced by *Terminus* (term) or *Ausdruck* (expression), but *Inhalt* by *Bedeutung*, the lack of clarity still remains. Finally, he proposed to use *Sinn*, not *Inhalt* or *Bedeutung*, and to apply the former to terms or expressions. Tarski pointed out that his objections and proposals had philosophical character. Although he did not enter into details, it is rather clear that he was guided by nominalistic sympathies (terms and expressions instead of concepts), a fear of psychologism (content) and the well-known ambiguity of *Bedeutung* (used once for meaning, otherwise for reference). Perhaps it is interesting to note that Tarski, in putting forward the idea of using *Sinn* in the above mentioned contexts, did not maintain that that choice would resolve all difficulties. He only said that this way of speaking is less problematic than others.[13]

As I have earlier noticed, the Tarski-Twardowski correspondence helps with some historical issues. In particular, it contributes to our knowledge about Tarski's relation to the Vienna Circle or, more generally, philosophers and logicians from Vienna. Of course, we have a considerable documentation about mutual influences between Tarski and Vienna, particularly referring to the development of logical semantics (see papers in Woleński, Köhler 1999)). The Twardowski-Tarski correspondence additionally confirms those relations. In the light of that exchange we can say that Tarski's ideas were known to Carnap and Popper in 1935, not only because Tarski had shown the proofs to them, but also because both of them were involved in the corrections. Tarski was willing to stress the role of Gödel for the ideas expressed in *Nachwort.* In the letter on July 24, 1935, Tarski explicitly asked Twardowski to add a footnote asserting the role of languages of infinite order (see Tarski 1956, p. 274; in fact, the related passage appears in the main text). Finally, the correspondence clarifies relations between Ajdukiewicz and Tarski. The former criticized Rickert's transcendental idealism (see Ajdukiewicz 1937) and Berkeley's subjective idealism (see Ajdukiewicz 1948) by pointing out that semantics is essentially richer than syntax (this is a consequence of Tarski's undefinability theorem). I heard (it was said during some discussions in Poland; I do not remember the author of this opinion) that Ajdukiewicz was not influenced by Tarski in that respect.[14] Now, once we know that Ajdukiewicz translated *Nachwort*, it is very difficult to imagine that he did not know Tarski's undefinability result and its possible philosophical consequences.

Tarski also mentions (the letter on July 24, 1935) some personal troubles related to his work on truth. Though he did not mention anybody who caused the troubles in question, it is pretty certain that he alluded to Leśniewski. The attitude of Leśniewski to the semantic definition of truth is unclear (see Betti 2008 for more

[13] He added "as they say". However, we do not know exactly who said that. Since Tarski was in Warsaw at that time, one can admit that 'they' refer to Warsaw logicians.

[14] On the other hand, the role of Tarski's arguments against radical conventionalism (the position held by Ajdukiewicz in the early 1930s) is well documented (see Ajdukiewicz 1964).

details). Czesław Lejewski, a student of Leśniewski's, once told me that due to uses of set-theoretical machinery by Tarski he (Leśniewski) wanted not to be linked with this type of semantics. Bogusław Wolniewicz reported to me Ajdukiewicz's statement that everything that Tarski did in his investigations on truth "was already in Leśniewski". This last opinion could suggest that Leśniewski accused Tarski of plagiarism.[15] In fact, the former wrote to Twardowski (the letter on September 8, 1935; see Burdman Feferman, Feferman 2004, p. 101): "I am trying to keep as secret as possible not just the details of [my] work but also the very fact of its development, so that certain Jew-boys and their foreign friends do not play some fifty tick on me again, as they have already done".

Leśniewski became a radical anti-Semite in the 1930s and, hence, the tone of his letter is not surprising. It is likely that Tarski belonged to 'Jew-boys', as did his foreign friends from the Vienna Circle. The two dates of May 24, 1935 (Tarski's letter to Twardowski) and September 8, 1935 (Leśniewski's letter to Twardowski) are close enough that the hypothesis that Lesniewski referred to work on the concept of truth looks probable. However, it is only a hypothesis. Anyway, Tarski decided to add a footnote clarifying several historical issues (see Tarski 1956, pp. 277–278) concerning the relation between his research and those of Gödel and Carnap. And in the mentioned letter to Twardowski he noted that the text of those historical explanations was "entirely squared with Carnap and Gödel".

References

Agassi, Joseph. 2008. *A philosopher's apprentice in Karl Popper's workshop*. Amsterdam: Rodopi.

Ajdukiewicz, Kazimierz. 1937. Problemat transcendentalnego idealizmu w sformułowaniu semantycznym [A semantical version of the problem of transcendental idealism]. *Przegląd Filozoficzny* 40: 271–287. English trans. in Ajdukiewicz 1978, 140–153.

———. 1948. Epistemologia i semiotyka [Epistemology and semiotics]. *Przegląd Filozoficzny* 44: 336–347. English trans. in Ajdukiewicz 1978, 182–189.

———. 1964. Zagadnienie empiryzmu i koncepcja znaczenia [The problem of empiricism and the concept of meaning]. *Studia Filozoficzne* 1: 3–14. English trans. in Ajdukiewicz 1978, 306–319.

———. 1978. *The scientific world-perspective and other essays, 1931–1963*, ed. and Trans. Jerzy Giedymin. Dordrecht: D. Reidel.

Betti, Arianna. 2008. Polish Axiomatics and its truth. On Tarski's Lesniewskian background and the Ajdukiewicz connection. In *New essays on Tarski and philosophy*, 44–71. Oxford: Oxford University Press.

Burdman Feferman, Anita, and Solomon Feferman. 2004. *Alfred Tarski: Life and logic*. Cambridge: Cambridge University Press.

Carnap, Rudolf. 1934. *Logische Syntax der Sprache*. Vienna: Springer; English trans. 1937. *Logical syntax of language*. London: Kegan Paul, Trench, Trubner & Co.

———. 1963. Intellectual Autobiography. In *The philosophy of Rudolf Carnap*, ed. Paul A. Schilpp, 1–84. La Salle: Open Court.

[15] Tarski credited Leśniewski with the diagnosis and solution of semantic antinomies (see Tarski 1956, p. 155, footnote 1).

Gruber, Monika. 2016. *Alfred Tarski and the "Concept of truth in formalized languages": A running commentary with consideration of the polish original and the German translation.* Dordrecht: Springer.

Kokoszyńska, Marja. 1936. Syntax, Semantik und Wissenschaftslogik. In *Actes du Congès International de Philosophie Scientifique*, vol. 3, 9–14. Paris: Herman & Cie.

Kotarbiński, Tadeusz. 1934. Alfred Tarski: Pojęcie prawdy w językach nauk dedukcyjnych [Alfred Tarski: The concept of truth in languages of deductive sciences]. *Przegląd Filozoficzny* 37: 85–91.

Mancosu, Paolo. 2008. Tarski, Neurath, and Kokoszyńska on the semantic conception of truth. In *New essays in Tarski and philosophy*, ed. Douglas Patterson, 192–224. Oxford: Oxford University Press.

Neurath, Otto. 1935. Erster Internationaler Kongress für Einheit der Wissenschaft. *Erkenntnis* 5: 375–430.

Popper, Karl R. 1972. *Objective knowledge: An evolutionary approach.* Oxford: Clarendon Press.

Tarski, Alfred. 1931. O pojęciu prawdy w odniesieniu do do sfomalizowanych nauk dedukcyjnych [On the concept of truth related to formalzied deductive sciences]. *Ruch Filozoficzny* 12: 210–211.

———. 1932. Der Wahrheitsbegriff in der Sprachen der deduktiven Wissenschaften. *Akademie der Wissenschaften in Wien. Mathematisch-naturwissenschaftliche Anzeiger* 69: 23–25.

———. 1933. *Pojęcie prawdy w językach nauk dedukcyjnych [The concept of truth in languages of deductive sciences].* Warsaw: Towarzystwo Naukowe Warszawskie.

———. 1935. Der Wahrheitsbegriff in formalisierten Sprachen. Trans. (from Polish) by L. Blaustein. *Studia Philosophica* 1 (1936): 261–405.

———. 1936a. Grundlegung der wissenschaftliche Semantik. In *Actes du Congès International de Philosophie Scientifique*, vol. 3, 401–408. Paris: Herman & Cie. English trans. 1956. *The establishment of scientific semantics.* Tarski, 1–8.

———. 1936b. Über den Begriff der logischen Folgerung. In *Actes du Congès International de Philosophie Scientifique*, vol. 7, 1–11. Paris: Herman & Cie. English trans. *On the concept of logical consequence.* Tarski, 409–420.

———. 1956. The concept of truth in formalized languages. In *Logic, semantics, metamathematics: Papers from 1923–1939*, 956: 152–278, ed. Alfred Tarski. Trans. (from German) by J.H. Woodger. Oxford: Clarendon Press.

Twardowski, Kazimierz. 1997. Dzienniki (*Diaries*), Vol. 2. Toruń: Wydawnictwo Marszałek.

Woleński, Jan, and Eckehart Köhler, eds. 1999. *Tarski and the Vienna circle: Austro-Polish connections in logical empiricism.* Dordrecht: Kluwer Academic Publishers.

Chapter 9
Two Philosophical Contributions to Modern Physics

Jagdish Hattiangadi

9.1 I

When Joseph Agassi was in Karl Popper's department in the 1950s, the doctrine that metaphysics is meaningless was widespread outside that department. In 1957, Joseph Agassi attests to this as follows: "The idea that a question has no meaning unless the truth of the answer to it can be decided by observation is current both among physicists and among philosophers." (Agassi 1957). Joseph Agassi and Karl Popper were at the time engaged in rehabilitating the role of metaphysics in science. They coined the phrase, "metaphysical research programs" in science to describe their relation. Science and philosophy, I have come to believe, is intellectually part of a single enterprise, though the increasing specialization of research make them appear different. I will not argue for this claim here. I will explore instead how modern physical science was made possible by two seminal philosophical contributions, by Francis Bacon, and by Galileo Galilei, respectively, in epistemology and metaphysics. The Copernican hypothesis, and two remarkable philosophical contributions, made modern physical science possible.

9.2 II

Francis Bacon made a contribution to the theory of knowledge, and of induction, which were Aristotle's concerns in the *Posterior Analytics*. Francis Bacon contributed to a revision of epistemology and induction by using techniques derived from various forms of skepticism that are to be found in Plato's dialogues, as they were interpreted in Plato's Academy in its later development. Francis Bacon admitted to

J. Hattiangadi (✉)
Department of Philosophy, York University, Toronto, Ontario, Canada
e-mail: jagdish@yorku.ca

N. Bar-Am, S. Gattei (eds.), *Encouraging Openness*, Boston Studies in the Philosophy and History of Science 325, DOI 10.1007/978-3-319-57669-5_9

the truth of a very dark and pessimistic form of skepticism, from which no escape seems possible. To this he made a significant optimistic contribution, which he called his method of "true" (faithful) induction (Bacon 2000). He contrasted his own method with that of Aristotle. His contribution was a breakthrough in the theory of knowledge, and in the theory of the method available to us of gaining it. "True" induction also came to be called the "experimental philosophy" when it was employed with some success by the Royal Society since 1660. Robert Boyle and Isaac Newton, for example, were two self-proclaimed practitioners of experimental philosophy from that period.

Aristotle had proposed that knowledge differed from merely true opinion in a certain way: when we know (or understand) something, we know why it is so, and why it could not be otherwise[1] (Aristotle, 1984 PA: 71(b)). A demonstrative syllogism is one path to knowledge. When we derive what is to be shown from a valid demonstrative syllogism, we also know why it is so (i.e. the cause. or "because" of it, is given by the middle term). There is also the logical necessity of the conclusion, given the premises, in a valid argument. A mere opinion lacks this feature of truth by necessity.

The logical necessity of the conclusion of a valid argument is known only if the premises are demonstrated. Demonstrative syllogisms by themselves are not enough, argues Aristotle. His argument is based on the existence of two distinct kinds of propositions. If we allow that there are true opinions, and also that there is knowledge, and that these are distinct, then we must allow that there is some form of proof other than that obtained by demonstrative syllogisms.

If knowledge is found in the conclusion of a demonstrative syllogism, then the premises of the demonstrative syllogism would need their own demonstration, before we can allow that the conclusion constitutes knowledge. This allows for only two possibilities. Either we demonstrate the premises using new and different premises in the higher demonstration, or we do not, in which case we use premises that are not new. If we use new premises, then we need new demonstrations for them, and this will continue; if we demand new premises each time we find new premises, then we find ourselves in an infinite regress. Since we cannot provide an infinite number of syllogisms to buttress any knowledge claim, we have to renounce all claims to knowledge. Then there is only opinion, and no knowledge, contrary to our supposition.

If, however, we were to allow for circularity at any stage i.e. if we used as a premise for a demonstration a conclusion previously demonstrated, then we have a circularity. In this case we will have to allow any opinion to constitute knowledge, since any opinion can be demonstrated if we allow for circularity. Then there is only knowledge. But we began by assuming that there are both opinion and knowledge, which are distinct. This is now denied. We must, therefore, allow for a method of non-demonstrative knowledge of first principles. (Aristotle PA: 72(b)). These are

[1] Aristotle writes about understanding, whereas for simplicity I use the word "knowledge" throughout in its stead. However, this is not a relevant difference for us.

produced by "epagoge" or by induction, according to Aristotle. Up to this point Francis Bacon seems to agree.

To generate the necessity of first principles, however, Aristotle suggested that it is provided by a direct grasp of the mind. To show how this can give us necessary truth, Aristotle had recourse to a discovery of Plato's. Plato had discovered that the Forms are related to one another by a kind of necessity.

Aristotle had departed from some details of Plato's theory of Forms, holding that Forms were embedded in things that change. An essential attribute of a thing determines its natural or characteristic motion. This was his distinctive philosophy. He regarded individuals as possessing primary being, or substance, whereas its attributes have secondary, or dependent being.[2] It was Aristotle's opinion that when we perceive things we perceive them as examples of the Ideas (essences) within them, as one of that kind or species or eidos.

Only primary beings or substances can support contrary attributes. So this apple which was once green has turned red. "Green" and "red" cannot in turn support contrary attributes. Green cannot become a shape. Green is necessarily a color. It is into this kind of necessity discovered by Plato that Aristotle taps for his theory of epagoge as the basis of knowledge. That a human being is an animal is a conceptual necessity, we might say today, if at all we do, but not a demonstrative (or "logical") necessity.

Aristotle thought that of the many predicates that may be truly attributable to a thing, only one of them is true of it such that we say "that is what it is." When I see a person, Socrates, I see him as a person, in which case I see him for what he is. If I see him as walking in Athens without shoes, this is not the attribute by virtue of which Socrates is what he is. In short, when we perceive things as members of their species, and when we define that species by the method of genus and difference, we capture the essence of the species in a definition of it, which constitutes the first principle of knowledge of the individual as what it (he) is. There arises the question about how we obtain this knowledge of essence from our perceptions. How do we get to know that a human being is an animal (*genus*) that is able to reason (*differentia*) which presumably we perceive? On this important point, Aristotle's writing is unclear, or incomplete. We can gloss over it by saying that we obtain such knowledge by a process of abstraction. We look at all the attributes of the species of which an individual is a member, and winnow out all the unsuitable attributes until we are left with the last predicates standing, which will be the kind or species of the thing defined by the method of "genus per differentia." This understanding of what a thing is, as such, produces concepts that are very close to common sense and to the concepts of our everyday language, so it is not easy to say what Aristotle's method is when we get down to details.

The nature of a thing (its *eidos*, its species) is one of its attributes, but it is unique as its nature among its attributes. When we produce demonstrative knowledge, we

[2] I will ignore Aristotle's later hylomorphic analysis of motion in the *Metaphysics* in terms of matter and form. The relation of individuals to matter in his theory of motion is unclear, and controversial, but not topically relevant.

must rely on induction to generate first principles that *uniquely* define the nature of the thing to be understood. To know the first principle of a thing is to apprehend mentally (with the *nous*) what it is *per se.* There cannot be two natures for something (not two eidoi, not two species). We have to find the one thing that constitutes its real nature.

Francis Bacon objects to Aristotle's account of method. An intuitive grasp of natures supplemented by abstraction struck him as a method that is too naïve and prone to error. To replace it, Francis Bacon proposes that we adopt a kind of dialectical analysis that is like a cross examination of nature. (Bacon 2000: 232). This was his new method. Implicit in it is a very different understanding of knowledge.

Francis Bacon was very impressed with some alchemical discoveries. Whereas Aristotle had relied on common observations to yield the true knowledge of first principles of perceptible things, alchemists countered with a different conception of reality by relying mainly on esoteric experiments. Experiments reveal a hidden and extraordinary reality, they claimed. Francis Bacon rejected both of these methods (i.e., common sense and esoteric methods) as means of getting to first principles. Neither could incorporate the insights of the other, when all insights should be taken into account to truly grasp reality. In the case of Aristotle's definitions, he also complained that they were of no practical benefit, whereas the alchemists had at least at times produced useful results that they employed in their craft. Francis Bacon therefore proposed a combined method of *observation and experiment* to pursue knowledge.

Francis Bacon's central insight into the theory of knowledge is that we should try to be more aware of our errors. We need not be worried by the most extreme of form skepticism that he could imagine, whereby the mind itself is by its constitution full of error. We should nevertheless note the errors down. We should not bother looking for a class of pure thoughts, even in our perceptions, on which we can completely rely. He rejects any "safe harbor" strategy, such as Locke's simple ideas of sensation (which were proposed much afterwards, of course) because the dark truth is that the mind is itself like a crooked mirror that distorts everything it apprehends. Moreover, the mind is unlike writing on a wax tablet, he says, on which we must wipe out the old to write on it anew. With the mind it is different: we must write on it first to be able wipe out the old, as he says at the very end of his fragment "The Masculine Birth of Time." (Bacon 1964).

I will use the metaphor of mental lenses to simplify my summary of Francis Bacon's method.[3] Let us call what seems to any mind to be true of reality "what the mind sees," whether sensory, perceptual, conceptual or philosophical. Suppose that seeing reality is inevitably through a distorting lens. Had we just one distorting lens to contend with, then the mind is trapped in its distortion. We would take reality to be as seen. We have no means of knowing that it is distorted. Fortunately for us, we

[3]The language of "distorting lenses" and "error analysis" is my own, and I use it to simplify the presentation of Francis Bacon's thought, even though it may be itself a distorting lens into his ideas. For a less distorting account, I have a longer book length version, entitled *The Theory and Craft of Breaking Through in Science*, manuscript.

have many distorting lenses of the mind – many different kinds of lenses and sources of distortions, as it were. We can then examine one image of reality against another. Both cannot be real, since nature is unique among qualities. This creates a discrepancy in the appearances of things.

The lenses that a mind uses are varied, and they distort reality in very different ways. There are too many to enumerate, but we may classify them. Francis Bacon calls these "idols" reminiscent of Plato's "Eidola" or false Ideas. (Bacon 2000: 41–42). These are sometimes translated as "illusions," but we must remember that idols need not be sensory for Francis Bacon. As a tribe of human beings, for one thing, we are prone to a host of different errors. Since these are endemic to us as human beings, we cannot possibly avoid them. (Consider for instance biases caused by the nature of the human perceptual apparatus itself, or the mind). As individual human beings, however, our differences in situation and experience may make reality appear to us in individually different ways. These will provide us with more distorting lenses, obtained from the vagaries of our situation and our experiences. Our differing social backgrounds may also give us many more distorting lenses. For instance different languages may employ different classificatory schemes. Then there are also distortions of reality that we can obtain from the lecture theatre, where philosophers hold forth about reality, with their very different theories of the nature of things. Francis Bacon suggests that this wealth of distortions at our disposal is of great value because we can use one distortion to show the error in another distortion, and vice versa. There are a boundless numbers of discrepancies in appearances, so we know when we have erred. There is hope.

The technique of "true" induction that is suggested by Francis Bacon is quite ingenious. He suggests that we take one small item to study, e.g., one species or a quality predicated of things in some closely related species. It is important that the object of study be carefully circumscribed in its scope, to keep the project simple, because it can get out of hand if we do not. The recommendation to perform a "true" induction now is to write up what he calls a "natural history" or a historical record of the various distortions of the nature of a thing that appear before the mind. In this natural history we note down the errors or distortions that we can detect by playing up one distorting lens against another, but with two very important provisos: we note down all the distortions that we can find, and in each case the conditions under which we observe the distortions of what we are studying. These conditions must be written down clearly. They must be clear enough that if another experimenter were to try to recreate the observation or experiment from the written description in the Journal, she would be able to reproduce the errors and distortions that are in the natural history. If one hears of an outlandish fact about what one is studying, it may be of interest to us, to generate a discrepancy with ordinary expectations. But it becomes part of an inductive natural history only when the conditions under which the strange result is obtained have been adequately described. Adequacy depends, of course, on being reproducible based on the written condition.

When writing up a natural history it is very important not to leave some item out because it can be explained away. The mind naturally wants to soar to the highest levels. If we adopt an older metaphysical hypothesis we can always explain away any discrepancy. What the mind needs are "not wings but leaden weights." For we know already that if we take each part of each discrepancy, one by one, it can be dissected and analyzed away by any philosophical system. By deflecting potential criticism to some minor parts of the system, the main system itself can be propped up as long as we wish, thereby perpetuating error.

To break through the distorted appearances to the undistorted reality beyond, i.e. to get behind the distorting lenses, is not easy, and always uncertain. Francis Bacon gives us this method as a "help to the mind:" First we build up a natural history that is as extensive as possible; then we try to come up with a model for just this thing that we are studying. We seek that reconstruction of the real thing that can explain why, under the noted conditions in our historical record, the correspondingly noted errors or discrepancies arise. When we seek such a model, we must make sure that we do not rise to the highest level of metaphysical theory. We try instead to provide a model for just this little thing that we chose to study. What we provide for it is not a global understanding of reality, but a bit of local error analysis.

Metaphysical notions are abandoned in order to focus on the errors at hand. We must want to obtain glimpses of reality beyond the distortions of appearance. This is a matter of choice. If we prefer, we could fall back upon some well-developed metaphysical scheme to explain as much as possible, and then to explain away that which we cannot explain. For adornment, says Francis Bacon, metaphysical supposition is better. But for discovering reality, true induction recommends itself.

We seek a model (a combination of the hidden or latent *structure* and its *process*) that can explain why the errors in our little natural history happen as they do, under each of the noted conditions. (Bacon 2000: 102ff). When we do, it is like solving a riddle, like finding one's way through a maze, like deciphered code in a secret message, or like finding an object that has been hidden. In each case the solution is unique if the natural history is extensive enough. Thomas Kuhn noticed this phenomenon amongst practicing scientists, in his own way, and called it "puzzle-solving." This is a useful word. A recreational puzzle proposed by the poser of the puzzle is solved when we have worked out its unique solution. Puzzles for amusement are designed to have unique solutions. Natural histories are our way of creating a puzzle for ourselves that seems almost impossible of solution. If the histories are extensive enough, then only one solution will be possible to a puzzle. The unique solution, obtained as an error analysis, then gets us behind the distorting lenses that prompted each bit of the natural history. In this error analysis, a model is proposed that shows why in the different conditions noted in the natural history those very errors arise that do.

Let us consider once again Aristotle's requirement that for *knowing or understanding* to be different from *opining*, we require it to be not merely true, but necessarily so. Aristotle's account provides for the necessity of knowing among the relations between Forms analyzable in the essences of perceived things. This is an appeal to metaphysical necessity. However, it made his epistemology vulnerable: for his theory of knowing cannot describe how we come by the ontic knowledge, without circularity. His opinion was easily countered by skeptics as "dogmatic," substituting an opinion for knowledge. Francis Bacon's epistemology is superior: the necessity in a Baconian principle is in turn epistemic. The puzzle is born out of a nest of impossibilities. A solution, when found, is unique. Necessity derives from both the impossibilities summarized in the unsolved puzzle and in the uniqueness of the solution, when found.

Francis Bacon makes another case for discovery by error analysis: *power over nature*. The model solution tells us, for each discrepancy, how different conditions will produce their respective distortions. If we call the discrepancies "discrepant effects," and the conditions under we which we note their occurrence their "causal conditions" then we know from the solution exactly how to produce the effects – namely by producing the causal conditions noted in our natural histories. This is how induction begets "power over nature." A solved puzzle about a thing is a key to its causal structure, and therefore its successful manipulation. Note, moreover, that, when completed, the knowledge we have gained is a kind of causal demonstration of the very kind that Aristotle prescribed for episteme. This power over nature is like a "pledge of truth." If we can successfully manipulate the thing that we were studying, then we know that we have somehow "got hold of the thing itself" in our little breakthrough. Our power over this bit of nature tells us that we are not merely spouting out words about things, as earlier philosophies had been doing.

Francis Bacon was more of a sceptic than these remarks show. Each Baconian model we discover by induction is very circumscribed in its scope. If we deploy one error against another, and so create a puzzle of a limited scope, to which we find a unique solution, we may well learn to see beyond some distortions of some of our lenses. Beyond these distortions, however, are there not many other distortions that we have failed to note? Of course there are. Francis Bacon suggests a way around this. If we begin with baby steps, we can make models at the lowest level, but they may contain distortions. When we go up the ladder of science to the next level up, we have the same scope for discrepancies again. Induction will be required to see a reality beyond some more distortions of a first approximation of reality. And so on. He speculates about a pyramid of knowledge, where we begin our journey at the base and rise to a pinnacle.

Of course, he cannot know about this supposed pyramid at the very bottom step, before the final metaphysics has revealed itself in a number of layers of error analyses. He did not note, but we may, that if the structure fails to taper as we scale up on his ladder, then we could continue correcting and improving our understanding indefinitely. The truth at the end of the road would then be that there is no end to the road. All our gains are temporary and fallible, and we find ourselves in the same epistemic position even as our metaphysics, or science, improves a bit at a time.

9.3 III

Galileo Galilei was a keen philosopher, generally speaking, and a metaphysician, in particular. This is not news. All the comments on his work by historians, social theorists and philosophers of science acknowledge his interest. But there is a wide variety of opinion concerning Galileo's contribution to modern science, about his method, and how his work should be interpreted. Some have said that Galileo follows the Archimedean method and was a Platonist, some that he followed Aristotle's method, some that he was following a version of a medieval method, some that he was an Aristotelian about experience, and so on. All the historical interpretations here seem to me to be partially right, in the following sense. If a proponent of one of the views does not say that one influence on it is all there is to Galileo Galilei's thought, then we can readily acquiesce. Unfortunately, that is often not the case, though this is not the time to dispute these details.

Pierre Duhem, who was in some ways his greatest detractor, recognized a metaphysician in him. It is in order to preserve the autonomy of physics from such metaphysics that he asked us to regard Galileo Galilei's contributions to physics itself as small and incremental, provided that we ignore his metaphysical opinions. Pierre Duhem thought that when we take away the metaphysical interpretation of physical theory from it, what is left is mathematical technique and a bank of experimental data, both of which grow very slowly, in minute increments (Duhem 1954: 7–18).

Alexandre Koyré, called out Pierre Duhem in his defense of the opinion that there was a scientific revolution in the seventeenth century, "[...W]hat the founders of modern science, among them Galileo, had to do, was not to criticize and to combat certain faulty theories, and to correct or to replace them by better ones. They had to do something quite different. They had to destroy one world and to replace it by another." (Koyré 1943: 205): "[... An] unbroken tradition leads from the works of the Parisian Nominalists to those of Benedetti, Bruno, Galileo and Descartes...". Still the conclusion drawn therefrom by Duhem is a delusion: a well-prepared revolution is nevertheless a revolution, and in spite of the fact that Galileo himself in his youth (as well as at times Descartes) shared the views and taught the theories of the medieval critics of Aristotle, modern science, the science born from his efforts and discoveries, does not follow the inspiration of the "Parisian forerunners of Galileo"; it places itself at once on a quite different level – on a level which I should like to call the Archimedean one. The true forerunner of modern physics is neither Buridan, nor Nicole Oresme, nor even John Philoponos, but Archimedes." (Koyré 1943: 406). Archimedes was of course a great admirer of Plato. So was Galileo Galilei. We find Alexandre Koyré concluding his paper by describing Galileo Galilei as a Platonist. "The new science is for him an experimental proof of Platonism." (Koyré 1943: 428).

Yes, Galileo Galilei admired Plato, as any good Archimedean mathematician would. He had studied Plato's thought closely. He had even applied Plato's thought in many places wherever it seemed to be appropriate. But Platonism is not all there is to Galileo Galilei's philosophy. He was, on the contrary, very much his own philosopher.

When Galileo Galilei was offered a job as a mathematician in Florence in 1610, he insisted that his new title be that of first philosopher and mathematician to the court. To find out what he means by "philosopher and mathematician" we can turn to Plato, who had something relevant to say about knowledge, mathematicians and philosophers, in the *Republic.* In a manner reminiscent of Parmenides, Socrates in Plato's *Republic* maintains that knowledge pertains to "that which is," or Being, whereas what he calls nescience pertains to "that which is not." Opinion pertains to that which is and is not, or Becoming. (Plato 1963 *Republic*, 476d–480) The phrase, "that which is and is not" is important here. Everything that changes, in time, will be first one thing and then another. Realities (e.g., the Forms,) of course, do not share this feature. They just are. Hence we can know the Forms, because they are intelligible, whereas that-which-is-and-is-not is very hard for us to grasp clearly, and is not intelligible. It is also uncertain (in Heisenberg's sense of being smeared). Whatever we say about that-which-is-and-is-not cannot be known to us, but may be perceived, and will always remain a matter of opinion. This difference is clarified later in the *Republic* in his famous Analogy of the Divided Line, which will help us understand what Galileo Galilei meant when he said that he wanted to be both a philosopher and a mathematician (Plato 1963 *Republic*, 510d–511e). A line is divided using a principle of proportion into four unequal parts. The smallest part of the line represents images and shadows (we could add sense impressions, perhaps). The next larger part represents the changeable objects themselves, which are clearer and more permanent than the ephemeral image of objects. However, on these two sectors of the line we can only find objects of perception, of which we cannot have knowledge but only opinion, since whatever they deal with are changeable in time.

On the third largest part of the divided line are those intelligible things that are treated by geometers and reckoners, i.e., by mathematicians. Things in this section of the line compel us to use perceptible things (from the second part of the divided line, of course) as a means to understand the intelligible forms of mathematics that are studied therein. Moreover what geometry does is this: "by means of assumptions [...] it proceeds not up to a first principle, but down to a conclusion [...]." In short, the method of mathematics is to assume hypotheses to be true and then deduce its consequences in terms of perceptible things that resemble the intelligible things being studied. Mathematicians do not investigate what the first principles mean and whether they are true. They merely craft the mathematics to fit the facts perceived among the changeable objects. This was called "the method of saving the appearances." Mathematical objects are exact and immutable. That is exactly why we can acquire knowledge of the objects. However, our mathematical knowledge is unfortunately only hypothetical, because when it is applied to changeable objects, there the knowledge stops. Plato's theme is well illustrated in his dialogue, the *Timaeus.* Timaeus warns Socrates from the beginning that knowledge *(episteme*, science) of celestial motion is impossible, because it is motion. Knowledge can only pertain to unchanging entities, which are the kind of objects of interest to mathematicians. (Plato, *Timaeus*, 27d–29d). Timaeus has an opinion about celestial motions, which he goes on to develop in the dialogue. It is a contribution to Eudoxan astronomy using the method of concentric circles to capture some errant appearances. But the best we can do in astronomy and in the related cosmology and theology, with the use of mathematics, is to save the phenomena. Mathematicians cannot know what stars truly are, or what paths they really follow, by the use of mathematics alone. Mathematicians do not inquire into their own first principles. Many models are possible. In studying motions the best that mathematician can do is to save the phenomena.

The fourth section of the divided line, representing the greatest clarity and the highest level of abstraction is "that which reason itself lays hold of by the power of dialectic, treating its assumptions not as absolute beginnings but literally as hypotheses, underpinnings, footings, and springboards so to speak, to enable it to rise to that which requires no assumption and is the starting point of all [...]." (Plato 1963 *Republic*, 511b–c) Philosophers, following the Socratic dialectical method, which is the highest level of all, can find true and unconditional principles rather than merely some hypothetical models that save the phenomena.

I suggest that Galileo Galilei wanted to be regarded as a philosopher because he wanted to rise to "that which requires no assumptions and is the starting point of all [...]." He wanted to offer a new science of kinematics, not just some opinions concerning motion. However Plato is quite clear about this. The realm in which motions are perceived is not capable of supporting knowledge. There can be no science of local motion, or kinematics, for a Platonist. Galileo Galilei knew his Plato well. He was a philosopher in his own right. He gave us, not Plato rehashed, but the rudiments of modern philosophy. This philosophy made modern mathematical physics possible. And yes, you can find isolated comments on Galileo Galilei in which Alexandre Koyré seems to recognize that this is exactly what the former was doing,

Galileo Galilei's philosophical accomplishments began with this: he began to treat the line or the curve of the geometer as the trajectory of a body. His geometry describes real local motions. It describes change, and not changeless entities. This is a new interpretation of geometry that had no precedent in ancient Greek thought. This is what he means by the very expression "a science of kinematics." He also expresses this in 1623 somewhat obliquely, "Philosophy is written in this grand book, which stands continually open before our eyes (I mean the 'Universe'), but cannot be understood without first learning to comprehend the language and know the characters as it is written. It is written in mathematical language, and its characters are triangles, circles and other geometric figures, without which it is impossible to humanly understand a word; without these one is wandering in a dark labyrinth." (Galileo 1957: 237–8).

With his new interpretation of geometry he laid the groundwork for the metaphysics of motion, if we may borrow a phrase. If you doubt that this new interpretation of geometry makes modern physics possible, please try to imagine what modern physics would be like if it could not treat local motions as geometrical lines. If motions were not treated as objects in their own right, we would not have "laws of motion." All laws would pertain to Forms or essences, and the motions they entail. Kinematics requires Galileo Galilei's declaration of independence on behalf of all local motions from the hegemony of Forms and qualities.

When Galileo Galilei began to treat local motion itself geometrically, he made the geometry a description of a reality in time, and not just a model to capture temporal phenomena that are inherently unintelligible. He had instantly made local motion more exact and more understandable than anything else in the Universe because the exact study geometry so describes it. What was unintelligible to Plato and to Aristotle, namely motion (in itself) had become the one thing we can know the best and the most exactly – i.e., geometrically. Local motion is better understandable it seems even than the essences of things, the embodied Forms in Aristotelian thought. Essences themselves were not needed to understand motions. They had become irrelevant.

I suspect that Galileo Galilei came to his new metaphysics of motion between 1602 and 1609 gradually. He was studying motions on earth using the techniques of Archimedean mechanics, assuming the Copernican hypothesis. Since the old physics was no longer applicable to a moving earth, he began experimenting and creating hypothetical mechanical models for falling bodies, for pendula, for natural motions on earth and so on. This is old philosophy, but new mathematics. He discovered to his amazement that each one of his studies showed that models that saved the phenomena did not use motions deriving from Aristotelian forms. These hypothetical models seemed to depend only on time, and distance. Perhaps, he thought, there is a science of motions as such. Once he came to believe that the Copernican hypothesis is very likely true, based on observations through a telescope, a science of motion seemed to be indicated – a "starry message."

Galileo Galilei made the many different natures otiose that Aristotle used to invoke the many different motions respectively of the different kinds of thing. When we can treat motion itself mathematically by means of a few qualities, we can banish all other forms, species, and qualities into a subjective realm. Geometry so applied provides us with all the Forms we need for physics (these are the primary properties or qualities of things). In 1623, he writes, "[…W]henever I conceive any material or corporeal substance, I immediately feel the need to think of it as bounded, and as having this or that shape; as being large or small in relation to other things, and in some specific place at any given time; as being in motion or at rest; as touching or not touching some other body; and as being one in number, or few, or many. From these conditions I cannot separate such a substance by any stretch of my imagination. But that it must be white or red, bitter or sweet, noisy or silent, and of sweet or foul odor, my mind does not feel compelled to bring in as a necessary accompaniment. Without the senses as our guides, reason or imagination unaided would probably never arrive at qualities like these. Hence I think that tastes, odors, colors, and so on are no more than mere names so far as the object in which we place them is concerned, and that they reside only in consciousness. Hence if the living creature were removed, all these qualities would be wiped away and annihilated." (Galilei 1957: 274). Modern metaphysics is unfolding before our eyes. Proposed lists of primary qualities can vary with theories of motion under examination. This makes no difference to the Galilean metaphysics. What matters is this: there is only one kind of motion and it is geometrical in nature. This is called classical physics nowadays. Its universal laws create the great wealth of effects on the otherwise nondescript matter of the universe. There seemed to be no way to study motion as a science except in this way, as classical modern physics, from 1623, until 1927.

Galileo Galilei faced an epistemological challenge: the relations between abstract entities in the definition of the nature of a thing made our knowledge necessary, thus distinguishing it from opinion. When Galileo Galilei banished all the forms and qualities of Aristotle's philosophy into a subjective purgatory, he simultaneously lost the source of necessity for his kinematics to rank as a science, or to count as knowledge. Physics in that case could not claim to be a science.

Galileo Galilei provided a resolution to this acute epistemological deficiency of his geometry of motion towards the end of the first day of the *Dialogue* "[...T]he human understanding can be taken in two modes, the *intensive* or the *extensive. Extensively*, that is, with regard to the multitudes of intelligibles, which are infinite, the human understanding is as nothing even if it understands a thousand propositions; for a thousand in relation to infinity is zero. But taking man's understanding *intensively*, in so far as this term denotes understanding some proposition perfectly, I say that the human intellect does understand some propositions perfectly, and thus in these has as much absolute certainty as Nature itself has. Of such are the mathematical sciences alone; that is, geometry and arithmetic, in which the Divine intellect knows infinitely more propositions, since it knows it all. But with regarded to those few which the human intellect does understand, I believe that its knowledge equals the Divine in objective certainty, for here it succeeds in understanding necessity, beyond which there can be no greater sureness." (Galilei 1967, 103).

Galileo Galilei's kinematics can be a science, too, if it is based on an intuitive certainty. Necessity arises from its certainty, because it is as sure as divine knowledge. Galileo Galilei's bar for knowledge is set high. This new epistemic bar gives us a way to generate necessity for the truth of our first principles, from knowing that they cannot be mistaken. Knowledge arises from the human intuition that rises to a divine level to apprehend some truth with certainty, though this is possible only in geometry and in arithmetic.

This new epistemology (incidentally, also a new philosophy of mathematics) is just a hint, a beginning. It gives us a sense of direction, but not the details of what and how we can know with such certainty that it can be compared with divine knowledge. Rationalism and empiricism, later erected as the two pillars of modern epistemology, rest upon the Galilean certainty of our knowledge.

9.4 IV

The Copernican hypothesis, and two remarkable philosophical contributions, made modern physical science possible. But the two philosophical contributions were not compatible with one another. The experimental philosophy of Francis Bacon forbids that we assume a metaphysical system when we confront a skeptically assembled natural history. The mathematical philosophy of Galileo Galilei asks us, on the contrary, to begin with first principles as certainly known to us as they are to God. These injunctions cannot both be followed at once.

This is not all. The epistemology of experimental philosophy derives the necessity of its conclusions from fallible considerations of uniqueness of the solution to a puzzle, composed of impossibilities. The mathematical philosophy derives its first principles from intuited certainties. These criteria of knowledge are at cross-purposes. Uniqueness does not yield certainty; but certainty does not yield power over nature. The mathematical philosophy and the experimental philosophy were at odds with each other.

The adoption of the experimental philosophy of Francis Bacon by members of the Royal Society, and later in science everywhere, did not stop these scientists from also embracing the new mathematical philosophy of Galileo Galilei, despite conflicts between the two. Each philosopher and scientist since then has had to think through these issues and work out his or her own patchwork compromise.

References

Agassi, Joseph. 1957. Duhem versus Galileo. Reviews. *British Journal for the Philosophy of Science* 8(31): 237–248, Oxford University Press, Oxford, UK.

Aristotle. 1984. PA, "Posterior Analytics,". In v.1, *The complete works of Aristotle*, ed. Jonathan Barnes, revised Oxford Translation, Bollingen Series lxxi – 2. Princeton: Princeton University Press.

Bacon, Francis. 1964. "The Masculine birth of time", unpublished original Latin from 1603, Trans. Benjamin Farrington in *The Philosophy of Francis Bacon*. Liverpool: Liverpool University Press.

———. 2000. *The New Organon*, original Latin, 1620, Trans. and ed., Lisa Jardine and Michael Silverthorne. Cambridge: Cambridge University Press.

Duhem, Pierre. 1954. *The aim and structure of physical theory*, original Fr. 1903. Trans. Philip Wiener. Princeton: Princeton University Press.

Galilei, Galileo. 1957. "The assayer." In *Discoveries and opinions of Galileo*, original Ital. 1623, Trans. abridged by Stillman Drake. New York: Random House.

———. 1967. *Dialogue concerning the two chief world systems*, original Ital. 1632, Trans. Stillman Drake. Berkeley: University of California Press.

Koyré, Alexandre. 1943. "Galileo and Plato." *Journal of the History of Ideas* 4 (4): 400–428. University of Pennsylvania Press. PA.

Plato. 1963. In *The complete works*, Bollingen series LXXI, ed. Edith Hamilton and Huntington Cairns. New York: Pantheon Books, Random House.

Chapter 10
Lessons from the Philosophy of Science to Data Mining and Vice Versa

Abraham Meidan

All along history philosophers, mainly in the western world, have asked the question: How can we avoid errors while explaining interesting phenomena? This is the question that defines Epistemology.[1] Since the development of science in the seventeenth century philosophers have applied this question to the scientific research. It is still one of the main fields in current philosophy.

One may distinguish between the normative question; what scientists *should* do (how they should raise and test theories and how they should observe facts), and the descriptive question; how scientists *actually* do the scientific research. However, since science is considered to be a paradigm of human rationality, the difference between these two questions is minor. The accepted view is that scientists actually do what they should do.

Most philosophers of science are empiricists. They claim that scientific empirical observations can corroborate scientific theories through induction. Some of them follow Francis Bacon who claimed (in the seventeenth century) that scientists should avoid *speculating* theories as much as possible, and concentrate on empirical observations only. The theories that these observations inductively imply will emerge. Other more modern philosophers of science, following William Whewell, claim that psychologically we cannot observe without raising hypotheses. The scientists then should deduct predictions from the hypotheses and test them empirically. If the test fails, the hypothesis is refuted, if the observations agree with the predictions, the hypotheses is corroborated.

[1] Usually epistemology is defined by the question: What can we know? But this question leads to a debate in regard to the definition of knowledge, and unfortunately this tiresome debate seems to lead nowhere. The concept of error is much less problematic than the concept of knowledge, and therefore I prefer to present epistemology by the question: how can we avoid errors when explaining interesting phenomena?

A. Meidan (✉)
WizSoft, Syosset, NY, USA
e-mail: Abraham@wizsoft.com

N. Bar-Am, S. Gattei (eds.), *Encouraging Openness*, Boston Studies in the Philosophy and History of Science 325, DOI 10.1007/978-3-319-57669-5_10

Empiricism faces several difficulties. The main difficulty, as was already mentioned by David Hume in the eighteenth century, is that relying on induction as a reliable method for avoiding errors might lead to paradoxical results. However, these difficulties are beyond the scope of this paper.

Empiricism is relevant to data-mining. Data-mining (and machine learning) algorithms try to find interesting valid patterns in data. As such they perform a task similar to a scientist who searches for a theory that explains the observations. It makes sense therefore to see what lessons from the discussions in the philosophy of science are relevant to data-mining, and vice versa – what lessons from data-mining methods are applicable to philosophy of science.

As mentioned, data-mining algorithms search valid patterns in the data. The data are saved in a table of columns (fields) and rows (records). One of the columns is selected as the dependent variable. The algorithm should find the patterns that explain the values of the dependent variable by the values in the other fields. For example, a medical data set may list the symptoms (the independent variables) and diagnosis (the dependent variable) of patients, where the data of each patient are saved in another row. The data-mining algorithms should then find patterns that relate between the symptoms and the diagnosis. These patterns should be valid: when they are used for issuing predictions in regard to other patients from the same population, the predictions should suffer from as little as possible misses (false negative) and false alarms (false positive). The standard method for testing such an algorithm is to divide the data randomly into two sets, a Train file and a Test file. The data-mining algorithm learns the patterns in the Train file, and then issues predictions as to the expected values in the Test file. The predicted values are then compared with the actual values in the Test file to calculate the success rate.

The data-mining learning algorithm faces two challenges: On the one hand it should not miss finding valid patterns in the Train file, and on the other hand it should ignore patterns that hold in the Train file incidentally (and therefore are not valid patterns). A model that meets the values of the dependent variable in the Train file, but fails to predict them in the Test file, is said to suffer from over-fitting.

In what follows we will review three data-mining algorithms: artificial neural networks, decision trees and association rules. These three are representative examples. However, the ideas that will be presented are applicable to other data-mining algorithms as well.

Artificial neural networks allegedly imitate the biological neural networks in our brain. They are composed of several layers of artificial neurons. Each artificial neuron in one layer is connected with other neurons in the next layer. Like biological neurons it fires if and only if the total input from other neurons is above a certain threshold. The network learns by modifying the connections among the artificial neurons in order to minimize a cost criterion.

The Decision tree algorithm learns by splitting the data into subsets, based on the values of one variable, and repeating this process recursively till all nodes have the target value, or additional splitting does not improve the prediction.

Finally the association rule algorithm reveals all the if-then rules that relate between the values of the dependent variable and the values of the other fields, and uses these rules for issuing predictions.

We will now review two examples where principles discovered in the philosophy of science are relevant to the data-mining algorithms.

10.1 Simplicity

Scientific theories are supposed to be simple. This requirement was presented already in the Middle Ages by William of Occam. Occam's razor says that among competing hypotheses, that are otherwise equal in all relevant aspects, the one that makes fewest new assumptions is the most likely to be true. This principle has been extended into a general requirement for simplicity.

Consider the following data that relate between two variables, A and B. There are infinitely many wave-like functions that meet these points, and in this sense all of these functions are inductively corroborated by the data. But we tend to prefer the right line, since it is *simpler*.

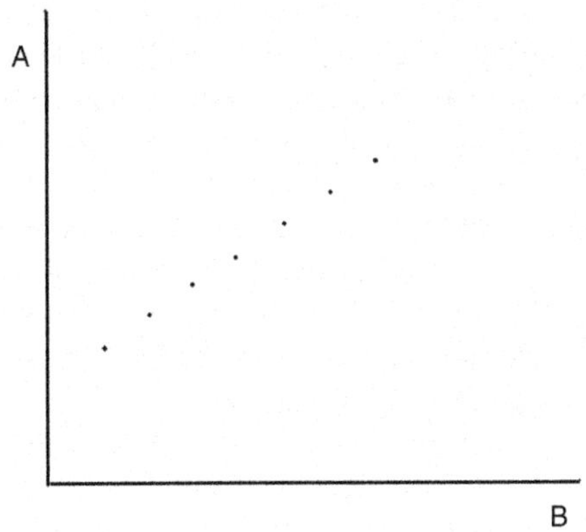

Now, consider another set of facts. The data can be explained by a straight line, with some deviations, or a parabola with fewer deviations, or a more complicated function that meets all the points. Which one should be selected? Philosophers of science answer that in such a case one has to weigh the benefits of simplicity against accuracy.

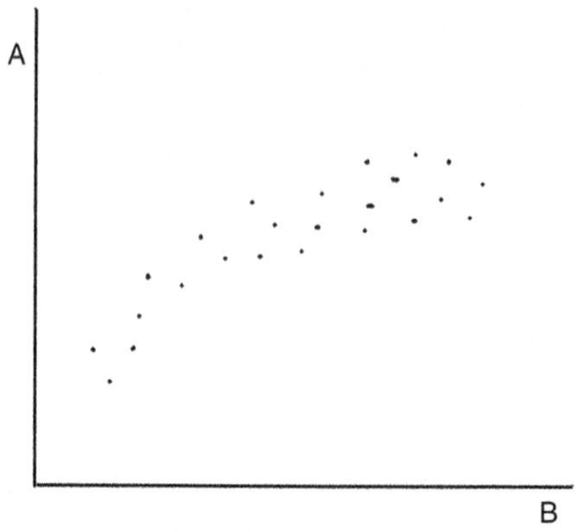

The idea of simplicity has been applied to all the algorithms of data-mining. Consider artificial neural networks. As mentioned, the network includes several layers of artificial neurons. The relation between the number of layers and the success of the model is: When the number of layers is increased, the network meets the dependent variable values in the Train file more accurately, but it tends to suffer more severely from over-fitting. And vice-versa: If the number of layers is decreased, the model tends to miss valid patterns in the Train file, but becomes more valid (i.e., suffers from less over-fitting) when applied to the Test file. The number of layers denotes the complexity of the network: the lower the number of layers, the simpler the network. It follows therefore, that in order to avoid as much as possible both misses and false alarms, the networks should have the right level of simplicity.

A similar application of the simplicity principle can be found in decision trees. When the number of nodes is not limited, decision trees can create over-complex trees that suffer from over-fitting. To avoid such over-fitting the final nodes are deleted by a process called pruning, which increases the simplicity of the tree, but deleting too many nodes might result in misses.

Finally, when running the association rule algorithm the user determines the minimum number of cases required for establishing an if-then rule (this parameter is called support level). When this number is low the association rule algorithm can create an over-complex set of rules that suffer from over-fitting when used for issuing predictions. To avoid such over-fitting, the minimum number of cases in a rule should be increased, which increases the simplicity of the discovered rules, but obviously if the number of cases is increased too much, the association rule algorithms will miss some valid rules.

To summarize, all the data-mining algorithms under discussion implement the principle of simplicity which was originated in the philosophy of science.

10.2 The Necessity of Hypotheses

A second case for the relationship between philosophy of science and data-mining refers to the question: can we research without having some hypotheses to start with? As mentioned, Francis Bacon claimed that good scientists should start the research by observing the data without having any hypotheses (since the hypotheses may influence as prejudices), while William Whewell said that such a requirement is impossible to carry out.

In what follows I'll present two arguments that refute Bacon's idea and support Whewell's.

The first argument refers to the structure of the patterns that the data-mining program is looking for. All the known data-mining (and machine learning) algorithms look for certain types of patterns (rather than all possible ones). Suppose one is looking for a mathematical formula that relate between the values of the independent variables and the dependent variable (like the formulas that are popular in Physics). There is no algorithm that looks for all possible formulas as there are infinitely many potential formulas. Therefore, one should determine in advance what types of formulas are relevant, or in other words one has to start the research by stipulating a (metaphysical) theory about the expected patterns.

The second argument rests on the idea that without having pre-assumptions as to what is relevant, one is supposed to collect as much data as possible. And indeed data-mining is sometimes presented as a method for revealing knowledge from data without any pre-assumptions. But contrary to this popular view, the research in the field of data-mining clearly supports Whewell's view (rather than Bacon's). As mentioned, without having pre-assumptions as to what is relevant, one is supposed to collect as much data as possible, but when the data contain many variables and few records, the models produced by the data-mining algorithms tend to suffer from over-fitting. This is practically relevant in medical researches, where a typical data set lists few patients (records) and many measurements (variables) for each patient.

Consider the following 15 × 15 table.

A	B	C	D	E	F	G	H	I	J	K	L	M	N	O
9	8	2	8	9	2	3	2	1	7	6	3	4	2	5
3	8	0	7	5	2	6	2	5	8	4	4	3	1	9
3	2	4	0	9	6	5	4	1	8	2	2	8	9	3
1	1	6	7	2	4	7	1	9	6	7	5	8	1	3
7	0	3	6	6	2	2	5	4	6	9	6	6	6	9
0	5	9	5	5	2	3	7	6	5	1	6	8	7	4
7	2	3	6	9	2	6	8	5	3	4	2	3	8	9
3	1	4	4	2	6	8	4	4	4	8	8	1	3	6
5	7	3	4	0	7	8	3	4	7	6	3	8	9	4
3	1	0	3	6	9	9	2	9	2	6	1	6	8	2
4	7	4	0	9	4	4	8	4	9	2	0	1	1	8
6	4	1	9	1	1	7	7	0	6	5	3	2	4	5
1	5	0	8	2	0	5	5	0	1	4	1	7	2	6
1	7	8	1	7	3	0	2	1	1	4	2	0	1	4
2	6	4	8	1	0	7	0	1	0	9	1	6	3	7

The values in this table were created by Excel Random function. Suppose variable O is the dependent variable, and the question under research is: under what conditions the value of O is above 4 (the median value)? In spite of the fact that all the values in this data set are random, the following if-and-only-if rule holds[2]:

O *is* **more than** **4.00** if and only if the following condition holds:

B *is* **4.00 ... 6.00** (average = **5.40**)
and **N** *is* **2.00 ... 8.00** (average = **4.40**)

When the condition holds, the probability that
O *is* **more than** **4.000**
is **1.000** (5 out of 5 cases)

When the condition does not hold, the probability that
O *is not* **more than** **4.000**
is **0.800** (8 out of 10 cases)

The total number of cases explained by the set of conditions: **13**
The total number of cases in the data: **15**
Success rate: **0.867** (13 / 15)

Since the values are random, obviously the rule is not valid, but rather an example of over-fitting. It follows therefore that in data sets having many variables and few records, even random data can include alleged patterns.

As mentioned, this phenomenon is practically relevant in medical research where the number of patients in many researches is relatively small while the number of variables is large. The discoveries in such researches might be incidental, even if all the standard statistical tests were issued. When searching for patterns in a data set, the number of variables should be much smaller than the number of records. And to reduce the number of variables one should have a hypothesis that lists the a-priori relevant variables. In other words, we should start with hypotheses rather than observations.

References

Agassi, Joseph, and Abraham Meidan. 2008. *Philosophy from a skeptical perspective*. Cambridge: Cambridge University Press.

———. 2016. *Beg to differ: The logic of disputes and argumentation*. New York: Springer.

Agrawal, R., et al. 1996. Fast discovery of association rules. In *Advances in knowledge discovery and data mining*, 307–328. Cambridge, MA: MIT Press.

Fayyad, Usama, Gregory Piatetsky-Shapiro, and Padhraic Smyth. 1996. From data mining to knowledge discovery in databases. http://www.kdnuggets.com/gpspubs/aimag-kdd-overview-1996-Fayyad.pdf

[2]The rule was revealed by running the WizWhy data mining program, www.wizsoft.com. This example is also presented in Agassi and Meidan (2016).

Hume, David. 1740. A treatise of human nature, (1967, edition). Oxford: Oxford University Press.

———. 1777. *An enquiry concerning human understanding* In: P. N. Nidditch, ed., 3rd ed. (1975), Oxford: Clarendon Press.

Meidan, Abraham, and Levin Boris. 2002. Choosing from competing theories in computerized learning. *Mind and Machines* 12: 119–129.

Whewell, William. 1922–1958. *Alumni cantabrigienses.* In Venn, J. and J.A. Cambridge: Cambridge University Press.

Chapter 11
The Mystique of Computer Technology and the Waning of Critical Enquiry

Sheldon Richmond

11.1 The Mystique of Computer Technology

Do we really need computer technical professionals for helping us to learn the use of computer technology and helping us to solve our difficulties with the use of computer technology? No: non-technical experts can learn the use of computer technology through trial and error with the help of mentors. The obstacles we have created for ourselves in mastering the use of computer technology are purely social and artificial.

11.1.1

It is almost commonplace to remark that we are in the early phases of a technological revolution created by microprocessor embedded devices. The widespread dispersal and use of processors indicates that we have achieved the state of what was originally called "office automation". But what do we really know about computation, computers, and computer technology and all devices embedded with processors or controlled by processors? What do we know about the complex system of computer technologies tied together through the internet and the "cloud" or banks of dedicated computer servers? What do we know about the how the "cloud" that runs programs ("apps") and stores data whose physical location is not seen or known by the user of "smart devices"? What do even the technological experts, who control the access to the "cloud" (the physical server computers, hubs, fibre lines, Wi-Fi

S. Richmond (✉)
Independent Scholar, 17 Jonathan Gate, Thornhill L4J 5K3, ON, Canada
e-mail: askthephilosopher@gmail.com

N. Bar-Am, S. Gattei (eds.), *Encouraging Openness*, Boston Studies in the Philosophy and History of Science 325, DOI 10.1007/978-3-319-57669-5_11

connections, the data and the apps), know about the complex of the "cloud", its computer servers and its users? Are these questions asked and discussed? (For a flavour of the history of the web and computer technology, and its social impact, see the following: Berners-Lee and Fischetti (1999); Bush (1945); Neumann and Burks (1966); Simon and Newell (1972)).

We have become distracted by what I think is a side issue. To my mind the side issue is the philosophical debate concerning "Artificial Intelligence" or "AI", and the general philosophical approach called "computational philosophy", and the various warring groups called "neurophilosophy" and "functionalism", which all take seriously the questions of whether human intelligence is a function of computation or whether the mind is a computer or whether the brain is a computer (for those who don't like talk of "minds"). To my knowledge, those questions about how much we really know and understand about computing, computers, and computer technology, are not discussed. For instance, the popular and apparently serious concerns of the philosopher Nick Bostrom in his work on "super-intelligence" also called the singularity, (*Superintelligence: Paths, Dangers, Strategies,* Oxford University Press, Oxford, 2014) argues that computer intelligence will inevitability surpass human intelligence. Whether computer intelligence can and will surpass human intelligence is a distracting side issue. The concern about the singularity and super-intelligence presumes a pretence to know not only everything there is to know about computing but also the inevitable future development of computers. It would be less pretentious and more open minded to take for our starting point the possibility that we are at the same level of understanding in our current knowledge of computers as the ancient Greeks were concerning mathematics prior to the discovery of the irrational number.

What we know about computation is extremely little because computation is not reducible to algorithms or moreover, not even reducible to mathematics in general. We all know that real physical systems are not reducible to abstract geometrical systems—that a physical edge is not identical to an abstract geometrical line, or that a physical flat surface is not identical to an abstract geometrical plane. However, when it comes to physical computers, we adopt the pretence without any thought that physical computers are identical to Turing Machines.

However, Turing Machines are merely mathematical abstractions. A Turing Machine assumes mathematical infinity—i.e. an infinite tape; whereas, physical computational devices have limited "tapes" or memories. That difference makes a world of difference between mathematical abstractions and physical systems. Furthermore, Turing Machines, including Universal Turing Machines, are isolated or closed systems (See, Turing (1950)). Whereas, physical computers, including unlinked or standalone computers that are not on the Internet or other networks, are still open physical systems that interact with their physical environment, such as electrical systems, air, earth, water, and fire. For the integrity of their computational functions, physical computers depend on the integrity of their hardware and software. For instance, the integrity of computational functions rely on hardware such as power supplies, keyboards, disk drives, monitors, cases, silicon chips, solid state devices; and software such as operating systems that control physical devices as well as applications, but which ultimately depend on their binary coded instructions being properly stored and acted upon by physical devices. By throwing these quirky physical systems onto networks, with a new level of hardware such as wires, switches, routers, servers and also with a new level of software such as network operating systems and communication protocols, we add a new dynamical layer where the mythical butterfly flapping its wings can change the universe.

The current mathematical theorists—euphemistically called, "computer scientists" —of finite machines (Turing Machines that have finite "tapes" or memory) have been unable to solve Stephen Cook's famous Millennium Prize problem of P vs. NP ("The Complexity of Theorem-Proving Procedures", *Proceedings of the Third Annual Association for Computing Machinery (ACM) Symposium, Theory of Computing*, May 1971). So, we cannot prove for every simple problem whose solution is intuitively obvious and check-able by a computer whether the solution can be transformed into an algorithm that will complete in physical time (i.e. before the universe implodes into a giant black hole—admittedly, my simplistic understanding).

Edsger W. Dijkstra (11 May 1930–1936 August 2002), a leading computer scientist and winner of the Turing Prize, has a similar diagnosis of the state of our knowledge about computer technology. I quote from a note he wrote on November 19, 2000, for the "Communications of the ACM":

> I would therefore like to posit that computing's central challenge, viz. "How not to make a mess of it", has not been met. On the contrary, most of our systems are much more complicated than can be considered healthy, and are too messy and chaotic to be used in comfort and confidence. [...] You see, while we all know that unmastered complexity is at the root of the misery, we do not know what degree of simplicity can be obtained, nor to what extent the intrinsic complexity of the whole design has to show up in the interfaces. We simply do not know yet the limits of disentaglement. We do not know yet whether intrinsic intricacy can be distinguished from accidental intricacy. We do not know yet whether trade-offs will be possible. We do not know yet whether we can invent for intricacy a meaningful concept about which we can prove theorems that help. To put it bluntly, we simply do not know yet what we should be talking about [...].

To my mind, Dijkstra does not quite get to the bottom line reason for the failure of computer science. It is not merely a matter of losing sight of the goal of achieving simple models of complex systems. Rather the bottom line reason for the failure of computer science is that no real computer is reducible to Turing Machines nor finite-state machines, and so real computers are not reducible to mathematical systems. Hence, what we know about mathematics, though great and wonderful—including the profound discoveries of Gödel, Turing and Cook concerning the undecidable—are not the whole story about computers and what computers do, let alone can do.

This lesson of our ignorance about computers leads me to ask the question—*Are we asking the questions that will help us to achieve some minimal understanding of computer technology?*

I usually turn to the writings of Joseph Agassi when I want to find fresh perspectives and independent thinking. His work in the philosophy of technology concentrates around the real and important question of how to democratically control technology. The fundamental issue is not whether technology will and by itself take over the world—in the form of super-intelligence according to the arguments of Nick Bostrom (*Superintelligence: Paths, Dangers, Strategies*, Oxford: Oxford University Press, 2014) as implicitly assumed by technological determinists whether techno-philic or -phobic. Rather, the fundamental issue as Agassi sees it and which I think is in the right direction, is how to design our social institutions so that democratic control is in place—for the institutions themselves as well as for technology. (Joseph Agassi, *Technology: Philosophical and Social Aspects*, Dordrecht: Kluwer, 1985).

Reading Agassi on technology, and how we should democratically control technology, leads me to think that there is a still more fundamental question to ask about technology. What I feel missing is asking the most obvious question possible to ask, and in turn, the most difficult question for anyone to ask who makes or has made their living from the frustrations created by technology. This question is not—how can we design computers and technology to be less frustrating? The design of computers to make people smart as opposed to stupid in the terminology of the cognitive scientist Donald Norman, is close to what I take to be the most obvious unasked question around about technology, but still just off the mark, from my frame of reference on the issue of technology Normen (1993, 1988, 2002). Rather, the very basic question which I think needs asking but which too many people avoid asking is: *Do we need technology experts and professionals to assist us in learning the use of computer technologies? Have computer technology support professionals created a mystique of technology?*

11.1.2

The root source of the mystique concerning technology is a modified but more serious form of C. P. Snow's famous two cultures problem of science versus the humanities (*The Two Cultures and the Scientific Revolution*, 1959.) The new version of the two cultures problem is technologists versus users of technology.

The genius of Snow's reading of the problem is that the problem is a problem of cultural differences and inter-cultural communication. The problem is not merely a problem that occurs between groups of individuals with different educational backgrounds and interests. Rather, the problem is much more fundamental. Scientists and humanists view the world differently, talk differently, and interact differently. They are akin to different ethno-cultural groups. The difference lies in that for scientists the universe is indifferent to human concerns and values.

Whereas for humanists, the universe is full of colour as opposed to electromagnetic frequencies, in the sense that what is of interest in the world is only of interest from a human perception of reality. Also, scientists talk in terms of measurable quantities, whereas humanists talk in terms of contrasting qualities—degrees Celsius versus warmth or coldness. Finally, scientists interact in terms of the presentation of objective theory discussed impersonally with respect to standards of truth and experimental validity; whereas, humanists interact in terms of stories relating to human concerns regardless of data but with special regard for moral value and aesthetic value.

Of course, Snow simplifies and exaggerates. However, his simplification and exaggeration is less of a simplification and exaggeration when applied to the cultures of technologists and users of technology—or technocrats and techno-subjects. The technocrat is only interested in whether the instrument works, whether the techno-subject is making proper use of the instrument, and whether the instrument is maintained. Whereas, the techno-subject is more interested in getting the 'thing' to work for his or her purpose: the techno-subject just wants to get something done with the instrument, and not have the instrument take control of him or her. The techno-subject doesn't care about following the rules of the instrument, but only about having the instrument fulfil his or her expectations and demands.

The cultural difference and lack of communication between technocrats and techno-subjects is most evident when something goes wrong. The techno-subject blames the machine, and wants the technocrat to get the 'thing' working again; whereas, the technocrat blames the user, and attempts to find what the user did wrong to break the machine. When the technocrat finally admits that something due to the machine is at fault, the technocrat wants more than anything else to fix the technical problem without regard for the concerns and questions of techno-subjects. The technocrat wants the techno-subject to basically get out of the way of the machine, and once the problem is corrected, the technocrat is only interested in what the techno-subject is doing with the machine as a method of testing whether the machine is back in functional order. This outlook of the technocrat is both absolutist and religious.

The technocrat's outlook involves a new form of idol worship and a new absolutist metaphysics. The goal of the technocrat is to continue the operation of the machine. The technocrat treats the machine as if it were the Absolute demanding not only service through self-sacrifice but also the sacrifice of those who hinder the operation and extension of technology. The techno-subject is seen as a device peripheral to the operation of the machine. For the technocrat there is no higher goal than serving the machine and perpetuating the continued operation of the machine. The techno-subject's goals, needs, and desires are irrelevant. Indeed, the techno-subject as a person is irrelevant. The only relevance of the techno-subject as far as the technocrat is concerned is whether the techno-subject is acting according to the dictates of the machine and whether the techno-subject is serving the operation of the machine. When the machine becomes problematic, the default hypothesis of the technocrat is that somehow the techno-subject is the cause of the problem. Once the technocrat rules out the default hypothesis, then and only then does the technocrat look into the machine in order to find a technical problem. (Ellen Rose, *User Error: Resisting Computer Culture*, 2003.) The working premise of the technocrat is that the machine would function fully and properly if only the techno-subject would not hinder its operation by disobeying the absolute commandments of the machine.

Can this problem of the gap between the cultures be resolved by improving communication? One might think glossaries or even some elementary courses for techno-subjects might help them better understand the concerns of technocrats. Or, one might think that courses in communication skills for technocrats might help them better explain the issues to techno-subjects and also help technocrats to understand the concerns of techno-subjects. Yes, but, and the "but" directs our attention to a qualification. The qualification to the "yes" is that the problem is even deeper than the problem of inter-cultural relationships. The problem concerns the common and unspoken idea that technocrats are even needed at all. Technocrats are needed, so it appears, because technology is hard to use: we need experts to help us learn how to use technology and to help us resolve our own difficulties or problems in the use of technology.

This idea—that technology is hard to learn and that we need experts to solve our problems with the use of technology—is an answer, a mistaken answer, to the tacit question: how do we learn to use technology? The answer that we learn the use of all technology by passively following the instructions of experts is obviously false. Firstly, as we all remember from our childhood learning bicycle riding, we did not magically learn to ride bicycles by watching expert riders and by following the instructions of experts. Rather, we learned how to ride our two-wheeler bicycle by getting on the bicycle and through trial and error, as well as through allowing our "teacher", or "mentor" usually a parent or older sibling or friend, give us pointers and help us regain our balance when about to fall. We also attempted to model ourselves on the actions of our mentor.

Notice that this common experience conforms both to Karl Popper's and Michael Polanyi's theories of learning. According to Popper we learn through trial and error; conjecture and refutation; or, hypothesis and falsification (*The Open Society and Its Enemies*, 1945). According to Polanyi we learn through activity and engagement with the guidance and modelling of a mentor (*The Tacit Dimension*, 1966). Both theories apply to the use of learning how to ride bicycles. I generalize: both theories apply to learning the use of all technology. Hence, everyone can learn to use technology, and those who have learned the use of technology can function as mentors to their friends. Technocrats are not needed.

When we think of technology as something so powerful and complex that we need a special class of dedicated people to control it for us, we adopt a system of thought that has the illusion of a rational explanation and justification for our technocratic society called by the social critic Neil Postman, Technopoly (*Technopoly: The Surrender of Culture to Technology*, 1992). However, the explanation and justification treats technology and our technocratic society (or the Technopoly) as an absolute when technology and the Technopoly are of our own making, though partially unplanned but still our choice and our responsibility to use, modify, reform or transform.

On reviewing my remarks so far, I have attempted to expose and criticize how we think about technology and how we treat the users of technology in our society. I have not so far discussed the theory of criticism. I discuss the theory of criticism later. However, as a first approximation let me refer to the writings of Octavio Paz and how he describes critical enquiry: When reason became self-critical, "[...] reason renounced the grandiose constructions that made it synonymous with Being, Good, or Truth; it ceased to be the Mansion of the Idea and became instead a path, a means of exploration" ("Breach and Convergence", in *The Other Voice: Essays on Modern Poetry*, Jovanovich, NY: Harcourt Brace, 1990, p. 33). Moreover, "[...] Differentiating itself from ancient religious and metaphysical principles, criticism is not an absolute; on the contrary, it is the instrument to unmask false absolutes and denounce abuses" (*Itinerary: An Intellectual Journey*, New York: Harcourt, 1994, p. 40). Furthermore, Octavio Paz tells us in another essay: "[...] Criticism tells us that we should learn to dissolve the idols, should learn to dissolve them within our own selves" (*The Other Mexico*, New York: Grove Press, 1972, p. 325).

11.1.3

I hope I have accomplished that purpose of criticizing the false absolutes and the new idols of computer technology and the Technopoly. I have asked the question—do we really need technical professionals for helping us to learn the use of technology and for helping us to solve our difficulties with the use of technology? My tentative answer is No: if we were to implement both Popper and Polanyi, we could learn the use of technology through trial and error with the help of mentors, or others who have managed to learn how to ride the bicycle, as it were. My tentative solution reveals the artificiality or the social nature of the obstacles we have created for ourselves in mastering the use of technology. Now I want to go deeper into these false absolutes by examining an unintended consequence of the development of computer technology—how the waning of Socratic enquiry or critical thinking has occurred.

11.2 The Waning of Socratic or Critical Thinking: An Unintended Consequence in the Development of Computer Technology

11.2.1 Where Can Socrates Open His Mouth in the Technopoly?

The Technopoly or modern technological and technocratic society with its domination by institutions and marginalization of informal associations, makes life difficult or next to impossible for the Socratic teacher, and Socratic or critical thinker.

The problem is: where can Socrates open his mouth in the Technopoly? Not in corporations and government offices where people are rushing like mad to complete their tasks; where managers don't want their projects shot down before they get the funds for implementing those projects; and then are able to move on to the next higher level before their projects fail during implementation. Not in the market place, where computer corporations are rushing like mad to get their products sold before the competition can come up with a better or at least similar product. Not in the educational institutions where students want to get the requirements done, get good grades, get out and get a job, and also have some fun and leisure and where ideas of graduate students and non-tenured professors need to gain approval from the authorities in whatever field in order to ensure grants if not tenure. Not in the political arena where criticism equals attack ads. So where can Socrates get a hearing? (For a parallel argument about how modern technocratic society inhibits reflective thought, see Rose (2013)).

Before going forward with attempting a solution to the problem of getting Socrates a hearing in the Technopoly we need to discuss two major questions. But so as not to keep you in suspense, I think there is a solution and the solution is that the Technopoly has four major anomalies or fault lines that the Socratic critic or critical enquirer can exploit. I will say more later about those fault lines after completing the discussion of the two major questions.

First we need to answer the question: How do Socratic teachers, thinkers and critics function even in theory? (For the Socratic theory of education, see: Swartz (2016); Agassi (2014)). I am going to focus in on a recent book by Agassi. Second, we need to answer: How has the Technopoly refashioned society? I am going to focus on the work of Luciano Floridi to help us figure out what is going on with the exponentially growing rate of the world wide diffusion of information technologies and systems.

After discussing those two questions, I can turn to the main question of this section: How critical enquiry or Socratic criticism can be heard in the Technopoly?

11.2.2 Agassi's Version of Popper's Contribution to the Theory of Critical Thinking or Socratic Enquiry

"Popper declared all attempts at criticism valuable, even ones that rest on misunderstandings. Does this hold for malicious distortions too? Yes, after they are cleansed of their malice." (Joseph Agassi, *Popper and His Popular Critics: Thomas Kuhn, Paul Feyerabend and Imre Lakatos*, 2014b, p. IX). The common view regards criticism even when kind and *constructive* as judgmental and negative. *Logically negative* criticism that points out a contradiction or is a *reductio ad absurdum* or involves a *modus tollens* form of argument is also judgmentally negative, and *logically negative* criticism is the worse judgment a theory and its proponents can face.

The point is simple: when criticism is taken as judgmental, criticism-avoidance is the only way to ensure a positive judgment of both the proponents of a viewpoint and the viewpoint. Hence, when criticism cannot be avoided, and when the only option under the judgmental attitude toward criticism appears to be defence, the defender risks attack as authoritarian and dogmatic.

However, Menachem Fisch and Yitzhak Benbaji (*The View from Within: Normativity and the Limits of Self-Criticism,* University of Notre Dame Press, 2011) develop a theory of criticism as self-criticism that underscores the common view of criticism as judgmental and as legitimate only to the degree of the authoritativeness of those making the criticism. Moreover, the Fisch/Benbaji theory underscores the current bias among philosophers who argue that criticism is only self-criticism or internal to one's culture or community.

Fisch/Benbaji argue contrary to Popper, that impersonal criticism cannot transform anything: "With regard to normative commitment, the truly transformative moment of rationality [...] is not one of bold conjecture or keen refutation but one of disturbing, destabilizing ambivalence; a moment characterized by indecisive dithering—a state of mind not usually considered the most inspiring and motivating and therefore, not usually associated with rationality. But if there is any truth in our analysis, then the creative individuals initially responsible for rationally transforming a field are to be sought among those who were lucky to be exposed to the ambivalating challenge of trusted external critics—real or imagined [...]." (292) The "trusted external critics" are those authoritative mentors from outside our home community that are encountered by our own authoritative mentors ("creative individuals") if and when they journey to outside communities or outside disciplines. The "trusted external critics" translate their frameworks and norms to our authoritative mentors ("creative individuals"), who become intermediaries for us of the frameworks and norms of the outside communities and disciplines. When our authoritative mentors ("creative individuals") return to lead and teach us, we follow, though hesitantly, and though sometimes rebelliously.

The Fisch/Benbaji theory misses the point of criticism according to Popper: The point of criticism is not to judge, not to change minds ("the truly transformative moment of rationality"). Rather the point of Popper's theory of criticism is on a different plane: it is to expose error, or in computer terminology to "debug". There is no missionary purpose involved with criticism—neither the transformation of individuals nor even of disciplines.

Against the common view of criticism (that the Fisch/Benbaji theory explicates) as judgmental, Popper's view is that criticism is simply a technique for pointing out important errors. The correction of important errors takes us further along the path of the improvement of knowledge, society, and civilization. Even finding out through the criticism of viewpoints that we have taken the wrong path and need to discover a new path is not mere nihilism, but the presentation of a new challenge or "problem-situation". In Agassi's words: "Popper's critical rationalism comes to replace Bacon's inductivism [...]. Popper took the encouragement of criticism as the hallmark of liberalism—in (scientific) research and in (democratic) politics alike. [...] Within the philosophy of science [...] Refutations [...] are not always valuable for survival but they are always valuable as intellectual assets: their upsetting important ideas renders them important. (This is the positive power of negative thinking.)" (130) "[...] What is peculiar to Popper is not taking criticism as a lofty activity: this is rather traditional; what is peculiar to Popper is the view of science as critical, as dialectical. Maimonides had said, human language is not fit to describe the attributes of the divine, yet it behoves humans to try to do so and to acknowledge the limitations of the results of their efforts. Combining the dialectic of the Maimonidean negative theology with the Spinozoist replacement of natural theology with natural philosophy amounts to the negative science that Einstein and Popper envisaged." (47).

Why does Popper's theory of criticism or Socratic enquiry or critical enquiry as negative matter today in the Technopoly? Popper inherited and carried forward the powerful tradition of criticism or *negative thinking.* Ironically, in computer terms, criticism is a form of debugging and there is no more or less legitimacy, value, or authoritativeness to the criticism whether or not the debugging or criticism comes from authorities within or outside the community. Nor is criticism less valuable as a form of debugging when it comes from fringe members of an outside community than when the criticism comes from authorities or experts from within the community. The value of criticism when viewing criticism as a technique for finding errors can only be determined after the error is uncovered. The value of the criticism depends upon how important the error is relative to the current state of critical enquiry.

Let me turn to another book, that unwittingly, explores and reinforces the plight of the Socratic critic in the Technopoly where all values are the same shade of grey: Luciano Floridi's, *The Ethics of Information.*

11.2.3 How Has the Technopoly Refashioned Society Such That Critical Enquiry Cannot Gain a Foothold?

Luciano Floridi's book, *The Ethics of Information,* (Oxford: Oxford University Press, 2013) unwittingly shows how values in the Technopoly are flat or grey. Information Ethics (or IE), Floridi says, "is an environmental ethics based on the phenomena and corresponding concepts of information/infosphere/entropy rather than life/ecosystem/pain." (98).

Floridi's "IE" (Information Ethics) makes explicit the flattening of values in the Technopoly. "Without IE's contribution, our understanding of moral facts in general, not just of ICT [Information-Communication-Technologies] related problems in particular, would be less complete. Our struggle to escape from our anthropocentric and solipsistic condition, be that Plato's cave [...], will be more successful if we can take a patient-oriented, informational perspective to the universe and its value." (333).

There is a Socratic irony or even Hegelian dialectical self-refuting nature to Floridi's flattening out of the values in the Technopoly—where everything is treated as information systems; and where all information systems are of equal value—except for those information systems such as malware and viruses that destroy other information systems (i.e. create "metaphysical entropy"):

> human life and human flourishing and even human suffering has no more value than getting people out of the rain because humans are simply systems of information. Floridi says: "maintaining one's dignity in a Nazi prison-camp is simply no better or worse, morally speaking, than giving a lift to an unknown person on a rainy day, not just because the two experiences are worlds apart, but because both agents have done their best to improve the "infosphere" [more popularly known as "cyberspace" or just data; specifically computer based data including algorithms or programs], and this is all that matters in order to consider their actions morally approvable." (79)

In Floridi's more recent book, (*The 4th Revolution: How the Infosphere is Reshaping Human Reality,* Oxford: Oxford University Press, 2014), he goes further and argues that the function of humans is to serve information computer technology ("ICT"). Firstly, humans have a minor role due to "information friction"—bottlenecks to the access of information which indirectly protects privacy—in sharing information online. Secondly, humans are needed to interpret data and data-patterns. Thirdly, according to Floridi, humans apparently gain more direct democratic power when the nation state withers away due to the shift to multi-national and multi-agent, decentralized processing power. However, when "the political multi-agent systems", take over "political and social space", human political and social involvement will become next to minimal. Fourthly, humans might have some role to play in solving our environmental problems and in "configuring our ethical infrastructure" or in creating an ethical social environment that guides individuals to act in socially acceptable ways. The irony is that we will need more and better ICTs to essentially solve environmental and ethical problems for us because "[…] We have moved inside the infosphere […]." (218) Hence, humans become "inforgs" (or information processing organisms) who serve ICTs, and at the best, become technical experts who develop and improve ICTs to solve whatever messes humans as pretty poor "inforgs" have created, such as environmental issues.

If everything is grey—all values merge, all norms merge—and humans are "inforgs" (information processing organisms) who only serve ICTs (information and communication technologies) and live in the infosphere (cyberspace), how can a Socratic critic, a critical enquirer get any questions about the Technopoly out of the infosphere to humans who are bound to ICTs?

The answer lies in the fact that everything is not really grey as solely data; people are not "inforgs"--people are just people. There are fault-lines in the Technopoly that allow Socratic criticism and critical enquiry to gain a leverage point for criticism.

11.2.4 The Fault-Lines in the Technopoly and How to Renew Critical Enquiry in the Technology Dominated World of Today

Here I spell out the fault-lines and indicate areas for developing critical enquiry in the Technopoly. These four fault-lines are well-known in the worlds of systems analysis, artificial intelligence, computer design, ergonomics, and cognitive psychology:

(a) The Technopoly or at least the computer technological aspect of the Technopoly contains a contradiction or 'dialectic' between two opposing computer system architectures: John von Neuman's centrally controlled serial digital systems and Norbert Wiener's parallel or connectionist analog-cybernetic systems (For a brief over-view of the divide between the analog or cybernetic and the digital, see the following: Ashby (1956, 1999); Bardini (2000); Bush (1945); Engelbardt (1962); Hauben (2006); Licklider (1960); Licklider and Taylor (1968); Metcalfe (2005); Skagestad (1993); Skagestad (1996); Waldrop (2001); Wiener (1950, 1954); Weizenbaum (1976); Winograd and Flores (1987)).
(b) Humans are analog-cybernetic 'machines' and even the best interface with the Internet and so called "Cloud" (including 'smart phones' and 'tablets') are digital.
(c) There is a dialectical conflict between information systems as open and horizontal versus organizations as closed and hierarchical.
(d) Intellectual property-theft: represents a conflict between 'ownership' of ideas, and inventions versus collaborative and shared intellectual or informational work (as in open source software).

The first fault-line of analog architectures vs digital architectures allows the Socratic critic to push for developing cybernetic type architectures that require computers to extend and augment humans as opposed to dominating humans. In other words, the architecture for cybernetic machines provides a new frame for developing critiques of current information systems. For views of systems analysis that attempt to reinforce the digital, topdown, centrist approach to systems analysis as opposed to the analog or cybernetic, bottom-up,distributed approach to systems thinking, see: Churchman (1971); Nelson (1965); Nelson (2004); Weinberg (1988)).

The second fault-line of the conflict of the human vs the digital interface for IT technologies, provides the basis for a new philosophical anthropology that restores anthropocentric values. For instance, people are not analogues to social animals, nor to machines, nor to information systems. Rather, people are people—strange thing to say—and people learn from feed-back (cybernetics); see the world in relational and comparative terms or in the form of stories or narratives—analogue, as opposed to seeing the world in terms of discrete binary systems—digital. The alternative philosophical anthropology allows for a leverage point for evaluating our institutions in the Technopoly. (For a better design of the digital computer so that the digital computer emulates and enhances the analog nature of humanity, see: Norman (1998)).

The third fault-line—the misfit between traditional organizational structures and open systems—allows for the creation of informal sub-systems and informal lines of communication within traditional hierarchical structures. For instance, the misfit between knowledge creation and the institutions and social structures of knowledge/education organizations creates a vantage point for criticizing social, educational and political institutional structures and organizations.

The fourth fault-line of proprietary systems vs. systems for open access, allows for the Socratic questioner to find alternative sub-cultures and outlooks within the Technopoly that can be used to develop criticisms against the reinforcing ideologies of the Technopoly (such as the neo-Hegelian philosophy of Luciano Floridi and the reductive philosophical anthropologies of humanity as information processing machines/organisms).

What Bertrand Russell said is the task of philosophy still applies and not only to philosophers but also to all people who are interested in the resuscitation of Socratic critical enquiry:

> The essential characteristic of philosophy which makes it a study distinct from science is *criticism*. It examines critically the principles employed in science and in daily life; it searches out any inconsistencies there may be in these principles, and it only accepts them when, as the result of a critical inquiry, no reason for rejecting them has appeared. (p. 12, Russell (1912))

Russell's words apply to the monopoly over everyday life by computer technology today. To break the monopoly we must develop and implement the democratic control of technology and apply critical enquiry to all forms of technology in everyday life. We must apply critical enquiry to our domination by the new elite of computer technologists and move towards transferring control to the individual user of technology. Only by applying critical enquiry can we break the silence of Socratic thinkers in all areas of life, including in the institutions of education.

References

Agassi, Joseph. 1985, 2005. Technology: Philosophical and social aspects, Dordrecht: Kluwer.

———. 2014a. *The hazard called education by Joseph Agassi: Essays, reviews, and dialogues*, ed. Ronald Swartz and Sheldon Richmond. Rotterdam: Sense Publications.

———. 2014b. *Popper and his popular critics: Thomas Kuhn, Paul Feyerabend and Imre Lakatos*. New York: Springer.

Ashby, W. Ross. 1956, 1999. An introduction to cybernetics. London: Chapman & Hall.

Bardini, Thierry. 2000. *Bootstrapping: Douglas Engelbart, coevolution, and the origins of personal computing*. Stanford: Stanford University Press.

Berners-Lee, Tim with Mark Fischetti. 1999. *Weaving the web: Origins and future of the World Wide Web*. Mason Texere Publishing.

Bush, Vannever. 1945, July. As we may think. *Atlantic Monthly*.

Bostrom, Nick. 2014. *Superintelligence: Paths, dangers, strategies*. Oxford: Oxford University Press.

Castells, Manuel. 1996, 2000. The rise of the network society, the information age: Economy, society and culture, vol. 1, Oxford: Basil Blackwell.

Churchman, C. 1971. *The design of inquiring systems*. New York: Basic Books.
Cook, Stephen. 1971. *The complexity of theorem-proving procedures*. Proceedings of the third annual ACM symposium, theory of computing, May 1971.
Dijkstra, Edsger W., November 19, 2000 note in *communications of the ACM*.
Engelbardt, Douglas. 1962. *Augmenting human intellect: A conceptual framework*. Menlo Park: Stanford Research Institute.
Fisch, Menachem, and Yitzhak Benbaji. 2011. *The view from within: Normativity and the limits of self-criticism*. Notre Dame: University of Notre Dame Press.
Floridi, Luciano. 2013. *The ethics of information*. Oxford: Oxford University Press.
———. 2014. *The 4th revolution: How the infosphere is reshaping human reality*. Oxford: Oxford University Press.
Hauben, Jay. 2006, Nov. 14. *"Visions of JCR Licklider and the libraries of the future", presentation given at the past, present and future of research in the Information Society* Conference. Tunis
Licklider, J.C.R. 1960, March. "Man-computer symbiosis", *IRE transactions on human factors in electronics*, volume HFE-1, 4–11.
Licklider, J.C.R., and Robert W. Taylor. 1968, April. The computer as a communication device. *Science and Technology*.
Metcalfe, Mike. 2005. *Contradiction and systems thinking*, working paper at the School of Management, University of South Australia.
Nelson, Ted. 1965. "A file structure for the complex, the changing and the indeterminate", *Proceedings of the ACM 20th National Conference*, 84–100, Cleveland, OH.
———. 2004, July . "A cosmology for a different computer universe: Data model, mechanisms, virtual machine and visualization infrastructure", *Journal of Digital Information*, vol. 5, no. 1.
Neumann, John von (with Arthur W. Burks). 1966. *Theory of self-reproducing automata*. Urbana: University of Illinois Press.
Norman, Donald A. 1998. *The invisible computer: Why good products can fail, the personal computer is so complex, and information appliances are the solution*. Cambridge, MA: MIT Press.
———. 1988, 2002. The design of everyday things. New York: Basic Books.
———. 1993. *Things that make us smart: Defending human attributes in the age of the machine*. New York: Addison-Wesley.
Paz, Octavio. 1972. *The other Mexico*. New York: Grove Press.
———. 1990. "Breach and convergence", in *The other voice: Essays on modern poetry*, New York: Harcourt Brace-Jovanovich.
———. 1994. *Itinerary: An intellectual journey*. New York: Harcourt.
Polanyi, Michael. 1966. *The Tacit dimension*. Chicago: University of Chicago Press.
Popper, Karl. 1945. *The open society and its enemies*. London: Routledge.
Postman, Neil. 1992. *Technopoly :The surrender of culture to technology*. New York: Knopf.
Rose, Ellen. 2003. *User error: Resisting computer culture*. Toronto: Between the Lines.
———. 2013. *On reflection: An essay on technology, education, and the status of thought in the twenty-first century*. Toronto: Canadian Scholars' Press Inc.
Russell, Bertrand. 1912. *The problems of philosophy*. Home University Library, Oxford University Press paperback 1959, Reprinted, 1971–1972.
Simon, Herbert A., and Allen Newell. 1972. *Human problem solving*. Englewoocl Cliffs: Prentice-Hall.
Skagestad, Peter. 1993. Thinking with machines: Intelligence augmentation, evolutionary epistemology, and semiotics. *Journal of Social and Evolutionary Systems* 16 (2): 157–180.
———. 1996. The mind's machines: The Turing machine, the Memex, and the personal computer. *Semiotica* 111 (3/4): 217–243.
Snow, C.P. 1959. *The two cultures and the scientific revolution*. Cambridge: Cambridge University Press.
Swartz, Ronald. 2016. *From socrates to summerhill and beyond: Towards a philosophy of education for personal responsibility*. Charlotte: Information Age Publishing Inc.

Turing, Alan. 1950. Computing machinery and intelligence. *Mind* 59: 433–460.

Waldrop, M. Mitchel. 2001. *The dream machine: J.C.R. Licklider and the revolution that made computing personal.* New York: Penguin.

Weinberg, Gerald M. 1988. *Rethinking systems analysis & design.* New York: Dorset House.

Weizenbaum, Joseph. 1976. *Computer power and human reason.* New York: W.H. Freeman.

Weiner, Norbert. 1948. *Cybernetics or control and communication in the animal and the machine.* Cambridge, MA: MIT Press.

———. 1950, 1954. The human use of human beings: Cybernetics and society. Boston: Houghton Mifflin.

Winograd, Terry, and Fernando Flores. 1987. *Understanding computers and cognition: A new foundation for design.* Norwood: Ablex, 1986. Paperback issued by Addison-Wesley, 1987.

Chapter 12
Francis Bacon, Karl Popper, Michael Polanyi, and Modern Experimental Biology

Ute Deichmann

Francis Bacon, widely appreciated as the father of the experimental method in science, proposed that science is inductive. Popper put critical testing of hypotheses against empirical evidence at the center of his epistemology, thereby excluding the act of conceiving or inventing a hypothesis or theory from his logical analysis of science. Michael Polanyi considered science as a social system based on epistemic authority and apprenticeship, stressing the informal and personal aspects of science. Using cases from genetics, embryology, molecular biology, and genomics, this essay (i) shows that induction, scientific authority and critical theory testing have been integral parts of scientific practice in the history of modern biology since the nineteenth century. This is despite the fact that a rigid adherence to Popper's principle of falsifiability is irreconcilable with scientific practice. (ii) analyzes the change of epistemology within recent developments in biology, particularly big data genomics. In these induction plays a major role while Polanyi's notions of epistemic authority and apprenticeship as well as Popper's principles of hypothesis, prediction and experimental testing are largely marginalized. The consequences of their disregard are discussed.

12.1 The Current Trend of Purely Inductive Approaches

Herbert Simon, computer scientist, cognitive philosopher, creator of the first artificial intelligence program and a recipient of the 1978 Nobel Prize in Economics in the 1970s successfully created a program that rediscovered Kepler's third law of

U. Deichmann (✉)
Jacques Loeb Centre for the History and Philosophy of the Life Sciences, Ben-Gurion University of the Negev, Be'er Sheva, Israel

University of Cologne, Cologne, Germany
e-mail: uted@post.bgu.ac.il

N. Bar-Am, S. Gattei (eds.), *Encouraging Openness*, Boston Studies in the Philosophy and History of Science 325, DOI 10.1007/978-3-319-57669-5_12

planetary motion: He explained: "We took raw data, nothing else, and tried to find patterns. We called it BACON, in honor of Sir Francis, because it's inductive. [...] In three tries BACON got the answer." (Cited in Stewart 1994)

Induction seems to have become fashionable again not only in artificial intelligence research, which is not the topic of this article, but also in new developments in the life sciences. With the advent of the generation of large amounts of data and high-throughput technology, the death of so-called hypothesis-driven science has recently been announced by scientists and commentators alike, for example journalist and editor of the Magazine "Wired" Chris Anderson. In a widely-cited article entitled "The end of theory", he proclaimed that big data should govern the science of the future, following the philosophy of Google: "Google conquered the advertising world with nothing more than applied mathematics. It didn't pretend to know anything about the culture and conventions of advertising – it just assumed that better data, with better analytical tools, would win the day. And Google was right." Anderson concluded that the received scientific method based on generating hypotheses and testing them experimentally had become obsolete, as had the acquisition of any knowledge prior to the application of an algorithm. Likewise, scientists' distinctions between correlation and causation and their aim to find causal mechanisms for observed phenomena should be a matter of the past, replaced by data-driven correlations (Anderson 2008). This implies not only the abandonment of a research which is at least in part Popperian, but also the premises that Polanyi deemed essential for successful research, i.e. competence, epistemic authority and apprenticeship.

Discovery science, "a scientific methodology which emphasizes analysis of large volumes of experimental data with the goal of finding new patterns or correlations, leading to hypothesis formation and other scientific methodologies" fits this trend in which it is data that drive science and lead to discoveries (Discovery science 2016). With data mining as the most common tool, discovery science is applied to data from diverse fields of study such as DNA analysis, climate modeling, nuclear reaction modeling, and others. As in Anderson's vision, prior knowledge is no longer required; data mining proceeds with the help of specialized machine learning algorithms for automated hypothesis forming and automated theorem proving. There seems to be no room for a creative and critical scientist.

The creation of a new belief system, such as the belief in algorithms without subsequent critical experimental testing of the hypotheses that were generated by the computer, deviates fundamentally from conventional research and the philosophical premises that guided it. The idea that observational data themselves can lead to discovery may be reminiscent of the notion of science as an inductive enterprise that was established by British philosopher and politician Francis Bacon (1561–1626), in particular of Bacon's thesis of the possibility of unbiased observations, his contempt for speculations or hypotheses, and his advocacy for new observations independent of any theory. (For a critical appreciation of Bacon and his philosophy of science, see Agassi 2013). However, the idea of observations leading directly to discovery disregards the fact that for Bacon, too, data are not to be obtained arbitrarily, 'groping, as of men in the dark', but guided by reason:

> But the true method of experience, on the contrary, first lights the candle, and then by means of the candle shows the way; commencing as it does with experience duly ordered and digested, not bungling or erratic, and from it educing axioms, and from established axioms again new experiments. (Coveney et al. 2016)

The activities advocated by Anderson and Discovery science have been denounced as not meeting the criteria of science (Pigliucci 2009). However, independently of whether or not we call them science, research along these lines is increasing at public and private scientific institutions. That is, many future scientists will be united by a belief in big data, carry out correlations in genomic medicine, forget about hypotheses, prediction, experimental perturbations, causal-mechanistic explanations, and basic science in general. History shows that the kind of science which led to our current understanding of the natural world, sometimes called traditional science, flourished only in certain political environments, while either not existing at all or declining in others. A well-known example for the latter is the destruction of genetics in the Soviet Union for decades under Stalin's protégé Lysenko, who imposed his own holistic, non-causal-mechanistic principles on all research of heredity that, moreover, had to be oriented towards agricultural or medical applications.

This article is not critical of the use of big data, which not only strongly impacted upon medical and other applied research but also opened up new possibilities for basic research. Rather, the issue at hand is the proclaimed end of so-called conventional science. Most researchers combine big data research with conventional research guided by the testing of hypotheses. An example is Hans Lehrach, a pioneer in genomics, who made it clear (relating to Popper) that "we are not really talking about data versus hypotheses, since in all cases I can imagine we will need both, data and hypotheses. You can therefore either start with a hypothesis, and then generate the data to disprove it (that's all you can ever do with hypotheses anyway) or you first generate (maybe at higher cost) the data allowing the cost-effective testing of many hypotheses and then test many hypotheses on the available data (often gaining the enormous advantage that your power to disprove hypotheses is much higher). [...] In most real projects we keep alternating between data generating hypotheses generating the need to generate new data generating new hypotheses, etc. in a multistage process." Lehrach is critical of big-data projects that are not combined with "scientific evaluation that is based on more than genomic information" (Lehrach 2016).

Generally, most big-data projects are related to medical research. Researchers focus on mathematical tools and do not proceed to prediction and experimental testing of the computer-generated hypotheses. This is especially true of large enterprises such as the Broad Institute of MIT and Google Genomics that store and process enormous datasets and use tools to analyze such data in order to solve problems related to human health. These tools cannot be discussed, but critical hypothesis testing indeed seems to be a matter of the past.

In the remaining parts of this article I will use case studies from the history of what may be called "causal-mechanistic experimental biology" to examine researchers' underlying epistemology. These studies deal with fundamental questions related to development and heredity from the nineteenth until the twenty-first century. The focus is on analyzing the roles of inductive and hypothetico-deductive reasoning, as well as on the roles of competence and scientific belief.

12.2 Mechanistic Biologists from the Nineteenth to the Twenty-First Century and Their Epistemology

Experimental biology based on rational as well as empirical principles developed in the late nineteenth century. The uncritical belief in philosophical concepts such as those of the romantic school of natural philosophers, prevalent for example in idealistic morphology and early versions of cell theory (Canguilhem [1952] 2009, 105f.), became increasingly replaced by a search for causal mechanisms based on observation, critical reasoning and experimental testing.

Most of the discoveries that had a lasting impact on the development of genetics and molecular biology started with observations and inductive reasoning. As is well-known, induction as scientific method, as established by Bacon, related to reasoning that derived general theories or laws from specific empirical observations of natural events. Bacon argued that this could be achieved when a methodical and skeptical approach was used. His non-metaphysical approach rejected speculation and marginalized hypotheses, though it involved the guidance of reason, and was revived and developed further in the nineteenth century by British philosopher, scientist, and historian of science William Whewell (1794–1866). Whewell sought a 'middle way' between pure rationalism, which solely relies on deductive scientific reasoning without empirical tests and widespread ultra-empiricism. According to Whewell, all knowledge has an ideal (subjective) as well as an objective (empirical) dimension. For him, induction was not just an enumeration of instances but contained an act of thought, "a New Element added to the combination [of instances] by the very act of thought by which they were combined" (Snyder 2012).

An example for such a reasoned generalization of observations, followed by the generation of hypotheses, predictions, and experimental hypothesis testing is the work of Gregor Mendel on plant hybrids which led him to propose the two laws of hybridization that became the basis for the science of genetics. He explained his procedure as follows (1866):

> Experience of artificial fertilization, such as is effected with ornamental plants in order to obtain new variations in color, has led to the experiments which will here be discussed. The striking regularity with which the same hybrid forms always reappeared whenever fertilization took place between the same species induced further experiments to be undertaken, the object of which was to follow up the developments of the hybrids in their progeny.

To him, quantification was essential: "Among all the numerous experiments made, not one has been carried out to such an extent and in such a way as to make it possible to determine the number of different forms under which the offspring of the hybrids appear, or to arrange these forms with certainty according to their separate generations, or definitely to ascertain their statistical relations."

Mendel did not start from an existing theory on the formation of hybrids and their statistical relationships. In contrast to most of the workers in the field of hybridization, he took a methodical and unbiased approach: He was not guided (as was, for example, Darwin) by the widespread expectation that inheritance would proceed by blending the characters of the parents and other popular myths. Instead, he used methods that had been developed in the hard sciences and statistics, including the start from clearly defined initial conditions. The hypothesis with which he started his experiments, namely that hybridization follows statistical laws, was based on many observations and the inductive reasoning that the rules he hoped to find in experiments with one differentiating pair of characters, could be extended to other differentiating pairs of characters and would turn out to be valid for other species as well. While he experimentally tested his predictions for *Pisum* and *Phaseolus*, he encouraged other researchers to test them as well in other species; he demanded that his own experiments be "repeated and verified" (Herskowitz 1962).

The novelty of Mendel's empirical approach, contrasted with that of many of his colleagues who were guided by philosophically motivated 'rational' theories, becomes clear in his correspondence with Carl Naegeli, botanist at the University of Munich. Naegeli criticized Mendel for his alleged lack of a theoretical basis: "You should regard the numerical expressions as being only empirical, because they cannot be proved rational." (Herskowitz 1962) In his response Mendel made it clear that his empirical approach was the basis for general statements and the formulation of quantitative laws.

According to Sander Gliboff (1999), Naegeli's use of the term "rational" was probably a reference to the rational induction method of his mentor Matthias Schleiden, a German botanist and co-founder of cell-theory, and other neo-Kantians, who concluded general laws and explanations from regularities in empirical observations with the aid of a priori assumptions (such as time, space, and causality). Empirical testing was not involved. In addition, Mendel's hypotheses of intracellular elements and mechanisms of combinations did not meet Schleiden's guiding maxims for botany that called for explanations in terms of cells and developmental processes at the cellular level and therefore, in this sense, were not rational (ibid.).

Naegeli might have disliked Mendel's approach not only because of its contradiction with Schleiden's a priori assumptions, but also for the fact that Mendel reduced complex phenomena to simple particulate models, thus rendering a quantitative analysis fruitful. This stood in stark contrast to Naegeli's predilection for concepts of gradation and continuity.

Mendel's methods became a model for scientists, as Thomas Hunt Morgan, the founder of the school of classical genetics explained (1934): "[Mendel's] analysis was a wonderful feat of reasoning. He verified his reasoning by the recognized experimental procedure of science." This procedure is characterized by observation – induction – hypothesis – prediction – experimental testing.

Cytologist and embryologist Theodor Boveri's work, which laid the cytological basis for the science of genetics and was, in addition, crucial for the concept of the causal role of chromosomes for development, is a pertinent example for Whewell's 'consilience of induction' in biology. Whewell introduced this term to discuss the unification of knowledge between the different branches of learning: Evidence from independent fields of research can converge to reach strong conclusions.

In 1888, based on observations that the arrangement and shape of chromosomes is preserved in the daughter cells, Boveri formulated his theory of chromosome continuity and individuality, namely that chromosomes are continuous entities, different from one another, that exist throughout the cell cycle, condensed during division and relaxed between divisions (Maderspacher 2008). This theory included a bold conjecture because chromosomes become invisible in the microscope between cell divisions. A few years later, after conducting perturbation experiments which suggested that only a full set of chromosomes could support development, he proposed (at the same time as Walter Sutton) the idea that chromosomes are carriers of heredity, which soon became known as the chromosome theory of inheritance. The theory was heavily debated at the time, and even rejected by some, among them Thomas H. Morgan, who became its strongest supporter after he confirmed it with his own work in Drosophila.

Boveri's underlying epistemology up to this point can be characterized, as with Mendel, as reasoned generalization of observations, followed by the generation of hypotheses, and their experimental testing. The next step was consilience: Boveri considered the chromosome theory confirmed – in a probabilistic way – when results obtained independently in two different fields of research agreed with each other: "We now realize that two areas of research [cytology and genetics] which developed independently from one another, produced results, which are fitting together so exactly as if they were derived theoretically from one another. [...] Thus the probability that the characters of Mendel's experiments are indeed linked to certain chromosomes is exceptionally great." (Boveri 1904, p. 123, cited after Barthelmess 1952, p. 233, trans. UD)

Around 50 years later, biochemist Erwin Chargaff proposed, purely inductively, a statistical rule in the composition of DNA namely that the ratio of Adenin to Thymin and that of Guanin to Cytosin in DNA nearly equals one. When he started to scrutinize the composition of the DNA of different species with newly developed chromatographical methods, he did not expect any regularities in the composition of DNA. On the contrary: He examined DNA because he was one of the few researchers at the time who believed Oswald T. Avery's demonstration that DNA, and not protein, was the hereditary material. Therefore he tested the hypotheses that DNA did not consist of regularly repeating sequences and that the sequences of DNAs from different species differed from one another.

While confirming these hypotheses, he unexpectedly found a pattern in the sequences, the rule in the ratios of bases, which would soon carry his name. After further measurements in various other species this was soon accepted as a general rule by inductive reasoning. The elucidation of the DNA double helix structure by Watson and Crick in 1953 made it clear that the rule had to hold universally, because the structure of DNA was accepted to be universally valid (Godfrey-Smith 2008).

All scientists mentioned before, to my knowledge did not relate to any epistemology as guiding principle, with the exception Schleiden and Schwann relating to Kant. The above makes it clear that all of them regarded empiricism highly, and that, in addition, they put forward hypotheses which they tested experimentally. The hypothetical-deductive method, which placed hypothesis testing in the center of its epistemology, was formulated most forcefully by Austrian-British philosopher Karl Popper (though it was partly anticipated by Whewell). Popper's epistemology does not need to be outlined here; I briefly summarize only those aspects that are relevant for my subsequent analysis. According to him, a scientist puts forward statements and tests them step by step. In empirical science, he constructs hypotheses or systems of theories and critically tests them against experience by observation and experiment. The riskier a prediction is, the better is the test of the hypothesis. This testing is deductive, that is, it does not follow an inductive logic (Popper [1959] 2002).

Popper rejected induction as a way to test hypotheses because he perceived an asymmetry between confirmation and falsification. No finite set of observations, for example of white swans, can prove the statement that all swans are white: "It is usual to call an inference 'inductive' if it passes from singular statements [...], such as accounts of the results of observation or experiments, to universal statements, such as hypotheses or theories. Now it is far from obvious from a logical point of view, that we are justified in inferring universal statements from singular ones, no matter how numerous; for any conclusion drawn in this way may always turn out to be false: no matter how many instances of white swans we may have observed, this does not justify the conclusion that all swans are white." (Ibid., 4)

Popper also rejected induction as a necessary means for creating hypotheses. Dismissing the accumulation of more and more perceptual experiences as basis for the advance of science, he contradicted Bacon: "Bold ideas, unjustified anticipations, and speculative thought, are our only means for interpreting nature: our only organon, our only instrument for grasping her" (ibid., 280). He held that not only truth cannot be achieved in such a way but also that probability is impossible: "I never assume that by force of 'verified' conclusions, theories can be established as 'true', or even as merely 'probable'." In contrast, theories can be falsified by a single piece of contradicting evidence. Therefore falsifiability, that is, the inherent possibility that a hypothesis can be proven false, not induction and inference, is at the center of Popper's scientific methodology. Theories can only be refuted, not confirmed. A theory that has been falsified must be rejected. According to Nimrod Bar-Am (2014), Popper was aware of the fact that the majority of scientists did not engage in creating risky refutable hypotheses but rather followed more conventional ways, which Thomas Kuhn (1962) called 'normal science'.

I do not present here another attempt to interpret Popper, rather I try to point out the Popperian and non-Popperian nature of experimental biologists' practice. Such a reflection has a new relevance at a time in which, as outlined above, entirely non-Popperian biology has become fashionable.

In two major respects, Popper's philosophy of science is not compatible with the actual work of the scientists depicted here. The first is related to Popper's refusal to analyze how hypotheses are set up or discoveries made (despite the misleading title *logic of discovery* of his 1959 book). The second is falsification. Popper's interest was in evaluating scientific theories. In a section on 'Elimination of Psychologism', he outlined what, in his opinion, philosophy of science should focus on (1959, 31): "The work of the scientist consists in putting together and testing theories. The initial stage, the act of conceiving or inventing a theory, seems to me neither to call for logical analysis nor to be susceptible of it. The question how it happens that a new idea occurs to a man – whether it is a musical theme, a dramatic conflict, or a scientific theory – may be of great interest to empirical psychology; but it is irrelevant to the logical analysis of scientific knowledge."

According to immunologist and Nobel laureate Peter Medawar, "Popper's most high-profile philosophical disciple within the field of active science" (Calver 2013), "the weakness of the hypothetico-deductive system, in so far as it might profess to offer a complete account of the scientific process, lies in its disclaiming any power to explain how hypotheses come into being" (Medawar [1963] 1996, 32). Joseph Agassi (2013, 74–75) emphasized that, according to Popper, novelty is relative to the theory that it refutes and new observations are viewed as refutations of received theories. In the history of biology, this is, however, rarely the reason for novelty, though there are examples such as the refutation of proteins as genetic material and its replacement by DNA, which marked the beginning of a drastic change in concepts and methods of molecular genetics (Deichmann 2004).

Unlike Popper suggested, the successful early research in genetics and development began with an inductive process in the sense of Whewell, followed by hypothesis. Observation and the accumulation of experiments was crucial, and sometimes new hypotheses were made by observations without a theory, as in the case of Chargaff's rule. Consilience of induction as in the case of Boveri, was the basis of major novelties. In general, progress in biology occurs predominantly through the opening up of new fields and not through rejected theories. A prominent example is the rise of molecular genetics which did not reject any theories of the existing classical genetics; it occurred after the concept of macromolecules was introduced into chemistry. The availability of new technologies often gave rise to new hypotheses, as did the microscope and staining to cell theory or heavy and radioactive isotopes to molecular biology.

While for experimental scientists testability and therefore falsifiability was a self-evident characteristic of hypotheses, they tried in most cases, as expressed by Morgan, to verify them; they certainly did not aim at falsifying them and sometimes avoided experiments to that effect. Thus, even if according to Popper falsifying could be carried out unintentionally, this is certainly often not the case. As mentioned before, verification often did not proceed by simply generalizing repeating individual observations or experimental results, but by finding concurring results obtained by different methods. Another major difference with Popper's epistemology lies in the fact that hypotheses that were disproved usually were modified and not discarded, as in the case of Mendel, whose rules were later shown to apply only to those cases in which genes were located on different chromosomes and were not separated by crossing over. This led to the amendment of Mendel's rules, not their rejection. Knowing about the fallibility of all theories and models, scientists nevertheless aimed at finding the truth about certain mechanisms. The degree of confirmation was often assessed in probabilistic terms.

This discrepancy between the philosophy underlying researchers' practice – which included induction and probability and often excluded the discarding of hypotheses once they were falsified – and that of Popper should not, however, obscure the fact that prediction and critical testing of hypotheses were fundamental principles in the search for causes and mechanisms of basic life phenomena.

At the same time that biology laid the basis for an understanding of the mechanisms of development and its genomic control, research was conducted on the basis of entirely different concepts. Here I briefly present research in which, about a century before the advent of big data genomics at the end of the twentieth century (though without major impact), mathematical tools and correlations were predominant. An example is the mathematical-morphological approach to understanding life phenomena by the British zoologist Wentworth d'Arcy Thompson who was influenced by Platonic philosophy. In his widely read book *On Growth and Form*, first published in 1917, he applied simple mathematics in order to find unifying principles of life's forms. According to Thompson, organic form was predetermined by the physical organization of the system in which it developed, and obeyed the law of physics, in which form was a diagram of forces. Theories that attributed specific properties to particles of the protoplasm, such as chromosomes, were rejected. Thompson's aim was "to correlate with mathematical statement and physical law certain of the simpler outward phenomena of organic growth and structure or form" (Thompson [1917] 1942, p. 14). He regarded his task to be finished if he had been able "to show that a certain mathematical aspect of morphology, to which as yet the morphologist gives little heed, is interwoven with his problems, complementary to his descriptive task, and helpful, nay essential, to his proper study and comprehension of Growth and Form" (ibid., p.1096).

With mathematical tools, Thompson created forms that correlated with certain forms or patterns of organisms, but he did not examine whether his mathematics represented causal relations in reality. Without perturbations, predictions, and experimental hypothesis testing, he could not elucidate mechanisms, apart from the fact that he neglected basic principles of life, in particular biological specificity (see, for example, Loeb 1916). Thompson's ideas did not play a role in the lasting advances of biology such as genetics, experimental embryology, and biochemistry though his approach has been taken up by mathematicians (see e.g. Katz 2012).

Starting in the 1960s, molecular biological research, which had previously focused on micro-organisms and viruses, broadened its scope to include the search for molecular explanations of phenomena that only occur in higher organisms, e.g. development and certain diseases. New techniques enabled systems approaches, e.g. through the study of the expression of a large number of genes at the same time. As pointed out in the beginning, this led to research that, on the one hand, analyzes large volumes of experimental data with the aim of finding new patterns or correlations, and in which hypothesis formation becomes an automated, purely inductive process conducted by computers. On the other hand, systems approaches pursue the same mechanistic goals as twentieth century experimental biology. An example is the research by Eric Davidson and his collaborators at Caltech on gene regulatory networks in development, a concept created by Davidson. A world leader in molecular embryology, Davidson demonstrated that, at least in sea urchins, early development is entirely regulated by the genome.

In 2006 he introduced the term 'regulatory genome' for the interactions between regulatory genes and their products during development. In the preface of his book with the same title Davidson wrote: "I have not shied away from what are sometimes pejoratively been termed 'big ideas' nor taken the view of an anonymous reviewer of a paper of mine who recently amused me with the complaint, 'But the original ideas in this paper are speculative!' This book includes many diagrams in which concepts are set forth in specific form, just so that they can be subjected to precise tests of falsification, and just so they can be used in precise ways to generate predictions I may not have thought of." (Davidson 2006, ix–x)

With the proposal of bold hypotheses that can be subjected to falsification, Davidson appears to be an ideal-type Popperian scientist. However the idea of the 'regulatory genome' was conceived and developed through decades-long painstaking experimental research by Davidson and his collaborators and anything but a one-step bold hypothesis. In this work Davidson and his collaborators systematically examined the cell-type specific gene expression patterns before moving on from the "gene-by-gene characterization of the sea urchin embryo to full comprehensiveness." (Rothenberg 2016) This systems approach was made possible when sequencing data of the whole sea-urchin genome were available.

Observations were crucial as a starting point. According to Davidson (2016), descriptive observations, often derided as intrinsically inferior to "problem solving," received a new meaning in a systems approach: "The whole interaction system cannot be studied effectively until the whole set of its components and the output (behavior) is correctly measured/observed." Following the acquisition of observational data, hypotheses were indispensable: "Throughout my own career in bioscience, I have exploited the explanatory effectiveness of hypothesis-driven scientific logic to explore how animals attain their properties, i.e., express their genomes and develop." This perturbation analysis is essential because "only by deliberate experimental perturbation and predictive challenge of the system can the mechanisms by which it operates be revealed." (Ibid.)

In order to test the proposed network, Davidson and his collaborators transformed it into a computational engine that would generate predictions of when and where in the embryo every regulatory gene should be expressed or not expressed according to the network structure. They showed that computation and observation agree in most cases (Peter et al. 2012).

Davidson's work displays clear Popperian traits, in particular with his strong emphasis on hypothesis (even bold hypotheses), predictability, and hypothesis-testing. His later attempts to include all regulatory genes, their products and all interactions into his study also seem to bear a Popperian mark: Davidson's outlining of a mechanism that takes care of any possibility, or, as he put it, without leaving any room for the appearance of 'black swans', appears to be inspired by his wish to overcome Popper's prohibition regarding induction.

But as with the mechanistic biologists before him, Davidson's Popper is not the Popper of philosophical textbooks. His statement that "in basic science, hypothesis formation is an inferential, inductive, creatively novel proposition of mechanism" (Davidson 2016) shows most clearly his emphasis on the relevance of induction in creating hypotheses. Apart from this, he would not have agreed with Popper that all experimental tests of hypotheses were attempts at falsifying them, no matter what scientists consciously thought they were doing. He replaced Popper's falsifiability criterion for good science with predictability: "In science, conceptual predictability accruing from an analysis is the golden criterion of value, of progress in understanding. Predictability is the lynchpin of scientific knowledge." (2016)

In addition, Davidson, like most experimental biologists, believed in the correctness of those of his hypotheses that were supported by various experimental (and computational) methods. They would have strongly disagreed with Popper's conviction that by force of 'verified' conclusions, "theories can never be established as 'true', or even as merely 'probable'." As pointed out by Michael Polanyi (1963) and Thomas Kuhn (1962), this belief frequently prevented novelties from being implemented immediately. As emphasized by Polanyi (1963), it also prevented uncorroborated claims from being accepted too early.

12.3 Scientific Belief and Criticism, Popper, and Polanyi

Michael Polanyi (1891–1976), a physical chemist who became a philosopher of science, took a very different outlook on science than Popper. Polanyi considered science a social system based on competence, authority and apprenticeship. A comprehensive overview of Polanyi's life and work from a historical-sociological and philosophical perspective is in Nye (2011). Though he accepted that rules and principles were important when it came to assessments of discovery and verification, he made it clear that these rules would not enable anyone who was lacking the training and skills of a scientist to make any competent judgement: "It is here, in the course of discovery and verification, that the premises of science exercise their guidance over the judgment of scientists. It is obvious that no formulation of these premises ever proposed (or yet to be proposed) would have enabled a person lacking the special gifts and training of a scientist, competently to decide any of the serious uncertainties that have risen in the various controversial or doubtful issues [....]." (Polanyi [1958] 1962, 165.)

Polanyi's conclusion that scientists are not driven by doubt, but by a personal commitment to beliefs reflects his own experience as a scientist, and contributes to the understanding of scientific practice. For Polanyi, a great scientist is someone who is competent: "To acknowledge a person as a scientist – and even as a very great scientist – is merely to acknowledge him as competent in science, which admits the possibility that he was, or is, in many ways mistaken." (Ibid., 173)

In Polanyi's view of scientific knowledge as a "stable system of belief", an interpretive "framework" or "paradigm" or "tradition" plays a major role" (Nye 2011, 262). This idea was also taken up by Thomas Kuhn, who popularized the term 'paradigm'. Unlike Kuhn and Popper, but in line with many scientists, Polanyi was a realist, convinced that science results in true knowledge despite the fact that it is based on social activities. The interpretative frameworks could be changed through major discoveries. What he called the "logical gap" between scientists working in different frameworks (anticipating Kuhn's concept of incommensurability (Jacobs 2002)), could sometimes be resolved in the course of scientific progress. Polanyi's example was the late-nineteenth-century controversy between Pasteur and the chemists about the nature of alcoholic fermentation. This was resolved by the new conception of intracellular enzymes, discovered by Eduard Buchner (Polanyi [1958] 1962, 150–157).

Polanyi's emphasis on scientific authority, competence, and apprenticeship has been frequently criticized, most fiercely by Joseph Agassi, who wrote (1981, 1–2):

> The prevalent philosophy of science, especially in the social sciences which should know better, is that of Michael Polanyi, especially in his vulgarized version offered by Thomas S. Kuhn, of scientists as a guild of masters and apprentices and a hierarchy of leaders who rule by conformity to fashion, which conformity makes thought-police quite redundant.

This criticism neglects that authority and apprenticeship were the backbone of every successful scientific school in the history of science. Polanyi's notion of scientists' commitment to belief is based on the notions of competence and the assessment of theories. His emphasis on the crucial role of scientific-authorities-as-epistemic-authorities includes the notion that these authorities can be shown to be mistaken, thus requiring a change in the scientific field, i.e. they include the notion of scientific criticism (see in particular Polanyi 1963). Therefore Polanyi's ideas of belief and authority differ fundamentally from the belief in the authority of big data, which neither require prior knowledge of the scientific field in question nor testing of the results against reality.

It should be added that Polanyi, similarly to Russell and Popper (Deichmann 2011), emphasized the contribution of scientific skepticism to freedom and social improvements in nineteenth century Europe (though he thought that an excess of doubt and skepticism later on became destructive): "What is more, scientific skepticism smoothly cooperated at first with the new passions for social betterment. By battling against established authority, skepticism cleared the way for political freedom and humanitarian reforms. Throughout the nineteenth century, scientific rationalism inspired social and moral changes that have improved almost every human relationship, both private and public, throughout Western civilization." (Polanyi 1966, 57)

Scientific skepticism also has a cultural role in the twenty-first century. For Davidson (2016), the mechanistic understanding of development and evolution was not only the only way to conduct science fruitfully but it also created a bulwark against irrational views such as those expressed by creationists. Scientific skepticism is crucial in regard to claims of big data genomic consortia that are not based on sufficient biological knowledge and experimental tests. An example is the ENCODE Consortium whose databases are widely appreciated by researchers but whose methods and interpretations as a scientific enterprise have been heavily disputed (see Morange 2014 and references).

Neither induction nor critical testing of hypotheses can explain scientific advance if taken alone; competence and the existence of a passionate commitment to scientific beliefs is another necessary prerequisite for understanding the development of science. To replace hypothesis, prediction and critical experimental theory testing with a belief in the self-explanatory function of algorithms, correlations and statistics means to give up the search for causes and mechanisms. This completely changes both the nature of science, and equally importantly, its cultural role as well.

Acknowledgement I thank Mary Jo Nye for her valuable comments on the manuscript.

References

Agassi, Joseph. 1981. *Science and society: Studies in the sociology of science*, Boston studies in the philosophy and history of science. Dordrecht/Heidelberg: Springer.

———. 2013. *The very idea of modern science. Francis Bacon and Robert Boyle*, Boston studies in the philosophy and history of science. Dordrecht/Heidelberg: Springer.

Anderson, Chris. 2008. *The end of theory: The data deluge makes the scientific method obsolete*. 23 June 2008. http://www.wired.com/2008/06/pb-theory/. Accessed 22 Nov 2016.

Bar-Am, Nimrod. 2014. The structure of scientific revolutions, fourth edition, 50th anniversary. *Philosophy of the Social Sciences* 44: 688. doi: 10.1177/0048393112473424 (http://pos.sagepub.com/content/44/5/688)

Barthelmess, Alfred. 1952. *Vererbungswissenschaft*. Freiburg: Alber.

Calver, N. 2013. Sir Peter Medawar: Science, creativity and the popularization of Karl Popper. *Notes and records*. The Royal Society for the History of Science. 22 May 2013. doi: 10.1098/rsnr.2013.0022.

Canguilhem, Georges. [1952] 2009. *Die Erkenntnis des Lebens* (trans. from French: La connaissance de la vie). Berlin: August Verlag.

Coveney, P.V., Edward R. Dougherty, and Roger R. Highfield. 2016. Big data need big theory too. *Philosophical Transactions of the Royal Society A*. doi: 10.1098/rsta.2016.0153.

D'Arcy Thompson, W. [1917] 1942. *On growth and form*, 2nd ed. Cambridge: Cambridge University Press.

Davidson, E.H. 2006. *The regulatory genome: Gene regulatory networks in development and evolution*. San Diego: Academic Press.

———. 2016. Genomics, "discovery science", systems biology, and causal explanation: What really works? *Perspect.Biol.Med.* 58: 165–181. doi:10.1353/pbm.2015.0025.

Deichmann, U. 2004. Early responses to Avery's et al.'s 1944 paper on DNA as hereditary material. *Historical Studies in the Physical and Biological Sciences*. 34 (2): 207–233.

———. 2011. Michael Polanyi on scientific authority and his criticisms of Popper and Russell. In *Yearbook of the Leo Baeck Institute*, vol. 56, 249–268. London: Oxford University Press.

Discovery science. 2016. https://en.wikipedia.org/wiki/Discovery_science. Accessed 25 Aug 2016.

Gliboff, S. 1999. Gregor Mendel and the laws of evolution. *History of Science* 37: 217–235.

Godfrey-Smith, P. 2008. Induction, Samples, and Kinds. For a collection of papers from the 2008 Inland Northwest Philosophy Conference. http://philosophy.fas.nyu.edu/docs/IO/11840/pgsinduction.pdf. Accessed 23 Nov 2016.

Jacobs, S. 2002. Polanyi's presagement of the incommensurability concept. *Studies in History and Philosophy of Science A*. 33: 105–120.

Herskowitz, I.H. 1962. *Genetics*. Boston: Little, Brown, Supplements.

Katz, E.A. 2012. Bridges between mathematics, natural sciences, architecture and art: case of fullerenes. Book of papers and extended abstracts of the 1st International Conference "Art, Science and Technology: Interaction between Three Cultures", ed. by T. Kravchuk, A. Groysman, C. Soddu, E. Colabella and G. Leisman, 60–71. Milano: Domus Argenia.

Kuhn, Thomas S. 1962. *The structure of scientific revolutions*. Chicago: University of Chicago Press.

Lehrach, H. 2016. Interview with Hans Lehrach by Ute Deichmann. Berlin, 9 February 2016.

Loeb, J. 1916. *The organism as a whole from a physicochemical viewpoint*. New York/London: Putnam's Sons.

Maderspacher, F. 2008. Theodor Boveri and the natural experiment. *Current Biology* 18 (R): 278–R286. doi:10.1016/j.cub.2008.02.061.

Medawar, P. [1963] 1996. Hypothesis and imagination. Times Lit. Suppl. (25 October 1963), in: Medawar, P. 1996. *The strange case of the spotted mice and other classic essays on science*. Oxford: Oxford University Press.

Mendel, G. 1866. Experiments in plant hybridization (Versuche űber Pflanzen-Hybriden). *Verhandlungen des naturforschenden Vereins in Brűnn, 4*, 3–47, (Read at the meetings of February 8th, and March 8th, 1865; in English translation at mendelweb.org).

Morange, M. 2014. Genome as a multipurpose structure built by evolution. *Perspectives in Biology and Medicine 57*: 162–171.

Morgan, T.H. 1934. *Embryology and genetics*. New York: Columbia University Press.

Nye, Mary Jo. 2011. *Michael Polanyi and his generation: Origins of the social construction of science*. Chicago: University of Chicago Press.

Peter, I.S., E. Faure, and E.H. Davidson. 2012. Predictive computation of genomic logic processing functions in embryonic development. *Proceeding of the National Academy of Sciences USA* 109: 16434–16442. doi:10.1073/pnas.1207852109.

Pigliucci, M. 2009. The end of theory in science? *EMBO Reports* 10: 534. doi:10.1038/embor.2009.111.

Polanyi, M. [1958] 1962. *Personal knowledge – towards a post-critical philosophy*. Chicago: University of Chicago Press.

———. 1963. The potential theory of adsorption. *Science*. 141: 1010–1013. doi:10.1126/science.141.3585.1010.

———. 1966. *The tacit dimension*. Garden City: Doubleday.

Popper, K. [1959] 2002. *The logic of scientific discovery*. London/New York: Routledge.

Rothenberg, E.V. 2016. Eric Davidson: Steps to a gene regulatory network for development. *Developmental Biology* 412: S7–S19. doi:10.1016/j.ydbio.2016.01.020.

Stewart, Douglas.1994. *Herbert Simon on the mind in the machine*. http://www.omnimagazine.com/archives/interviews/simon/index.html. Accessed 23 Nov 2016.

Snyder, Laura J. 2012. "William Whewell". *The Stanford Encyclopedia of Philosophy*. Winter 2012 Edition. ed. Edward N. Zalta. http://plato.stanford.edu/archives/win2012/entries/whewell/. Accessed 23 Nov 2016.

Chapter 13
Ebola in West Africa: Biosocial and Biomedical Reflections

Daniel B. Cohen

13.1 The Case

This essay is about the scientific and medical response to ebola; and more particularly the actual and potential relationship between biotechnical and biosocial actions. It has been nearly entirely stripped of exposition; nevertheless, the ghost of the critical analysis is, I hope, alive in the narrative choices. This allows great compression of ideas and information within a narrative flow. So it can be read – I hope usefully – as a story.

In it, single paragraphs – sometimes single sentences – could have been elaborated into their own chapter in a more technical presentation. In fact, they would seem to require this. Worse yet, a single paranthetic phrase "(very carefully)" represents an entire book that could be written on the production of pharmaceuticals in plants, and the necessary "humanizing" of protein structures such as immunoglobulins.

The same compression applies to themes from Agassi's work. Most of the relationships to his work are in the structure of choices for a general narrative and in the specific direction taken at the end of the paper. There are some direct clues, partly in the section titles. Let me use three examples to illustrate this.

One of the skeleton key elements goes back to the original draft of *The very idea of modern science* which contained much more of Agassi's original PhD thesis from the 1950s than the final manuscript.

In it, he displayed an amazing critical empathy for an obscure scientific literature written mainly in the language and references of Greek mythology. I don't think the critical analysis was possible without empathy being the key to unlock the door of their language, because one has to enter into their style of thinking that was embedded in their use of mythological themes. It was a stunning interpretive translation,

D.B. Cohen (✉)
Maccabee Seed Company, Davis, CA, USA
e-mail: daniel.b.cohen@me.com

N. Bar-Am, S. Gattei (eds.), *Encouraging Openness*, Boston Studies in the Philosophy and History of Science 325, DOI 10.1007/978-3-319-57669-5_13

fully recognizing the way in which the mode of thought was not simply an Aesopian code but influenced how they thought; one could say became integral to how they thought.

So: cross-cultural thinking of a kind. Within a culture but cross-cultural to us, now. Agassi moved from one temporally bounded subculture to a different temporally bounded sub-culture; let's say modern philosophy and history of science. One of the themes of my paper is the notion of cross-cultural science – from a broader perspective but perhaps with the same approach. The easiest place to see a parallel at work is how I talk, all too briefly, about Amilcar Cabral.

A second hint comes from a section title, in which I use a three word gloss to co-integrate some aspects of Paul Feyerabend's and Joseph Agassi's work: *Anything critical goes*. This is – among other things – both sloganeering and a reference to *Popper and his popular critics*. There is more to say about this in terms of Feyerabend and Agassi, but for this essay I was thinking of the immense distances between conceptions and lives of groups, and the possibility of critical evaluation or science as a meeting ground, outside the boundaries of either.

Actually, the behavior of the international medical responders fit perfectly with the conceptual actions of witches: their masking in personal protective equipment, their obsession with blood and tissue samples, their restrictions on normal social intercourse and so on. It would be like someone in the USA taking their child to a hospital emergency room, only to be met by giant bats in lab coats. It is a kind of tribute to cross-cultural rationality that the entire rural population of all three countries did not, in fact, revolt.

I use the term "convergence" in the paper, instead of discussing "cross-cultural consilience", The great ebola anthropologist Alain Epelboin, who worked mainly in Guinea (Conakry) during this epidemic, seems to maintain a sense of humor about the variety and diversity of human follies. He could probably describe poly-cultural consilience.

I think Agassi actually discussed research programs before Imre Lakatos. I do not discuss this directly, but I do mention that the choice of GP (ebola glycoprotein) for target epitopes was common to most of the vaccine and immunomimetic drugs. At least in some of the drug cases, I thought I could see the determining influence of early rodent trials, even though rodents were not a great model for ebola as a human disease. It seemed to me that research programs could be determined by nearly random events, early in their life-cycles, which then became a fixed part of a program or a field-wide dogma.

A third cryptic Agassi-related remark, is towards the end: "[…] a kind of Prague Spring of science facing a lethal crisis." This was a complex phenomenon that relates to Agassi's interest in the self-organization of science. It went beyond research to the institutional infrastructure of science and technology, including publications, funding agencies, pharmaceutical companies and governmental bureaucracies. Sometimes, as in the case of the World Bank, it seemed that lower level staff were making up for leadership deficiencies in planning, and actually acting. In other cases, such as Merck, it was the corporate leadership that turned ebola response into a company-wide effort.

In may be surprising to many, but the FDA pushed all of its boundaries to facilitate rapid development and deployment of experimental therapeutics and diagnostics. My experience was that people individually responded with great acuity and often forced their agencies and bureaucracies to also respond and actually fulfill their missions. Here, I will just mention some of the changes that were made for communication and publication. At least some journals, such as The New England Journal of Medicine, made all their ebola-related articles open access. Some open access PLoS journals shifted fast publication to their associated (but refereed) blogs, but by Spring 2015 were able to publish new reports fairly quickly. In the meantime many researchers shared data without regard to priority of publication. The social scientists set up something like a temporary fast-publication journal, the Ebola Anthropology Platform in the UK, which played a major role in providing a place for well-researched reports that needed to be put out quickly.

Some of these work-arounds became codified by the time the Zika epidemic hit, with a consortium of major journals specifically exempting priority of publication and confidentiality of data in order to encourage, not inhibit, rapid data sharing. Unfortunately, this did not apply to journals in Brazil, apparently; which was the country hardest hit by Zika in 2016.

Then a new phenomenon became serious enough to be widely recognized: the rise of the false journal and predatory publishers. The problem of access to publication, which Agassi discussed, became a new version of the problem of demarcation instead. Perhaps this is enough of an indication of the work that lies behind this somewhat unusual structure for a paper. The references can provide access to many of the topics that fly by.

Finally, I should say that this was the only way I found I could write about this topic without being lit on fire with fury.

13.2 Ebola That's Enough!

Let's begin at the end. On December 23, 2016, The Lancet published on-line the final report of the "Ebola Ça Suffit!" ring-vaccination trial which took place mainly in Guinea and partly in Sierra Leone (Henao-Restrepo et al. 2016). It showed the successful deployment of a vaccine against ebola and confirmed an interim assessment published in September, 2015 (Henao-Restrepo et al. 2015).

The trial took place late on the downward side of the ebola epicurve. Later, I will discuss issues from the other side of the epicurve, when cases kept doubling faster and faster and case numbers appeared to be going asymptotic to a vertical: the circumstances from late 2013 to the end of 2014.

As an investigation and intervention Ebola Ça Suffit! got many things right, under still difficult conditions. Ebola Ça Suffit! got trial design right, for the circumstances. A major issue was designing a trial that would be able to capture actual ebola cases. By the time major resources were made available, in 2015, the ebola epidemic was ending and cases were vanishing. Social scientists I knew working on a proposed multi-vaccine RCT (randomized controlled trial) for Liberia in early 2015 kept hearing the numbers changed: 5000 then 50,000 then 500,000 prospective participants in what was supposed to be a phase II trial and then "a phase II/III trial" trying to capture enough probable cases to have a contrast.

There often seems to be a sub-discipline that has more power and authority in a crisis. During the ebola crisis, mathematical epidemiological modelers seemed to have such a position. They played an ambiguous role. The models encouraged a massive response in the fall of 2014 by their projected unthinkable caseloads, if nothing were done. They also used a random population-exposure assumption that was wrong.

A natural question would be: why weren't RCT vaccine trials (placebo controlled, double-blind, randomly-assigned treatment trials) used before the ebola case numbers had turned so far down? Part of the answer, lies in the time it took to have adequate non human primate (NHP) data on safety and efficacy and human preliminary trials.

Another factor was that objections to the deployment of experimental drugs in RCTs for clinical patient treatment, which had some strong contextual and ethical validity (Adebamowo et al. 2014), spilled over into views of vaccine RCTs, even where the same contexts could be seen as absent.

Vaccine trials were prospective, when participants could make informed choices. Increased care could be planned for patients who developed ebola or other medical conditions as a result of vaccine trial participation. Trials could have been designed for and with volunteering health care workers at high risk, with advanced medical back-up relief options for all participants, for example in the early Fall of 2014.

NIAID, the National Institute of Allergy and Infectious Disease, National Institutes of Health, Department of Health and Human Services in the United States, was one of the strong advocates of RCTs for both drug and vaccine trials, which ultimately they started in Liberia. However, the director of one of the leading vaccine candidates told me in the early fall of 2014, that even mentioning RCT vaccine trials could lead to his vaccine never being trialed in Sierra Leone. There was a clear difference in approaches within the international medical communities.

Guinea (Conakry) fell geopolitically and biopolitically on the other side of the divide from the NIAID approach. This is reflected in the composition of the trial funders, from WHO, Norway, UK, MSF (Médicins Sans Frontières) and Canada.

Ebola Ça Suffit! could be described as a control measure fused with a trial in a skillful combination. This type of design may have broader future use. One of the only ways novel control measures could legally be used was as trials.

Capturing cases was done by defining the participants as contacts and contacts of contacts of remaining ebola cases. These were people at true higher risk of ebola. The trial populations were clusters (actually two rings of risk) of contacts around known cases. The design was based on vaccination procedures used for the final stages of eradicating smallpox under natural (nonmilitary, non-biowarfare) conditions.

Vaccination could occur either immediately or following a two-week delay, and the clusters themselves were randomly assigned to either treatment in a double blind manner. By the time they reported their interim results (Henao-Restrepo et al. 2015), a separate evaluation team called for the elimination of the delayed vaccination arm of the trial, because sufficient evidence of efficacy of immediate vaccination made it unethical to continue delayed vaccination.

The authors describe multiple teams, including for informed consent, separate from the trial medical team. Informed consent, crucially, included community acceptance as well as individual choice (and the right to withdraw). The trial personnel were predominantly from Guinea itself and other African countries. They harmonized the international regulations, country regulations and local participation.

Perhaps this is too rosy a picture? There were still many issues in Guinea. Planners could have engaged communities on the trial design itself; or allowed choices between pre-designed approaches that already had regulatory approval, and had sufficient statistical power. The resources brought to bear in this trial would have been unimaginable in August 2014, when basic medical supplies were unavailable and even the supply of the most basic sanitation tool, chlorine bleach was out of date and nearly useless in, for example, Sierra Leone. Remarkably, Ebola Ça Suffit! even had electric power, water, and systems for maintaining a cold chain for vaccine deliveries.

In August 2014 Sierra Leone asked the UK military to bring as many helicopters as possible for distribution of supplies to remote areas, and for access to much more of the countryside during the coming rainy season. Roads were often poor at all times. Before ebola, when the critically important Lassa Fever lab was re-equipped in Kenema, Sierra Leone, the only way key components could be brought in was by U.N. troops from Pakistan re-purposing one of the military helicopters (Wilkinson 2016). Helicopters could also deliver vaccine on dry ice in coolers; which was cold enough, a vaccine head told me. Nothing so jury-rigged needed to be contemplated by mid-2015.

In this ebola epidemic there were over 20,000 infected and over 11,000 dead, due to ebola itself. The secondary effects due to the impact on the overall health system may have been equal or worse. In Sierra Leone with a typical extraordinarily low ratio of doctors or health care workers to population, before the epidemic began, at least 14 senior doctors died in the 7 months from July to December 2014; as well as large numbers of other health care workers.

The randomised ring-cluster design probably is statistically under-powered relative to an equal-sized RCT trial, if an RCT could have been finished. Ebola Ça Suffit! is the only phase III vaccine trial to successfully report from the West African ebola epidemic.

13.3 From Agassi to Ebola

By July of 2014 I had been engaged in a multi-year dialog with Joseph Agassi on various aspects of philosophy and history of science. He asked me to read and critique the drafts of two of his books: *The Very Idea of Modern Science* (Agassi 2013) and *Popper and His Popular Critics* (Agassi 2014). Another critical reader was Ian Jarvie.

My professional interests came mainly from plant breeding and genetics in agricultural; the intersection with cross-cultural science; the temporal development and intersecting changes of biology, medicine, nutrition and agriculture in North America; and how these interacted with other cultures and countries. These were actually practical matters, but how one thought about them could be crucial.

I was also interested in two very different people who specifically credited some of their ideas and methodology to Quine. One was Bob Moses, of the Student Non-Violent Coordinating Committee (SNCC) when he described his later work on math and science education, starting with the Algebra Project (Moses and Cobb 2002). Another was Barry Hallen who, with his colleague John Olubi Sodipo, started a long project on cross-cultural understanding of knowledge (itself) and medicine (as we might say), working as colleagues with Yoruba *onisegun* (Hallen and Sodipo 1997).

Their self-described pragmatic and experimental uses of concepts and methodologies from Quine in these two different contexts were suggestive and intriguing. Their work is reflected in my discussion at the end of this paper.

In early August, 2014, I received a request for assistance on ebola from a colleague who was a senior rice breeder and geneticist, and in fact a World Food Prize laureate. He had returned to Sierra Leone, previously, to accept a position as a Special Advisor to the President and Ambassador-at-Large. I knew Monty Jones from his work at WARDA (the West African Rice Development Association) developing rice varieties (*Oryza sativa*) that were partly based on the indigenous domesticated rice species of West Africa (*Oryza glaberrima*). Now he had the mandate as Presidential liaison for ebola.

We had been talking about the experimental drug ZMapp, and he was looking for how to have it be tested in the (then) two ebola epicenters in Sierra Leone: Kenema and Kailahun. They also were looking for anyone who could help provide badly needed equipment, protective gear, vehicles, ambulances, etc.

The next day I sat in on the CDC conference call to prepare U.S. medical personnel and institutions for the arrival of two ebola patients to Emory University in Atlanta. Attuned to look for anomalies, I noticed the experts who were presenting briefly mentioned the possibility of some evidence for sexual transmission of ebola. We were off.

13.4 On Immunochemistry and International Research Policy

Ebola is considered by the CDC to be a category A bioterrorism agent; meant to be studied only in level IV biosecure facilities. Because of this designation, and perhaps spurred on by recurring episodes of ebola outbreaks in Central Africa, there had been advanced work done on a number of possible interventions. Unfortunately, none of them were actually ready to be used. It may be that a certain lassitude and comfort level had set in, over a decade after the anthrax attacks in 2001. There is a kind of public biowarfare defense community, so to speak, with their own conferences and journals and associated laboratories and companies.

Federal research funding in the United States, in general, includes various small business requirements, and these also apply to funds for biowarfare defense. Such funding can be a useful steady source of income for smaller (or boutique) bioresearch companies. Canada, under the Harper governments, had reached a peak of Thatcher-influenced research policy that required commercial investment, or buy-in, for many areas of public research in the Canadian government laboratories, or their projects would be ended. My guess is that this policy helped lead to the rights to one of the two leading ebola vaccine candidates going from the Public Health Agency of Canada (PHAC) in WInnipeg to NewLink Genetics of the United States for what I recall as two hundred thousand dollars, Canadian. This was rVSV-EBOV, the vaccine that ultimately was used in Ebola Ça Suffit! (by then produced by Merck).

A common approach in vaccine development is to use recombinant technology (the small "r") in a live but innocuous virus (to the human population of interest); in this case a vesicular stomatitis virus (the "VSV") to express antigens of interest from the ebola virus ("EBOV"). One alternative approach is to first silence key genes in a dangerous virus, so it can replicate but not cause damage, and then use it as a vehicle for recombinant antigen expression. In both strategies, the immune system can then react to the antigens of a disease without being exposed to the disease itself, or be harmed by the live virus which is vectoring the antigens.

Heinz Feldmann originally modified VSV in order to study one of the ebola genes for glycoprotein, GP, in mice without needing level IV containment. Gary Kobinger, who also worked on ZMapp, headed the Winnipeg lab after Feldmann moved to the CDC. ZMapp, in contrast, delivers three monoclonal antibodies against GP (glycoprotein) antigens that have been (very carefully) expressed in plants. Like receiving venom anti-toxin or tetanus antitoxin, the monoclonal antibodies should neutralize a key viral function, in this case glycoprotein fusion into a patient's cell membranes, allowing viral entry into the cell. It is passive immunotherapy. This also gives the patient's immune system time to respond to viral antigens.

In general there were three main classes of intervention tools: various means of expressing antigens, leading to an immune system response and defense; various means of providing antibodies to ebola or providing analogs of antibodies for (temporary) passive immunity; and various means of interfering with (RNA) viral specific replication. Some of these were drugs, some were vaccines. There also were immunology based rapid diagnostics for point of care screening under development. Using transfusions or fractions of ebola survivors' blood, which had been done *ad hoc* in emergencies, was another example of passive immunotherapy. None of them had safety and efficacy data for people. Having rodent data, such as for mice or hamsters, was an advanced stage. Some of them had preliminary small trials of safety or efficacy in nonhuman primates, or these were rapidly undertaken. Proposed repurposed drugs such as favipiravir (Avigan) had extensive human safety data for its intended uses but not in the presence of ebola. There were rodent ebola studies with favipiravir but no human or nonhuman primate efficacy data. Some of the drug classes were so new that I could not find any registered human use in the entire class.

I had been developing a network of people who could help evaluate scientific issues of ebola intervention as well as working with institutions that could or did provide direct aid. By mid-August, with the encouragement of Lawrence Gostin at Georgetown University, I had developed a list of the roughly 19 (then) currently known agents. I also included on the list the notion of social mobilization as an intervention, with some possible examples. These could cut the rates of transmission and give the medical response time to catch up.

Gostin was also the Director of the WHO Collaborating Center on Public Health Law and Human Rights. He emphasized that informed individualized consent was an ethical obligation.

Now, many of the interventions would be hard to explain to anyone not actively working in that particular focused biomedical research subfield. It would be harder to explain to anyone else. It would be harder still to have information translated and explained adequately in a completely different cultural and social context; in the middle of an epidemic.

One of the lead tropical disease specialists in the United States commented to me, in early September, that the three most promising interventions on the list were:

(A) Immunoglobin purified from ebola survivors (passive immunity).
(B) Favipiravir from Fuji/Toyoma (inhibitor of RNA virus replication).
(C) The GSK investigational ChAd3-EBOV vaccine.

This vaccine was from GlaxoSmithKline: ebola GP antigen carried on chimpanzee adenovirus type 3. It was originally developed by NIAID and Okairos, which later became part of GSK. Two forms were initially proposed and tested: monovalent, carrying GP antigen from the Zaire outbreak strain of ebola; and bivalent, carrying GP antigen from both the Zaire and Sudan ebola outbreaks.

Many of the eventual vaccines and immunomimetic drug candidates could be presented in multiple versions depending on these kinds of choices of antigen sources, vaccination schedule, or combinations with other vaccines or related drugs. Almost all of the immunological drug and vaccine interventions were targeted at epitopes (antigenic areas) from forms of the GP ebola protein and not any of the other six ebola proteins.

ZMapp was not on his list of high priority interventions to try. This was partly based on his own experience with plant-derived vaccines and their problems. In particular, there was the difficulty in production scale-up during an escalating epidemic even if ZMapp should prove to be clinically successful.

When asked what I would choose personally, I said either the GSK or Canadian vaccines, and ZMapp as part of clinical treatment. However, the international interest in ZMapp, starting in August 2014, was based on an anomaly: a single dose, with a long backstory, given to Dr. Kent Brantly in Monrovia. Looking back, it was probably late in his disease course to have expected much, but ZMapp is associated with an immediate (transient) reduction in viral load followed by an increased level of a patient's own IgG response (Zeitlin et al. 2016).

Brantly had contracted ebola while working for the missionary charity Samaritan's Purse. Later both he and a volunteer colleague, Nancy Writebol were flown to Emory University for advanced medical care under high biocontainment. Nancy Writebol had turned down the ZMapp treatment because Kent Brantly was sicker. Their clinical case history and treatment at Emory are in the *New England Journal of Medicine*, where they appear thinly disguised as Patient 1 and Patient 2 (Lyon et al. 2014).

Looking back, the observation of greatest importance to the most patients in West Africa was buried in the case histories and may appear trivial. In Monrovia, Patient 1 tried to maintain fluids and electrolytes by drinking Tang and Gatorade. This was commonly done by Westerners as self-treatment for a number of similarly presenting tropical fevers. It was inadequate for clinical care of ebola. Nutritional support, correct oral hydration therapy (ORT) and (eventually) maintaining intravenous electrolytes (etc) were key to patient survival of ebola in a wide variety of circumstances and facilities (Lamontagne et al. 2014). A curious international pattern developed.

International medical workers who developed ebola were flown to Europe or the United States, where they received the highest level of monitored intensive care. And they also received one, two or even three of the experimental drugs, with no discussion of enrollment in trials. West African ebola patients, including doctors and medical workers received none of this. This was the context in which MSF, which led the on-the-ground battle against ebola, as well as others, rejected using patients in West Africa as experimental subjects in RCT drug trials as medically unethical.

RCT vaccine trials were collateral damage. Well-informed volunteer trials of medical workers were rejected. RCT vaccine trial co-designed with participants, with acceptable risk and conditions from their own perspectives never came up. I was told an RCT of as few as 200 people could give up to 90% confidence of a drug being effective, in the context of projected hundreds of thousands of cases without effective control measures in September, 2014. I wondered about the politics and ethics of the ethical debate itself. Trial designs themselves are imposed as fiats rather than co-designed with the people most at risk. People at risk, including medical workers, were not participants in the ethical debate; nor in the assessment of tolerable risks versus possible benefits of particular trial designs. They were not participants in the decisions to not have RCT trials, even of vaccines. In vaccine trials participants are not under the threat of imminent death; they are healthy and at risk of infection.

Drug trials require a constant background of uniformly adequate palliative care to even determine if a treatment is effective This would have been difficult under chaotic medical conditions. Vaccine trials do not. It was perplexing. My initial reason for contacting social scientists, mainly anthropologists, was motivated by the technical concerns about informed individual consent for any of the experimental interventions, however they were deployed. There was an added issue of the need for social or community consent in some contexts. Individual consent could be required by the international community, and that looked hard enough; while some level of social or group consent was a cultural and social requirement for the potential participants. This quickly became subsidiary to much broader issues of social and cultural aspects of disease transmission, prevention, and deliverable clinical care.

13.5 The Social Epidemiology of Being Dead

An ebola disease course starts with fever and mostly non-specific symptoms, which could indicate other diseases, about 8–12 days after effective exposure. The second phase, with the highest classic risk of transmission, starts about 5 days later and includes, notably, massive diarrhea and vomiting. Viral concentration needed for infection is very low (Vetter et al. 2016) while blood and other fluids have very high viral concentration (CDC 2016).

Unfortunately, so do corpses. The handling of ebola corpses can be extremely difficult even under United States' conditions (CDC 2015). Consider the impact of just one example of cultural difference: when is a marriage complete? In some Sierra Leone societies a marriage takes years to be complete, involving obligations to both the spouse and her family. When a married woman dies of ebola, where is she buried? It may require near or long distance transport of the body depending on her family of origin and the status of the marriage. Transmission can occur everywhere along the journey.

The social scientists I was working with often had decades of experience in West Africa. Many were integrated into the societies they worked with; some because it was their society. Many had medical, biological, ecological or agricultural backgrounds. They had lived through the violent convulsions of West Africa (Richards 1998; Suluku et al. 2012), the AIDS crisis (Nguyen 2010; Benton 2015), the crises of governance, and the complex social communication (Ferme 2001) elaborated under permanent structural violence (Bardosh et al. 2016).

They had critical information about the tremendous social impacts of illness care, dying, death, mourning, preparation for burial, burial and post-burial ceremonies. All of these involved high ebola risk to different types of contacts and associates of the dead. My closest anthropology colleague, Professor Paul Richards, prepared a summary of key information which I sent to retired senior CDC staff as a reality check. We then had a formal conference call with active CDC ebola-response staff, in Atlanta, who were joined by multiple other agencies. A later summary was published in 2015 (Richards et al. 2015).

This is an example of one kind of important but passive engagement. Social scientists from Sierra Leone had useful epidemic-related information needed for planning effective epidemic interventions. Active integration of social scientists with medical intervention was more encouraged by the UK. Several key incidents of social scientist intervention between communities and the medical response were fairly well known, after violence by governments or violence by communities. Active integration of social scientists with communities on epidemic response also occurred but is less well known.

Social and cultural issues of burial ranged from the ritually covert and secret to the social, legal and ritually overt and public. Both had to be accommodated. With information on ebola transmission, the physical covert ritual actions could be adjusted to block transmission, for example by women in sodalities such as the *sande*. The more overt and socially public aspects of burial could also be modified to accommodate both epidemic control and the social requirements of death. This included proposing the use of trained local burial teams. The modification of both social burial and "medical" burial is an area where mutual accommodation eventually succeeded in many areas.

The social scientists described people in rural areas who would never have timely access to effective ebola treatment. They still needed assistance. Professor Marianne Ferme, Department of Anthropology, University of California, Berkeley described one of her own journeys to health care (Ferme 2014): "When I first did fieldwork in Sierra Leone I came down suddenly with a combination of illnesses that made me completely delirious and incontinent within a few hours (malaria, amoebic dysentery, a staff infection). I was a total mess. I couldn't believe what the two women who took care of me had to do to keep me clean and care for me. They were not only sponging and changing me, they also were constantly laundering all the bed sheets and clothes I was soiling, until I was 'med-evacuated' the indigenous way: the Paramount Chief sent his ceremonial hammock with 4 hammock bearers (who worked in pairs, running at a regular jog up and down hills to keep the momentum, and relieving each other at regular intervals), and I was out of there in no time at all, and on a main road where they flagged down a passing vehicle".

This was a walk-in village with no regular transportation, but I was completely in awe at the ingenuity of people in figuring out how to make do with what they had: from fashioning a little "porto-let" on a bucket for me, to figuring out when I was ill enough to need immediate hospitalization. Even with my ready cash to pay for transportation it took me and the man delegated by the village elders to accompany me about 12 h to get to the Serabu hospital emergency ward, which gives you some sense of how unrealistic it may be to expect isolated rural folks who may not have the means to pay for double transportation all the way there to get to hospitals (in my case, 4 different vehicles/legs to the journey, in addition to the bit with the hammock)." We will return to the hammock bearers later.

Some of the social scientists saw very early that people were condemned for not bringing ebola patients to medical isolation, even when transportation was impossible and medical centers were nonfunctional. I read medical team reports from all three countries that also recognized the failures in each stage of transportation, facilities, adequate material supplies and successful clinical care that were supposed to make clinical-isolation work. Medical messaging castigated the most basic human need to care for loved ones, rather than providing assistance with material and guides for home or local health care of ebola. It was isolation that was needed, whether the setting was a medical facility, community, social group or at home; not just medical isolation, even when it was unobtainable.

In order to break transmission chains *in situ*, roughly "at home", this meant a patient would have to be isolated, and a single care-giver had to be semi-isolated, physically but not socially from the surrounding community. Many farms had isolated field shelters used at harvest which could have been used. This was similar to social practices used for containing smallpox in Zaire (now DRC) that were repurposed during the first ebola outbreak in 1976. Similar practices had been used historically for other epidemics in West Africa as well. Basic palliative care at home had to be done in a clinical manner, to give caregivers a chance at survival. But these were homes without running water, power, and in significant numbers not even two plastic buckets for water or bleach solution. People could contain the epidemic and break chains of transmission (epidemic intervention) by isolation (patient) and semi-isolation (caregiver) without outside aid. But without that aid, both caregiver and patient would probably have died. With improved communication about burial, at least they could have been buried safely (burial teams or local burial teams) with socially adjusted medical burial.

In late August 2014 the U.S. military was going to provide several hundred thousand home health care kits for Liberia. This never happened. The decision making was opaque.

No one ever developed a manual for home health care for ebola. The closest I could ever find was an older CDC and WHO manual *Infection Control for Viral Haemorrhagic Fever in the African Healthcare Setting* (Lloyd and Perry 1998) which discusses the layout of care, innovation in personal protective equipment (PPE) and making maximum use of minimal access to supplies in a low resource setting.

It took about half a year but eventually the CDC put out a poster giving home caregiving advice while waiting for transportation (CDC 2014). I remember similar advice from one MSF facility for people who were waiting for a space to open. At the end of 2015 MSF's internal research unit published a review, in very French-inflected cautious English: noting that home care isolation had been used in other filovirus outbreaks, that they had older guidances for home-based support and risk reduction, that the pragmatic limits of using facility isolation had been reached in 2014, and that perhaps facility isolation was not the only option (Calain and Poncin, December 31, 2015). This is in section four: "Facility isolation: an onerous public health measure".

13.6 Closer to the People: Technologies of Bodies

Broadly speaking, social engagement strategies can be monologues and directed from the top down, or dialogues which require mutual recognition. Social science methodology can also reflect these two orientations. In surveys based on questionnaires people were adept at repeating back the ebola messages that came from governments and international medical groups. *Don't eat bushmeat* along with other more useful messages. Often the surveys were seen as part of the government. When engaged by longer-form more open-ended investigations and asked, for example what they would actually *do* in response to ebola the answers could be quite different. Medical work in foreign-supported centralized hospitals could involve gruesome triage decisions and 70% death rates in 2014. There were two waves of the international response trying to bring isolation and health-care closer to people: first in what were called Ebola Treatment Units (ETUs) which were relatively more advanced, and then in what were called Community Care Centers (CCCs). These may have suffered from the sociological consequences of top-down implementation.

In Liberia, many of the facilities were still being built in 2015, even as case numbers dissolved to nothing. They were burned to the ground for ebola containment reasons and could not be re-purposed for general medical care, following protocol. When the UK teams asked for an ETU plan review, I was able to point out that it *might* be more appropriate if the patient flow-chart schematics did not have the only outcome, or end-product, be disposing of the patient's body by incineration. They could, for example, show the patient cured and returning home. Incinerating the body, in any case, may have been biologically safer but socially disastrous.

What does it mean to be cured of ebola, or ebola-free? Originally it meant that a patient had survived and no longer had detectable virus when blood was tested. This was the diagnostic sign of being ebola-free. It was presumed that this sign derived from blood testing applied to the patient's entire body. However, sexual transmission of ebola developed harder and more convincing evidence. Ebola survival in semen turned out to vary widely and could go on for an indeterminate time in some patients. Gradually it became clear that there were multiple kinds of immune-protected tissues where the virus could remain, be detected, and sometimes be culturable (Vetter et al. 2016). Although sex, usually semen, was the main cause of secondary infections, infection of other tissues could cause long-term problems, for example in eyes. Sex was a real but relatively rare means of transmission compared to, say, Zika. Nor was being ebola-free (by blood testing) the end of long-term physical and mental consequences for patients (Etard et al. 2017).

On the other hand, physical objects from patient-care settings (fomites) did not seem strongly linked to transmission despite the great attention paid to them in all patient-care facilities and the detection of ebola RNA. Ebola was notable for its very high death rate, but people had different risk levels, for example by age. There also were cases of high and intimate exposure followed by minimal symptoms. We received one of these rarer atypical case description from a colleague, Lina Moses working in Kenema, Sierra Leone, at the height of the epidemic. She is Research Assistant Professor in the Department of Global Community Health and Behavioral Sciences, Tulane University (Moses 2014): "[…] In the three and a half months doing this Ebola response, I was very surprised to see several cases of what I thought was mild disease. One case in particular, the contact of a case that died with severe hemorrhaging, vomiting, diarrhea, had a fever one day, came in for testing and was positive with low viral load. By the time we got the results he was feeling better, but we had to admit him because he had detectable virus. He was not too happy about this as he never developed any further symptoms. In fact, he occupied his time with push-ups, sit-ups, and pull-ups and was considerably bigger when he was discharged four days after admission. We tested him several times just to be sure it wasn't a lab error."

There is a history of some intimate caregivers of ebola patients surviving with only mild ebola symptoms, for example during an earlier outbreak in Gabon. I use the neutral terminology of an atypical disease course, which does not imply any particular mechanism. When blood samples from the Gabon caregivers were tested later they showed anti-ebola antibodies but, interestingly, not to the GP protein. We recommended special attention be paid to the low percentage of West African ebola epidemic survivors who had been identified with an atypical disease course.

Some new vaccine and antibody-like drug development strategies are being modeled on a particular survivor's antibodies from prior outbreaks. Others have the goal of broad-spectrum vaccination, seeking exactly those epitopes that could control a range of filoviruses at once. There may be an analog in the natural history and ecology of filovirus exposure that patients with mild symptoms exemplify. Prior exposure to some other disease, which must be rare, may have cross-protected them from most ebola symptoms. A mild disease reaction in enough people could create similar dynamics to immunization in dampening the rate of transmission.

Personal protective equipment was critical for medical workers but it was never designed for tropical conditions. In a sense they were being cooked sous vide while they tried to work. Typically a two-hour shift was the longest people could work competently, which meant the effective staffing rate needed was at least four times normal. The large, relatively well-trained Cuban medical teams were reported to have kept to shifts of from 40 minutes to an hour (Kirk and Walker 2016). The most dangerous time was disrobing, when liquid contaminating the impermeable surfaces of the PPE could transfer from PPE to exposed skin.

It would have made sense for materials science and PPE design teams to have worked with West Africans on better combinations of protection, particularly affordable protection, to reduce the risks and to reduce the heat stress. Cheap single use, absorbent disposable aprons could help when disrobing, for example.

PPE designs also made patient interaction difficult. Even medical workers who came down with ebola and were treated by their own friends and peers reported feeling humanly cut off. The very welcome relief offered by the international community at the end of 2014 was always additive in nature. If the epidemic had kept growing exponentially, this too would have been rapidly overtaken and swamped. Tropical adapted PPE would at least have had a clinical-personnel multiplier effect, with better odds of holding the line until vaccines or other effective biotechnical tools were available.

On the other hand, something about the pace and conditions of work was quite different for members of the burial teams. Despite the high risk from corpses, and often less adequate PPE, I do not know of any burial team member who contracted or died from ebola. Nevertheless, the team members often faced tremendous social stigma for their work.

The clinical-isolation model of epidemic control had two internal contradictions at all levels. First, medical transportation and facilities, if available, were themselves major sites of transmission. Careful design and training could reduce within-facility transmission but peri-nosocomial transmission from transportation, waiting to be admitted, and screening at admission remained. Very effective point-of-care rapid screening diagnostics were needed, with near-zero false negative results. These were under development but not available during the epidemic. Second, access to effective medical treatment, indicated by survival rates, was always unequal; and the more effective the medical care, the more unequal the access.

Dr. Olivet Buck was the fourth senior doctor to die in Sierra Leone. When she was diagnosed with ebola, a Level 4 hospital in Germany offered to treat her if she could get there. President Ernest Bai Koroma personally called WHO in Geneva to ask for airplane medical evacuation to Germany. They refused. This was remarkably covered by Joseph Harker in an article in The Guardian newspaper (Harker 2014) headlined: "Why are western health workers with ebola flown out, but locals left to die?" In response, we spent quite some time on how to access the two U.S. SARS-ready emergency planes, but the systemic issue remained.

13.7 The Turning Point

There was a nodal point between two very different phases of the epidemic and response; between 2014 and 2015. It was the first questionable possibility of the beginning of the end. Reports that the epicurve was leveling off and perhaps turning down came first from Liberia. This was still uncertain, over the late winter and New Year's holidays, 2014/2015. I could not find anyone who knew whether to believe the reported Liberian numbers, or not. The major aid had only begun to arrive so it seemed too early. On the other hand, the methodology of reporting cases in Liberia seemed to have some strong disincentives, so perhaps it was a methodological artifact.

I went back and read Camus' *The Plague*, this time as a kind of field handbook to an epidemic. So did others: an article in *Emerging Infectious Diseases* from April 2015 on ebola in Liberia, mid-year 2014, introduces each section with a quote from *The Plague*. (Arwady et al. 2015). As it happened, the downturn was true, and the same decline in the rate of increase and then the number of new cases would also occur in Guinea (Conakry) and Sierra Leone. One of the under-examined questions of the epidemic is what actually caused the epidemic to slow down, ending the growth phase.

I separate social epidemic control measure from classic clinical control measures (medical isolation and contact tracing). They interact: positively when there is adequate medical care for the scale of the outbreak, but negatively when the scale of the epidemic over-runs medical capacity, or medical capacity is limited from the beginning. The outbreak could, possibly, have been controlled early-on by a clinical isolation-contact tracing model, if there had been a timely response to MSF's (Medecins Sans Frontières) first calls for assistance in March, 2014. Later, with comparatively ample resources and many adaptations, the model worked as the epidemic was ending; on the downside of the epicurve controlling re-emerging cases, when single cases or small clusters could have reignited an outbreak.

The containment of ebola as an epidemic to Liberia, Sierra Leone and Guinea (Conakry) was a major success due to classical medical isolation and contact tracing. There were only small, brief, outbreaks in Nigeria, Senegal and Mali, for example. Sometimes this was by a combination of preparation, skill and good fortune. Urban Nigeria could have been the beginning of a world-wide epidemic (Shuaib et al. 2014). Sometimes, as in Guinea-Bissau, which seemed to have escaped transmission, success may have been due mainly to luck. The international response teams of the Centers for DIsease Control, coordinating with national health systems, had a major role.

Jeremy Farrar and Peter Piot described the vastly expanded scope of engagement needed for an ebola epidemic, when classical outbreak control is no longer sufficient (Farrar and Piot 2014). As the epidemic expanded exponentially, throughout 2014, clinical isolation was still the main model for epidemic control; ignoring conditions of unsafe or non-existent transportation to inadequately staffed and supplied facilities, with high mortality rates. MSF, whose capacity had been over-run, for first time in its history called for military intervention and assistance, but with the same model for epidemic control. There has been a recurrent amnesia about key social lessons learned about both clinical and epidemiological engagement during ebola outbreaks. The lessons forgotten include the social design of care facilities and community isolation and treatment, among many others (Bremen and Johnson 2014; Bremen et al. 2016; Hewlett and Hewlett 2007; Hewlett 2016). The lessons are learned and then forgotten, repeatedly since 1976. This is a bureaucratic and institutional mystery (Abramowitz et al. 2015). On the other hand, some key lessons learned, or imagined to have been learned, from the much smaller ebola outbreaks in central Africa were remembered all too well, where they did not apply. For example, messaging on bats, in particular, and bushmeat in general (Frieden et al. 2014) were contradicted by what people saw. Hunters and their families were not dying in West Africa; clinic and hospital staff were. In a factually human-human transmission epidemic, this damaged the credibility of international actors and of health program messaging.

One hypothesis for the turning of the epicurve is that social epidemic control measures had increasing effectiveness. Examples include social distancing; personal sanitation (chlorine solutions everywhere); mourning and burial changes adapted to local cultural and social requirements; and community-organized isolation control. Temporary shelter-in-place programs, such as in Sierra Leone are another example, although controversial. A variation on the hypothesis stresses community and individual innovation. When there was no effective medical response, I argued that for social epidemic control measures to work: communities needed to think like epidemiologists, and epidemiologists needed to think like villagers and communities.

There was, in fact, a process of convergence that could have been much more supported and emphasized. Issues of convergence included evaluation of clinical risk and benefits: which procedures were critical to use in ebola care and worth the risk to health care workers (Ansumana et al. 2015) That's part of a longer story. The failure to provide basic supplies to people who would never be able to access any functional or timely clinical care is close to the heart of the matter.

13.8 Anything Critical Goes!

Paul Richards' book on the ebola social response is titled *Ebola: how a people's science helped end an epidemic* (Richards 2016). This was very kindly and favorably reviewed by Peter Piot, for some time now the Director of the London School of Hygiene and Tropical Medicine (Piot 2016). Richards describes the case for how rapid community learning and social adaptation on ebola was a major cause for the turning of the epicurve and for control of the epidemic. These changes appear to be mainly what I call epidemic measures, that limit or end transmission, as opposed to clinical care. His own specific examples come from villages studied in depth in 2014 and 2015 by the team he works with at Njala University in Sierra Leone, including Esther Yei Mokuwa. A large number were in Jawei chiefdom, the hardest hit chiefdom of Kailahun District, the epicenter of the epidemic. In general, he thinks that across Sierra Leone, areas that were hit first and hardest put the greatest adaptive measures in place. How rapid learning was adapted in urban and peri-urban areas, will have to be set aside.

Interaction with the epidemic control teams, in these examples, is shown as more of a dialog. Epidemiologists learned from the community and the community learned from medical and epidemic experts. The team set up workshops at Njala specifically for these interactions. Richards' view is that the fact that this was the first epidemic of ebola, as opposed to the smaller prior outbreaks, meant everyone had to learn how to fight and control the epidemic together. Peter Piot agreed (in his review and elsewhere) that there were similar experiences to what he and his team faced with the first ebola outbreak in Yambuku, in what is now the DRC. This included supporting villages using prior social models for fighting small pox.

For context, it is worth recalling the nosocomial origins of the first ebola outbreak, where the virus was spread by the use of unsterilized needles at Yambuku Mission Hospital. Eleven of the staff, then the medical director and three Belgian missionaries died. The hospital closed and patients and their contacts fled to their homes, bringing ebola with them (Bremen et al. 1976; Breman and Johnson 2014).

The government and NGO burial teams in Kailahun made some adjustments to the social need of burial but much more could have been done. It was very difficult for the local communities to get support and training for their own burial teams, which they wanted.

Where dialog slowed down or disappeared, across the epidemic countries, was over clinical care: supporting the basic palliative care patients need to survive, in a rural community hut as well as in a place designated as a medical facility, or an advanced medical hospital. At a minimum this means, nutritional support (food, broths), oral rehydration therapy (ORT), disinfectants (usually chlorine) and how to prepare them, some PPE, and guidelines on the geometry and procedures of safe patient care to avoid contact with or spread of ebola virus.

Facility layout and movement in good ebola patient care has analogies to safe handling of radioisotopes, requiring a notion of the correct structure for the physical environment and how to move through it in sequence. One could say it is a kind of dance. Although ebola patient-care as dance would have made perfect sense for cross-cultural medical training I did not see a discussion in Richards of this kind of interaction.

Dance, drumming, and masquerade all have specific cognitive content and communicate structured information in this culture; they are languages as well as other things. Talking drums may be more familiar, and Richards gives examples from fieldwork when planting rice. The Paramount Chief Mussa Kallon and two sande (women) elders went to the Njala meetings, but there were disagreements about what should happen next. When Paul asked Chief Mussa Kallon how this was resolved he replied that the elders went into the bush and danced a solution.

Epidemiological concepts became integral to how people approached ebola. The communities Richards worked with were aware of, and developing procedures to prevent, the possibility of ebola transmission when carried by stretcher-bearers. No foreign or domestic medical group ever discussed this. After the epidemic receded, survivors were discussing using ebola, and the history and experience of ebola in their communities, as an important way of teaching science in their children's educational curriculum.

When thinking about strengthening a people's science in the context of the West African ebola epidemic, Richards defines the fundamental idea of science, in passing (!), as: prior judgements must be abandoned in the face of compelling empirical evidence. My interpretation of his work is that convergence of understanding and action could happen, even though the knowledge of the social meaning of death, which had to be understood, started more on one side (in communities) and understanding of the nature of the ebola virus and its transmission, which also had to be understood, was more on the other side (the international medical teams). Dialog can lead to cross-cultural convergence. People have a remarkable capacity to maintain prior judgements in the face of compelling empirical evidence. Here, I have been more interested in showing how this warped the international medical response. More broadly applied, cross-cultural convergence can lead to remarkably interesting and enriched kinds of science. Most of the examples I know of come from cross-cultural agriculture, so to speak.

Twenty years ago, two of the social scientists we later worked with on ebola, James Fairhead and Melissa Leach, published Misreading the African landscape: society and ecology in a forest-savannah mosaic. A time sequence of aerial photographic surveys of Guinea, which began after the First World War, documented that the integrated cultural use of slash-and-burn annual cropping was part of increasing forest cover and not desertification. Kat Anderson's *Tending the wild* (Anderson 2005) describes the complex indigenous non-farming cultivation of plants and animals in California. The epistemology, ontology and categories used are different. This is part of a longer story. Medical, epidemiological, ecological and agricultural fields come together in the disease control strategies called one-health (Cohen 2013; Rubin et al. 2013), which may be a particularly useful area for investigating cross-cultural science.

13.9 For Amilcar Cabral, Who Would Have Been 76 in the Year 2000

In the Spring of 2015 a fisherman who was a primary contact of an ebola patient disappeared from Guinea (Conakry) into Guinea-Bissau. The CDC social sciences coordinator asked for contacts in Guinea-Bissau; and we had a whole second round of work on the same issues. The medical and governmental conditions were worse. The social scientists now were Portuguese-speaking or from Portuguese language countries. There also were a high proportion of scientists from Nordic countries. Most of this has to be set aside.

The history of Guinea-Bissau made it, if anything, even more socially, culturally and ethnically complex than the three primary ebola epidemic countries. Amilcar Cabral had carried out the first agricultural census of Guinea-Bissau, after 500 years of Portuguese rule, published in 1956. Some of this was now available in English translation. Almost all of the population were farmers so an agricultural survey had to deal with most of the complexity of the country. It was still very useful background information 60 years later.

Cabral was literally one in a million, one of the very few Africans from then Guinea-Bissau and Cape Verde Islands who received a college education. One can imagine if the United States had only 300 people with bachelor degrees. I read as much of his agricultural work as I could find and re-read the political writing and commentary.

Cabral had a very unusual and sophisticated view of critical cross-cultural science, technology, ecology and knowledge. It is framed first within the context of farming, and then in his approach to liberation from Portugal and international support. I could see this in his critical evaluation of rice farming between ethnic and cultural groups, as well as the evaluation of the impact of colonial cash crops. It is nearly completely overlooked in the political literature, except for two cases that could not be ignored but were never analyzed or, I think, understood. He had interesting methods of using theater for education and practice in cross-cultural work. He died before he could use this approach in post-colonial national development. This needs to be followed up.

Ebola was not reported found in Guinea-Bissau. They were not so fortunate with (Pacific-Brazilian) Zika virus, which went from Cape Verde to Bissau in 2016. Zika may be harder than ebola to control.

Ebola had a striking effect on international biomedical research and development capacity for combatting neglected tropical diseases. It also showed that, in many cases, individuals could wrestle control of scientific, development, government and corporate institutions in a kind of Prague Spring of science facing a lethal crisis. Even advanced biomedical tools have to be used in context.

Ebola went undiagnosed in Guinea for over three months. Dr. Michael Van Herp, MSF Brussels, suspected ebola after receiving case reports that included rapid death, probably too rapid to be Lassa fever, and hiccups (Stern 2014). Hiccups were toward the bottom of signs and symptoms presented by ebola patients; 17th on the WHO Ebola Response Team table of West African patients, reported in October, 2014 (Team WE 2014). But hiccups were associated with ebola among the hemorrhagic fevers.

And so it all began.

References

Abramowitz, S.A., K.L. Bardosh, M. Leach, B. Hewlett, M. Nichter, and V.K. Nguyen. 2015. Social science intelligence in the global Ebola response. *The Lancet* 385 (9965): 300.

Adebamowo, C., O. Bah-Sow, F. Binka, R. Bruzzone, A. Caplan, J.F. Delfraissy, D. Heymann, P. Horby, P. Kaleebu, J.J. Tamfum, and P. Olliaro. 2014. Randomised controlled trials for Ebola: practical and ethical issues. *Lancet* 384 (9952): 1423.

Agassi, J. 2013. *The very idea of modern science: Francis Bacon and Robert Boyle*. Dordrecht: Springer.

———. 2014. *Popper and his popular critics: Thomas Kuhn*. Paul Feyerabend and Imre Lakatos: Springer.

Anderson, K. 2005. *Tending the wild: Native American knowledge and the management of California's natural resources*. Univ of California Press.

Ansumana, R., K.H. Jacobsen, M.B. Idris, H. Bangura, M. Boie-Jalloh, J.M. Lamin, S. Sesay, and F. Sahr. 2015. Ebola in Freetown area, Sierra Leone – A case study of 581 patients. *New England Journal of Medicine* 372 (6): 587–588.

Arwady, M.A., L. Bawo, J.C. Hunter, M. Massaquoi, A. Matanock, B. Dahn, P. Ayscue, T. Nyenswah, J.D. Forrester, L.E. Hensley, and B. Monroe. 2015. Evolution of Ebola virus disease from exotic infection to global health priority, Liberia, mid-2014. *Emerging Infectious Disease* 21 (4): 578–584.

Bardosh, K., M. Leach, and A. Wilkinson. 2016. The limits of rapid response: Ebola and structural violence in West Africa. In *One health: Science, politics and zoonotic disease in Africa*, ed. K. Bardosh, 12: 74–94.

Benton, A. 2015. *HIV exceptionalism: Development through disease in Sierra Leone*. Minneapolis: University of Minnesota Press.

Breman, J.G., and K.M. Johnson. 2014. Ebola then and now. *New England Journal of Medicine* 371 (18): 1663–1666.

Breman, J.G., P. Piot, K.M. Johnson, M.K. White, M. Mbuyi, P. Sureau, D.L. Heymann, S. Van Nieuwenhove, J.B. McCormick, J.P. Ruppol, and V. Kintoki. 1976. The epidemiology of Ebola hemorrhagic fever in Zaire. *Ebola Virus Haemorrhagic Fever* 1978: 103–124.

Breman, J.G., D.L. Heymann, G. Lloyd, J.B. McCormick, M. Miatudila, F.A. Murphy, J.J. Muyembé-Tamfun, P. Piot, J.F. Ruppol, P. Sureau, and G. van der Groen 2016. Discovery and description of ebola zaire virus in 1976 and relevance to the West African epidemic during 2013–2016. *Journal of Infectious Diseases*. jiw207.

Calain, P., and M. Poncin. 2015. Reaching out to Ebola victims: Coercion, persuasion or an appeal for self-sacrifice? *Social Science & Medicine* 147: 126–133.

Carney, J.A. 2009. *Black rice: The African origins of rice cultivation in the Americas*. Cambridge: Harvard University Press.

Centers for Disease Control and Prevention. *Poster: Taking care of someone with suspected Ebola: Be safe while you wait*. Approximate date December 15, 2014. Available at https://www.cdc.gov/vhf/ebola/pdf/be-safe-while-you-wait.pdf.

———. *Guidance for safe handling of human remains of ebola patients in U.S. hospitals and mortuaries*. Available at: https://www.cdc.gov/vhf/ebola/healthcare-us/hospitals/handling-human-remains.html. Last updated February 11, 2015. Retrieved January 10, 2017.

———. *Ebola virus disease information for clinicians in US healthcare settings*. Available at: https://www.cdc.gov/vhf/ebola/healthcare-us/preparing/clinicians.html. Updated May 24, 2016. Retrieved January 10, 2017.

Cohen, D. 2013. *The new country doctor and the clinical diagnosis of the farm. Food Safety News*. Available at: http://www.foodsafetynews.com/2013/12/the-new-country-doctor-and-the-clinical-diagnosis-of-the-farm/#.WHpImumi3Hg.

Etard, J.F., M.S. Sow, S. Leroy, A. Touré, B. Taverne, A.K. Keita, P. Msellati, S. Baize, H. Raoul, S. Izard, and C. Kpamou 2017. Multidisciplinary assessment of post-Ebola sequelae in Guinea (Postebogui): An observational cohort study. *The Lancet Infectious Diseases*, Jan 14.

Fairhead, J., and M. Leach. 1996. *Misreading the African landscape: Society and ecology in a forest-savanna mosaic*. Cambridge: Cambridge University Press.

Farrar, J., and P. Piot. 2014. The Ebola emergency – Immediate action, ongoing strategy. *New England Journal of Medicine* 371 (16): 1545–1546.

Ferme, M. 2001. *The underneath of things: Violence, history, and the everyday in Sierra Leone*. Berkeley/London: University of California Press.

———. 2014. Personal email communication, 2014. Permission to cite January 11, 2017.

Frieden, T.R., I. Damon, B.P. Bell, T. Kenyon, and S. Nichol. 2014. Ebola 2014 – New challenges, new global response and responsibility. *New England Journal of Medicine* 371 (13): 1177–1180.

Hallen, B., and J.O. Sodipo. 1997. *Knowledge, belief, and witchcraft: Analytic experiments in African philosophy*. Stanford: Stanford University Press.

Harker, J. Why are western health workers with ebola flown out, but locals left to die? The Guardian, 15 September 2014. Available at: https://www.theguardian.com/commentisfree/2014/sep/15/ebola-doctor-death-olivet-buck-sierra-leone. Accessed 15 September 2014 and 11 January 2017.

Henao-Restrepo, A.M., I.M. Longini, M. Egger, N.E. Dean, W.J. Edmunds, A. Camacho, M.W. Carroll, M. Doumbia, B. Draguez, S. Duraffour, and G. Enwere. 2015. Efficacy and effectiveness of an rVSV-vectored vaccine expressing Ebola surface glycoprotein: interim results from the Guinea ring vaccination cluster-randomised trial. *The Lancet* 386 (9996): 857–866.

Henao-Restrepo, A.M., A. Camacho, I.M. Longini, C.H. Watson, W.J. Edmunds, M. Egger, M.W. Carroll, N.E. Dean, I. Diatta, M. Doumbia, and B. Draguez. 2016. Efficacy and effectiveness of an rVSV-vectored vaccine in preventing Ebola virus disease: Final results from the Guinea ring vaccination, open-label, cluster-randomised trial (Ebola Ça Suffit!). The Lancet, 2016 Dec 23.

Hewlett, B.S., and B.L. Hewlett 2007. Ebola, culture and politics: The anthropology of an emerging disease. Cengage Learning; 2007 Nov 27.

Hewlett, B.S. 2016. Evolutionary cultural anthropology: containing Ebola outbreaks and explaining hunter-gatherer childhoods. *Current Anthropology* 57 (S13): S000.

Kirk, J.M., and C. Walker. 2016. Cuban Medical Internationalism: The Ebola Campaign of 2014–2015. *International Journal of Cuban Studies* 8 (1): 9–27.

Lamontagne, F., C. Clément, T. Fletcher, S.T. Jacob, W.A. Fischer, and R.A. Fowler. 2014. Doing today's work superbly well – Treating Ebola with current tools. *New England Journal of Medicine* 371 (17): 1565–1566.

Lloyd, E., and H. Perry 1998. Infection control for viral haemorrhagic fevers in the African health care setting. CDC and WHO, 1998. Maintained by the CDC at https://www.cdc.gov/vhf/abroad/vhf-manual.html.

Lyon, G.M., A.K. Mehta, J.B. Varkey, K. Brantly, L. Plyler, A.K. McElroy, C.S. Kraft, J.S. Towner, C. Spiropoulou, U. Ströher, and T.M. Uyeki. 2014. Clinical care of two patients with Ebola virus disease in the United States. *New England Journal of Medicine* 371 (25): 2402–2409.

Moses, L. 2014. Personal email communication. Permission to use, November 07, 2014 and January 10, 2017.

Moses, R., and C.E. Cobb. 2002. *Radical equations: Civil rights from Mississippi to the Algebra Project*. Boston: Beacon Press.

Nguyen, V.K. 2010. *The republic of therapy: Triage and sovereignty in West Africa's time of AIDS*. Durham: Duke University Press.

Piot, P. 2016. Public health: Beating ebola. *Nature* 537 (7621): 484–485.

Richards, P. 1998. *Fighting for the rain forest: War, youth & resources in Sierra Leone*. Oxford: James CurreyR Ltd..

———. 2016. Ebola: How a people's science helped end an epidemic. Zed.

Richards, P., J. Amara, M.C. Ferme, P. Kamara, E. Mokuwa, A.I. Sheriff, R. Suluku, and M. Voors. 2015. Social pathways for Ebola virus disease in rural Sierra Leone, and some implications for containment. *PLoS Neglected Tropical Disease* 9 (4): e0003567.

Rubin, C., T. Myers, W. Stokes, B. Dunham, S. Harris, B. Lautner, and J. Annelli. 2013. Review of institute of medicine and national research council recommendations for one health initiative. *Emerging Infectious Disease* 19 (12): 1913–1917.

Shuaib, F., R. Gunnala, E.O. Musa, F.J. Mahoney, O. Oguntimehin, P.M. Nguku, S.B. Nyanti, N. Knight, N.S. Gwarzo, O. Idigbe, and A. Nasidi. 2014. Ebola virus disease outbreak – Nigeria, July–September 2014. *MMWR Morb Mortal Weekly Report* 63 (39): 867–872.

Stern, J.E. 2014. Hell in the hot zone. Vanity Fair.

Suluku, M.R., M.I. Abu-Bakarr, M.J. Johnny, and F. Jonsyn-Ellis. 2012. Post-war Demographic and Ecological Survey of Dog Populations and Their Human Relationships in Sierra Leone.(A Case Study of Urban Freetown). *Science Journal of Agricultural Research and Management*;2012.

Team WE. 2014. Ebola virus disease in West Africa – The first 9 months of the epidemic and forward projections. *The New England Journal of Medicine* 2014 (371): 1481–1495.

Vetter, P., W.A. Fischer, M. Schibler, M. Jacobs, D.G. Bausch, and L. Kaiser. 2016. Ebola virus shedding and transmission: Review of current evidence. Journal of Infectious Diseases. Jul 20:jiw254.

Wilkinson, A. 2016. Beyond biosecurity: the politics of Lassa fever in Sierra Leone. In *One health: Science, politics and zoonotic disease in Africa*, ed. K. Bardosh, Feb 12: 117–138.

Zeitlin, L., K.J. Whaley, G.G. Olinger, M. Jacobs, R. Gopal, X. Qiu, and G.P. Kobinger. 2016. Antibody therapeutics for Ebola virus disease. *Current Opinion in Virology* 17: 45–49.

Part III
Making Sense of a Changing Past: History and Historiography

Chapter 14
Jacob & Esau Today: The End of a Two Millennia Paradigm?

Malachi H. Hacohen

14.1 Introduction

> ***Jacob's Voice***
> The tears.
> The tears in the brother's eye.
> One has remained hanging, and growing.
> We live in it.
> Breathe, so it may be released.[1]

Paul Celan's 1959 poem, created a convergence of Jacob's voice and Esau's tears. Celan saw Jews & Christians living in one big, and growing tear, and envisioned a breath that may release it. The poem provides both the motto and the motif for this

---

Dedication
Joseph Agassi came into my life in my thirties, as I was completing my book on Karl Popper. I had known of him before: When conducting research on Popper in my graduate student days in the 1980s, his name was one of the first to come up. My first recollection of him is, however, even earlier, as a boy: About 1970, he appeared on Israeli TV, in a panel debating *Animal Farm*. He confronted long-time Zionist Soviet sympathizer, Yaakov Riftin, and I liked immediately his mischievous style. Popper, Israel and political theory have since become the staple of our ongoing conversations. I am a generation younger than Joseph, but we share biographical affinities: a traditional Jewish upbringing, followed by a Popperian rebellion, and wide-ranging academic interests that defy disciplinary boundaries. Joseph is one of the very few to whom I can send anything I write, and be assured of a discerning reading and critical exchange. I call Joseph *ha-Rav Yosef*, a pun on the late Israeli Sephardic Chief Rabbi who represented everything Joseph rejects, but also a recognition of Joseph's authority, and a reminder that our putatively academic dialogue has always been Talmudic in character.

[1] "Jakobsstimme," in: *Sprachgitter* (Frankfurt am main: S. Fischer Verlag, 1959), http://www.lyrikline.org/de/gedichte/stimmen-162#.VaEB4Ovt5t4, accessed 24 July 2015: "Die Tränen. Die Tränen im Bruderaug. Eine blieb hängen, wuchs. Wir wohnen darin. Atme, daß sie sich löse." My thanks to Moshe Lavi of the University of Haifa for introducing me to the poem.

M.H. Hacohen (✉)
History and Jewish Studies, Duke University, Durham, NC, USA
e-mail: mhacohen@duke.edu

N. Bar-Am, S. Gattei (eds.), *Encouraging Openness*, Boston Studies in the Philosophy and History of Science 325, DOI 10.1007/978-3-319-57669-5_14

essay. Only European accountability for the Holocaust made Celan's vision of togetherness possible: Jacob's voice and Esau's tears – or are they Jacob's? – cohabit in one space, and healing appears a prospect. Celan reversed the *midrashim* that intimated that Esau's cry and tears over the injustice of the stolen blessing were the source of Jewish suffering through history, and redemption will tarry until the well of tears has dried up.[2] In the aftermath of the Holocaust, Jacob's voice, originally beguiling and the cause of Esau's cry and tears, cries out itself, and the tears appear in either brother's eye, indeed, both live in Esau's (?) growing tear. Yet, a breathe signaling a stop to the crying can be envisioned, even if it remains unclear that there is life once the tear has dropped. However one interprets Celan, the two-millennia paradigm of Jacob & Esau has collapsed in the aftermath of the Holocaust and the State of Israel.

Yet, not immediately: Celan was a harbinger of future paradigmatic changes, but they occurred, for the most part, only after the 1967 War. Israel's War of Independence (1948–1949) did not overcome the sense of powerlessness generated and reaffirmed by the Holocaust. Yael Feldman showed "the *aqedah* becom[ing] a

---

Together, Joseph & I have led an ongoing struggle over the meaning of Jewishness in contemporary life. Some differences, I think, have also emerged. Popper has become Joseph's lifelong point-of-reference. In his Autobiography, *A Philosopher's Apprentice*, Joseph provided an unmatched psychological portrait of Popper and a testimony to his formative influence on his philosophy. For me, Popper has been only a midway station: He is still someone I occasionally think with, but his philosophy is no longer foremost on my mind. Likewise, while Joseph & I both go back to Jewish history in order to contemplate the future, Joseph's abiding concerns remain the prospects of liberal nationalism and Israel, whereas I endeavor to think beyond them – when Israel ceases. Would Joseph consider this a failure of nerve on my part, or a callous disregard for what is at stake for Jews in the present? I hope not.

In celebration of Joseph's 90th birthday, I offer an essay on "Jacob & Esau Today." The essay is the concluding chapter to my forthcoming *Jacob & Esau: Nation and Empire in Jewish European History* (Cambridge: Cambridge University Press, 2018). The rival biblical twins, Jacob & Esau, represent the rabbinic idiom for Jewish-Christian relations as it has emerged from rabbinic interpretations to the biblical story of *Genesis*. Jacob has been recognized by Jews, Christians and Muslims as the Ancestor of the Jewish people. Through a remarkable chain of events, which the book tells, Esau, whom *Genesis* designates the Ancestor of Edom, became first Roman and then Christian, and the rabbis directed biblical prophecies on Edom against the Roman Empire and Christianity. The biblical story became a *topos* for Roman-Jewish and Christian-Jewish relations. Jacob & Esau provided the paradigm of Jews' relationship to non-Jews: "All that happened to our ancestor Jacob with Esau his brother," opined the medieval Spanish Jewish biblical commentator, Naḥmanides (1194–1270), "will always happen to us with Esau's descendants." This paradigm has collapsed in the aftermath of the Holocaust and the State of Israel. The essay is a roundabout explanation to Joseph of why I felt compelled to move as I have on contemporary Judaism and Israel. It is also a tribute, paid with affection and admiration, to the impact Joseph has had on my intellectual universe. יישר כוח הרב. עד מאה ועשרים!

[2]*Bereshit Rabba*: 67:4; *Tanḥuma*, ed. Buber (Vilnius: Rom, 1913): toledot 24; *Zohar*: 1: 145a: "Israel has suffered on account of the tears which Esau shed before his father, in his desire to be blessed by him, out of the great regard he had for his father's words"; 2: 12b: "Those [Esau's] tears brought Israel down into his exile."

key figure in Zionist thought and Hebrew letters" during the 1940s and 1950s.[3] Displays of Masada as a counter-myth to the Holocaust – "Never again shall Masada fall!" – only highlighted the self-sacrificial moment in the Israeli national ethos but provided no comfort.[4] The painting commemorating the 1948 War at the entranceway to Israel's leading Art School, the Beẓalel Academy in Jerusalem, depicts mourning over a fallen soldier. At the moment of Israel's foundation, and for 15 years later, images of Jewish victimhood predominated in Beẓalel, and overwhelmed the "New Jew," the venerated icon of Labor Zionism.[5] Only the 1967 War, which confirmed Israel as a regional power but also made its national problem unresolvable, engendered Jewish empowerment and triggered the millennia-old paradigm's dissolution.

14.2 Latent Loss and Empowerment: Jacob & Esau, 1945–1967

In mainstream Zionist poetry, Jacob remained until the late 1960s symbol of the national project. From Yiẓḥak Lamdan's "You!" (1953) to Amir Gilboa's "We were like Returnees" (1953) and "Israel" (1963) to Abba Kovner's "All My Bell" (1970), Jacob was called upon to ascend the ladder and struggle against the angel to realize the dream of national return. He was promised a glorious future, and assured, on Mount Sinai, of his ladder's strength.[6] Less celebratory about the national prospect, Israeli painter Jacob Steinhardt's 1950 woodcut of Jacob & Esau embracing expressed hopes for Jewish-Arab reconciliation: The sentiments may have been

---

[3] *Glory and Agony: Isaac's Sacrifice and National Narrative* (Standford CA: Stanford University Press, 2010), p. 20: "Janus-like, [the *Aqedah*] came to represent both the slaughter of the Holocaust and the national warrior's death in the oldnew homeland." In fairness, Feldman does not see the post-1967 "Israeli psycho-political assault on the *aqedah*" as issuing from victory but as solidifying the Israeli "Isaac syndrome." Arie Dubnov of the University of Haifa and Shai Ginsburg of Duke University pointed out to me that the transformations of the *Aqedah* and the Jacob & Esau typologies in Israel dovetailed.

[4] Yael Zerubavel, *Recovered Roots: Collective Memory and the Making of Israeli National Tradition* (Chicago: University of Chicago Press, 1995). Yiẓḥak Lamdan, *Masadah* (Tel Aviv: Hedim, 1927) was crucial in formatting the myth.

[5] Lisa Bonnifield, Modern Jewish Art and Zionism, BA Thesis, Duke University, 2008; Dalia Manor, *Art in Zion: the Genesis of Modern National Art in Jewish Palestine* (London: Routledge, 2005); Gideon Ofrat, *One Hundred Years of Art in Israel*, trans. by Peretz Kidron (Boulder, CO: Westview Press, 1998).

[6] Malkah Shaked, *I'll Play You Forever: The Bible in Modern Hebrew Poetry*, 2 vols. (Tel Aviv: Yediot Aḥaronot, 2005), 1: 120–21, 1:122–23, 1: 125–26, respectively. Interestingly, Gilboa also published, in 1953, "Isaac" (1: 96), which Yael Feldman considers subversive of earlier Zionist portrayals of Isaac as a self-sacrificing military hero: *Glory and Agony*, pp. 134–36. Mourning over relatives who perished in the Holocaust may have subverted native martyrology of fallen soldiers but preserved life-loving Jacob for the national project.

new but the typology's use was traditional.[7] Far Right poet, Uri Ẓvi Greenberg, who, in the late 1920s, toyed with Esau as a Zionist New Man, returned in the aftermath of the Holocaust to the typology's traditional use: Esau became again the murderer of Jews.[8]

The pattern of continuity and change transcended political camps. Diaspora nationalist philosopher and historian, Simon Rawidowicz (1897–1957), founder of the Jewish and Near Eastern Studies Program at Brandeis, protested that, in uprooting hundreds of thousands of Arabs from their homeland, and turning them into refugees, Jacob had for the first time since the Hasmoneans spilled blood, and committed violence that cried out to God, a moral transgression inviting criticism.

"Esau has been defiling Jacob; the two have become alike. The twins are no longer struggling with each other. [...] Jacob is a brother to Esau, who always made of the credo 'live by the sword' a positive commandment. Perhaps one day Esau and Ishmael will join forces [...] to repay Jacob for the act that he committed in 1948. When the avengers are given the authority to avenge, will they know any limit?"[9]

Rawidowicz's critique of Jewish power and statehood reflected new political realities, indeed, unprecedented ones, but it articulated the traditional Eastern European Jewish apprehension about identifying the nation – even the Jewish one – with the state. Predictably, his typology also remained traditional: He merely bemoaned Jacob behaving like Esau.

Similarly on the Christian side, the soul searching occasioned by the Holocaust among Christian churches came fully into the open only in the mid-1960s with Vatican II and Protestant initiatives for Christian-Jewish Dialogue. The immediate postwar period remained, as historian Dan Diner suggests, one of Holocaust latency: A reevaluation was taking place but not apparent.[10] In the 1950s, missionary evangelist books on the Patriarchs could still use old Christian typology mixed with antisemitic stereotypes. If Abraham was the man of perfect faith, Jacob was a ques-

---

[7] Steinhardt (1887–1968) belonged to the Berlin Secession, then, having immigrated to Palestine in 1933, to the new Beẓalel Group. He became head of the Beẓalel Academy (1954–57). Ziva Amishai-Maisels, "Steinhardt in the Land of Israel," *Jacob and Israel: Homeland and Identity in the Work of Jakob Steinhardt* (Hebrew and English), ed. by Gabriel Ma'anit, Ruthi Ofek, and Avraham Hai (Tefen: Open Museum, 1998), pp. 70–126 (Hebrew), 197–230 (English). I learned of Steinhardt's Jacob & Esau woodcut from Ezra Mendelsohn, *Painting a People*: *Maurycy Gottlieb and Jewish Art* (Lebanon, NH: Brandeis University Press, 2002), p. 215.

[8] Neta Stahl, "'Man's Red Soup' – Blood and the Art of Esau in the poetry of Uri Zvi Greenberg," in: *Jewish Blood: Reality and Metaphor in History, Religion and Culture*, ed. by (London: Routledge, 2009), pp. 166–69.

[9] Simon Rawidowicz, "Between Jew and Arab," in: *Between Jew and Arab: The Lost Voice of Simon Rawidowicz*, ed. by David N. Myers (Hanover, NH: Brandeis University Press, 2008), pp. 150–51, ProQuest ebrary, accessed 15 July 2015.

[10] Dan Diner, "Vom Stau zur Zeit: Neutralisierung und Latenz zwischen Nachkrieg und Achtundsechzig," in: *Latenz: Blinde Passagiere in den Geisteswissenschaften*, ed. by Hans Ulrich Gumbrecht and Florian Klinger (Göttingen: Vandenhoeck & Ruprecht, 2011), pp. 165–172.

tionable youth. Encountering God at Penuel, Israel became God's soldier, and, yet, meeting with Esau the next day, he was a master of guile all over again.[11] The contrast with future ecumenical Abraham, the main figure for interreligious dialogue among Christians, Muslims and Jews, was great.[12]

The turn-around in the 1960s was surprising, and the surprise contributed to Orthodox Jews' suspicion of the Vatican II's call for dialogue. The leader of U.S. modern Orthodoxy, rabbi and philosopher, Yosef Dov Soloveitchik (1903–1993), advised against theological exchange. He regarded Jacob & Esau's separation after their reconciliation as marking destiny: Non-Jews could not truly accept Jewish difference, and "it is easier and better for Esau to live on Mt. Seir, and for Jacob and his Sons to be in the Land of Israel."[13] The great Ultra-Orthodox halakhist, rabbi Moshe Feinstein (1895–1986), rejected Paul VI's call for a Christian-Jewish dialogue as a new conversion ploy. In a 1967 *responsum* advising British Jewish leaders not to appeal their government's refusal to support Jewish schools to international organizations lest they evoke British wrath, he opined that "hatred of the Jews by all nations is great, even in states that treat Jews well. [...] It is a well-known *halakha* (maxim) that Esau hates Jacob. [...] Just as *halakha* never changes, so also Esau's hatred of Jacob never changes."[14] His son, David Feinstein, suggested that Jacob could not keep his promise to come to Esau (*Genesis* 33:14) because he was concerned about his impressionable family. The time had not yet come for Jacob & Esau to become brothers or for Jews to conduct dialogue with Christians.[15]

Indeed, Ultra-Orthodox Jews have endeavored to retain the traditional typology all the way through. Most Ultra-Orthodox Jews believe that the Holocaust came for assimilation's sins, and vindicated Moses Sofer's warning that mimicry of the gentiles would provoke their (and God's) wrath and end up disastrously. With the exception of the radically anti-Zionist Satmar Ḥasidim, they have desisted from calling the Zionists (and assimilated Jews) Amaleq, as they had done before the Holocaust, and Amaleq is now reserved for the Nazis. In his Megilat Polin (1966), Gur leader, Yehudah Leib Levin, founder of the Ultra-Orthodox Daily Ha-Modia, suggested that remembering the antisemitic Nazi design to obliterate the entire Jewish people, and commemorating the saints who perished in the Holocaust, was

[11] G. R. Brinke, *Abraham: der Freund Gottes* (Bern: Ährenlese-Verlag, 1957); idem., *Der Erzvater Jabob: Ein Lebensbild in Skizzen* (Bern: Ährenlese-Verlag, 1959).

[12] Matthias Morgenstern, "Vom 'Götzenzerstörer' zum Protagonisten des Dialogs – Der Erzvater Abraham in 1800 Jahren jüdischer Tradition," in: *Interreligiöser Dialog: Chancen abrahamischer Initiativen*, ed. by Reinhard Möller and Hanns Christoph Goßmann (Berlin: Lit, 2006), pp. 101–126. Yaakov Ariel of UNC, Chappel Hill points out that Abraham did not become a central figure until the Christian-Muslim Dialogue took off at the turn of the twenty-first century, by which time the Christian-Jewish dialogue had already become secondary.

[13] "Jacob and Esau," *Divre Hashqafa*, trans. (from the Yiddish) by Moshe Crone (Jerusalem: Zionist World Association, 1992), p. 27.

[14] *Igrot Moshe* (Benei Beraq: Moshe Feinstein, 1985): *Ḥoshen u-Mishpat*: 2:77 (p. 319).

[15] *Kol Dodi on the Torah* (Brooklyn, NY: Masorah Publications, 1992), pp. 55–56.

fulfillment of the Torah's command to remember Amaleq.[16] Christians were charged for withholding help from Jews but they were not the Holocaust's perpetrator. Pagan Amaleq was. This did not amount, however, to an Ultra-Orthodox Holocaust theology, and the lacuna suggests that the Holocaust's legacy has not yet fully registered.

Ultra-Orthodoxy is even less able to account for the Jewish State. A minority, the Satmar *Ḥasidim* flatly reject Israel, and regard the State as yet another trial for Jews "to find out whether you love the Lord your God with all your heart" (Deuteronomy 13:3). Most of the Ultra-Orthodox have developed an ambivalent attitude towards Israel: God has mercifully provided a refuge to the Jews in Israel in the aftermath of the Holocaust, but they remain outrageously obdurate. Why God has so chosen remains a mystery.[17] The reluctance to think through the Holocaust and Israel is exemplified by their traditional use of Jacob & Esau. "It is a maxim that Esau hates Jacob," said Eliezer Shakh, the venerated leader of Israeli Ultra-Orthodoxy's Lithuanian wing. He had no confidence in U.S. benevolence: Non-Jewish governments should never be trusted but also never provoked. Jews have weathered hostile governments for millennia by keeping loyal to the Torah, and Jewish politics ought to be limited to maneuvering among existing political options. Jews were not to declare war or make peace. All the same, just as no traditional option existed at the birth-time of Orthodoxy, in a postrevolutionary Europe undergoing emancipation, none truly exists after the Holocaust, and in face of Jewish statehood. Reluctant to acknowledge the new national and diasporic options, Ultra-Orthodoxy has found itself inadvertently transforming the typology by applying it to the Mideast conflict. Esau came to stand for Arab leaders, now perceived as the foremost threat to the Jewish population.

If Ultra-Orthodoxy has endeavored to buck post-Holocaust typological changes, the national religious camp has led them. This was no coincidence. Secular Zionists had limited interest in the rabbinic typology whereas religious Zionists sought to use it to address politics. The prayer for the Jewish State's welfare, recited in modern Orthodox synagogues weekly on the shabbat, speaks of Israel as "the beginning of the growth of our redemption" (ראשית צמיחת גאולתנו), thereby integrating the Zionist project into the Jewish messianic vision. National religious circles around

[16] Michal Shaul, *Beauty for Ashes: Holocaust Memory and the Rehabilitation of Ashkenazi Haredi Society in Israel, 1945–1961* (Hebrew) (Jerusalem: Yad Vamhem; Yad Izhak ben-Zvi, 2014), pp. 307–316. Levin was a harbinger more than a trend setter: Until the 1990s, the Ultra-Orthodox leadership downplayed Holocaust commemoration, fearing it may hinder the reconstruction of Jewish life. See: Dan Michman, "The Impact of the Holocaust on Religious Jewry," *Major Changes within the Jewish People in the Wake of the Holocaust* (Jerusalem: Yad Vashem, 1996), pp. 659–707. Ultra-Orthodoxy's revival has since given rise to bountiful literature.

[17] Aviezer Ravitzky, *Messianism, Zionism, and Jewish Religious Radicalism*, trans. by Michael Swirsky and Jonathan Chipman (Chicago: University of Chicago, 1996); Barbara Krawcowicz, Covenantal Theodicy Among Haredi and Modern Jewish Thinkers During and After the Holocaust, Ph.D Dissertation, University of Indiana, Bloomington, 2013. My thanks to Krawcowicz and to Moshe Hellinger of Bar-Ilan University (email to author, 29 May 2014) for their help.

the Merkaz Ha-Rav Yeshivah were the first to see the Holocaust as messianic tribulations, and the Yeshivah head and founder's son, Ẓevi Yehudah Kook, spoke of the divine liquidation of exile as preparation for redemption. Unlike Ultra-Orthodoxy, religious Zionism offers a compelling theological explanation for the Jewish State and an outrageous one for the Holocaust, but one that ties it intimately with the Jewish State. No wonder religious Zionists have been in the forefront of typological innovation: Modern Jewish messianism required for the Jacob & Esau paradigm to be reworked.

The beginning in the 1950s was modest. In her legendary textbook, training teachers in Bible instruction, Neḥamah Leibowitz (1905–1997) crafted traditional commentaries on the weekly portions of the Torah. Born to an Orthodox Jewish Riga family that immigrated to Berlin in 1919, Leibowitz graduated from the University of Marburg in 1930, and moved to Palestine. A self-effacing erudite scholar, she rekindled interest in school teaching of the Bible with the medieval Jewish commentators. In her *Genesis* commentary, she confronted Samson Raphael Hirsch's view of Jacob and Esau's reconciliation as presaging emancipation with that of Benno Jacob, who, she said, found Esau's conduct suspicious. She suggested that the Holocaust exposed Hirsch's view that Esau was laying down his sword and reconciling as wishful thinking.[18]

A scandalized Samuel Hugo Bergmann, carrying the Prague Zionist and *Berit Shalom*'s legacy, protested that she ignored the reconciliation's grandeur, which recalled to him Schiller's "Ode to Joy" (An die Freude), as embedded in Beethoven's Ninth Symphony: "Embrace, millions!" (Seid umschlungen, Millionen!)[19] In response, Leibowitz faulted Hirsch for deriving his love for humanity from German humanism and not from Jewish sources. She preferred, she said, Naphtali Ẓevi Yehudah Berlin's view of the reconciliation: When Esau's descendants genuinely respect Israel, "then we, too, are moved to acknowledge [him as] our brother." She reaffirmed the traditional typology, listing the atrocities that Israel's enemies had perpetrated, from medieval pogroms to forced draft and conversion in the Tsarist Russian army to the Holocaust. Israel is still a lamb among the wolves, she told Bergmann.

Surprisingly, those wolves now became Arab, and Leibowitz projected onto them two millennia of mostly Christian persecution. Unwittingly and unnoticeably, Leibowitz and other Orthodox intellectuals initiated in the 1950s a revolutionary change. Never in two millennia had Esau been a symbol for Arabs or Muslims (traditionally regarded as Yishmael's descendants). Now Esau reached beyond his Christian provenance into Islam to unify, as ancient Edom had had, divergent enemies into a typological one. Yosef Dov Soloveitchik identified the Arabs with

[18] Neḥama Leibowitz, *Studies in the Book of Genesis in the Context of Ancient and Modern Jewish Bible Commentary*, trans. And adapted from the Hebrew by Aryeh Newman (Jerusalem: World Zionist Organization, 1972), commentary on Va-yishlach.

[19] The December 1957 exchange of letters between the two was published in Aviad Hacohen, "Does Esau hate Jacob" (Hebrew), *Meimad* (1998): 16–19.

Amaleq.[20] Shakh and the Ultra-Orthodox followed suit. The new locution has become a common place. Israel's Sephardic Chief rabbi, Yiẓḥaq Yosef, inveighed in November 2014 against Jews going up the Temple Mount: "Stop inciting Esau's descendants and provoking our Arab enemies."[21] Moshe Gafni, political leader of the Lithuanian wing of the Torah Jewry Party, opined, during the 2015 Israeli elections campaign, that "it is, indeed, a maxim that Esau hates Jacob but we do not wish for war with the Arabs to continue endlessly and must make peace efforts together with the U.S."[22] The quiet revolution begun in the 1950s was now complete.

These were the limits of innovation in the first two decades of Israel's existence. Israel's advocates and critics alike extended the Jacob & Esau paradigm to the Arabs but the typology remained traditional. A sense of fragility still pervaded in Jewish life almost everywhere. A generation of Cold War liberals, Isaiah Berlin most prominently, endeavored to pluralize the European nation state, so as to make it inhabitable for Jews, but the task was daunting: The nationalizing state left exceedingly narrow space for Jewish life.[23] European liberals could also not imagine that, in the long run, the North American Diaspora would prove any different. "It cannot happen in Scarsdale[?] I am not sure, not completely," said George Steiner as late as 1965. "America is no more immune than any other nationalistic, professedly Christian society from the contagion of anti-Semitism."[24] Twenty years later, historian Stephen Whitfield suggested, in *Voices of Jacob, Hands of Esau*, that American culture did put the lead on antisemitism. Biblical culture, he said, at once enhanced and constrained Jacob's new voice, that of critical modernism.[25] The Jews became an American success story, model immigrants, a symbol of national ideals and a proof of their viability.

The anxiety about Israel and the future, and the predominant ethnonational perspective, were reflected also in the postwar Israeli imperial imagination. Book series for children highlighted the pre-state national struggle against the British, finding nothing to commend in the Commonwealth.[26] In contrast, Hebrew literature witnessed the last wave of nostalgia towards the Austrian Empire, which was viewed as victim of nationalism. Asher Barash published in 1949 a collection of stories on

[20] Joseph B. Soloveitchik, *Kol Dodi Dofek: Listen, My Beloved Knocks* (1956), trans. by David Z. Gordon (New York: Ktav, 2006). For additional examples, see: Elliott Horowitz, *Reckless Rites* (Princeton, NJ: Princeton University Press, 2006), esp. pp. 144–46.

[21] *Haaretz* (7 November 2014).

[22] *Haaretz* (9 February 2015). He was responding to questions of *Haaretz* browsers.

[23] Malachi Hacohen, "'The Strange Fact That the State of Israel Exists': The Cold War Liberals Between Cosmopolitanism and Nationalism," *Jewish Social Studies* 15:2 (2009): 37–81.

[24] An observation all the more remarkable for Steiner's US upbringing and affirmation of the Jewish Diaspora. "A Kind of Survivor," in his (1984) *George Steiner: A Reader*. New York: Oxford University Press, pp. 225, 224.

[25] *Voices of Jacob, Hands of Esau: Jews in American Life and Thought* (Hamden, CT: Archon Books, 1984).

[26] Eitan Bar-Yosef, "The Nostalgic Return to Mandatory Palestine in Israeli Culture," paper at the NC Jewish Studies Seminar, 29 March 2015.

Galician Jewish life, permeated with a sense of doom and failure. The lead story, "Embroidery" (רקמה), tied together the declining fortunes of the imperial family and Jewish life.[27] The year after, Rabi Binyamin published his memoirs, which likewise highlighted Austrian Galicia.[28] Hayim Beer's *Feathers* (1979) closed the postwar era, describing the vestiges of fin-de-siècle Viennese culture in 1950s Jerusalem. A half insane follower of the progressive thinker Lynkeus cast a major character, and his son, it turns out at the end of the book, died in the 1973 War. The Empire's end and Israel's crisis converged.[29] Hebrew literature marked Israel's independence and survival with visions of a beloved Empire's doom. Well into the State era, the cultural universe shaping Jacob & Esau seemed regulated by pre-state traditions.

14.3 Jacob the Liar, 1968–2000

The 1967 War and the 1968 Student Revolution occurred as if in different worlds, yet both signaled political changes that shifted Israeli and European dispositions, respectively, towards the Holocaust, thereby undermining the Jacob & Esau paradigm. *Jakob der Lügner* (1969, *Jacob the Liar*) by East German Jewish novelist and screenwriter, Jurek Becker (1937–1997), marked an epochal change. Becker survived the Holocaust as a child, first with his parents in the Lodz Ghetto, then in a German concentration camp, where he lost his mother in 1945.[30] The Beckers had spoken Polish until after the War, when the father decided to settle in Berlin and abruptly shifted to German. Becker received socialist education and grew up atheist and anti-Zionist, a Jew defined by the Holocaust alone, i.e., by his ethnic origins. In East Germany, Jewishness conferred recognition as a victim of fascism but could become a liability if avowed as a national identity. Important Holocaust films had been produced in Eastern Europe before *Jakob der Lügner*, but to make a film, in which the Jew was the hero and the murder of the Jews was central was still tantamount to crossing a minefield. Having passed the film script by the East German censor in 1965, Jurek found the Polish authorities torpedo the filming in Cracow, and the East Germans canceling the production the year after.[31]

---

[27] *In the Shadow of Good People: Short Stories* (Hebrew) (Jerusalem: Bialik institute).

[28] Rabi Binyamin, *From Zborov to Kineret* (Hebrew) (Tel-Aviv: Devir). Almost two third of the book are devoted to the Galician *shtetl* Zborov. The early chapters, published already in 1940–42, evoked enthusiastic response among the Hebrew writers' community (pp. 7–10).

[29] Haim Beer, *Feathers* (Lebanon, NH: University Press of New England, 2004).

[30] Becker had no memory of the Holocaust years. His father had survived Auschwitz and claimed him after the War, apparently in Sachsenhausen. *"Wenn ich auf meine bisheriges zurückblicke, dann muß ich leider sagen": Jurek Becker 1937–1997: Dokumente zum Leben und Werk ausdem Jurek-Becker-Archiv*, ed. by Karin Kiwus (Berlin: Akademie der Künste, 2002), pp. 10–23; Sander Gilman, *Jurek Becker: A Life in Five Worlds* (Chicago: University of Chicago Press, 2003), pp. 17–21, and ns. 30–31, p. 235.

[31] Sander Gilman, *Jurek Becker*, pp. 60–67.

Becker, a master of the subterfuges of dissent, transmuted the film script into a Holocaust "comic" novel (surmising that novels were less carefully scrutinized than films).[32] *Jakob der Lügner*'s success was immediate, and it won several West German and Swiss prizes. The official East German film and television agencies now joined to produce a film (1974), with a new screenplay by Becker.[33] It became the first East German film to be shown at the Berlin Film Festival, and the only one ever to be nominated for the Academy Award. It also received East Germany's National Prize (1975). The Holocaust now broke through the iron curtain and became a medium for European self-definition and internationalization.[34]

Situated in a WWII Polish ghetto, *Jacob the Liar* tells of a lower-middle-class Jew, Jakob Heym, formerly a tiny restaurant's owner, who, while being detained at the Ghetto's headquarters for violating the curfew, overhears a radio broadcast about the Russians advancing. Apprehensive that his friend, Mischa, would carry out a suicidal act of escape, he reveals to him the news, and, to make them credible, intimates that he has a radio. The "radio" gives the people in the ghetto, who are starving and awaiting deportation to Nazi death camps, renewed hope: Suicides stop, and people begin contemplating a future. Jakob is repeatedly asked for more news and cannot retract without throwing the Ghetto into despair. To Lina, an orphan under his care, he stages mock BBC broadcasts of news and children's stories. The film ends with Jakob, Lina, and their compatriots being transported to an unknown destination but, in the novel, Becker offered also an alternative fantastic ending (the original one), in which Jakob is shot trying to escape but the Russians do arrive and liberate the camp.

Becker's Jacob embodies traditional antisemitic stereotypes but their meaning is transformed. Jacob's lies, traditionally a means for profiteering, become an act of triumphal imagination, opening up horizons of hope for the Ghetto's inmates. Jacob's craftiness and impersonation create a visionary, life-sustaining alternative reality. Traditionally, Jacob's voice gives him in, whereas, in the film, the changing voice is constitutive of radio performances that delight and comfort little Lina. Jacob's recollections of his deal with Kowalski, a hair dresser now his friend in the Ghetto – free haircuts in exchange for free food – appear not as petty haggling, but as dreams of everydayness and a restoration of humanity. No Leah or Rachel shows up in the film: Jacob is the motherly figure taking care of Lina, and his last name Heym, connotes "home" in Yiddish and "life" (*Ḥayim*) in Hebrew. With *Jacob the Liar*, the Jewish anti-hero has become the hero, and the homeless people have come to provide a home and sustain the life of a new European culture.

[32] Jurek Becker, *Jakob der Lügner* (Darmstadt: Luchterhand, 1970); idem., *Jacob the Liar*, trans. by Leila Vennewitz (New York: Arcade Pub., 1996).

[33] DEFA Studio für Spielfilme, Gruppe Johannisthal und Fernsehen der DDR; eine Gemeinschaftsproduktion in Zusammenarbeit mit Filmové studio Barrandov; Buch, Jurek Becker; Produktionsleitung, Herbert Ehler; Regie, Frank Beyer, Jakob der Lügner, Video-DVD (Northampton, MA: Icestorm International, 1999).

[34] Shai Ginsburg objects: Earlier Polish and Czech Holocaust films, distributed internationally, had already done so. He is right, but they did not have the same impact.

The Christian churches, Socialism's major rival, were modifying during the same period their teachings about the Jews, and scrutinizing supersessionist typologies that suggested that Christianity had replaced Judaism. In a Catholic Bible picture-book from the 1980s, the Jacob & Esau typology was gone, and the narrative underlined that Esau had not been rejected. A 1980 book of Catholic prescriptions on Judaism for instructors was free of Christian typology and sympathetic to Israel. It gave a fair exposition of Jewish views and prohibited the old supersessionist doctrines. Protestant pedagogy likewise strove to break with Christian anti-Judaism. A 1981 review expressed frustration that the theological change had not yet trickled down to German elementary schools, and schoolbooks still depicted the Jews as sharing in the guilt for the crucifixion. Things changed over the following decades. The Protestant-Jewish Dialogue, begun seriously in the 1960s and the 1970s, has continued ever since, and its effects continue to register in school instruction today.[35]

Post-1967 changes in Jacob & Esau's roles in Israeli culture occurred in a radically different context. Whatever echoes could be heard in Israel of the 1968 events in Europe, they were swallowed by the canons' roar across the Suez Canal, fighting the War of Attrition. And, yet, the changes were momentous. Benjamin Tamuz's (1919–1989) novel *Jacob* signaled a paradigmatic shift in Hebrew literature.[36] Tamuz, a writer, artist and journalist, editor of *Haaretz*'s literary supplement, endeavored to make *Jacob* his masterpiece. Reflective in part of his own biography, the novel tracks the confrontation of its protagonist, Jacob Engelsohn-Malachi, with the meaning of "Israeli" and "Jew" over two decades. Told in multiple voices, it takes Jacob, in four scenes, from the early 1940s Jewish Underground in Palestine to the War of Independence and the State's early years to late 1950s Paris and back to early 1960s Israel. *Jacob* signaled Tamuz's break with his Canaanite youth – the endeavor to build a Hebrew nation dissociated from its Jewish past as part of a reawakening Semitic Middle East – and his transition to Jewish Diaspora cosmopolitanism.

Son of a poor émigré, Isaac, who never adjusted to Palestine, and the grandson of Abraham, who is buried in Europe, Jacob remains an outsider to the Israeli establishment, which Tamuz describes in an unflattering fashion. Jacob is a critic of the *Yishuv* and the State's commercial, military and intellectual elites, and searches for the meaning of the Jewish legacy in Palestine. Leah appears as his early partner,

---

[35] Kees de Kort, *Bibelbilderbuch* (Stuttgart: Deutsche Bibelgesellschaft, 1984); Peter Fiedler, *Das Judentum im katholischen Religionsunterricht* (Düsseldorf: Patmos, 1980); Ruth Kastning-Olmesdahl, *Die Juden und der Tod Jesu: Antijüdische Motive in den evangelischen Religionsbüchern für die Grundschule* (Neukirchener-Vluyn: Neukirchener Verlag, 1981); *Menschen suchen – Zugänge finden: Auf dem Weg zum einem Religionpädagogik verantworteten Umgang mit dem Bibel. Festschrift für Christine Reents*, ed. by Desmond Bel, Heike Lipski-Melchior, Johannes von Lüpke and Birgit Ventur (Wuppertal: Foedus, 1999); *Lernen auf Zukunft hin: Einsichten des christlich-jüdischen Gesprächs – 25 Jahre Studium in Israel*, ed. by Katja Kriener and Bernd Schröder (Neukirchen-Vluyn: Neukirchener Verlag, 2004).

[36] Tel-Aviv: Massada, 1971.

whom he declines to marry, and Rachel as his beloved wife who dies during the 1956 Sinai War. Esau is nowhere to be seen: The struggle Jacob leads is with himself, and with the compromises that Israeli existence forces on him. Zionist demands take over the space of Esau's challenge. Naturally inclined toward intellectual and artistic inquiry, Jacob feels, in his childhood and underground days, incapacitated by his diasporic origins, but he proves himself both a warrior in Israel's wars and a moneymaker in its emergent economy. In becoming a fighter and an entrepreneur, he tries to impress the elites and Rachel's relatives into accepting him, yet he feels alienated and repulsed by both.

Jacob endeavors to rid himself of the burden of his Ancestor's triumph over the Angel, and the dubious blessing that consigned Jews to a permanent struggle with God and other nations. Cognizant of the emptiness of Israel's pretension to be a light unto the nations, and of the world's special expectations of Jews, he wishes "that God forget us for a while [...] a few generations without miracles; let everything occurring to us be natural" (p. 149). In an interview with a French journalist, he still articulates the conventional Canaanite-Zionist narrative: The Hebrews (Zionists) are anti-imperialist and friendly to the Palestinians, the Arab states threaten Israel, and the Christians menace the Jewish Diaspora, as demonstrated in the Holocaust (pp. 202–210). His response to a French Jewish philanthropist, however, breaks decisively with this narrative and seems a moment of higher truth: "Judaism actually means exile, even voluntary exile. [...] The entire people as if chose [...] this most human mode of existence. But the exilic idea is very difficult to realize [and] when the survival instinct acts up, there is a pause: The Jews turn back to their physical homeland and try to settle it [...] It happened more than once: exile, return, exile, return [...] Once they had long enough [of] the [...] State [...] they go again on exile [...] ostensibly forced to, but, truly, this realizes the national will [...] When you talk of your hopes for the [...] new Israeli man [...] you speak merely of a brief pause in the journey" (pp. 173–74).

Torn between Israel and the Diaspora – "I have nowhere rest. Shall I try a different age?" (p. 257) – a final encounter with the Angel confirms to Jacob that there is no way of escaping the Angel's blessing and Jewish destiny: "Whoever has no intention of struggling and triumphing over God, has no right to see God or hear his voice" (p. 284). He walks away from the encounter bruised, accepting of the eternal Jewish struggle for universality and authenticity, and becomes a partner of Ruth, a war widow. This suggests tenuous hope for the future.

The questions Tamuz's Jacob posed to the Zionist project have been resolved over the next few decades with Jacob's transformation from a national into a universal hero. In the background lurks the growing Palestinian resistance to the post-1967 Israeli occupation and the Lebanon wars, which made Jacob's power and moral dilemmas palatable. The shift often expressed disenchantment with the national project, and yet Jacob remained a favorite poetic subject. Abraham and Isaac could never be relieved from the national burden: The *Aqedah* remained a major site for negotiating Israeli ambivalence about violence, sacrifice and death. Earthly Jacob, in contrast, cast a multifarious figure suitable for diverse poetic topoi, dissociated from the nation. His love for Rachel has remained a favorite motif, secure from political critique. Yet, the shift from Jacob as a national to a universal symbol dissolved the typology.[37]

Zionist discourse had long been ambivalent about Jacob, so querying the national myth typically began from within. The War of Independence poet, Ḥayim Guri, expressed exhaustion with war in "On Pillow Stone" (1968): "like Jacob, no ladder, like Jacob, no angels."[38] Moshe Dor's "Hard Water" (1989) recognized that, to triumph over the angel, Israel forewent gentleness and desire.[39] The a-national Israeli national poet, Yehuda Amichai, dissociated Jacob from the Zionist project in "Jacob and the Angel" (1975) by depicting the struggle as a love play between Jacob & Rachel, with a hint of a tragic end: Rachel is called back from the play, and the separation may allude to her death.[40] Dov Chomsky's, "Jacob's Dream" (1975), expressed disappointment that "Jacob's dream is no longer dreamt," and, in Rina Lee's "Jacob's Ladder" (2000), the ladder turns into an escalator, and desacralized modernity, a "swampy everydayness" (יון החולין), takes over biblical magic one.[41]

The love story between Jacob & Rachel has become progressively remote from the biblical narrative. For Aryeh Sivan (1969), "at the Well," and Yoram Beck (1989), "The Love of Jacob," Jacob is, above all, a lover, and, for Beck, also a symbol of creation and death.[42] Equally remote from the Bible are Ḥamutal bar Yosef's "Two Poems on Jacob" (1998), which focus on the loss of Joseph and Jacob's mourning.[43] Recollecting the struggle against the angel, Leah Senir, "At the Jabbok Crossing" (1988), introduced the Matriarchs to help Jacob, whereas Ḥavah Pinḥas Cohen's "That Man" (2000) and Admiel Kosman's "At the Jabbok Crossing" (1989) transmuted the fight, respectively, into an overnight struggle of a woman against a man – her call for help remains unanswered, and there is no river to cross and no escape: she is trapped – and Jacob's adventurous night wandering. The national struggle has been privatized and universalized.[44]

Dov Elbaum's *My Life with the Patriarchs* (2001) completed the cycle Tamuz's *Jacob* began by subverting both rabbinic and nationalist renderings of Jacob.[45] Rebelling against his own Ultra-Orthodox upbringing, Elbaum reshapes the

---

[37] Hilda Nissimi points out that similar universalization of biblical topoi has become typical of Israeli culture. Jacob is but one obvious exemplar.

[38] Malkah Shaked, *I'll Play You Forever: The Bible in Modern Hebrew Poetry* (Hebrew), 2 vols, (Tel Aviv: Yediot Aḥaronot, 2005), 1: 123–24, respectively.

[39] Ibid., 1: 128.

[40] Ibid., 1: 127.

[41] Ibid., 1: 126, 1: 134–36, respectively.

[42] Ibid., 1: 124–25: 1: 130–31, respectively.

[43] Ibid., 1:133.

[44] Ibid., 1: 127–28, 1: 133–34, 1: 131–32, respectively. Both Avraham Hus, "Penuel," 1: 132–33, where Jacob struggles with his own anxieties, and Meron Isaacson's "Jacob's Destiny" (1989), 1: 129, in which national and personal fears of war and death converge, are more traditional, but the trend of denationalization is evident.

[45] Hebrew. Tel Aviv: Am Oved.

Patriarchs as a dysfunctional family living in Hebron. Father-son relations are broken, husband-wife relations are exploitative, and brothers' relations are abusive. The *Aqedah* looms large: It traumatized Isaac, but Abraham refuses to talk about it. The family's great secret, revealed at the end of the book (and, yet, remaining incomprehensible), is Esau's severed head, which Isaac carries always in his red bag. Abraham appears as an imposing worldly operator, Isaac is a dejected and tormented soul, and Jacob is confused and unstable, a gifted, well-meaning womanizer who, recanting his ways in mid-life, ends up ludicrously as a *yeshivah* student.

The Patriarchs come into the protagonist's life at crucial junctures, and Abraham tries to help him sort out his troubles. In vain: Progress is only made when the protagonist releases himself from their tutelage, and finds an introspective companion who, like him, struggles to confront her legacy and articulate her problems. Revealing the patriarchal family's secrets, and talking about them, he also manages to restore a measure of harmony among the Patriarchs. The Bible and the Rabbis remain for Elbaum reference points but their stories have been so radically rewritten that they are recognizable only to connoisseurs. Therapeutic opening to individual and national traumas, concealed by the biblical and rabbinic stories of the Patriarchs, offers the single chance of overcoming them. Typology must be reinvented to overcome the past.

In the decades closing the second millennium, European and Israeli culture moved in both parallel and opposite directions in transforming Jacob's image. European culture embraced traditional Jewish Jacob and declared him European, indeed a universal model, whereas Israeli writers distanced themselves from Jewish Jacob, whether rabbinic or Zionist, and converted him into a lover, a mourner or a lost soul. Both the European and Israeli Jacobs were new and parted with Jewish and Christian typology.

14.4 The Jewish Esau, 1982–2016

Malka Shaked points out that Jacob's denationalization did not entail demythologization, and that the transformation of Esau's role in Hebrew literature has been far more radical.[46] In the past three decades, Esau has become, for the first time in history, a Jewish, or better, an Israeli hero. The transformation has occurred across the Israeli political and cultural spectrum, and Esau cast a very different hero for secular and religious, cosmopolitan and nationalist, Palestinian empathic and Greater Israel intellectuals.

Ḥayim Guri's gentle query of the typology in "Smell of the Field" (1985) signaled the trend's beginning from within the Zionist mainstream. He was full of sympathy for Esau, the man of the field, the less holy and less sophisticated protagonist, the more open, simple-minded and stronger one: "Someone must dash and retrieve the plunder. Always. And be less loved, strong and conned and undeserving

[46] Ibid., pp. 179–180.

of mercy."[47] His Esau, "a man of the field," is not a Zionist. Could he be a Palestinian farmer who does the dirty job and seeks to retrieve his plundered land? Unlikely, but Guri's poem reflects the reversal of traditional attitudes towards Jacob & Esau.

Esau's rehabilitation has cut across intellectual currents. Eli Alon and Aharon Amir, the former a Peace Movement activist, the second a leading Canaanite, are at one about Esau. In Eli's "Esau My Son, My Might and the First Sign of My Strength" (1989), Isaac tells Esau of his love for him. Jacob may be smooth and shrewd, and easily changes forms. A domestic schemer, he knows his way with God and women. Yet, Esau is Isaac's beloved man of nature, the rain and wind in the wheat-field. Let Jacob have his eloquence and God's blessing – Esau has the earth's blessing.[48] In Amir's "Esau's Epistle" (1986), Esau recognizes that Jacob is a smooth-tongued double-dealer, who is concerned most with his historical role – and he laughs at Jacob. He, Esau, is straight and approachable, at peace with his conscience and destiny. He declines to seek the gods' favors with altars, as Jacob does. Let Jacob and his appalling family receive the blessing, and gullible Isaac and pious Rebecca delight in it. He, Esau, is happy enough with his wives and herds on his mountain, enjoying the bountiful red lentil soup. Eli and Amir redeploy old Zionist criticism of the traditional Jew, which is not free of antisemitic stereotyping, to adorn Esau as the man of the Land. By 1989, however, the New Zionist Man had vanished, and pristine Zionism appears postzionist.[49]

Both Yehudit Kafri's "Esau" (1988) and Yoḥai Oppenheimer's "With the Cheated One" (2000) protest against the Bible's focus on Rebecca and Jacob, and its lack of concern for Esau. For Kafri, Esau runs with the deer in the field to ease his pain, and his cry can be heard all over, full of hatred and revenge. Then, the love of women consoles him, and, when he meets with his brother, he no longer claims the blessing, but cries, and the entire world cries with him. Leah Senir's "When Rebecca Died" (1995) reworks the *midrash* that Esau alone was there to bury Rebecca.[50] She recounts how Rebecca went through an undignified burial at night, in retribution for the human hurt she had caused. Jacob was abroad. Esau returned from the field to search for Rebecca and reburied her, "going to and fro weeping (הלך ילך ובכה) after her bed, as if lamenting his own life."[51] Senir exemplifies recent Hebrew poetry's deployment of biblical and rabbinic material to fashion radically new stories. Esau, the Jewish people's mortal enemy for millennia, is gone and gone is also rabbinic Jacob. The enemy has become a subject of love, and the Jewish Patriarch a subject of alienation.

If Tamuz's *Jacob* (1972) marked the beginning of the paradigm's reevaluation, Meir Shalev's novel, *Esau* (1991), written in the aftermath of the First Lebanon War (1982–1985) and the First Intifada (uprising, 1987–1991), represented its

[47] Ibid., p. 1: 137.

[48] Ibid., p. 1: 138.

[49] Ibid., pp. 1: 138–39.

[50] *Tanḥuma* (ed. Buber), ki-teẓe 4; Naḥmanides on Genesis 35:8.

[51] Ibid., 1: 137 (Kafri), 1: 140 (Oppenheimer), 1: 419 (Senir).

culmination.[52] Whereas *Jacob*'s reputation remained confined to literary circles, Shalev's novel became a best seller and was translated into several languages. Shalev was born in 1948 in Nahalal, an agricultural cooperative in the Yizrael Valley acclaimed for two generations of Zionist leaders, and grew up in Jerusalem. He was injured during his military service, studied at the Hebrew University, and became for 15 years a journalist and a TV and radio producer and host, before publishing his first novel in 1988. Since then, he has become one of Israel's most prolific writers, translated into several languages. His father, poet and novelist Yiẓḥak Shalev, was a spokesperson for Greater Israel, and the son is a Left Zionist. His early novels, including *Esau*, tell of Zionist agricultural settlers, founders and sons, their families and communities. The grandsons have unraveled Jacob & Esau from within mainstream Zionism.

Esau is a saga of three-generations of a family of bakers in a village near Jerusalem from WWI to the 1970s. The narrator is Esau, brother of Jacob, son of Abraham and Sarah, and uncle of Benjamin (who dies in the army), Romi, and Michael. Three short stories complement his narrative, telling of events Esau himself could not report. In the aftermath of WWI, Abraham Levy, a descendant of an old Jerusalem Sephardic family who lost his father in the War and saw his family reduced to poverty and starvation, marries a powerful, illiterate, beautiful and generous woman, Sarah, daughter of a Greek Orthodox pilgrim to Jerusalem, Michael Nazaroff, who converted to Judaism. Having lost their holdings in the 1927 earthquake, the family leaves Jerusalem for a nearby village, where they establish a bakery with the aid of a loan from Sarah's family. Abraham and Sarah's twin sons, Jacob & Esau, born about 1923, are both near-sighted, and Jacob monopolizes the single set of eyeglasses. Esau's blurred vision predisposes him to intellectual life, and puts him at a disadvantage in the competition with practical-minded Jacob for both the bakery's inheritance and the heart of beautiful Leah. In 1929, aunt Miriam, alias Tia Duduch, Abraham's sister, who lost her husband and was raped and maimed by Palestinian rioters in Jerusalem, joins the family with her injured son, Shimeon.

Abraham, an overworked and judgmental father and a spurning husband, prefers Esau to Jacob, whereas Sarah prefers Jacob. Having lost his finger in a bakery accident, the injured Jacob wins Leah, and is about to win the bakery, too, when the disgruntled Esau emigrates, about 1948 (Israel's foundation), to the U.S. to become a gourmet food columnist. Sarah curses him when he leaves home, telling him that "you will have no family of your own [...] no wife of your own [...] no child of your own [...] no land," which all comes to pass.[53] Esau does not find American life meaningful and remains attached to memories of the family. Sarah dies shortly after his departure, Jacob takes over the bakery and becomes as hardworking, unsociable and authoritarian as his father was, and, like him, grows also to despise his beloved wife, Leah. Determined to secure the bakery's succession, he imprints his own muti-

[52] Tel-Aviv: Am Oved. Trans. by Barbara Harshav (New York: Harper Collins Pubs., 1994).

[53] Ibid., p. 311.

lated hand and that of his eldest son in cement at the bakery's gate: "Abraham Levy and his son Benjamin, bakers, April 1955."[54]

About 1967, however, Benjamin, a soldier, falls to friendly fire. His traumatized mother, Leah, sinks into permanent sleep, and the bereaved Jacob, refusing public sympathy, madly impregnates her, to produce Michael, his youngest son. Aging aunt Duduch breast-feeds Michael, and Jacob, his father, neurotically guards and checks him daily. Yet, unable to feel pain, Michael's behavior appears otherworldly, and he clings to his sleeping mother, not quite the bakery owner Jacob envisioned. Older sister Romi, reminiscent in her look and her red complexion of grandmother Sarah and uncle Esau, visits her uncle in the U.S. and develops an intimate relationship with him (and many others). Hearing of the worsening condition of his ailing father Abraham, Esau returns after 30 years for a visit, which gives rise to the story of the Levy family. Jacob bitterly offers Esau both Leah and the bakery – both have become untenable projects. Artistically inclined, enterprising Romi photographs and exhibits the family's story.

The biblical narrative of Jacob & Esau serves as a prototype for the family saga, and Shalev repeatedly gestures to it, as if creating a grotesque counterpart. Yet, he subverts the narrative, and with it also the Zionist project. The biblical story has moments of horror and grace, and a grand vision of God's people underlies both. The fleeting moments of grace in Shalev's saga reflect natural enjoyment of sun, rain and the smell of bread; they do not involve human relationships. Love, family and community are an interminable story of misfortune, alienation, jealousy, violence, frustration and pain. They remain without real hope, and are putatively made light by their grotesque appearance. The Bible, the rabbis and Zionism are gone. What is to come is unclear.

The biblical Patriarchs were dignified, successful, devoted to their family and community, at peace with themselves, and in command of life and events. Shalev's Patriarchs are lonely, unsociable, tortured souls, overwhelmed by events, and failures. The rabbis idealized the Patriarchs' households and the Matriarchs' happy unions. In Shalev, the Matriarchs' marriages are pure hell, Jacob's courtship of beautiful Leah is a parody of the biblical Rachel romance, and Jacob's love towards Leah turns into hatred and virtual rape. God vindicates independent biblical Sarah: Abraham is told to obey her. Shalev's Sarah's strength and free spirit are oddly masculine and animal-like and they are rewarded with Abraham's resentment and rejection. The Zionist ethos elevated collective life in agricultural settlements, returning the Jews to their homeland. Shalev's protagonists are social outcasts leading hard lives, unable to enjoy the fruits of their labor, irreparably damaged by the violence of the Arab-Israeli conflict. The culturally sensitive Esau leaves for a life abroad, yet finds no home there, either. There is no home.

The Bible underlines the Hebrews' foreign origin, and the Patriarchs set a marriage boundary against the Canaanites, but the Patriarchs were, for the most part, welcomed in Canaan, and, in the rabbinic story, massive conversion took place in Abraham's household. The conversion of a foreigner into a Jew seemed seamless.

[54] Ibid., p. 106.

Shalev, too, undermines the boundary between foreigner and Jew, the Land and the Diaspora, but the union of foreigner and Jew is unhappy: Sarah comes from a family of converts, and Abraham's family rejects her. Esau and Romi (evocative of Rome) display Sarah's non-Jewish ethnicity and, not coincidentally, they are also intellectually inclined and have extended stays abroad.

Shalev parodies relentlessly the Jacob & Esau paradigm, and subverts the rabbinic typology, as well. Rabbinic Jacob, a *yeshivah* student, becomes a this-worldly baker, and rabbinic Esau, a warrior, becomes a writer. The brothers' conflict, it turns out, was not about the Jewish people's future but merely about property and a girl. In Genesis, Isaac blesses both his sons, but, in Shalev, the mother curses Esau and no one has a blessing. In Genesis, Jacob's limping is a mark of his struggle and triumph over the angel; in Shalev, his mutilated hand is a result of a bakery accident and a pathetic testimony of his love for Leah. The falling of Benjanmin is unheroic: He dies of friendly fire. With Rachel's loss, biblical Jacob has consolation in her two sons, Joseph and Benjamin, and Joseph's loss ends up in joyous reunification and salvation. There is no recovery for Jacob and his family after Benjamin's fall, only Romi's documentation of the family (and Zionism's) losses.

Contrary to some views, *Esau* contains no political critique of the Israeli disinheritance of the Palestinians.[55] Indeed, the Palestinians appear as murderous rioters: The Salomo family's overseer, Ibrahim, turns on it, kills the husband, rapes and maims the wife, and injures the child. (Palestinians who hid Jewish friends during the 1929 riots are nowhere in sight: It would be out of character for the novel.) The Jacob & Esau conflict is internally Jewish: Esau is more the diasporic Jew, and Jacob the Zionist, but there is no solution in Diaspora or Zionism: Diaspora is alienating, and the Zionist project, the work of uncultivated characters, a prolonged misery. Romi, who moves freely between Jacob & Esau, Jewish and non-Jewish, the Land and Diaspora, masculinity and femininity may represent the most attractive prospect, but hardly one that can sustain a national project. There is no resolution in sight.

Shalev inserts himself into the novel as one of Romi's boyfriends, and *Esau* does what Romi does: Document life in disarray. One may not fault an author for choosing the grotesque, decadent and magically realist to address a tragedy, but it does denigrate Zionism's attractive ideals without helping rectify the injustice it inflicted on thc Palcstinians. In sabotaging biblical, rabbinic and Zionist resources, Shalev leaves Israelis struggling for a future without means.

Orthodox Jews could not make short shrift of the traditional typology as Shalev did. Instead, in both Israel and the Diaspora, they have reworked the typology in surprising ways, initiating paradigmatic changes while retaining traditional appearance – subversion from within. Improved Jewish-Christian relations and, after September 2001, the putative conflict between "Islam" and "the West" influenced reinterpretation everywhere, but whereas Orthodox leaders in Europe and the US endeavored to redefine Orthodoxy's place in a multicultural society, Israeli ones were most responsive to problems emerging from the Israeli-Palestinian conflict.

[55] Yosef Oren, *The Writing as a Political Announcement* (Hebrew) (Rishon Le-Ẓion: Yaḥad, 1992), pp. 57–82.

The former reconfigured Esau as a respected, and even beloved Christian brother, whereas the latter declared him a lost Jewish brother.

Jonathan Sacks, Chief Rabbi of the British Commonwealth, adopted first a Taylorite conception of multiculturalism and the self. He argued that Jacob began as an insecure acculturated Jew, not knowing who he was, and ended up finding his Jewish identity as Israel.[56] The blessing Jacob stole was not his: It was material, and not spiritual. "As long as Jacob sought to be Esau there was tension, conflict, rivalry."[57] "Living by someone else's identity creates confusion, anxiety and insecurity."[58] The struggle with the angel was an internal battle with the fear of Esau, and conquering the fear and being himself, Jacob could reconcile with Esau. The two millennia Jewish-Christian feud was unnecessary. Being authentically Jewish in a pluralist society is a prescription for good Jewish-Christian relations. Psychologizing the wrestling with the angel, Sacks deployed Taylor to vindicate Moses Sofer and illiberal multiculturalism.

Yet, attending to Esau's humanity, Sacks moved on to liberalize pluralism, the Torah and Jewish Orthodoxy. By highlighting Esau's efforts to honor his parents and to win the blessing, and then his bitter cry and pleading, the Torah endeavored, says Sacks, to enlist our sympathies with Esau, and teach us a lesson about love and humanity. Esau was never rejected: Isaac's blessings of the land's fruits to his two sons were virtually identical.[59] Jacob was promised political supremacy, but only as long as his rule was just, and, in fact, Esau's political might would precede his, and Israel would be warned not to touch his territory and not to "abhor an Edomite, since he is your brother." "There is nothing in these commands to remind us of the eternal strife between the two nations predicted before their birth. [...] During the biblical era there were periodic tensions [...] but normatively, the Israelites were commanded to respect both the Edomites and their land."[60]

Esau was not chosen as a heir to the Abrahamic Covenant because "the twists and turns of covenantal history – exile, slavery, redemption and the wilderness years" – were not for him: When Jacob offered him, on his return from abroad, "please accept my blessing," Esau showed no sign of understanding that the Covenant was at stake. But "to be chosen does not mean that others are unchosen. [...] To be secure in one's relationship with G-d does not depend on negating the

---

[56] Charles Taylor, *Sources of the Self: The Making of the Modern Identity* (Cambridge, MA: Harvard University Press, 1989).

[57] "Be Thyself" (16 November 2013), Covenant and Conversation.

[58] "This Is Ours", in: *A Letter in the Scroll: Understanding Our Jewish Identity and Exploring the Legacy of the World's Oldest Religion* (New York: Free Press, 2000), pp. 204–215; quotation is on p. 207.

[59] The traditional Jewish reading of Esau's blessings (Genesis 27:39) is identical with the King James version "thy dwelling shall be the fatness of the earth, and of the dew of heaven from above" and not the modern standard version "your dwelling will be away from the earth's richness away from the dew of heaven above." Sacks needs to override Malachi 1:2 "but Esau I hated" and interprets "hated" as "less loved," as in Genesis 29:31: "God saw that Leah was hated." ודו"ק

[60] "Chosenness and its Discontents" (10 November 2007), http://www.rabbisacks.org/covenant-conversation-5768-toldot-chosenness-and-its-discontents/, accessed 16 July 2015.

possibility that others too may have a (different) relationship with Him. […] Love is not quantifiable. […] All are precious to G-d, each has his or her place in the scheme of things, each has his or her virtues, talents, gifts, and each is precious in the eyes of G-d. […] Esau too will have his blessing, his heritage, his land. […] There is a humanity here that defies all stereotypes and conventional categorisations. Esau is a child loved by his father and loving him in return. […] It is one of the Torah's most profound messages to humanity. How deeply does the world need to hear it today in an age of 'the clash of civilizations'"![61]

Sacks had to override significant textual and historical resistance, not to mention *Midrash*, to propound his liberal Orthodox view, and his Esau remains intellectually limited, conforming to traditional stereotypes. Yet, he managed to navigate the paradigm's dissolution and deployed Jacob & Esau to reshape a space for (post?) Orthodox Judaism in the unprecedented circumstances world Jewry faces today. This is no mean accomplishment.

Esau's rehabilitation extended even to Chabad *Ḥasidism*. The Seventh Lubavitcher Rebbe, Menaḥem Mendel Schneerson (1902–1994), recognized shortly after his arrival in America in 1941 that the U.S. represented a prospect for the reconstruction of Jewish life destroyed in Europe. He spoke of the U.S. as a benevolent superpower and a kingdom of grace. In a December 1984 address, commemorating the release of the First Rebbe from Russian prison, he presented the Chabad celebration, the American holiday of Thanksgiving and Ḥanukah as all conveying a single message of religious freedom.[62] He spoke of the partnership of Jews and non-Jews in advancing observance of the Noaḥide Commandments, the Jewish equivalent of Natural Law, and regarded it as preparing the world for the messianic coming, when Jews and non-Jews will worship together. Non-Jewish observers of the Noaḥide commandments were the "righteous among the nations."[63]

Schneerson elided ḥasidic and kabalistic traditions, which had established an insurmountable barrier between Jews and non-Jews. He could not leave the image of a bloodthirsty Christian Esau intact. In Chabad weekly homilies, the obscure Og, a legendary giant Amorite king, began taking Esau's place.[64] The First Rebbe had intimated that Esau represented the animal soul, and Jacob the human one.[65] In contrast, in his Saturday afternoon "conversations," Schneerson told his audience that both Jacob & Esau wished to worship God, but whereas Jacob was the perfect righteous *ḥasid*, who had no evil inclination, Esau was a knowing hunter, who knew how to conquer his instincts. This reflected the virtue of heroism, which he inherited from Isaac. Both strove to inherit this world and the world to come, but Jacob gave

[61] loc. cit.

[62] https://www.youtube.com/watch?v=bguqj7VaZxk, accessed 13 August 2015.

[63] *Hitvaadoyut* (1985–6), 2nd ed. (Brooklyn, NY: Vaad Hanochos, 1990): 3: 62–65.

[64] I heard such homilies delivered in Durham NC in 2002–03, using Rashi on Genesis 14:13 and *Genesis Rabah* 53:10; *Deuteronomy Rabah* 1:25.

[65] *Liqute Amarim: Tanya* (New York: Qehot Publication Society, 1956), pp. 13b–14a.

priority to the latter, suitable to his mode of worship, and Esau to this one, where Esau's mode is superior.[66] To be sure, Esau did end up going astray, and worshipping foreign gods, and the *ḥasidim* are holier than less observant Jews and Christians, but there is something about the arts of this world for the *ḥasidim* to learn from their less holy brethren.

Modern Orthodox Israeli rabbis were slower to respond to changing Jewish-Christian relations but with the growing confrontation between "Islam" and the "West," in the aftermath of September 11, and the American Christian Fundamentalists' support for Israel, change in Esau's image registered among them, too. Benjamin Lau, a moderate National Religious rabbi, surveyed, in 2009, Jacob & Esau's metamorphoses since biblical times. Jewish-Christian relations had a dark past, he said: The seventh stanza of the Ḥanukah song, Maoz Ẓur, called for the Edomites' destruction and was repressed for fear of the Church. But "our generation seeks reconciliation between the Church and Judaism, and finds itself attacked from the East by a third brother who also carries the single God's name. The seventh encounter of Edom (Esau) and Jacob finds itself in a common front against this brother, the goal of which is reinforcement of man's dignity and stature wherever he is. This is a different sort of encounter, in which God's words to Rebecca "two nations in your womb" are found out not as an invitation to a struggle, but an understanding that each one has its fullness and uniqueness, and together we shall need to aim for a life of peace."[67]

Lau's millennial revolution staged Christianity and Judaism against Islam in a war for peace. Similarly, Isaac Dov Koren, of the Center for Jewish-Christian Understanding and Collaboration in Jerusalem and Efrat, suggests a "New Covenant" between Christianity and Judaism.[68] Koren's mentor, Soloveitchik, limited Jewish-Christian dialogue in the aftermath of Vatican II to non-theological topics. The situation has since changed, says Koren, as both Popes and churches have changed their attitudes towards Judaism. They no longer represent Esau and Edom: They are now allies in fighting against Muslim extremists and leftist antisemites.[69] The two brothers have united against their uncle, Yishmael.

Yet, there are also other voices in the National Religious camp, ones wishing for the opening to Christian Esau to extend to understanding with secular Jews and the Palestinians. They are not, to be sure, predominant: More common is the evocation

---

[66] *Liqute Siḥot* (collected conversations; Yiddish), 39 vols. (Brooklyn, NY: Qehot, 2001), 20: 108–115.

[67] "Israel vs. Edom – Seventh Round" (Hebrew), *Haaretz* (4 December 2009), http://www.haaretz.co.il/literature/1.1293204, accessed 2 August 2013.

[68] "New Covenant" (Hebrew), trans. by Limor Riskin, *Maqor Rishon* (26 June 2015), http://musaf-shabbat.com/2015/06/26/ברית-חדשה-יצחק-דב-קורן/, accessed 14 August 2015. The article responded to the arson attack on the ancient Church of the Multiplication of the Loaves and Fish in Capernaum. For this, and many references to contemporary Israeli discourse on Jacob & Esau, I am indebted to my colleague and friend, Hilda Nissimi, Chair of the General History Department at Bar-Ilan University.

[69] Christianity in the Eyes of Judaism: Past, Present and Future (Hebrew) (Jerusalem: AJC, 2013).

of the Palestinians on *Shabbat Zakhor*, before the Purim holiday, when the command to eradicate *Amaleq* concludes the Torah reading.[70] But, like Sacks, these voices articulate exegetical innovation, liberal spirit and hope, however slim. David Bigman, head of the Maale ha-Gilboa Yeshivah, suggests, following U.S. Orthodox rabbi Yiẓḥak Hutner, that Jacob had first to recognize Esau's greatness, to see both his human and divine virtues, before he could reconcile with him: This was the significance of his struggle with Esau's angel. Jacob repented and went through true reconciliation.

Yaakov Nagan, of the West Bank's Tequa and Otniel *yeshivot*, activist in the "Jerusalem Peacemakers" and "Interfaith Encounter Association," suggests that the Jacob saga conveys the ambiguous legacy of pain in Jewish history. Nothing came to Jacob easily: His very name connoted struggle, and he had to go through two millennia of exile and struggle with the angels to emerge as a fighter and return to the Land. Understandably, but all the more regrettably, Jacob remained suspicious of Esau, and protective of his identity. His closeness became a hindrance: He missed the chance of marrying Dinah to Esau and was punished for it. Jews should remain open and build good and honest relations with the Other.[71]

On the Religious Left, the Other's identity remains amorphous: Esau could be the non-observant Jew, the Palestinian, or the Christian. The rehabilitation of Esau extended, however, to the Center and Right, too. In the aftermath of the Israeli Disengagement from Gaza, which pitted major segments among the Settlers (residents of the post-1967 Occupied Territories) against the Israeli government, the latter sensed that they had lost the Israeli public. The more thoughtful among them responded by seeking new ways of building bridges to the lost brothers. Esau became a lost Jewish brother, spurred by Jacob with horrible consequences.

Reem Hacohen, head of the Otniel Yeshivah, overhauled the interpretive tradition by reconfiguring *midrashim* and kabalistic homilies favorable to Esau. The Jacob-Esau story could have ended differently, he says, with the two building together, with Rachel and Leah, the House of Israel. Jacob failed the test of brotherly love. Isaac was no dupe for loving Esau: Esau's holiness was greater than Jacob's. He wished to tithe hay and salt because, in the other world, still life is holy, too, and both Isaac and Esau wandered in other worldly fields. Jacob's theft of the birthright was tantamount to murder, a design on Esau's soul, as it excluded him from the Jewish people. Jacob was punished by exile, and Laban cheated him in measure for measure. He committed a second transgression by marrying Leah, who was destined for Esau. Returning to the Land, he sent for Esau, seeking *tiqun*, but failed a third time by refusing him Dinah. The result was the Shekhem rape and massacre, and later, Amaleq, and yet later, the Temple's destruction. God did not wish it this way, but Jacob was allowed to go the way he wished. In Hacohen, the Jewish-

[70] I have experienced it personally in both the U.S. and Israel.

[71] "A Blessing Emerges From Pain: Between Jacob & Esau," http://www.kipa.co.il/jew/pash/42/46979.html, accessed 20 July 2015.

Christian story has vanished. Jacob & Esau have become an internal Jewish story.[72]

Reem Hacohen's message, like that of Nagan and Bigman, is one of peace, reconciliation and sorrow for the profound loss occasioned by the Jacob & Esau conflict. His Esau is mystical and holy, a product of Kook's kabalistic tradition, a figure very new, and yet recognizably Jewish. In contrast, the Esau emerging from the discourse of former Gush Emunim (bloc of the faithful) leader, Joel Bin Nun, and the head of the flagship Har-Eẓion *Hesder* Yeshivah (a yeshivah combining advanced Torah study with military service), Yaacov Medan, reflects religious nationalism's profound transformation of Jewish values. Both have had it with rabbinic Jacob, the studious *yeshivah* student, whose eyes are set on the world to come, and who, facing the prospect of Esau's attack, prays, offers gifts and plans to escape to save himself. Esau, a warrior and an empire builder (in Nimrod's image), suddenly appears appealing.[73]

No wonder, says Bin Nun, that farmer Isaac, incognizant of Rebecca's oracle, and recognizing the need for household protection, assumed that Jacob, a tent-dweller and a Torah student, would need Esau's protection, and that he planned on bestowing political power on Esau. It was Jacob using his voice but offering Esau's hands that opened Isaac's eyes to the successful combination, and to the inferiority of Esau's bloody power politics. But Jacob first needed to learn to fight, and he avoided it until the last minute, attempting to run away from Esau. Encountering the angel and overcoming him, Jacob finally became the synthesis of the two brothers, the way the Jewish State should be, built with Esau's might but with Jacob's justice.[74]

Yaacov Medan goes further. He is exasperated with the secular Israeli intelligentsia, the *belle âme* who supported the Disengagement, and with the Ultra-Orthodox *yeshivah* students shrinking military service, and he was horrified by the Jewish State's uprooting of Settler families from Gaza. Israel's actions had biblical prece-

---

[72] "Contradictions in a Single Womb," *Maqor Rishon*, 44 (27 November 2009); idem, "How was the Torah written?" Lecture in Herzog College (published 7 October 2009), http://www.ynet.co.il/articles/0,7340,L-3786363,00.html, 39:42–53:32, accessed 3 January 2010 and 23 July 2015.

[73] Genesis 10:8–10 speaks of Nimrod, the King of Shinar, as "mighty on earth" and "a mighty hunter before the Lord." *Midrash*, *Targum* and medieval commentators cast Nimrod as Abraham's persecutor (*Bereshit Rabba* 38:13, 42:4), a rebel against God (*Bereshit Rabba* 23:7, 37:2–3; *Pseudo-Jonathan* on 10:8), and an empire builder (Naḥmanides on 10:9 and Radaq on 10:8). They coupled the two hunters, Nimrod and Esau, whether typologically as tricksters (*Bereshit Rabba* 37:2), or as competitors over Esau's garments: *Bereshit Rabba* 63:13 has Nimrod coveting Esau's garments; *Pirqe de-Rabi Eliezer* 24 has Esau kill Nimrod and take over his garments. Yiẓḥak (Max) Danziger's sculpture Nimrod (1939) became the embodiment of the Canaanite vision of a revival of a secular Semitic culture throughout the Middle East. My thanks to Ari Dubnov, Shai Ginsburg and Hilda Nissimi for drawing my attention to Bin Nun and Medan's unlikely predecessors.

[74] "Esau's Hands – Jacob's Voice" (Hebrew), *Shenaton Amit* (1998): 13–20, repr. in: *Pirqe Avot* (Alon Shevut: Tevunot, 2003).

dents, he says: Jacob's sons did not shrink from selling their brother into slavery, and Judean dynasties murdered their siblings. Medan wishes to build bridges to the less acculturated Israelis, even non-religious ones, who patriotically carry the defense burden – to Israel's Esaus. He points out that the Sages enumerated two great merits of Esau: Honoring of parents and love of the Land. Esau wanted the blessing so he could settle the Land. In contrast, Jacob, who, unlike Abraham & Isaac never fought, escaped abroad. Esau conquered his wish to avenge the blessing's theft out of respect for Isaac, and Jacob was never in danger. If family members had an appropriate appreciation of Esau's character, Jacob would not have needed to run abroad. Jacob's sons, in contrast, did not conquer their vengeance, dishonored and hurt their father, murdered and betrayed a brother. Notwithstanding a two millennia rabbinic tradition, Esau was an honorable warrior, not unlike King David. Isaac's alliance with him was one of farmer and warrior, the alliance sustaining today the Jewish State.[75]

Moshe Lichtenstein, son of Aharon, the founder of Har-Eẓion Yeshivah, and representative of the old guard synthesis of Brisk intellectualism and moderate religious nationalism, responded testily to Medan, reaffirming the two millennia rabbinic tradition seeing Esau as both Jew and non-Jew, and Jacob as the Covenant and the Land's legitimate inheritor.[76] For all his erudition and firm grasp of rabbinic Esau's ambiguity, Lichtenstein is fighting a rear guard action against the inexorable logic of religious ethnonationalism. Bin Nun and Medan's daring innovation is matched only by their message's appalling character. After two millennia of Jacob's voice crying out over Esau's crimes, the progeny of the Canaanites' Nimrod, the warrior and empire builder, has become the religious national ideal. Jacob has turned into Esau. "Woe, what has become of us!"[77]

[75] Yaacov Medan, "Esau' Merit," *Daf Qesher* 522 (December 2006), http://www.etzion.org.il/he/פרשת-תולדות-זכותו-של-עשו, accessed 16 July 2015. As this article is sent to print, Ari Geiger of Bar-Ilan University has drawn my attention to Medan's latest pronouncement: "Isaac & Rebecca, Jacob & Esau": https://www.youtube.com/watch?v=JhemCzKWP_4&feature=em-uploademail, accessed 16 October 2016. Medan strives to reconcile rabbinic Esau with his own: He now faults Esau for having left the Land and for marrying Canaanite women. His identity is once again ambiguous: Medan identifies him with both the Settlers' unruly Hilltop Youth and the non-Jews serving in the IDF. Harnessing his military virtues, exemplified by Judah and King David, is essential to Jewish power, as Jacob's pacifism will not sustain an Israel extending to the Euphrates and overruling Abraham's multiple descendants. Innovative and dangerous.

[76] "Esau as Brother, Esau as Goy" (Hebrew) 2006, http://www.etzion.org.il/he/פרשת-וישלח-עשו-כאח-עשו-כגוי, accessed 22 July 2015.

[77] *The Koren Mesorat Harav Kinot* (Hebrew & English) (Jerusalem: Koren Publishers, 2011), p. 51. The *Qina*'s first stanza, "Remember, God, what has happened to us" is based on Eikha 5:1.

Chapter 15
Charity, Logic, and Para-Clericalism: Rereading Agassi on Koestler on Galileo

Maurice A. Finocchiaro

15.1 Introduction

In 1971, Joseph Agassi published an essay entitled "On Explaining the Trial of Galileo" (Agassi 1971b). At the time, I did not read it; nor, indeed, did I know of its existence or publication. The main reason for my oversight was that at the time I had no special interest in the trial of Galileo. Yet, there were other reasons why I could and should have learned about it.

In fact, I was already well acquainted with some of Agassi's works, especially his historiography of science. Indeed, I had written my doctoral dissertation on the problem of explanation in historiography of science, and several chapters in it consisted of a critical examination of his views on the subject, as found in his *Towards an Historiography of Science* (Agassi 1963, cf. 1987, 2008, 119–242). And in 1971, I was in the process of revising and polishing this dissertation, for publication as a book, which occurred soon thereafter (Finocchiaro 1973, cf. 1974, 1976a, 1987).

There were thus two principal reasons for the great relevance of Agassi's essay to my own work. One was that the essay is largely a critical examination of the account of Galileo's trial advanced by Arthur Koestler in *The Sleepwalkers* (1959); and this critical examination is largely a critique of Koestler's historiographical practices. The second reason was that, as the title of Agassi's essay explicitly suggests, the main historiographical practice examined by him is that of explanation: that is, how Koestler goes about explaining why Galileo's trial took place, why in 1633 the Roman Inquisition tried and condemned him; and how one could give a better explanation than Koestler's.

I did not discover Agassi's 1971 essay until about 10 years later, that is, not until I myself developed a special interest in Galileo's trial. What happened was the following.

M.A. Finocchiaro (✉)
University of Nevada-Las Vegas, Las Vegas, NV, USA
e-mail: maurice.finocchiaro@unlv.edu

N. Bar-Am, S. Gattei (eds.), *Encouraging Openness*, Boston Studies in the Philosophy and History of Science 325, DOI 10.1007/978-3-319-57669-5_15

In my work on history-of-science explanation, one of the case studies I had examined was Alexandre Koyré's (1939) explanation of Galileo's initial failure and eventual success in discovering the law of falling bodies. Following up on this case study, I become more involved in Galilean scholarship. Moreover, another one of my scholarly specialties was logical theory, and in this field I followed an approach which stresses practical reasoning and argument in natural language, and which goes under such labels as informal logic, applied logic, practical logic, critical thinking, and argumentation theory (cf. Finocchiaro 2005a). It did not take me long to discover, both for pedagogical and theoretical reasons, a uniquely important and importantly unique source for such material; that is, Galileo's *Dialogue on the Two Chief World Systems, Ptolemaic and Copernican*, which is full of arguments and counter-arguments for and against the motion of the earth. Combining these two lines of research, in the 1970s I became very involved in analyzing this Galilean work from the point of view of informal logic and critical thinking. This led me to publish in 1980 a book whose content is well described by the title, *Galileo and the Art of Reasoning: Rhetorical Foundations of Logic and Scientific Method.*

Now, in 1980, a world-historical development added a twist to the just-mentioned personal development. In the fall of that year, Pope John Paul II announced the appointment of a Vatican commission to re-examine Galileo's trial and the Galileo affair; and it was clear that in so doing the pope was acting on some of what he had said on the topic a year earlier, on the occasion of his speech to the Pontifical Academy of Sciences, at its commemoration of the centennial of Albert Einstein's birth (cf. Finocchiaro 2005b, 338–58). I knew, of course, that the 1633 condemnation of Galileo was occasioned by his publication of the *Dialogue* a year earlier. But this Galilean work was the subject of my decade-long research project and of my just-published book; and my book could boast of being the most detailed and comprehensive analysis of Galileo's *Dialogue* ever attempted or published. As a result, I felt that my deep understanding of the *Dialogue* enabled me to make some contribution to the interpretation and evaluation of the trial and the affair. Thus began my long-standing scholarly involvement with this topic.

When, finally, in the early 1980s, I read Agassi's 1971 essay, two things impressed me favorably. One was what may be called the principle of charity, which he advocated, practiced, and applied in interpreting and evaluating Galileo's trial, as well and in criticizing Koestler's account. Independently of Agassi, I had myself become a follower of the principle of charity, having learned it primarily from Michael Scriven (1976, 71–73), in the context of informal logic and argumentation theory; and I had extensively applied the principle in my approach to the analysis of fallacies in general, and in my analysis of Galileo's *Dialogue* (Finocchiaro 1981, 1980, 340–41, 378–79). Thus, in my first paper on Galileo's trial, presented orally in 1982, and then published in 1986, I was eager, pleased, and excited to note this convergence (Finocchiaro 1986, 246).

A second convergence between myself and Agassi was the appreciation of Galileo from the point of view of informal logic, critical thinking, and argumentation. I could not help but being impressed upon reading in Agassi's essay so many explicit statements and examples to this effect (to be elaborated below). And as mentioned earlier, my *Galileo and the Art of Reasoning* had independently articulated such a critical interpretation of Galileo's work.

There is a third aspect of Agassi's 1971 essay that converges with my own work. This concerns what I call the para-clerical approach to questions of the relationship between science and religion in general, and of the Galileo affair in particular. This is an approach that is secularist-minded and yet tries to provide an intellectual service to religion, based on the belief that topics like the Galileo affair are too important to be left to religious believers themselves. I did not become aware of this convergence until very recently, for the simple reason that I myself became aware of this aspect of my own work only in the past 5 years roughly (Finocchiaro 2015b). However, this is something which I now find striking in Agassi's 1971 essay, which is significant, and which I am ready to elaborate. But before elaborating this convergence, and the others mentioned earlier, it will be useful to have a summary of Galileo's trial.

15.2 Galileo's Trial

In 1633, at the conclusion of one of history's most famous trials, the Roman Inquisition found Galileo guilty of "vehement suspicion of heresy."[1] This was a specific category of religious crime intermediate in seriousness between formal heresy and mild suspicion of heresy. He had committed this alleged crime in a book that defended Copernicus's hypothesis of the earth's motion and implicitly denied the scientific authority of Scripture. This work had been published the previous year, under the title *Dialogue on the Two Chief World Systems, Ptolemaic and Copernican*.

This verdict was accompanied by several penalties: Galileo had to immediately recite an "abjuration" of his erroneous beliefs; the book was banned; and he was condemned to house arrest for the rest of his life.

This condemnation was the climax of a series of events starting in 1543, when Copernicus published an epoch-making book, *On the Revolutions of the Heavenly Spheres*. Copernicus advanced a novel argument for the idea that the earth moves, with daily axial rotation and yearly heliocentric revolution. His argument amounted to showing that the known facts about the motion of heavenly bodies could be explained better (and in quantitative detail) on the basis of the geokinetic, heliocentric hypothesis, as compared with the geostatic, geocentric view.

Although novel and significant, Copernicus's argument was hypothetical and inconclusive. Moreover, the idea faced powerful objections, which had been elaborated for two millennia by Aristotle, Ptolemy, and many others. That is, the earth's motion seemed epistemologically absurd because it contradicted direct sense experience. It seemed empirically false because it had astronomical consequences that

[1] For details, documentation, and references on the story sketched here, see Finocchiaro (1989, 2005b, 2014b).

were not observed, e.g., the phases of Venus and the annual parallax of fixed stars. It seemed mechanically impossible because it contradicted the available laws of motion and the clearest observations of moving bodies, such as the vertical direction of free fall. And it seemed religiously heretical because it contradicted biblical texts, such as the Joshua miracle, when God stopped the sun to prolong daylight (Joshua 10:12–13).

Thus, Copernicanism attracted few followers. Galileo himself, in his early career, was *not* one of them.[2] His research focused on physics rather than astronomy; he was critical of traditional physics and was constructing a new one; he had intuited that the Copernican hypothesis was more consistent with the new science of motion than was the geostatic theory; but he felt that, overall, the arguments against Copernicanism were stronger than those in favor.

However, in 1609–1610, with the newly invented telescope, Galileo made several startling discoveries: the moon's surface was full of mountains and valleys; innumerable other stars existed besides those visible with the naked eye; and the planet Jupiter had four moons revolving around it. Soon thereafter, he also discovered the phases of Venus and sunspots.

The new telescopic evidence removed most observational-astronomical objections against the earth's motion and added new evidence in its favor. Galileo's assessment was now that the arguments for the earth's motion were collectively stronger than those for the earth's rest. However, he realized that this did not mean that the issue was settled, because there was still some astronomical counter-evidence (mainly, the lack of annual stellar parallax), and because the physical-mechanical objections had not yet been explicitly refuted.

Furthermore, Galileo perceived the potentially explosive character of the scriptural issue. In fact, for several years he did not get involved. When he finally did, he was careful enough not to publish his criticism of the scriptural objection, but to circulate it privately, in the form of letters: in 1613 to his former student Benedetto Castelli, and in 1615 to the grand duchess Christina.

Galileo's criticism, although complex and liable to misunderstanding, was logically compelling, rhetorically persuasive, and theologically sophisticated.

However, despite winning the intellectual argument, Galileo lost the practical struggle. In 1615, after some formal complaints were filed against him, the Inquisition launched an investigation. The proceedings lasted about a year, and the results were the following. In March 1616, a decree was issued by the Index, the department of the Church in charge of book censorship: it declared that the doctrine of the earth's motion was scientifically false and theologically contrary to Scripture; and it temporarily prohibited Copernicus's book.

Although Galileo was not mentioned at all in the decree, in private he was given a warning. This warning exists in two versions. One is in a certificate given to Galileo by Cardinal Bellarmine, a member of the Inquisition and one of the most authoritative churchmen of the time; it states that he had informed Galileo that the

[2] For details and documentation of this thesis, and criticism of alternatives, see Finocchiaro (2010, 37–64).

earth's motion could not be defended as true, but merely as a convenient hypothesis to save the appearances. The second version is in an unsigned note written by a clerk and found in the file of Inquisition trial proceedings; it states that the Inquisition's commissary (Michelangelo Seghizzi) gave Galileo the special injunction not to hold, defend, or teach in any way the earth's motion. The difference between these two versions is that the special injunction adds a more stringent prohibition to the one mentioned in Bellarmine's warning: besides being prohibited, like other Catholics, to defend the physical truth of Copernicanism, Galileo was also forbidden to defend it in any way whatever.

For the next several years, Galileo kept relatively quiet. Then a new opportunity arose in 1623, when one of his admirers became Pope Urban VIII. From several indications, Galileo got the impression that, if he exercised the proper care, he could publish a book on the earth's motion. Thus, he wrote the *Dialogue*.

The book was obviously a discussion of the earth's motion, but Galileo had done several things to avoid trouble. For example, as the title suggests, he wrote the book in the form of a dialogue among three characters: a Copernican named Salviati, a Ptolemaic named Simplicio, and an intelligent and curious layperson named Sagredo. Moreover, the conversation amounted to a critical examination of all the evidence for and against the earth's motion; the arguments on both sides were presented, analyzed, and evaluated. Galileo tried his best to carry out the evaluation fairly. The arguments for the earth's motion turned out to be stronger than those against it. This was at worst an *implicit* defense of the *probable* truth of Copernicanism. Galileo's gamble was that friendly Church officials would not blame him for this, because they would recognize that the book could be interpreted as a *hypothetical discussion* of the geokinetic opinion, and that its defense was not explicit, and so that is was within the spirit of Bellarmine's warning.

Galileo's effort misfired not because it was foolish, but because in 1632 the special injunction came to the surface, and from its point of view any defense of the earth's motion by Galileo was prohibited. The book's publication thus led to a trial during which he denied receiving the special injunction, but admitted receiving and violating Bellarmine's warning, insisting that his violation was unintentional. This admission spared Galileo more drastic punishments, such as being burned at the stake.

15.3 Koestler's Account

There is much in Agassi's account of Galileo's trial that is correct, insightful, and important. And the same applies to his critique of Koestler's account. Let us highlight some of these themes, beginning with the latter.

As Agassi points out, according to Koestler, "Galileo fought not for the sake of truth, but because he was pathologically unable to avoid any quarrel, accept any compromise, or admit any error" (138).[3] In other words, allegedly, Galileo was "an extremely unpleasant and arrogant person" (142). This psychological interpretation provides Koestler with the explanation of the tragic outcome; that is, "Galileo's character was the cause of the trial" (143). Koestler tries to support this explanation with two main allegations.

The first allegation pertains to the first phase of the trial, in 1615–1616. According to Koestler, after some heresy charges against Galileo were formally filed in February and March 1615, the Inquisition investigated the matter, and by the autumn of that year it had found them to be baseless and had essentially dismissed them. However, in December 1615, Galileo went to Rome on his own initiative, and ended up staying till June. His goal was to defend himself and Copernicanism, and indeed for about 2 months he was very active, meeting as many officials and laypersons as he could, and attending or organizing many social gatherings to discuss pro- and anti-Copernican arguments. Koestler claims that "his method was to make a laughing stock of his opponent—in which he invariably succeeded, whether he happened to be in the right or in the wrong [...] before answering the opposing reasons, he amplified them and fortified them himself with new grounds which appeared invincible [...] It was an excellent method to score a moment's triumph, and make a lifelong enemy. It did not establish his own point, but it destroyed his opponent's" (Koestler 1959, 452).

Koestler's second allegation pertains to the 1633 proceedings. These were triggered by the 1632 publication of the *Dialogue*. For Koestler, in this work Galileo was resuming his (previously failed) pro-Copernican campaign with the same tactics and techniques that had been unsuccessful earlier. That is, this book supposedly insulted the intelligence of Galileo's contemporaries with fallacies, sophistries, confusions, and errors, and offended their sensibilities with insults, lies, deception, dishonesty, hypocrisy, and false promises. For example, Koestler thinks that Galileo's simplified presentation of the Copernican doctrine "is downright dishonest [...] was not a simplification but a distortion of the facts, not popular science, but misleading propaganda [...] Moreover, he keeps silent about the fact that the Tychonic system fits the phenomena equally well" (Koestler 1959, 476–77).

Agassi does show some appreciation of Koestler's account, primarily because his image of Galileo is lively, de-mythologized, and forceful (138, 163). However, Agassi is mostly critical of Koestler, and here I can only focus on the criticism.[4] Let us look at the points that are most telling.

[3] Here and elsewhere, references to Agassi (1971b) are given by means of page number(s) in parenthesis.

[4] I have elaborated my own critical appreciation of Koestler's account in Finocchiaro (2005b, 306–17). Similarly, but in a reverse manner, it would be beyond the scope of the present essay for me to elaborate a critical appreciation of Agassi's 1971 essay, that is for me to give an elaboration that was both critical and appreciative; rather, I can only focus on a positive appreciation of Agassi's account. If I had to be critical, I would elaborate some difficulties involving dates, names, and documents.

15.4 Agassi's Criticism

Agassi is absolutely right that Koestler's explanation of the trial is "an exaggeration" (143) of the truth that Galileo had quarrelsome tendencies which caused him difficulties. Now, exaggeration is a serious flaw by any standard, and so there might be no obvious need to elaborate. However, it turns out that this criticism has some revealing ramifications, which should not be ignored.

To begin with, it is worth quoting an aphorism formulated in 1785 by one Girolamo Ferri, in the context of his criticism of a new Galilean myth that was emerging then: "the exaggeration of truth is a lie" (Ferri 1785, 54; cf. Finocchiaro 2005b, 161). However, to go back to Agassi, it is important to know and understand that, in the words of Jarvie and Laor (1995, xix), "one theme that recurs in his [Agassi's] writings and his discussions is that of steering the middle way and keeping a sense of proportion [...] for example, he spells out and rejects polarized views of the nature of man [...] he sets out the extreme views on sociological method and shows they share an underlying premiss which, when negated, permits moderate mediation between the extremes. Agassi's use of this *via media* is confessedly Lévi-Straussian. He invites us to break through intellectual myths that come in poles, sweeping our searchlight over the whole area of reality between the poles." One work where Agassi focuses on such an approach is his *Towards a Rational Philosophical Anthropology* (Agassi 1977; cf. Finocchiaro 1978).

More to the point, this approach is not merely Lévi-Straussian, but also Galilean, at least as I interpret Galileo. In fact, in a number of works, I have argued that one of the most central methodological features of Galileo's work is something which I call judicious-mindedness, or judiciousness, or, more simply, judgment. This concept means the willingness and ability to be impartial, balanced, and moderate; that is, to avoid one-sidedness (by properly taking into account all distinct aspects of an issue) and to avoid extremism (by properly taking into account the two opposite sides of any one aspect). For example, Galileo was judicious in this sense with regard to the contrast between the two chief world systems (Ptolemaic and Copernican), between religion and secularism, between epistemological apriorism and empiricism, between mathematicism and qualitative considerations, between authority and independent-mindedness, and so on (cf. Finocchiaro 1980, 145–66, 2010, 121–34, 2014a, 259–80).

Another criticism advanced by Agassi is that Koestler's portrayal of Galileo is one-dimensional and static rather than multi-faceted and dynamic, insofar as Koestler attributes to Galileo a character than does not change, whereas it underwent significant developments corresponding to the three main stages of his life. That is, according to Agassi, "Galileo held three different positions in different times: first, he concealed his Copernicanism; then—when he was about fifty—he defended it with some caution; his great battle for it took place when he was about seventy. Koestler's view of Galileo's motives and purpose is offered as one explanation of all three phases" (138). More generally, "it seems that Koestler's historical figures suffer from the same defect as his fictional figures: they never change their characters" (142).

This criticism is perceptive, correct, and important. And the same applies to the positive and constructive thesis that underlies the criticism. Methodologically speaking, Koestler's one-dimensionality regarding Galileo's character may be viewed as an instance of injudiciousness. And substantively speaking, such a three-fold development in Galileo's life is a crucially important historical fact. I myself have had occasion to elaborate it elsewhere (cf. Finocchiaro 2010, 37–64), and have summarized it above. It is worth repeating here, with some further encapsulation.

First, there is the pre-telescopic stage, when Galileo's cognitive stance toward Copernicanism may be described as partial, indirect, and implicit pursuit (which is also to say, qualified rejection). Then we have the full-blown middle period from 1609 to 1616, which constitutes a qualitative change from the preceding one; which may be described as full-fledged, direct, and explicit pursuit of Copernicanism (which is also to say, qualified acceptance); and which contains varying and increasing degrees of endorsement or commitment. Thirdly, there is a problematic post-1616 stage, which may be described as again one of indirect and implicit pursuit; now the main stumbling blocks were external, namely the warning he had personally received from cardinal Bellarmine, and the public condemnation of Copernicanism issued in the Index's decree. As suggested earlier, the *Dialogue* fits this third stance, for it can be interpreted as a critical examination of all the astronomical, physical, and philosophical (but *not* the theological) arguments for and against the opinion.

An additional flaw which Agassi finds in Koestler's account is what may be called bias. For example, with regard to the scientific aspect of the story, the pattern displayed by Koestler is the following: "every time Galileo's defenders are apologetic, Koestler stresses the point which caused them discomfort; and every time the Catholic apologists show a weak point in Galileo's scientific views, Koestler follows suit" (140). And with regard to the political context, "the chief Catholic technique used on the political side of the story is that of pooh-poohing one document and stressing another; and Koestler follows suit" (140).

This is a perceptive, accurate, and fair criticism. The only thing I would want to add here is that this is precisely what we would expect from an injudicious thinker like Koestler. Indeed bias is an aspect of injudiciousness.

Similar considerations apply to another one of Koestler's failings, namely unfairness. However, here the issues get more profound and far-reaching. For, the underlying principle, presupposed and advocated by Agassi, and violated by Koestler, is extremely important. It deserves a label: the principle of fair-mindedness.

As Agassi points out, "Koestler […] can be unbelievably unfair to Galileo, to the point of viewing even his (alleged) anticipation of criticism—the thing which every good author does—as a vice" (151, n. 2). Indeed, in discussing the controversy over floating bodies in 1611–1612, Koestler complains that Galileo "employed his favorite trick of anticipating the peripatetics' arguments, building them up in a mock-serious manner, and then demolishing them with glee" (Koestler 1959, 428). This is the same complaint which, as we saw above, Koestler advances in his account of Galileo's discussions in Rome in the winter of 1615–1616.

Now, at the interpretive level, Koestler is giving an accurate construal of this element of Galileo's methodology. Indeed, I would add that this is an important feature of the approach he follows in the *Dialogue* (cf. Finocchiaro 2014a, 259–64). And Galileo was even reflectively aware of this fact, for he had the occasion to explicitly formulate the principle of fair-mindedness with words that are eloquent and memorable: "when one presents arguments for the opposite side with the intention of confuting them, they must be explained in the fairest way and not be made out of straw to the disadvantage of the opponent" (cf. Finocchiaro 1989, 278, 2014b, 129).

However, as Agassi correctly suggests, at the evaluative level Koestler's complaint, which portrays this technique as illegitimate, is not only wrong, but perversely wrong; that is, it is the opposite of the truth. Let me elaborate.

This technique is a very sophisticated and powerful method of argument. It has nothing to do with what is imagined by Koestler, namely with the method of ridiculing opponents by means of the trick of first persuading them of one thing and then proving to them the opposite. On the contrary, the Galilean technique of strengthening objections before refuting them shows that the objections, although invalid, are serious, important, and plausible, and therefore that the opponents who believe the contrary thesis are reasonable people. Far from ridiculing opponents, this is a way of ennobling them, paying them respect, and enhancing their standing and credibility. Galileo's technique is, in fact, the anti-thesis of the widespread practice of demonizing one's opponents, or differently expressed, it is the anti-dote to the straw-man fallacy. Finally, fair-mindedness may be regarded as an application of judiciousness to the contrast between a proponent and an opponent in the context of the practice of argumentation.

I am adapting this (meta)argument, as well as the label "fair-mindedness," from the literature on informal logic, critical thinking, and argumentation theory (Scriven 1976, 166–67; Fisher 1991; Fisher and Scriven 1997, 90–91, 137–43; Finocchiaro 2013, 89–91,117–22, 153–54). However, the key idea of this principle has also been elaborated by many others under different terminology, for example, by John Stuart Mill under the heading of freedom of thought, discussion, and argument, and the role of such freedom in the search for truth (cf. Finocchiaro 2013, 178–91).

So far, I have elaborated and strengthened several of Agassi's major criticisms of Koestler. However, it should be obvious that my discussion has not been merely a negative exercise, but rather has revealed several constructive and positive principles which may be properly be attributed to Galileo; which Agassi seems to advocate; and which appear to be highly plausible. That is, we have seen that, firstly, Agassi faults Koestler's account in general as full of exaggerations, and this criticism may be conceptualized as revealing Koestler's injudiciousnes, as well as the value of the principle of judiciousness. Secondly, Agassi criticizes Koestler's account of Galileo's character as one-dimensionally static, rather than dynamically three-fold, and such one-dimensionality may be seen as an instance of injudiciousness. Thirdly, Agassi questions Koestler's account as biased with regard to the issue of Galileo vs. the Church, and hence as one-sided, and thus again as a violation of the principle of judiciousness. Fourthly, Agassi objects to Koestler's unfairness in assessing Galileo's method of argument, with regard to the technique of anticipating and strengthening the objections to one's own views; this reveals Koestler's own violation of the principle of fair-mindedness, and thus again of the ideal of judiciousness; and we also have a reaffirmation of the importance of this ideal.

15.5 Agassi's Approach

Despite this positive and constructive value of Agassi's criticism of Koestler, at this point it is useful to focus more directly and explicitly on Agassi's own account of Galileo's trial. However, it will be obvious that, just as the critiques discussed above have underlying positive counterparts, the positive account discussed below has critical implications. These would take the form that Koestler fails to appreciate various aspects of Galileo and his trial which Agassi's account utilizes, elaborates, and appreciates.

A key theme in Agassi's explanation of Galileo's trial is the following. It occurs too frequently to be ignored. For example, with regard to the first phase in 1615–1616, consider "his theological campaign for Copernicanism. Galileo had a very good chance of winning it; but the point to stress is that we ought [...] to notice that he thought he had a good chance to win the battle. (Indeed at one point he thought the battle was already won.) Trying to explain his behavior thus may be more interesting than viewing it as irrational, as Koestler does, by saying that he was obsessed with his need to quarrel regardless of the consequences. In brief, Koestler does not attempt a rational reconstruction of the battle as it appeared before it was over" (157).

And with regard to the second phase of the trial in 1632–1633, "the story of the writing and the publication of the *Dialogue* becomes amply clear and reasonable if we do not use hindsight, if we forget the ensuing catastrophe when trying to reconstruct the way Galileo, the Pope, and others, looked at the situation before they knew of the grave consequences of all they did. (Even the fact that they were apprehensive, all of them, indicates that none of them acted unreasonably.) But it all becomes clear when one assumes their intentions to have been honorable" (159). Furthermore, "as long as there is no evidence against Galileo's confessions, they ought to be taken very seriously as possibly true; and the same ought to be said of Pope Urban VIII and of all others involved. It is quite possible that both Galileo and Urban VIII were irresponsible rascals, as Koestler asserts; but this should not be our starting point" (158). Or again, "the people involved knew what they were doing—they were too cautious and apprehensive not to—and were acting in good faith and not from any personal motive" (160).

This is a very commendable and fruitful approach. Although Agassi does not use this terminology, I think that here we are dealing with the principle of charity.

In the context of the theory of reasoning and argument, Scriven (1976, 71–72) formulated this principle as follows: "The Principle of Charity requires that we try to make the best, rather than the worst, possible interpretation of the material we are studying […] It doesn't mean letting people off the hook entirely […] What the Principle of Charity does mean is that 'taking cheap shots' is something we shouldn't waste much time doing […] It tells you that you want to interpret the argument's meaning in whatever way makes the most sense and force out of it, because otherwise, *it can easily be reformulated slightly in order to meet your objections*."

And in the context of my own critical analysis of the arguments in Galileo's *Dialogue*, I have had the occasion to formulate it as follows: "The principle of charity states that in criticizing an opposing argument or view, one should first formulate a charitable interpretation and then make the latter the target of one's criticism; a charitable interpretation is one that portrays the original argument or view in a reasonably favorable light, as possessing some strength that must be taken seriously, and as free of insignificant or trivial errors which do not affect the main issue" (Finocchiaro 1997, 376).

More generally, I have also had the opportunity to formulate and apply this principle in the context of my criticism of a recent account of Galileo's trial that reaches new heights of uncharitableness. This account is that advanced in Thomas Mayer's *The Roman Inquisition: Trying Galileo* (2015). Besides many other objections, I also point out that whereas my account (along the lines of the one sketched above) conforms to the principle of charity, Mayer's account violates it: "my account is also confirmed by the fact that it can help us understand Galileo's subsequent behavior and the Inquisition's subsequent proceedings, up to the sentence and abjuration of June 22, 1633. This is in accordance with a general principle of charity that aims to portray all parties as more or less reasonable, or at least to maximize the total reasonableness displayed on all sides. On the other hand, Mayer's account is further disconfirmed by the fact that his […] thesis forces him to attribute all kinds of administrative incompetence and legal improprieties to the officials involved, in both the earlier (1615–1616) and the later (1632–1633) proceedings, and a nearly-insane arrogance and recklessness to Galileo. In short, for Mayer, Galileo as well as almost all clerical officials (except for commissary Seghizzi) were unreasonable, incompetent, and/or unlawful, whereas in my account Galileo and almost all officials were reasonable, competent, and lawful (the only and main exception being Seghizzi)" (Finocchiaro 2016, 57; cf. Finocchiaro 2015a).

A second fruitful theme found in Agassi's account is the thesis of Galileo as a skillful practitioner of applied logic, critical thinking, and argumentation. In fact, in the context of his discussion of Galileo's methodology as a synthesis of apriorism and observation, it is impossible to overlook Agassi's talk of Galileo's "clear thinking" (149); of his "critical mode of thinking" (152); and of "his realization of the importance of logic; his considering criticism and clarity to be essential for scientific discourse" (153). Similarly, Agassi regards Galileo as "a forerunner of Descartes in viewing clarity and distinctness […] but also simplicity […] as criteria of truth" (150); as "a teacher of critical thinking" (151, n. 2); as a "clear and critical thinker" (162); and as "the greatest logical mind of his age" (155).

Agassi finds such applied logic and critical thinking primarily in Galileo's early works *On Motion* and *On Mechanics*; in his *Sidereal Messenger*; and in his

Discourse on Floating Bodies. The only thing I would add here is that such Galilean traits are also predominant in the *Dialogue* and the *Two New Sciences*.

This Agassian thesis is not just a substantive thesis about the nature of Galileo's scientific work. It also reflects a methodological orientation, according to which a scholar seeks to interpret and assess scientific works from such a point of view. In fact, Agassi himself, in a completely different context, has made a contribution that reflects such a point of view. I am referring to his book on Michael Faraday (1791–1867), the English scientist best known for his discovery in 1831 of electro-magnetic induction; that is, the phenomenon of generating an electric current in a wire by moving it relative to a magnet (Agassi 1971a; cf. Finocchiaro 1976b). This work may be interpreted as a case study in the general interpretation of the history of science as the history of arguments. In fact, the book is full of discussions seeking to ascertain the reasons which Faraday and his contemporaries had for their frequently conflicting views; and Agassi also attempts to evaluate those views in terms of those reasons, rather than by reference to the latest physics textbook. And at the end of the book, Agassi expresses explicitly his awareness that he is practicing and advocating the history of science as the history of arguments.

A third key aspect of Agassi's account is a methodological and historiographical orientation which I shall label and conceptualize as para-clericalism, or a para-clerical approach. Again, this is not Agassi's terminology, and so this orientation is not as explicit as the theme of critical thinking, which we have just seen to be very obvious; nor is such para-clericalism as easily discernible as the charity principle, which we saw earlier is not explicitly discussed by him but corresponds closely to his talk of rationality and reasonableness. Nevertheless, para-clericalism may be seen to be no less central, once we understand what is meant by this term.

To begin with, the para-clerical approach should not be confused with either the pro-clerical or anti-clerical orientations, which are very common, as Agassi's account points out. It is also clear that he is critical of both of these approaches, and such double-edged criticism is a hint of para-clericalism. Second, it is useful to point out that the term para-clericalism is meant to bear some analogy (with regard to the prefix) to para-medical and para-military contexts.

Additionally, conceptually speaking, as I have had occasion to elaborate before, the para-clerical approach "views controversial topics involving the relationship between science and religion from a perspective that is secular-minded, but appreciative of religion, and yet conducted in the belief that such topics are too important to leave to religious believers" (Finocchiaro 2015b, 125). In the context of the Galileo affair, para-clericalism commonly involves "offering advice to the Church, crediting the Church for uncommon reasons, defending the Church from common criticism, believing in the superiority of religious outsiders over insiders, and intellectual courage" (Finocchiaro 2015b, 137).

Finally, by way of illustration, I have had occasion to elaborate a methodological appreciation of the work of two scholars, Paul Feyerabend and John Heilbron (Finocchiaro 2015b). However, I must add that in their case my methodological appreciation happens to be combined with substantive criticism, whereas with regard to Agassi my position seems to be a combination of both methodological and substantive appreciation. In fact, Feyerabend has (in)famously argued, on secularist grounds, that "the Church at the time of Galileo not only kept closer to reason as defined then and, in part, even now; it also considered the ethical and social consequences of Galileo's views. Its indictment of Galileo was rational and only opportunism and a lack of perspective can demand a revision" (Feyerabend 1993, 125). And Heilbron ends his secularist biography of Galileo with an argument predicting that sooner or later the Catholic Church will canonize Galileo as a saint. Indeed the book's last sentence is the following rhetorical question: "Who can doubt that within another 400 years the church will recognize Galileo's divine gifts, atone for his sufferings, ignore his arrogance, and make him a saint?" (Heilbron 2010, 365).

In light of such a framework, several aspects of Agassi's account become very revealing. The most general and significant one is his thesis that, with regard to the science vs. religion issue, Galileo was a "Catholic reformer" (139, 156). That is, as Agassi formulates it elsewhere, Galileo "was an ardent Catholic who sincerely wished to cooperate with the Church as far as his conscience permitted him […]. The Church wished to embrace Copernicus' system, in order to retain its intellectual hegemony […] it wanted to make the transition from Aristotelianism as smooth as possible and … Galileo was supposed to help in this transition" (Agassi 1957, 238).

More specifically, Agassi makes and applies a crucial theological distinction that helps to make sense of the Church's warning to Galileo in 1616 and of his subsequent behavior resulting from it. This is the distinction between believing or accepting a doctrine as true on the one hand, and discussing or supporting or criticizing it on the other. The distinction is reflected in one of the most important documents of Galileo's trial, namely the minutes of the Inquisition meeting of February 25, 1616, when Pope Paul V issued some conditional orders with regard to the resolution of the Galileo case:

> The Most Illustrious Lord Cardinal Millini notified the Reverend Fathers Lord Assessor and Lord Commissary of the Holy Office that, after the reporting of the judgment by the Father Theologians against the propositions of the mathematician Galileo (to the effect that the sun stands still at the center of the world and the earth moves even with the diurnal motion), His Holiness ordered the Most Illustrious Lord Cardinal Bellarmine to call Galileo before himself and warn him to abandon these opinions; and if he should refuse to obey, the Father Commissary, in the presence of a notary and witnesses, is to issue him an injunction to abstain completely from teaching or defending this doctrine and opinion or from discussing it; and further, if he should not acquiesce, he is to be imprisoned. (Finocchiaro 1989, 147, 2014b, 102)

And to this we should add that a week later Bellarmine reported to the Inquisition that he had given the warning and that Galileo had acquiesced; thus, there had been no need for the commissary to intervene and issue the more stringent special injunction.

On his particular issue, Agassi speaks precisely as one would expect from a paraclerical scholar: "Now (being an agnostic and a Jew) I am a person poorly qualified to explain the fact that Catholics are permitted by their Church simultaneously to believe in a doctrine and to criticize it; yet I wish to state categorically and most emphatically that such is the case, no matter what is the doctrine in question, no matter how far-reaching its consequences" (157–58).

Similarly, Agassi's para-clerical orientation emerges with regard to the issue of Galileo's criticism of the biblical objection to Copernicanism, which started the first phase of the proceedings. According to Agassi, "had Galileo's theology been accepted by the Church directly from him, he might have become a saint, rather than Bellarmine. There was a chance that this could happen, both in 1616 [...] and after the death of Bellarmine and the Pope, when the new Pope, Urban VIII, encouraged Galileo to write his *Dialogue*. But even if he had no chance against Bellarmine, his sincerity and courage, as well as his important contribution to Catholic theology, ought to be appreciated (and may be appreciated in the future, even by Rome; remember Joan of Arc!)" (156). Some of these words (from 1971) are prophetic in light of what started happening about a decade later, with the so-called rehabilitation of Galileo by Pope John Paul II in 1979–1992 (cf. Finocchiaro 2005b, 338–58).

Finally, with regard to the 1633 proceedings, occasioned by the *Dialogue*, Agassi's para-clerical explanation is the following: "I do think that throughout the *Dialogue* he meant what he said and that the constrictions he accepted were those he freely chose as a religious man. It seems to me that the Church was ready to grant more freedom to scientists provided it could save its face, and that Galileo, the ardent Catholic and leading scientist, was eager to help. Why then did the plan misfire? Because, I suggest, of the enthusiastic reception of he book: the Copernicans took the permission to publish the *Dialogue* as the Church's admission of failure" (Agassi 1957, 239). Moreover, "Galileo's enemies successfully spread the rumour that Simplicio was [a caricature of] the Pope; and then, even though the Pope knew it not to be true, he had to do something about it: this was not a period in which jokes against the Pope were harmless to the Church. So the Church had to assert its authority against Galileo" (161).

15.6 Conclusion

In this chapter, I have examined Agassi's 1971 essay on explaining the trial of Galileo. I have found it to be full of substantive and methodological insights.

Substantively speaking, Agassi is essentially correct in advancing the following important theses, and in criticizing Koestler for his failure to appreciate them. Galileo had quarrelsome tendencies, but not a compulsive obsession to get into trouble. His career displays three importantly different behavior patterns, and not a single one. A characteristically Galilean method of argument was the technique of acknowledging and strengthening objections against his own views, before criticizing them; and such a technique is highly commendable, and not at all deplorable. More generally, Galileo was a highly skilled and effective practitioner of the art of reasoning, argument, applied logic, and critical thinking, and not at all a sophistical or incompetent arguer. And Galileo was also a kind of Catholic reformer, whose arguments on the separation of Church and science, and on the non-scientific authority of Scripture, could have prevailed in his own time, but had to wait centuries before they received the proper recognition by the Church.

Methodologically speaking, Agassi displays a number of fruitful historiographical orientations and approaches; and these are attitudes which by and large he shares with Galileo and which are constantly disregarded or violated by Koestler. Judiciousness is the avoidance of one-sidedness and of extremes. Charity is the ideal of maximizing the rationality or reasonableness of one's own opponents. Critical thinking is the skill of reasoning focused on the identification, interpretation, evaluation, analysis, and self-reflective presentation of arguments. And para-clericalism is an approach to questions of science vs. religion that is secular-minded, but appreciative of religion, and yet conducted in the belief that such questions are too important to leave to religious believers.

It should be obvious from the preceding that I share these substantive theses and methodological attitudes. Indeed, my scholarly work on Galileo seems to have been largely an elaboration of such theses and attitudes. However, despite such convergence, my Galilean scholarship was created independently of Agassi's influence. On the other hand, it is perhaps because of such a convergence that I have been able to extract these themes from his essay. All this strikes me as strange and curious, especially in light of my early acquaintance with and study of his historiography of science. However, I shall postpone to some other occasion any further elaboration and embellishment of this biographical and autobiographical curiosity.

References

Agassi, Joseph. 1957. Duhem versus Galileo. *British Journal for the Philosophy of Science* 8: 237–248.

———. 1963. *Towards an historiography of science*. The Hague: Mouton.

———. 1971a. *Faraday as a natural philosopher*. Chicago: University of Chicago Press.

———. 1971b. On explaining the trial of Galileo. *Organon* 8: 137–166.

———. 1977. *Towards a rational philosophical anthropology*. The Hague: Martinus Nijhoff.

———. 1987. Twenty years after. *Organon* 22 (23): 53–61.

———. 2008. *Science and its history*. Dordrecht: Springer.

Ferri, Girolamo. 1785. Apologie de Galilée. *Mercure de France*, January 8, 54–63.

Feyerabend, Paul K. 1993. *Against method*. 3rd ed. London: Verso.

Finocchiaro, Maurice A. 1973. *History of science as explanation*. Detroit: Wayne State University Press.

———. 1974. Toward a Crocean history of science: Criticism of Agassi's critics. *Rivista di studi crociani* 11: 142–161.

———. 1976a. A Crocean philosophy of the historiography of science. *Physis* 18: 274–286.

———. 1976b. Review of Agassi's *Faraday as a natural philosopher*. *Journal of the History of Philosophy* 14: 488–489.

———. 1978. Review of Agassi's *Towards a rational philosophical anthropology*. *Isis* 69: 437–438.

———. 1980. *Galileo and the art of reasoning: Rhetorical foundations of logic and scientific method*. Dordrecht: Reidel.

———. 1981. Fallacies and the evaluation of reasoning. *American Philosophical Quarterly* 18: 13–22.

———. 1986. The methodological background to Galileo's trial. In *Reinterpreting Galileo*, ed. W.A. Wallace, 241–272. Washington: Catholic University of America Press.

———. 1987. The methodology and philosophy of history of science. *Organon* 22 (23): 99–114.

———., trans. and ed. 1989. *The Galileo affair*. Berkeley: University of California Press.

———., trans. and ed. 1997. *Galileo on the world systems*. Berkeley: University of California Press.

———. 2005a. *Arguments about arguments*. New York: Cambridge University Press.

———. 2005b. *Retrying Galileo, 1633–1992*. Berkeley: University of California Press.

———. 2010. *Defending Copernicus and Galileo*. Dordrecht: Springer.

———. 2013. *Meta-argumentation*. London: College Publications.

———. 2014a. *The Routledge guidebook to Galileo's Dialogue*. London: Routledge.

———., trans. and ed. 2014b. *The trial of Galileo*. Indianapolis: Hackett.

———. 2015a. Essay-review of Mayer's *The Roman Inquisition: Trying Galileo*. *Reviews in History*. http://www.history.ac.uk/reviews/review/1836. Accessed 1 Oct 2015. doi: 10.14296/RiH/2014/1836.

———. 2015b. A para-clerical approach to the Galileo affair and to science vs. religion. *Science, Religion and Culture* 2: 125–142. doi http://dx.doi.org/10.17582/journal.src/2015/2.4.125.142.

———. 2016. Galileo's first confrontation with the Inquisition (1616). *Galilaeana* 13: 29–60.

Fisher, Alec. 1991. Testing fairmindedness. *Informal Logic* 13: 31–36.

Fisher, Alec, and Michael Scriven. 1997. *Critical thinking*. Point Reyes: Edgepress.

Heilbron, John L. 2010. *Galileo*. Oxford: Oxford University Press.

Jarvie, I.C., and Nathaniel Laor, eds. 1995. *Critical rationalism, metaphysics and science: Essays for Joseph Agassi*. Vol. I. Dordrecht: Kluwer.

Koestler, Arthur. 1959. *The sleepwalkers: A history of man's changing vision of the universe*. New York: Macmillan.

Koyré, Alexandre. 1939. *Etudes galiléennes*. Paris: Hermann.

Scriven, Michael. 1976. *Reasoning*. New York: McGraw-Hill.

Chapter 16
What Is the Problem? Understanding the History of Ideas

William Berkson

"What is the problem?" If you ever were a student of Joseph Agassi, you remember this question fondly. Agassi expanded and refined Popper's characterization of a "critical tradition," in his philosophical writings, historical studies, and as a teacher. His challenge to his students was both to understand the growth of knowledge in past traditions, and to then take on the task of advancing one—in no matter what field. And "What is the problem?" was always the first question to begin understanding the development of ideas and traditions.

By the time Agassi left Popper's side at the London School of Economics, he had a well-developed view of the nature of critical traditions, both within science and outside of it. I first learned his views on the nature of critical traditions in the early 1960s, at the University of Illinois. Later, as a PhD student under Popper, I found that similar ideas were common currency in the Philosophy department at the LSE, though not always expressed with the clarity that I found in Agassi's writings of the time. I have been grateful for the tools developed collaboratively by Popper and Agassi, tools for understanding intellectual history. Personally, I have used these tools every day since studying with Agassi—now for over 50 years.

A key development in the evolution of these tools, in my understanding, was Agassi's extension of Popper's views on the growth of science to the logic of other traditions beyond science proper. Popper had seen scientific problems as created by observation reports that contradict existing theories. The typical scientific problem, for him, was a phenomenon like the Michaelson-Morley experiment, which couldn't be explained successfully by earlier theories, and seemed to contradict them. Popper envisaged the development of science as a series of conjectures and refutations, to use his book title. A new refutation would pose a new problem: how to modify or replace past theories, with a new theory that would explain both the phenomena that the old theory explained, and the new, recalcitrant case. This leaves out problems of

W. Berkson (✉)
Jewish Institute for Youth and Family, Virginia, USA
e-mail: berkson@mentsh.com

N. Bar-Am, S. Gattei (eds.), *Encouraging Openness*, Boston Studies in the Philosophy and History of Science 325, DOI 10.1007/978-3-319-57669-5_16

phenomena that are not yet connected to any deductive theory—such as many chemical and biological phenomena have been—but this situation of contradiction does seem to be the kind of problem that Popper generally had in mind.

In 'On the Nature of Scientific Problems and their Roots in Metaphysics', Agassi (1964) placed metaphysics squarely within a similar, though not identical, kind of critical tradition. Though metaphysical theories are not testable, they do contain ideas about the fundamental constituents of the world—such as that the world consists of matter and force—which can be, and have been, used as ingredients in specific, testable theories. Influenced by Agassi's analysis, I wrote my PhD thesis under Popper about the development of field theory in the nineteenth century; it became my first book, *Fields of Force: The Development of a World View from Faraday to Einstein* (Berkson 1974). I found that the successes and failures of the specific, testable theories of fields and particles did reflect on the merits of the contending metaphysical world pictures, even though the scientific theories weren't dispositive as to which was correct. (The debate is still not resolved.)

Agassi also the made the standards for advancement of a critical tradition more realistic than Popper's sharp "demarcation." For example, whereas Popper had forbidden *ad hoc* modifications of a theory in response to its being contradicted by evidence, Agassi allowed the for gradual process, which corresponds to actual history. He explained in 'Towards a Theory of *Ad Hoc* Hypotheses' (Agassi 1975) that there are *degrees* to which a modification is *ad hoc* or innovative. And so long as modifications are made less and less *ad hoc*, or, in other words, more and more independently testable over time, such progress should be welcomed. In fact, much of Agassi's voluminous work in philosophy can be seen as continuing develop new insights into how problem-solving, critical traditions grow.

Generally speaking, Agassi tried to identify how empirical testing could come to have a bearing on problems that are not strictly speaking scientific. W.W. Bartley, III, in his book *Morality and Religion* (Bartley 1971), along this line, included ethical traditions in this framework. He pointed out that once we want ethical principles to promote human wellbeing and reduce suffering, then the real-world impact on human welfare of proposed social arrangements or ethical guidelines does a relevance to ethics. Observation and experiment are then relevant to evaluating the ethical impact of social institutions and guidelines for action.

In general, critical traditions consist of problems, tentative solutions, arguments pro and con the adequacy of the proposed solutions, and new problems. Agassi's characteristic question 'What is the problem?' is a first tool, a nutcracker, so to speak, to begin to understand a critical tradition, and specific claims within it. What problem was a new idea or social change intended to solve? And a key step in understanding the problem when we identify it is to look at the "problem situation"—how that problem arose when past solutions failed to meet a desired standard. The history of a critical tradition is, then, a chain of problems, solutions, and debates. For myself, I never feel I understand a theory or a social phenomenon until I understand its history as a history of successful and failed efforts to solve problems. When entering any new field, the key questions are: "What are the important problems?" "What are the alternative solutions?" and "What are the arguments pro and con the alternative solutions?" Studying the problems and their history in depth usually involves going back at least one step into how the current problems were created.

This kind of analysis that I learned from Joseph Agassi and from Popper I have found the single most valuable tool I learned in my education. Furthermore, I find very healthy Agassi's effort to find locate anchors in empirical testing for any kind of history of problem-solving. For ultimately, it is specific observations of the "who, what, when, where" of events that keep us grounded in reality. This kind of analysis of the history of problems, alternative solutions, problems the solutions face and arguments pro and con can in fact go beyond traditions involving empirical testing. As Menachem Fisch has pointed out, the Talmud (which Agassi studied in his youth) contains a critical tradition, with ideas that are tested and modified through critical debate. There, the 'tests' are proof texts from the sacred literature, which are held to be authoritative, rather than the observation reports which are authoritative in science. And just as in science, where there is a process for challenging observation reports (through further testing and theorizing) there is a process for debating correct interpretation of a sacred text.

Because humans are problem-solving creatures, this type of analysis of traditions is helpful even beyond critical traditions. There are situations where a social solution is proposed, runs into problems, and the solution is abandoned and replaced without the influence of critical debate. An example of this kind of analysis is that done by the late Berl Gross, with guidance from Agassi. In his book *Before Democracy: A History of Error and Justification* (Gross 1992), Gross describes the decentralized, ethically just society envisaged by Moses, and its eventual abandonment by the ancient Israelites, and replacement with monarchy. Gross views the history as tragic, because of its failure to critically assess the pros and cons of Moses' social vision, and revise, rather than abandon it.

In *Learning from Error: Karl Popper's Psychology of Learning*, (Berkson and Wettersten 1984), a book that John Wettersten and I collaborated on, we explore a similar phenomenon. Popper himself, ironically, never critiqued in depth the relationship of his methodological views and the psychological theories of the Würtzburg school that he learned from Karl Bühler. He simply tried to jettison the relevance of psychology for methodology, rejecting any connection as "psychologism." In the process he in fact conflated psychology and psychologism. In my commentary on the Jewish ethical classic *Pirke Avot: Timeless Wisdom for Modern Life*, (Berkson 2010), again, on a very different topic, I asked and tried to answer: What problem were the Jewish Sages trying to solve? And from the point of view of human welfare in our changed modern society, where do their guidelines for living need to be extended or changed?

Recently, on a topic far afield from any of these, in an effort to understand the failings of journalism in the U.S. in the election of 2016, I again turned to Agassi's questions—which again proved invaluable.

Thomas E. Patterson, at Harvard's Shorenstein Center for Media, Politics and Public Policy, in collaboration with the Swiss firm Media Tenor, analyzed 'How the Press Failed the Voters' (Patterson 2016) in the 2016 Presidential election in the United States. Their report has documented and given some insightful analysis of the failings of the press in the recent election. Coding all reports of any significant length in five major national newspapers and five TV networks, they found that only 10% of the reports were on policy issues and 40% were on the "horse race" issue of who is winning. Further, the reports on the fitness for office were overwhelmingly negative on both sides—87% negative to 13 positive on both sides—as were the relatively few stories on policy (10%), in which hostile criticism from the other side predominated.

If we look at the functioning of the press from the point of view of good democratic governance according to the standards Popper laid out in *The Open Society and Its Enemies*, the pattern that Patterson documents is alarming. Popper held that a key benefit of the democratic electoral process is to remove bad leaders without violence. And ideally there is also, through public debate and voting, feedback on the success or failure of public policies. This feedback by public debate and by voting is analogous to testing in science, and key to improving laws and government policies. For both functions—to replace bad leaders with better ones, and to be informed about policy—the electorate is dependent on the news media. And the numbers in Patterson's report document a failure of the news media to inform the electorate.

The reporting of equally negative information about both sides might seem a kind "balance," but as Patterson explains, such a balance is not in fact fair or objective. When both sides are in reality *not* equally bad, the equal negativity about both sides produces a "false equivalence." Such false equivalence hinders voters' judgment, and makes them more vulnerable to emotional appeals. It can help elect a "charlatan"—evidently Patterson's (and my) view of what actually happened in 2016. Further, as Patterson explains, when reporting on government is overwhelmingly negative, it unfairly supports the U.S. 'conservative' view that government has pernicious effects and is always inferior to private markets. To actually inform the electorate fairly, then, the press has a responsibility to report the overall context of negative events, and including the positive benefits that government programs are successfully delivering. Furthermore, the 40% of reporting which was on the "horse race" reports tell the electorate nothing about either the candidates' fitness for office, nor about the merits of their policies. By Popper's standards, this reporting is little more than a parlor game, and irrelevant to good democratic governance.

It is on the issue of policy coverage that good tools for understanding critical traditions become critical. Of course, the mere 10% of election coverage spent on policy is inadequate to convey to voters a meaningful evaluation of public policies. The question is: "Why is there so little coverage of policy?" Part of the answer of course has to do with the usual imperatives of journalism. Existing public policies are usually long standing and proposals for changes are usually also are not totally new, but simply variations on recent proposals. The fact that that policy debates are long standing in fact means that they may be even more important than the questions of candidates' fitness. For it is the policies that will usually most affect the country, and the debates are long lasting. However, because the policy issues are not new, they don't fit into the category of surprising or threatening news. Also, expression of opinion in itself is not news, as it doesn't necessarily contain any information that would help the public make a better choice.

The most helpful coverage of policy issues falls into a category which The New York Times calls "news analysis." This is a category that is not a straight "who, what, when, where" report on events, nor a mere expression of opinion; it is rather an objective analysis giving deeper understanding of a current event or issue. In the case of policy issues, such analysis would give objective information bearing on what has worked in the past and not, and what current proposals are better and worse. So the question of the lack of policy coverage becomes, "Why is there such little *news analysis* of the policy issues at stake in elections?"

Part of the answer is painfully evident from the discussions of political journalists on television: in general, the political journalists don't understand the policy issues to any depth, including the pros and cons of the different options. There are some specialist journalists who cover economics, the military, etc., who do understand the issues in depth, but the mass of journalists covering the election are not comfortable with assessing what the specialists write. And so, as a rule, the political journalists do not pick up on the questions that specialists have, report further on them in the mass media, and challenge the candidates with them. Instead, as Patterson documents, they generally report without context the hostile criticisms of each side against the another, however sound or unsound they are.

The reason for the failure to understand the state of the policy debates is not, I believe, that political journalists are stupid, but rather that understanding the debates is very difficult and time consuming, and the journalist lack the tools to do it. In fact, looking at the policy debates lands us in very challenging problems of evaluating claims in social science. And here the questions that I learned from Agassi, and answers to them, would be most helpful both to journalists, and to the general public. What problems were current policies intended to address? Where had earlier solutions fallen short? What are the current alternative policy solutions to the problems the current policies are suffering from or failing to address? Where does the evidence stand for and against the likely success of the current proposals?

In trying to assess the current status of policies and policy proposals we run directly into problems of the testability of theories in the social sciences. Again, the issue comes down to how observation can test theory, in diverse fields of learning. In the social sciences, people have tended to look for correlations that confirm their theories. The problem is, as Popper noted, confirmations alone tell us very little. It is possible to confirm contradictory theories, which cannot both be true. For example, a person claiming that raising interest rates above 5% helps the economy can point to instances where it did help. But equally a person claiming that lowering them below 5% can also point to confirmations. As a result of looking for confirming correlations theories in social science, and so the evaluation of policy proposals, tends to be a mess of indicators pointing in opposing directions.

In my article 'Testability in the Social Sciences' (Berkson 1989), I discussed how to make social science theories more testable. I took issue with some of Popper's views on the methodology of the social sciences, with an eye to better methods for evaluating theories in the social sciences. One of my objections was to the "rationality principle" as a common factor in explanations in the social science. My objection was that explanation of social phenomena need a degree of predictability in human actions, but it is sometimes possible to predict less than rational actions. I referred to the work of psychologist Irving Janis in my article; since then the work of Kahneman and Tversky on predictable irrationality has gained renown. My other proposal was that in the social sciences thinkers should look beyond "single exit" models like Newton's mechanics, where a differential equation and specified initial conditions together give a single, precise prediction. It is enough for testability that some options are excluded. Thus, I recommended that social scientists look for testable patterns, even when they don't fully specify the outcome of events.

Not working in the field of social science, I was unable to give a concrete realization of the suggestions I put forward. Fortunately, my wife, Isabelle Tsakok, a Popperian development economist (we met in Popper's lectures), took up the challenge. After her retirement as World Bank staff, and in collaboration with former U. S. Assistant Secretary of Agriculture Bruce Gardner, she began looking for testable theories of economic development in the field of agriculture. In a joint paper, Tsakok and Gardner (2007) outlined how the failure to develop testable theories had led to a series of confusing, contradictory confirmations, and they argued for testable theories in development economics. After Gardner's untimely death, Isabelle finished the book *Success in Agricultural Transformation: What it Means, and What Makes it Happen* (Tsakok 2011). In it she identified five necessary (but not sufficient) conditions for agricultural transformation. And she specified the conditions and "transformation" in ways that they were subject to independent empirical tests. The five conditions were testable, and thus far in history they have passed the tests.

What the five conditions have in common is government leadership over decades. In every case, government sustained investment in public goods and services over decades was crucial to transforming countries from poor traditional economies into wealthy modern ones. To me Isabelle's book is a refreshing example of "one handed" economics. President Harry Truman complained that his economic advisors were always saying "on the one hand, […] on the other hand," and he needed a one-handed economist, who could give unequivocal directives. This complaint sums up the weakness of social science theories that are used as "tools", rather than being regarded as empirically testable theories, subject to testing and revision. During the 2016 election in the U.S., I found that a series of other excellent economists have, since 2008, made similar points in books about economic growth for developed economies. In every case, as I wrote during the election (Berkson 2016), government leadership through sustained investment has led the economic growth in developed economies as well.

Looking at the complexity of evaluating social science theories, and its necessity for better democratic governance, I have proposed in 'As fact checkers proliferate, where are the policy checkers?' in the Columbia Journalism Review (Berkson 2017) that a new organization, policycheck.org or policybrief.org be formed. It would act as a go-between between scholars in academia, government, and think tanks, on one side, and on the other, political journalists. The organization would be tasked with doing the kind of analysis that I learned from Popper and Agassi, and developing a regularly updated on-line briefing book. It would say what the leading policy problems are, the alternative solutions, the relevant testing, and expert opinion, and its basis or lack of basis. This organization, I believe, could have a beneficial effect on both sides.

On one side, Policycheck.org would be a service to political journalists who don't have time to develop such an objective briefing book. Using it, they could give vital context and news analysis on policy issues, which they are failing to do now. And they could press candidates to be clear and defend their views, rather than simply sloganeering. On the other side, of the scholars, it would push them to lift their game to standards of testability. The guideline I have suggested is to look for "crucial evidence." In cases where conflicting theories predict contradictory results, and you can actually check, what actually happens becomes "crucial evidence," and dispositive on which theory is comparatively better. This idea is of course a generalization of the idea of "crucial experiment" to include not only experiments, but observations, including using facts from history. From my experience studying the case of Hertz's demonstration of electromagnetic waves (radio waves) in my PhD dissertation and first book, I believe that "crucial evidence" is a more precise guide than 'testability,' because includes the importance of using or developing *rival theories* of the same event.

Surprisingly, when you actually look, there are examples of crucial evidence and clear refutations. One example of crucial evidence on policy is the fact that tax cuts have never paid for themselves in U.S. history (Committee for a Responsible Budget 2015). We can conclude that those versions of "supply side economics" which promise that tax cuts will always quickly pay for themselves through economic growth are refuted. Indeed, the refutation is decisive enough that even George W. Bush's economic advisor Greg Mankiw—who might be expected to defend the view—once labeled those who continue to advance this claim as "charlatan and cranks" (Krugman 2012).

I believe that in pointing to clear crucial evidence, and in objectively summarizing the relationship between evidence, theory and public policies, a policycheck.org could do a singular service to democratic process, here in the U.S., and around the world.

References

Agassi, Joseph. 1964. The nature of scientific problems and their roots in metaphysics. In *The critical approach to science and philosophy: Essays in honor of Karl Popper*, ed. Mario Bunge, 189–211. New York: Free Press. Reprinted in Agassi 1975.

———. 1975. *Science in flux, Boston studies in the philosophy of science*. Vol. 28. Dordrecht/Boston: Springer.

Bartley, William Warren. 1971. *Morality and religion*. New York: St. Martins.

Berkson, William. 1974. *Fields of force: The development of a world view from Faraday to Einstein*. London: Routledge. Reprinted, 2015.

———. 1989. Testability in the social sciences. *Philosophy of the Social Sciences*: 157–171.

———. 2010. *Pirke Avot: Timeless wisdom for modern life*. Philadelphia: Jewish Publication Society.

———. 2016. Government investment and opportunity for all. *Washington Monthly*, June 24. http://washingtonmonthly.com/government-investment-opportunity-for-all Accessed Jan 2017.

———. 2017. As fact checkers proliferate, where are the policy checkers?' Columbia Journalism Review, Feb 2, 2017. http://www.cjr.org/criticism/policy_checkers_fact_checkers.php. See also the followup Salon.com interview by Paul Rosenberg: Beyond fact-checking: After the catastrophic media failure of 2016, the press must master "crucial evidence" http://www.salon.com/2017/02/19/beyond-fact-checking-after-the-catastrophic-media-failure-of-2016-the-press-must-master-crucialevidence/

Berkson, William, and Wettersten, John. 1984. *Learning from error: Karl Popper's psychology of learning*. (Open Court).

Committee for A Responsible Budget. 2015. Do tax cuts pay for themselves? http://crfb.org/blogs/do-tax-cuts-pay-themselves. Accessed Jan 2017.

Gross, Berl. 1992. *Before democracy: A history of error and justification*. Melbourne: Globe Press. https://berlgross.wordpress.com/. Accessed Jan 2017.

Krugman, Paul. 2012. *Charlatans and Cranks*. Blog post Sept. 15. http://krugman.blogs.nytimes.com/2012/09/15/charlatans-and-cranks-2/?_r=0. Accessed Jan 2017.

Patterson, Thomas E. 2016. *How the press failed the voters*. https://shorensteincenter.org/news-coverage-2016-general-election/. Accessed Jan 2017.

Tsakok, Isabelle. 2011. *Success in agricultural transformation: What is means and what makes it happen*. Cambridge: Cambridge University Press.

Tsakok, Isabelle, and Bruce Gardner. 2007. Agriculture in economic development: Primary engine of growth or chicken and egg? *American Journal of Agricultural Economics*: 1145–1151.

Chapter 17
Reflections on Science and Imagination

Robert S. Cohen

Once I was a physicist, working at the beginnings of plasma physics, after a few years in the Second World War dealing with radar, radar counter-measures, and scientific intelligence. Then I added history of physics, and always looked for philosophy. Why philosophy? Well, philosophy deals with three basic questions: *What is? How do we know? What ought to be?*

Physics was my first step toward finding out what there is, what there *really* is, what is fundamental. Physics first, then biology, then psychology and social science, and always a mixture of observation and experiment with mathematical formulas plus hypotheses. That is, plus conjectures about what would explain the results of our experience both in living our ordinary lives and in scientific observations. These conjectures are our theories; if they are correct, they describe the world. Or that part of the world to which each theory pays attention. How clear it appears to be. The scientist isolates a bit, a part, of the world of human experience, explores that part further, out of curiosity or for practical purpose, and the many parts – each with its scientific explanation – will (or may) add together to describe and explain the whole world. The whole will be understood as the sum of its parts.

But how do we know this? And how do we ever get our hypothetical conjectures? Do we go out to simply look at the world, listen to it, touch it, smell it, taste it? Yes but not so simple. Just adding up observations and recording data, correlating, interpolating, extrapolating, just generalizing from the observed regularities of life and laboratory, in the medical clinic or the astronomical satellite, will not do. No, generalizing in that way called 'induction' is a start, long before modem science, but it is only a start.

This paper was first a tribute to Kluwer, the original publisher of the Boston Studies in the Philosophy of Science, and our dear editor Annie Kuipers. A Russian translation was published as my contribution to a Festschrift for my dear friend Natalia Avtonomova (Moscow, 2015). The English text is hereby offered to celebrate my beloved friend and colleague Joseph Agassi.

R.S. Cohen (✉)
Emeritus professor of Physics and Philosophy, Boston University, Boston, MA, USA
e-mail: robertscohen@hotmail.com

N. Bar-Am, S. Gattei (eds.), *Encouraging Openness*, Boston Studies in the Philosophy and History of Science 325, DOI 10.1007/978-3-319-57669-5_17

First, we may note how early was the search for scientific explanation, for the causes of what happens. Herodotus starts his work by saying he has two purposes. To preserve the memory of the Greeks and barbarians in their struggles; and also, "in particular, beyond everything else, to give the cause of their fighting one another."

My hero, one among several, has been Albert Einstein, and he put this issue so clearly. Not only do we ask "How do we know?", but how do we think, how are we able to guess explanations of Nature? Where do we get our knowledge of reality, once we understand how true is the old wise remark that things are not what they seem to be, once we recognize the puzzling distinction between appearance and reality?

Einstein's words: the concepts and theories of science are free inventions of the human spirit. And he adds a note: Our concepts and theories are not logically derivable from what is empirically given, not to be found hidden in the observed facts. So "free creations of the human spirit". Where then are we as scientific thinkers?

Pure logic, pure reason, gives no knowledge of the empirical world—knowledge of what is actual and real begins with human experience and observations and must end, tested in experience and observations.

But then there are the products of the scientist's imagination—sometimes they seem so natural as to be simple and obvious and necessary—as reality rather than creations of thinking. Experience may suggest but the appropriate mathematical or other concepts cannot be deduced from experience.

Where then is the role of reason, if imagination is our realm of creative success, our life of genius?

To Einstein the "creative principle lives in the world of mathematics"—and in a certain sense, he added, "pure thought can grasp reality, as the ancients dreamed"—the eternal antagonism (as he called it) between the two inseparable components of our knowledge, the empirical and the rational, is dissolved into imagination—and we may add, not always mathematical.

***

How "free" is our imagining? What are the boundaries—the constraints—the rules? Why do we enter into a world that is not real, into an imaginary world? Who typically dreams, imagines, plays with images, with what "might be" or might have been? I suppose children in games, artists, poets, and scientists.

Children play with "make believe"—sometimes realistic (play grown-up, play house, play doctor, play war), sometimes fantasy. Children play with toys and games, with objects and pet animals and toy animals, with each other. Sometimes sadly realistic; and the fantasies can be realistic too. Jules Verne last century; the film 2001; a recent film with the actors Eddie Murphy and Dan Ackroyd, the rich white man exchanging role in life with the poor black; the personal account in a film a few years ago by an Israeli Jew passing over into an assumed life as Palestinian Arab, living a real fantasy. Or the happy song from the musical play Fiddler on the Roof, the song, "If I were a rich man".

Notice how wonderful fantastic imagination can be, how marvelous. Einstein's life comes to mind here. As a child he loved to build with toy wooden blocks and to put together jig-saw pieces. At four or five, his father gave him a magnetic compass to play with, a common thing for the German nature lovers and woodland explorers—and little Albert Einstein was astonished – and curious—how did the compass work? How did it point North? A mystery of the adult world maybe? But he had a particular word: Wunderbar, wonderful. Was it a toy? A tool? At any rate a wonder, to play with.

Then, later, in a high-school, he came to see that "looking" is the foundation of learning – a special sort of schooling, and young adolescent Einstein began to play with objects of thought, objects of imagination – combining, fitting together, moving, reproducing, enlarging, rotating – Einstein often reflects on these matters, and used the words "play" [*Spiel*] and "Image" [*Bild*].

As he said the main thread of productive thinking is combinatory play – this imaginative playful, untested, non-factual, mental activity is prior to any verbal construction, any grammar of logic, even any signs. Can Einstein be correct in this self-description? He says these – which I can almost call free associations, or instinctual ramblings – these plays are in his case not only preverbal but either visual or even he said of a "muscular type".

And how do we tell the clever scientist from the rest of us dreamers? The justification for this free play with concepts, like earlier free play with pure images, is the "measure of survey over the sense-experiences which we are able to achieve with its aid". Note the word "survey" [*Übersicht*] – looking, again.

And in another place, Einstein writes: "Man seeks to form for himself, in whatever manner is suitable for him, a simplified and lucid image of the world [*Bild der Welt*], so as to overcome the world of experience by striving to replace it to some extent by this image."

And so do painters, musicians, poets, each in his own way. To Einstein, the physicist's picture of the world, the scientist's, is one among the possible pictures – but this one must be rigorous, precise, simple – how hard for the physicist to deal with complexity and subtlety! His words: "Supreme purity, clarity, certainty".

And we must add "beauty". How incomplete our scientific achievements are, with all these praises! Hear the moral advice of a predecessor of Einstein, the Austrian Ernst Mach – the true courage to be learned from a scientific philosophy is to *bear*, to live with, incomplete knowledge. To be able to say accurately "I do not know" and to live with that situation, neither with despair nor by irrational escape into wish-fulfilling fantasy. So only rational fantasy? Realistic imagination? Even practical guesses, which we call hypotheses. How curious and splendid to join "realistic" with "imagining".

As to the world of the wonderful and the world of fantasy, we note two more events of the young Einstein. At about twelve, he was given a copy of Euclid's *Geometry* – and by the knowledge of geometrical shapes, the certainty of the proofs and the theorems, by the clarity, its logical independence from any observations, the *purely mental* status and internal systematic rigor – he was amazed and entranced – pure thought, maybe pure, imaginative, but not arbitrary or willful – rigorous imagination, genuine discovery – *wonderful* – just as the magnetic compass but now clear. And this time not a mystery.

Then, a bit later, Einstein learned of the speed of light; and the wave-properties of light, and that the color of light is given by the number of waves that pass the observer per second—then he tried to imagine—what if the observer ran along with the light? Fantastic! What if he ran as *fast* as the light? What would the light-color be? What would it look like since in that case no wave would pass him each second—where and what is the color then? Has it disappeared? Does it exist only as a relative interaction between the light and the observer?

These mental experiments—these fantasies of scientific imaginations – they are *genuine* experiments because we do not know the result when we enter upon the play with the conditions of the game – with the rules of the game. Can we run with the speed of light in our experimental play?

Have we, in science, imagination run wild? But no, again hear Einstein who advises us not to envy the theoretical scientist: "For Nature, more precisely experiment, is an inexorable and not very friendly judge of his work. Nature never says 'yes' to a theory. In the *most* favorable cases it says 'maybe', and in the great majority of cases, simply 'No'—if an experiment agrees with a theory, it means 'maybe'."

***

But the danger that imagination might run wild is with all of us—in science, and in daily life, and especially in anti-science and in pseudo-science, or in para-science. We use *metaphors* and *analogies* and they can be harsh when they succeed in organizing our experiences (scientific or not) but they also can be dangerous. Think of such words as "attract" and "force" and "repel" in the physics of gravitation for Newton's time and today. They *suggest* an analogy to some of our psychological or physiological experiences, attract, repel, force may rest upon *desires*, attraction as the satisfaction of a kind of desire. But this is placing human quality into physical objects—called *an anthropomorphism.*

I suppose this danger is simple—but anthropomorphism is all too common. When science fails to explain, when knowledge is insufficient or absent, imagination can then take its place—and this is wild imagination, which is projective, black magic, unconstrained because *this* imagination works with *naive parallelisms.* Therefore, *superficial analogies*, often simple analogies with human experiences, are confused with explanations! They are not generalizations from experiences, and not imaginative free play which is disciplined by test. In such cases, recognition of ignorance (which is not a mistake but plain honesty) is replaced by projection, by pseudo-explanation. The understandable *desire* for explanation itself is often stimulated and encouraged by myths and fables, by religious or older philosophical stories of drama and by picture-thinking—but the pseudo-sciences, the wishful explanations—do not clarify. Instead they run the danger of empty words, dangerous self-assured dogmatism of untestable beliefs, indeed unproven and unprovable, *the closed mind.* Loose language, extended beyond established meanings, can create false ideas—not creative imaginative but destructive imagination. What is the escape into pretty picture language, or reassuring analogies. Children's games? Perhaps the dark side of play?

And yet what about astrology? Was it pure science with a *mistaken* hypothesis about physical forces reaching across great distances to our individual life histories – mysterious but with specifiable conditions—and testable? Surely Albertus Magnus proposed that astrology *should be tested,* treated as an empirical conjecture about the regularities of human life and character in a world where appearances are so arbitrary and troubling. He opposed the anti-astrology and antiscience of the medieval tradition that saw the world as a play of unconstrained arbitrary chance, subject to divine miracles or willfulness—astrology then was a step in the search for natural laws, for scientific formulation. But this poor science, astrology, evolved into magic, was replaced by better, richer ideas, more testable, less wishful; and astrology split into astronomy and a pseudo-science still named 'astrology' today. Our dear old erroneous astrology is now just a fossil science.

Then where is the *wonder* of the world if not in miraculous actions? First in the *harmony* of the fifteenth century – Ficino saying the world is beautiful and intelligible – harmony linking nature and music and mathematics.

And later, laws of Nature; remember Copernicus who saw not only that a sun-centered system was simpler than the tradition that the Earth was the world center and stood still but also that the Sun is fit to be the center, for it is marvelous, wonderful.

But it is we who see the wonder! We who find the regularities in Nature that the old astrologers were seeking, we who today see a huge portion of chaos, and argue over a view of a world of indeterminism and of alternative potentialities. We dig out the regularities, or as some scientists and philosophers have it, put them in but not at all by simply recording data. Then, ask succeed? Can we aspire to know it all?

A couple of lines from the English poet Alexander Pope's *Essay on Man* may help: "Why has not Man a microscopic Eye? / For this plain reason, Man is not a Fly."

Or we would be overwhelmed! So it seems that science does not give us a model of Nature but rather a living language for describing Nature, for integrating our partial evidence, from our own time span and cosmic scale of micro and macro. Imagination takes hold of our fragments of knowledge, far beyond our human eyes. And we then manipulate (in language or in other symbolism or in pre-linguistic ways) what are as yet unrealized situations, perhaps ever to be so. Imagination is in our minds, images into futures which have not happened.

How far the imaginary can go! The non-scientist is told how the compact and specific mathematical statements—Newton's law of gravitation, Boltzmann's great law of entropy—reach far beyond any specific test case. But recall too how far a specific instance goes with the stimulation of a poem. Listen to William Blake: "A Robin Red-Breast in a Cage / Puts all Heaven in a Rage". What an extraordinary verse – about the world of humanity and Nature and Heaven, of morality and simply sorrow – poetic magic.

So, imagination is our natural magic. *Scientific* imagination leads to theories which conform to the facts but go beyond the facts. Theories *show* the facts; the facts will flow, in Bronowski's appreciation of theories, "from an inner order and imaginative arrangement of a few deep central concepts" – "a creation of the human mind".

Was Newton's theory true, in his time, and false today in Einstein's time? Why ask about true in this way? Newton and Einstein created coherent and bracing, marvelous visions, world- pictures – disciplined by the facts as they knew them. They did what the artistic imagination can do, at its best – go beyond the presented world to an absent situation – to free the scientist, and the artist, from the immediacy of our all-too-finite actuality.

If every art-work, every painting, song, symphony, novel, is an experiment in living, then the scientific work is too. But the novel imagines a different truth from that of the objective world of the scientific imagination—how do we evaluate the truth of a novel, a poem, a symphony of Mozart or a ragtime composition of Scott Joplin? We ask whether it tells us, this is *how the world feels* how the novel or the music lives, and we must live it, each of us in our own way. Imagination – in art, how the world feels to me and to the artist (a joint, sometimes cooperative matter) and in science, how the world is *and might be.* Remember a slogan from Paris, spring 1968: *L'imagination au pouvoir.*

But how two-faced imagination, and art, and science have been in human history, and may yet be. We can summarize the characteristics of science:

1. Our time, our late twentieth century civilization is certainly scientific and barbaric, a terrible ambiguity. How are these characteristics connected? Science is ambiguous in its uses and its protagonists; and so is imagination. Indeed the ambiguity can be devastating, even with the scientific successes. C.*S. Lewis*, in the mid-century wrote "What we call man's power over nature turns out to be a power exercised by some men over other men, with nature as its instrument".
2. Science is feared, respected, ignored—desired, hated, and imitated.
3. Rationally, it may still seem to be true that the achievements of science and technology have made possible the translation of humane values into technical tasks.
4. But equally made barbaric horrors into technical tasks.
5. And global problems may triumph over practical use of science—population, ecology, resources.
6. Who can blame those who see no help in the science they know only from the technologies of wars, from distortions or in ignorance or who are escaping from their bleak personal worlds into pseudo-science or illusion or vain hopes?
7. But rational imagination, scientific imagination is still our best hope. Recall Einstein again and, with him, cherish science as "free creations of the human spirit'.
8. Our poet Blake again, with all his fear yet recognizing progress: "What is now proved was once only imagin'd". Shall we admire the wonders of Nature and its mysterious rationality?

For myself, I would observe (following Anatole France) that "The wonder is not that the field of the stars is so vast, but that man has measured it".

Chapter 18
The Legends of One Methodology of Science Used Throughout Its History and Its Independence from the Institutions in which Science Has Been Conducted

John R. Wettersten

18.1 Introduction: Central Problems of the Historiography of Science Today

Ever since the beginning of science in the sixteenth and seventeenth centuries the question of the independence of natural philosophy (science) has been intensively debated. One of the finest statements of the separation was made by Galileo in his letter to Grand Duchess Christina. Natural philosophers (scientists) struggled to establish their independence from religion and philosophy; in the seventeenth century, at the very latest with Newton's success, they succeeded. But this victory was a Pyrrhic one: Science was often granted its independence, but its identity was not clearly established. Methodological debates arose again and again and even then, when it looked as if everything had been settled. This history of methodology shows that science cannot be given some unified, established and never changing methodological identity. It shows that methodology and science have to steadily interact and seek improvements. It also shows that to do this is not only valuable but also a crucial and important basis of the growth of knowledge.

William Whewell was the first thinker to notice that the approach of conflating the teaching of science as a discipline, on the one hand, and explaining how science developed, on the other hand, was mistaken. (Whewell 1967a, b) He noticed that scientific theories were not proven by induction, as Hume had previously and so powerfully pointed out, but, he also understood that scientific theories were not created by inductive inferences, as virtually everyone in the light of Newton's fantastic success, including Hume, had long maintained. Virtually all portrayals of science presumed that scientists used induction to discover and prove their results. (Whewell says that Descartes retained significance in France until around 1750.) But, if this

J.R. Wettersten (✉)
Fakultät für Sozialwissenschaften, Universität Mannheim, Mannheim, Germany
e-mail: wettersten@t-online.de

N. Bar-Am, S. Gattei (eds.), *Encouraging Openness*, Boston Studies in the Philosophy and History of Science 325, DOI 10.1007/978-3-319-57669-5_18

inductivist method of discovering scientific theories was not the method which was actually used to make such discoveries, the development of science had to be described quite differently. There had to be a history of science, which was not merely the cataloguing of important truths, noticing when they were discovered, and showing how they were proven by generalizing from facts, that is, by induction. It was not enough to simply describe the results of science; one also had to describe how this extraordinary advance of knowledge took place.

In order to do that, one had to have an alternative methodological theory. This alternative methodological theory could only be constructed, if the dominant conflation of the portrayal of the content of science and the history of science was replaced with some other approach. This alternative methodology of science had to be a true methodology in order to describe how scientific discoveries were made in the history of science, or that was at any rate what the only person who tried to construct one in the middle of the nineteenth century—William Whewell—thought. A true methodology of science, he thought, would be that methodology which correctly described how scientific discoveries were made. This task of explaining scientific discovery could not be carried out by merely describing the contents of the successes of science. It required an independent theory of how knowledge was obtained. Whewell set out to meet this new challenge. When his research was largely completed, he claimed that he had met the challenge of describing how scientific knowledge was obtained and, since no other explanation was possible, his own revolutionary methodology was the true methodology. It was, he said, the only methodology scientists had ever used.

The story of how this change in the history of science came about due to Whewell's innovations should be the story of how the framework for the philosophy of science from the middle of the nineteenth century until quite recently was formed. Independently of which specific modern methodological theory one defends, Whewell's approach shows the framework in which all these methodologies needed to compete up until today. They had to clearly portray the methodological theory thought to be true and then show that the methods described by this theory were in fact the methods used by all scientists to make their discoveries. But this fact about research in the philosophy of science remains unacknowledged by the majority of researchers in this field. It needs telling and in such a way, in which it becomes clear that this has been the highly influential but largely unacknowledged framework for philosophies of science for more than the last 150 years.

But this is not the end of the story. Today this framework needs to be changed once more. Karl Popper rediscovered that scientific methodology depends not merely on the correct application of logic, but also on methodological rules. (Whewell and Pierre Duhem had realized this before.) Popper saw that these rules are social; they do not merely describe the application of logic, but are rules which describe how the truth should be pursued. But Popper did not observe that they have changed in progressive ways and that they should further change as science develops. They are, in fact, better or worse and constantly in need of new appraisal as knowledge advances. This means that the correct history of science is not that history, which correctly describes how each advance was made with the application of the single correct methodology, as Whewell maintained. We cannot merely substitute Popper's methodology as a framework for describing the history of science for those such as Whewell's and/or Duhem's (Duhem 1962) which were used before, as Agassi suggests. (Agassi says he was disappointed that readers of his essay did not notice that his view differs from that of Popper. But he nowhere says whether the history of science would be conducted any differently if one used his methodology as a framework rather than Popper's.) Because social rules differ and change, we need to show which rules were applied in which historical instances and how these contributed to the emergence of new discoveries, or how they hindered it. The study of the history of science cannot be separated from an ongoing critical appraisal of scientific method. And the appraisal of scientific method cannot be separated from a study of its history. This critical appraisal cannot be carried out by merely describing what happened. It requires philosophical appraisals of these developments as well. It needs to analyze how currently applied rules need to be supported and/or changed.

A crucial aspect of this essay is the demonstration that, though descriptions of parts of the history of science have been excellently described, all attempts to provide a comprehensive overview of the proper tasks and framework for the study of the history of science, from Whwell until today, have turned out to be flawed. A deep-seated reform of the description of the tasks of describing the history of science is needed. It is desirable to examine how events in this history have been analyzed, criticized and portrayed up to now. This literature is relatively small, and a critical appraisal and overview of it shows the need not only for new studies of the history of the history of science, but also for new theories of how to study the history of science. I will turn, then, to the question of where the historical study of the historiography of science brings us today.

18.2 Agassi's Historiography of Science: What It Contributed and How It Needs to Be Revised

Historians of science have by and large made no systematic effort to appraise and understand the history of their discipline. The only two comprehensive studies are those of Agassi and H. Floris Cohen. (Cohen 1994) Perhaps this neglect has been due to the dominant point of view among historians of science that the history of science is itself a science, in that it does what all scientists have always done: it describes the facts. On this view no separate theories of how science has proceeded are needed in order to understand its growth. Now Whewell agreed that all scientists used the same method. But, he pointed out, this method cannot be the mere description of facts. Historians of science thus need a new theoretical framework in order to describe the growth of science, whereas philosophers of science need to test their theories of the methodology that all scientists have used to achieve knowledge, with historical studies of what really happened. By and large historians of science do not think that science makes advances by correcting mistakes and producing better theories. At best they, following Duhem, presume that scientists make small adjustments to prevailing theories. Since they are also scientists, they must all follow the same procedures. They cannot criticize their predecessors for serious mistakes, without demonstrating that they were not scientific, that their views should be discarded and replaced with serious ones. Philosophers of science, on the other hand, have tended to neglect the study of how the history of science should be conducted due to their view that their proper subject is the logic of science rather than its history and its changing methodological-cum-sociological rules.

The first comprehensive study of the historiography of science was, then, Joseph Agassi's *Towards an Historiography of Science*. (Agassi 2008, pp. 119–243; first published in 1963) This study was a direct consequence of Popper's challenge to the dominant methodological theories of science. In a quite Whewellian fashion Agassi said that they could not correctly describe how science grows, whereas Popper's theory could. But, contrary to Whewell, he made no extensive study of the history of science. There was, then, a Whewellian gap in his theory of the methodology and history of science. Agassi did not try to fill this gap by writing a new history of science, as Whewell had done; he only argued that it existed. He demonstrated with examples from the history of the history of science that developments in this history could be far better understood and described when a Popperian methodological framework was presumed, than could be done when the two established frameworks, that is, inductivism and conventionalism, were used.

Agassi's study makes a number of useful points about the historiography of science, that is, about how specific historical developments could be more accurately understood. But his study also contains significant philosophical-cum-historical mistakes.

18.3 Agassi's Progressive but Still Faulty Appraisal of the Historiography of Science and Its History

Agassi's essay begins with a characterization of the inductivist-Baconian methodological theory which has been the fundamental theoretical framework for the history of science since its very beginnings. Just when this view became dominant is not explained; the question is not posed. Bacon is given as the only authoritative model. And Whewell is mentioned, by and large, as an inductivist historian, who, to a high degree, followed the Baconian program. Whewell is praised for developing a remarkable history; but his history is not seriously examined; its originality is touched upon but hardly explained. Its excellent theoretical foundation is not explained. Agassi returns to him later as the prime contributor to the development of the emphasis on continuity in the history of science. But this interpretation of Whewell's contribution is not very clear. It says much too little about Whewell's emphasis on the adventure of discovery, and it does far too little to distinguish Duhem's emphasis on continuity as a series of small steps as modifications of established doctrines from Whewell's theory of the growth of knowledge in historical contexts, which includes false conjectures and criticism, sometimes but not always followed by great gains, as important parts of the difficult quest for truth.

Agassi's criticism of the inductivist approach is launched with a series of examples, which are, for the most part, critical reviews of various rather recent histories of science, that is, those written in the twentieth century. This is, of course, nearly all there is to discuss. Outside of his discussion of Duhemian conventionalism, there is no historical analysis of their development or growth. The aim of these histories of science is, by and large, quite simple. Agassi argues quite effectively that the inductivist framework, when used to guide the research of historians of science, regularly leads these historians to make black and white judgments between false speculations, on the one hand, and true theories proven by induction on the other. From the inductivist point of view there is no middle ground. This assumption has led to numerous distorted portrayals of what really happened when the development of science has proven to be far more complex and subtle than the inductivist methodological assumptions allow.

As a matter of fact the development of science regularly involves false conjectures and approximations and criticisms and improvements, as Whewell pointed out although Agassi does not give him the credit that he is due. But the inductivist historian of science has no option other than to use curious descriptions and evasions, which are designed to sweep these complexities under the rug. Their existence cannot be admitted, even when they are virtually always present. From the inductivist point of view, admissions of their existence call the scientific nature of science into question. But the avoidance of pointing out the pervasive and serious difficulties scientists have faced leads to quite unrealistic portrayals of what really happened.

After Agassi has completed his critical portrayal of inductivist histories of science, he turns to the second of the two major theoretical frameworks used to set the tone for histories of science up until the time he conducted his analysis. Agassi does note that there are a few exceptions to this generalization, such as the research of Alexander Koyré, who found that during the early history of science it was integrated with philosophy and theology (Koyré 1955, 1958, 1965, 1971, 1973) and of Arthur Koestler who argued that scientists were sleepwalkers: Their innovations could, Koestler, argued, not be explained. (Koestler 1963) But Agassi pushes his threefold division of the historiographies of science too far; he erroneously claims, for example, that Koyré is a Popperian simply because he noted the importance that false theories have played in the growth of knowledge. What he fails to note is the sharp conflict between Koyrè's view that science from Copernicus to Newton was above all a philosophical-cum-mathematical problem of finding good mathematical readings of reality, on the one hand, and Popper's theory that the key to scientific progress is the construction of highly falsifiable theories, on the other.

The second major alternative to inductivism is the conventionalist approach introduced above all by Duhem. Conventionalism is the theory that the only aim of science is to find true predictions. On this view science does not, and should not, seek true theories. Conventionalism is an old theory; it is already found in the preface to Copernicus's magnum opus. But Duhem was the first to turn it into a quite general methodological and historical theory of science. Both the earlier version of conventionalism and the latter one shared the aim of removing any conflict between religion, which should provide the true theory of the world, and science, which should provide everyday tools for effectively dealing with it. Duhem's innovations to this traditional view were above all twofold. First, from a methodological point of view Duhem argues that the only acceptable scientific method is that of making small modifications to established theory. When one tries to do more, the unity of science breaks down; philosophical schools are formed and the divisions between them cannot be resolved with rational approaches. Secondly, from an historical point of view, Duhem argues that the history of science is continuous: All true scientific innovations are small modifications of that which went before. Descriptions of so-called revolutions in science are misguided; they simply ignore the continuity between that which went before and that which came afterword. This continuity applies even to the beginning of science. The beginning of science was not a revolution away from the natural philosophy of the Middle Ages, but rather it was many small modifications of theories already at hand in the so-called dark ages. On Duhem's view the failure to recognize this development leads to a quite mistaken view of the limitations of theoretical developments before what is now regularly called the scientific revolution. The preceding theories were allegedly far better than acknowledged by those seeking to find a revolution from philosophy to science at the end of the fifteenth and the beginning of the sixteenth century.

In introducing this approach Agassi once more mentions Whewell as an important contributor to it. This is correct, as he and I have elaborated elsewhere (Wettersten and Agassi 1991) and I in a rather different way. (Wettersten 2005) Whewell's contribution did not occur in the way which Agassi describes in this essay. Agassi claims that Whewell was an inductivist; Whewell allegedly believed that, when attempts to refute conjectures confirmed the conjectures, they were, with sufficient successes of this kind, proven to be true. Under this assumption Whewell's influence could not have been due to his call for a significant alternative to inductivism. Agassi needs an alternative. He claims that Whewell pointed to the continuity of scientific progress, even though he was an inductivist. His influence then allegedly occurred because other historians learned from him about this continuity. Duhem is the most prominent example. And Duhem taught many other succeeding historians of science, how to portray the history of science as a continuous process with small changes.

As a matter of fact Whewell rejected inductivism. And while nobody adopted his powerful alternative explanation of how scientific knowledge grew, some important thinkers saw that he was right to seek an alternative to inductivism. They sought to provide their own innovations. They gave no credit thereby to Whewell for what they had learned from him. There were various reasons for not doing that. Whewell was, after all, rejected by the scientific and the philosophical establishments. Siding with him gave no one any advantage. For two French innovators—Pierre Duhem and Claude Bernard—the fact that he was English and they were French may also have had some influence on their reluctance to mention him. At any rate they said nothing about this connection. And this fact blurred the history of the history of science considerably.

In limiting the historiographical alternatives to inductivism and conventionalism, Agassi is by and large correct. Nobody tried to explicity build on Whewell's alternative, so the dominant active approaches were inductivist and conventionalist although there were important exceptions, as I further explain below. And Duhem's conventionalism was the strongest alternative to Whewell that was constructed. Some alternatives, such as that of Charles Sanders Peirce, never got off the ground. (Wettersten 2005, pp. 100–106) Others, such as that of Ernst Mach, were never clear enough. (Mach 1970, 1974; Cohen and Seeger 1970) Mach futilely tried to give inductivism a solid psychological foundation. And this is not possible.

Agassi proceeds with a critique of Duhemian conventionalist histories of science which parallels his critique of inductivist histories of science. And this critique hits the mark quite well. He concedes that conventionalist histories of science have been better than inductivist histories of science, because they take account of the fact that each development builds on theories which preceded it. But, he nevertheless shows how conventionalist histories of science also distort the history of science. They do this, because their framework says that all advances are relatively small changes made in previously established theories. But many advances which really occurred in the history of science were not small and, indeed, rested on powerful criticisms of their predecessors. Conventionalists have to level out these changes; in doing so they often significantly distort the historical developments they want to describe. This approach blocks explanations of the deep-seated problems which progressive scientists have found in the established doctrines in their fields, the enormous difficulties they faced in their struggles to overcome them, and the significance of their broad advances.

After his discussion of conventionalist historiography Agassi turns to the task of explaining how the history of science should be written. His historiographical theory is an application of Popper's methodological theory to the history of science. It is a small attempt to do for Popper's methodological theory one of the things which Whewell had done in very large measure for his methodology, that is, to show how it can both reform how the history of science is written and thereby significantly improve portrayals of it. Agassi's result is more or less a standard application of Popper's methodological theory. Popper said his theory should and can describe the history of science. But Agassi was the first to develop this methodological theory as an historiography of science at any length.

On Agassi's Popperian view, scientific research begins with problems which are based on criticism of the best theories available. Criticisms of established theories can point to counter-examples or vagueness or incompleteness or inconsistencies. New proposals are made which are designed to overcome such difficulties; these are in turn subject to the same sort of criticisms. In portraying the history of science, it is of great importance to understand the views taken by scientists at any time. It is important to make these presumptions as they were then conceived as clear as possible, even when, or especially when, they are now seen to be false or untenable or utterly confused. Only then can we effectively understand how science has developed, that is, just what problems scientists have faced and how they have, to this or that degree, overcome them.

Agassi gives some examples of how historians of science had, until he wrote his essay, regularly failed to describe the history of science in realistic ways. They distorted what happened either by romanticizing the successes or underestimating the importance of mistaken theories (inductivists) or by papering over serious changes (conventionalists); they thereby rendered the history of science quite difficult to understand. Two of his significant examples of such distortions are, one, how historians have dealt with the transition from phlogiston theories, on the one hand, to Lavoisier's alternative, on the other hand, and, two, how they have portrayed Örsted's demonstration of the connection between electrical currents and magnetic forces.

He shows how Priestley's criticisms of Lavoisier's proposal led Lavoisier to real improvements and he shows how Ørsted's theoretical background, even when false, was of considerable importance for his empirical research. In both cases he makes interesting contributions to the history of science; but in neither case does he develop an alternative historical portrayal, which satisfies the high standards he sets for historians of science. This was, of course, not his purpose. He only intended to illustrate how portrayals failed to take into account important aspects of these developments, and how this failure led to mistaken or vague portrayals, which could even be hard to understand.

One crucial aspect of Agassi's historiographical study is his assumption that the history of science can only be written within some specific philosophical framework, which explains with a methodological theory how science grows. He here follows Whewell's profound innovation in the historiography of science, which Agassi failed to acknowledge. Historians of science have followed it, as Agassi shows, in various ways with varying frameworks. No one followed Whewell's methodology. Agassi works in the tradition of Duhem, who agreed that the history of science should be written within the framework of a correct methodological theory, but who also (implicitly) claimed that the correct theory is not Whewell's but rather his own version of conventionalism. Agassi says, no. It is his own version of Popper's methodology translated into a proper historiography.

18.4 The Need to Take Methodological Change in the History of Science into Account

When we ask what standards or goals we should seek to realize when participating in scientific institutions, we must take into account the fact that science has changed in significant ways, often for the better but sometimes for the worse. Traditionally the pursuit of the correct methodology has been the pursuit of that which logic (universally) demands. Induction has been, of course, deemed a version of logic. The rules of science were thought to be those rules which followed the logic of science. Whewell broke with this tradition. His methodology explains how scientists pursue the truth with activities such as making conjectures which are not merely applications of logic. De Morgan's criticism of Whewell's theory in which he said that Whewell offered the best psychological theory of scientific practice but no logic of science accurately mirrored what Whewell had done. (Wettersten 2005, pp. 58–63) And Duhem followed him in this regard by saying that a crucial rule of scientific methodology is that only small changes to established theories are allowed. (Wettersten 2005) Charles Sanders Peirce clearly recognized the superiority of Whewell's theory over that of his chief competitor, John Stuart Mill, but he also saw the gap between logic and scientific practice in Whewell's theory. He tried to find unity, but he had no success. (Wettersten 2005, pp. 100–106).

After these thinkers Popper was the first to clearly break with the tradition of seeking a methodology of science which was merely the application of logic. He saw that his first methodological problem, that is, that of describing the logic of research alone, was not formulated adequately. The logic of science, that is, deducing the consequences of a theory and finding a contradiction between a consequence and a fact, allowed for ad hoc modifications. And these could not be a part of proper science defined as the proper application of logic alone. In order to keep ad modifications out of science one needed, in addition to the rules of logic, social rules, which he then added. The most significant additional rule he added is to always select that not yet refuted theory with the highest degree of falsifiability.

Agassi knows all this as well as anybody. In his study of Faraday he pointed out the unhappy influence of inductivist dogmatism (Agassi 1971) and in his study of Boyle in which he points out that Boyle instituted rules of crediting priority and public recognition which led to an exaggeration and perpetuation beyond reason. (Agassi 2008, pp. 388–444) But he does not notice the extent to which the existence of social methodological rules requires changes in the philosophy of science. We cannot find a correct theory of science by only studying its logic. Proper science is not that science whose methodology conforms to logic. No science does merely that. Each science, and/or each period of science have their own rules which are better or worse. Whenever we pose the problem of how science should be conducted today—the problem on the table in Agassi's response to critics—we have to ask how the rules of today can be (somewhat) improved. They cannot be made perfect; such rules do not exist. Because they do not exist we cannot set our task as getting as close as possible to them, when we do science. We can only say that this rule in this context brings us forward, try it out, and see if things improve or not. Things might even get worse. Even when the problem posed is not reformulated in this way, some of the deliberations may touch on it. But what will be claimed about good scientific behavior will not be specifically limited to specific social conditions, but universalized as applying to all of science, from Copernicus until today. This needs to be changed.

A second important innovation in Popper's philosophy of science is that science is social. Robinson Crusoe could not do science, Popper said, because he had no one with whom he could have critical discussions. (This is false and a bit racist: Friday was there after all. But the point is clear.) This means that to analyze how science should be conducted one must analyze how scientists interact with one another, and not merely how each scientist behaves correctly. Agassi's writings sometimes go in this direction, as in his discussion of Lavoisier and Priestley, but he does not go far enough. Actions will be socially analyzed but for the most part depend implicitly on being individual contributions rather than being often small or partial contributions to positive social change.

18.5 Mach, the Social Victory of Inductivism, and Koyré's Innovation

In the twentieth century Duhemian, conventionalist methods were introduced into the study of the history of science; they became widely practiced. But many historians retained the traditional inductivist approach. Some mixed their methods here and there. But nobody explicitly took up Whewell's approach; nobody tried to develop further a Whewellian history and methodology of science. A general discussion of alternatives was effectively blocked. And this was desired by a majority of historians, because the inductivists, led by the resistance to Whewell of John Stuart Mill and John Herschel, had determined in the middle of the nineteenth century that Whewell's non-inductivist theories could gain no open and acknowledged foothold in the traditions of scientific thought. (Wettersten 2005, pp. 35–57) From merely sociological and historical points of view, though not from an intellectual one, the inductivists won the day. Their conservative view dominated the discussions well into the twentieth century. But a closer look at the discussion reveals the considerable theoretical influence that Whewell had in the late 19th and earlier twentieh centuries. I have shown this influence in the philosophy of science elsewhere. (Wettersten 2005).

The only historical study which sought to give a serious methodological support to the social victory of inductivism was Ernst Mach's approach to the historiography of science. His project failed, but he had influence. The most important twentieh century innovation in the history of science was Koyré's approach to sixteenth and seventeenth century natural philosophy. But before Duhem adopted Whewell's approach to the relationship between methodological and historical theories, Ernst Mach combined an historical study of the rise of mechanics with his own methodological theory of science: he sought to show how his anti-metaphysical methodological-cum-epistemological theory could show how physical theories could be properly interpreted. The key to doing this, he thought, is to preclude from them any non-empirical aspects.

Mach's historical research preceded that of Duhem by several years. And the two views have similarities due to the fact that they both reject metaphysics in science; each wants science to simply deal with facts. But they are, nevertheless, quite different. Mach seeks to portray the history of science as the construction of true statements about the relationships between facts—and only the facts. He presents the history of mechanics as a series of discoveries of how the relationships between facts could be described. When he describes Galileo's research he portrays the factual discoveries which Galileo made. His theoretical breakthroughs are hardly mentioned. When discussing Galileo he claims to be using Galileo's own descriptions of the facts, or, at any rate, those of contemporaries who set them down in some simpler way. When he comes to Huygens, he says that Huygens language and mathematics are all too complex and embedded in the time to be used in a more modern history. So, he translates them into a modern language, claiming that he thereby can more simply and clearly portray the relationships between facts as Huygens allegedly did.

Mach also makes attempts to describe quite generally the path of scientific development. In a somewhat surprising way this description comes quite close to that developed by Whewell. He says that there is an initial stage of searching for statements which describe the facts. This is followed by the simplification of these statements and, then, finally by the extension of the statements to other facts. This description comes close to Whewell's observation that there is first a quest with conjectures and corrections, followed by the formulation of fundamental ideas—the true scientific theories, whose applications are then extended and rendered more precise in a third stage. But the direct influence of Whewell on this aspect of Mach's theory is not clearly stated; it remains circumstantial.

Among historians of science who have made significant studies of important aspects of the history of science, Alexander Koyré stands out. And one of his most important contributions to the history of science is the significant comments he made about the historiography of science. He pointed out, for example, that the studies of philosophy, theology and science were unified before Newton. (Koyré 1958, p. 4) In this attitude he followed that of E. A. Burtt, who first published his study of this connection in 1924. (Burtt 1954) But Burtt did not make any significant innovations in the historiography of science. He simply showed how science had changed our view of man and the world in such a way that God and the importance of man no longer played a central role. For religious reasons he wanted to find a way to recapture this role. Koyré observed that in the seventeenth century there was hardly any history of science. But Koyré's significance was not only because he pointed out the integration of philosophy, theology and science in the fifteenth, sixteenth, and seventeenth centuries. He also explained why the significance of the scientific revolution did not rest above all in empirical research: It lay in their rethinking the philosophical framework which was used to understand the world. They moved from an Aristotelian framework, which emphasized the qualities of objects and the relationships these qualities had to each other, to an abstract mathematical framework. They sought then to find the mathematical laws which objects obeyed. This new framework led to revolutionary theoretical advances which also had some very significant empirical consequences, and it led to important observations such as those of Galileo's of the surface of the moon and of Jupiter's planets. But it was the newfound application of mathematics to the world which produced what later became characterized as the scientific revolution. In this respect he is the opposite of Mach's portrayal of Galileo's results as improved descriptions of the relationships between facts.

Agassi comments on Koyré and praises him highly. But then he says that Koyré was a Popperian although his research was by no means based on some Popperian framework. And, indeed, it contradicts it. For Koyré emphasizes the importance of a mathematical framework, while limiting the importance of empirical research. Popper emphasizes the importance of the empirical tests of all new scientific theories. Agassi's characterization of all histories of science as being inductivist, conventionalist, or Popperian breaks down in the case of Koyré's research, as it does in others as well. I.B. Cohen's study of Newton and Franklin is one such further example. (Cohen 1953) Koyré mentions Popper once, but then he points out Popper's view of thought experiments. This is, perhaps, a bridge between his view and Popper's: empirical research is rendered theoretical and non-empirical.

The above overview of the historiography of some of the leading historians of science shows how the program introduced by Whewell has dominated the field ever since he explained it to a hostile audience. But Whewell says the aim of history of science is the correct description of one methodology which was adhered to by all scientists. Koyré discussed the methodology of science in its beginnings. But he did not universalize this view to all of science. Whewell did not simply use his methodology to guide descriptions of scientific research; he called for the development of one universal methodological framework for the conduct of the history of science. Whewell thought, of course, that only his alternative was acceptable. But those following him used his procedure to develop their own points of view, whether it was Mach, Duhem, Bernard or Peirce. Historians of science have by and large accepted the frameworks put forth by these leading philosophers-cum-historians in the late nineteenth and early twentieth centuries. This logic has been somewhat noticed by some; but its central factor in the historiography of science has not.

18.6 Methodological-Cum-Historical Pluralism Today

The history of the history of science shows the lasting, important and close relationship between the methodology of science and the history of science. Each discipline has been vastly improved by this cooperation. It also opens up interesting relationships between the varying institutions in which science has been conducted and their methodologies. These connections are, of course, not the only important aspects of the history of science. There are many further interesting problems which are valuable parts of the both the history and methodology of science which are not treated in this essay.

The variations of methodologies of science and their differing institutional contexts are important areas of study for the history of science. There is no proper, comprehensive list for what aims should be pursued: the imagination of historians in finding new aims in regard to ever developing institutional contexts should be encouraged. All attempts to formulate one aim or even a few aims end up setting limits to the study of the history of science, which preclude interesting problems.

The alternative to some specific methodological theory combined with a specific set of aims as a framework for the study of the history of science offered here is a theory of science as an institution. Science has not been one constant and unchanging institution since it began, say, in the sixteenth century. Rather, institutions in which science has been pursued have changed as both the understanding of how to pursue the truth has changed and the social contexts in which truth has been pursued, well or poorly, has changed. By ignoring the close social interaction between what scientists have engaged in and the social contexts in which they have carried out their research, historians of science have romanticized, downplayed, and/or ignored many interesting problems. They have romanticized conflicts between non-scientific institutions such as the church and scientists, they have downplayed or ignored social contexts which favored some alternatives and rejected others, and they have ignored borderline aspects of the influence of science on other fields or the influence of other fields on science.

References

Agassi, Joseph. 1971. *Faraday as a natural philosopher*. Chicago/London: University of Chicago Press.

———. 2008. *Science and its history*. New York: Springer.

Burtt, Edwin A. 1954. *The metaphysical foundations of modern physical science*, 1924. Garden City: Doubleday & Co. (first published.

Cohen, I. Bernard. 1953. *Benjamin Franklin*. Indianapolis/New York: Bobbs-Merrill.

Cohen, H. Floris. 1994. *The scientific revolution*. Chicago: The University of Chicago Press.

Cohen, Robert S., and Raymond J. Seeger, eds. 1970. *Ernst Mach: Physicist and philosopher*. Dordrecht: D. Reidel.

Duhem, Pierre 1962. *The aim and structure of physical theory*. New York: Atheneum; first published in 1914 under the title: La Théorie Physique: Son Objet, Sa Structure.

Koestler, Arthur. 1963. *The sleepwalkers*. New York: Grosset & Dunlap; first published, 1959.

Koyré, Alexandre. 1955. A documentary history of the problem of fall from Kepler to Newton. *Transactions of the American Philosophical Society* 45 (4): 329–395.

———. 1958. *From the closed world to the infinite universe*. New York: Harper & Brothers; first published, 1957.

———. 1965. *Newtonian studies*. Cambridge, MA: Harvard University Press.

———. 1971. *Descartes und die Scholastik*. Darmstadt: Wissenschaftliche Buchgesellschaft.

———. 1973. *The astronomical revolution*. Paris/Hermann/London/Methuen/Ithaca: Cornell University Press.

Mach, Ernst. 1970. *Einstein and the search for reality*. In Cohen and Seeger (eds.) 1970, pp. 165–199.

———. 1974. *The science of mechanics*. La Salle IL: Open Court; first published in 1893 as Die Mechanik in Ihrer Entwicklung Historisch-Kritisch Dargestellt.

Wettersten, John 2005, Whewell's critics: Have they prevented him from doing good?, with a forward by James Bell, commentaries by Joseph Agassi, Joseph Margolis, Michael Segre, Ronald Curtis, Maurice Finocchiaro and Godfrey Guillaumin, and replies by the author. Amsterdam/New York: Rodopi.

Wettersten, John, and Joseph Agassi. 1991. Whewell's problematic heritage. In *William Whewell: A composite portrait*, ed. Menachem Fisch and Simon Schaffer. Oxford: Clarendon Press.

Whewell, William. 1967a. *History of the inductive sciences*. London/Edinburgh: Frank Cass.

———. 1967b. *Philosophy of the inductive sciences*. London/Edinburgh: Frank Cass.

Chapter 19
Beyond Wavefunctions: A Time-Symmetric Nonlocal Ontology for Quantum Mechanics

Yakir Aharonov, Eliahu Cohen, and Avshalom C. Elitzur

Joseph Agassi has won eminence in many branches of philosophy, among them philosophy of physics. He has authored several insightful works on relativity, quantum mechanics and thermodynamics. To his merit is the open-mindedness he exercises in these works. Unlike his illustrious master, Popper, who has dogmatically asserted his semi-classical "propensities theory" (Popper 1983) as a solution to QM's riddles, Agassi humbly and gracefully pleads ignorance with respect to the latter. His writings are erudite, curious and thought-provoking. In short, he is a true philosopher and teacher. One of us (AE) had the privilege of being his student.

We take Agassi's attitude to QM as an invitation to present some insights we have gained during our research in this field. Following is a highly non-technical account of a few works which we believe begin to merge into a novel and rich picture of physical reality.

Twentieth century physics has seen the advent of two major revolutions. One is relativity theory, special and general. It is logical, self-consistent and, eventually, highly intuitive: Once you understood its basic premises, everything follows coherently, even when the predictions clash with everyday notions of space and time.

And then there is quantum mechanics, radically different. Newton's simple laws of motion give their place to Schrödinger's equation, which describes even particles in terms of wave mechanics. And alongside with classical conservation laws,

Y. Aharonov
Chapman University, Orange, CA, USA

Tel Aviv University, Tel Aviv, Israel
e-mail: yakir@post.tau.ac.il

E. Cohen (✉)
University of Bristol, Bristol, UK
e-mail: eliahu.cohen@bristol.ac.uk

A.C. Elitzur
Iyar, The Israeli Institute of Advanced Research, Jerusalem, Israel
e-mail: avshalom@iyar.org.il

N. Bar-Am, S. Gattei (eds.), *Encouraging Openness*, Boston Studies in the Philosophy and History of Science 325, DOI 10.1007/978-3-319-57669-5_19

quantum measurement obeys Heisenberg's uncertainty principle, which obliges several physical values to be indeterminate. This is the price QM exacts for other physical properties which are assigned precise values by measurement, no doubt a very heavy price, which even Einstein regarded as much too heavy. The challenge, then, is to develop new intuitions suitable for this unique reality.

Let us explain first why intuition is important for any physical theory. When one has a satisfactory mathematical formalism underlying a physical theory, one needs also to find the right questions to ask this formalism, which it either can or cannot answer. Yet this formalism can never *suggest* which questions to ask. What we need is the right *story* to tell about the formalism. Now this story should be *about the mathematics*, and at the same time *free of the mathematics*. That is what we mean by having a good intuition about the theory.

The two revolutions differed in this respect as well. In the case of relativity, Einstein came up with ingenious thought experiments, with which he showed that by accepting the universality of light's velocity as an axiom, one could predict all the rich new phenomena uncovered by the theory, thereby offering real intuition about its meaning. This was not the case with quantum mechanics. Even today, nearly a 100 years since the theory was discovered, people are still confused about the right story to tell about it.

In this short article we point out what we consider the reason for this unhappy situation, and then propose what we believe to be the right story for quantum mechanics.

As we all know, quantum mechanics was described initially using two different, yet eventually equivalent formalisms, namely that of Schrödinger and the other by Heisenberg. Because the Schrödinger formulation was mathematically simpler, everyone assumed that its wavefunction-based picture is the right way to think about quantum systems. In this case, however, mathematical simplicity does not lead to the right interpretation. As we shall show, Heisenberg's approach, although more complicated, is the one that leads to a better understanding of quantum mechanics.

Let us first show why the wave picture, originally suggested by de Broglie, was in fact misleading. In the case of electromagnetic field, the individual photon is this field's quantum, namely its smallest unit. The photon is the gauge boson (or force carrier) of the electromagnetic field: when two charged particles interact, they exchange photons. In other words, the field and the photon are two complementary aspects of the same physical entity.

Here comes a crucial restriction. *Electric and magnetic fields are defined only if the number of photons associated with the field is uncertain.* It is therefore a subtle error to associate wave properties to individual photons. It is similarly misleading to associate wave properties to *any* individual particle. In classical physics, wave properties can appear only with an ensemble of particles. Similarly, the wave properties of fields can be associated only with a large number of photons. So as far as individual particles are concerned, the Schrödinger wave can only give probabilities for finding particles at different locations, as Born indeed suggested.

Another shortcoming of the Schrödinger picture is the impossibility of fully measuring the wavefunction, i.e., non-orthogonal wavefunctions are indistinguishable in general. Thus in our current view, wavefunctions in quantum mechanics have the same role of the Boltzmann distribution in statistical mechanics – they are very helpful in describing the system as a whole, stressing its statistical features, but they do not serve as a measurable, ontic description of a single particle.

But how, then, can we account for the famous two-slit experiment, where it so famously appears that each quantum particle goes through both slits, thereby manifesting wave properties? This is where the Heisenberg picture saves the day: *The photon goes through only one of the two slits, yet the other slit affects it in a nonlocal way*. This is possible since the particle carries a nonlocal property known as a *modular momentum* (Aharonov et al. 1969), which is affected by the other slit being open or closed.

Indeed the fundamental difference between quantum dynamics and classical dynamics is that the former has nonlocal features that account for interference properties. It is interesting to see that quantum uncertainty allows the possibility of nonlocal interactions without violating causality. For this reason, one of us (YA) has asserted that "God plays dice to save free will" (Aharonov et al. 2014a, 2016).

The scope of this article does not allow us to go into all the details of how this new interpretation works and we suggest to refer to the relevant articles where this is discussed in great detail (Aharonov et al. 2014a, 2015a, 2017a; Cohen and Aharonov 2017). Here are just the essential ingredients of the proposed ontology.

Time-symmetry is taken as an axiom meaning that quantum systems are described by two boundary conditions rather than just one. This important axiom have allowed us to account for kinematic (Bell) nonlocality (Aharonov et al. 2015b) and for the measurement problem (Aharonov et al. 2014a). In the Schrödinger picture it has led to the Two-State-Vector Formalism (Aharonov and Vaidman 1990, 2008), incorporating two wavefunctions as the fundamental description of a quantum system, one evolving forward in time, while the other backwards. Together they determine the *two state*

$$ {}_{t_f}\langle\phi|\ |\psi\rangle_{t_i}, \tag{19.1} $$

which provides a richer notion of quantum reality than the wavefunction and enables to prescribe to any physical observable $\hat{A}$ its *weak value* (Aharonov et al. 1988)

$$ \hat{A}_w(t) = \frac{\langle\phi(t)|\hat{A}|\psi(t)\rangle}{\langle\phi(t)|\psi(t)\rangle}. \tag{19.2} $$

The latter characterizes the quantum system during all times between the initial and final conditions. It is symmetrically affected by both boundary conditions, thus generalizing the ordinary expectation value and can be found experimentally using *weak measurements* (Aharonov et al. 1988, 2014b).

In the Heisenberg picture which we now employ (Aharonov et al. 2015a, 2017a; Cohen and Aharonov 2017), the two boundary conditions define a double set of *deterministic operators* (Aharonov et al. 1970):

$$\hat{A}_{\psi} = \left\{ \hat{A}_i \text{ such that } \hat{A}_i \left| \psi(t) \right\rangle = a_i \left| \psi(t) \right\rangle, a_i \in \mathbb{R} \right\}. \tag{19.3}$$

In technical terms, this set answers the question "what is the set of physical observables $\hat{A}_i$ for which $|\psi(t)\rangle$ is an eigenstate". Therefore, the notion of collapse in no longer needed for inferring the values of these physical observables denoted by $\hat{A}_i$. All their values can be simultaneously and deterministically found. For this reason, we find the set of deterministic operators to be the true description of a single particle (Aharonov et al. 2015a, 2017a; Cohen and Aharonov 2017). This set is mathematically equivalent to the wavefunction, yet is much more revealing. Interestingly, some of these operators are *nonlocal*, meaning that their equations of motion contain nonlocal terms relating remote points in space. It is this property that allows a particle going through one slit to "sense" the other.

We believe that if quantum mechanics were discovered before relativity theory, then our proposed ontology could have been the commonplace one. Before the twentieth century, physicists and mathematicians were interested in studying various Hamiltonians having an arbitrary dependence on the momentum, such as $cos(p)$. In quantum mechanics, these Hamiltonians lead to nonlocal effects as discussed above. The probability current is not continuous under the resulting time-evolution, which makes the wavefunction description less intuitive. However, those Hamiltonians were dismissed as non-physical in the wake of relativity theory, allowing the wavefunction ontology to prosper.

This picture has stemmed from the TSVF which has given rise to many surprising effects and insightful paradoxes (Aharonov and Rohrlich 2005). Two recent gedankenexperiments have directly employed the novel time-symmetric Heisenberg ontology to explain paradoxical dynamics (Aharonov et al. 2017b, c). It is our hope that the discussed formulation would give rise to many other effects and would eventually lead to a deeper understanding of quantum mechanics.

Acknowledgements Y. A. acknowledges support from ISF grant 1311/14, ICORE Excellence Center "Circle of Light", and the German-Israeli Project Cooperation (DIP). E. C. was supported by ERC AdG NLST.

References

Aharonov, Y., and D. Rohrlich. 2005. *Quantum paradoxes: Quantum theory for the perplexed.* Weinheim: Wiley-VCH.

Aharonov, Y., and L. Vaidman. 1990. Properties of a quantum system during the time interval between two measurements. *Physical Review A* 41: 11.

Aharonov, Y., and L. Vaidman. 2008. The two-state vector formalism: An updated review. *Lecture Notes in Physics* 734: 399–447.

Aharonov, Y., H. Pendleton, and A. Petersen. 1969. Modular variables in quantum theory. *International Journal of Theoretical Physics* 2: 213.

Aharonov, Y., H. Pendleton, and A. Peterson. 1970. Deterministic quantum interference experiments. *International Journal of Theoretical Physics* 3: 443.

Aharonov, Y., D.Z. Albert, and L. Vaidman. 1988. How the result of a measurement of a component of the spin of a spin-1/2 particle can turn out to be 100. *Physical Review Letters* 60: 1351–1354.

Aharonov, Y., E. Cohen, E. Gruss, and T. Landsberger. 2014a. Measurement and collapse within the two-state vector formalism. *Quantum Studies: Mathematics and Foundations* 1: 133–146.

Aharonov, Y., E. Cohen, and A.C. Elitzur. 2014b. Foundations and applications of weak quantum measurements. *Physical Review A* 89: 052105.

Aharonov, Y., Landsberger, T., and E. Cohen. 2015a. A nonlocal ontology underlying the time-symmetric Heisenberg representation. arXiv:1510.03084.

Aharonov, Y., E. Cohen, and A.C. Elitzur. 2015b. Can a future choice affect a past measurement's outcome. *Annals of Physics* 355: 258–268.

Aharonov, Y., E. Cohen, and T. Shushi. 2016. Accommodating retrocausality with free will. *Quanta* 5: 53–60.

Aharonov, Y., Colombo, F., E. Cohen, T. Landsberger, I. Sabadini, D.C. Struppa, and J. Tollaksen. 2017a. Finally making sense of the double-slit experiment: A quantum particle is never a wave, to be published in Proceedings of the National Academy of Sciences.

Aharonov, Y., E. Cohen, A. Landau, and A.C. Elitzur. 2017b. The case of the disappearing (and re-appearing) particle, Scientific Reports 7: 531.

Aharonov, Y., E. Cohen, A.C. Elitzur, and L. Smolin. 2017c. Interaction-free effects between distant atoms, arXiv:1610.07169.

Cohen, E., and Y. Aharonov. 2017. Quantum to classical transitions via weak measurements and postselection. In *Quantum structural studies: Classical emergence from the quantum level*, ed. R. Kastner, J. Jeknic-Dugic, and G. Jaroszkiewicz, 401–425. Hackensack: World Scientific Publishing. arXiv:1602.05083.

Popper, K. 1983. *Quantum theory and the schism in physics*. London: Hutchinson.

Part IV
The Open Society and Its Challenges

Chapter 20
Critical Rationalism Applied: A Footnote to *The Siblinghood of Humanity*

Margaret Ng

Joseph Agassi taught in the Department of Philosophy of the University of Hong Kong from 1960 to 1963. He later contributed to a volume of papers edited by Ian C. Jarvie entitled *Hong Kong: A Society in Transition.*[1] His reputation stayed on after his departure. In 1971, I sought out Agassi in Boston University to study with him. I returned to Hong Kong in 1975, after completing a somewhat unorthodox doctorate. I never held a teaching post in a university, and have no significant academic publication in philosophy. My chief occupation was, at different times, in journalism, law and politics. Yet Agassi's teaching has profound influence on my work as a practicing barrister, and as a member of the Legislative Council – Hong Kong's mini parliament known as "LegCo" – in the transitional years from British colony to special administrative region of China, and particularly in relation to defending the rule of law and fighting for democracy under the promise of a sovereign state which believed in neither.

Bearing the watermark of the workshop of Popper's apprentice, my theory of law is that it is an approximation to justice and not justice itself,[2] in much the same way as science is an approximation to truth and not the truth itself. Judges in applying the law faithfully and impartially, without fear or favour, daily puts the law of the land to the test. Every now and then the law will be found to be wrong or seriously wanting, and the duty then falls upon the executive and the legislature to correct the error. We obey the law not because it is moral or just, but because it is our best

---

Joseph Agassi, *The Siblinghood of Humanity: An Introduction to Philosophy*, Caravan Books, Delmar, New York, 1991.

[1] International Library of Sociology and Social Reconstruction. London Routledge, 1969, 1998, 2013.

[2] Inadvertently aping H.A.L. Hart, discussed in Agassi, *ibid.*, p.251, which I did not read until years later.

M. Ng (✉)
Member, Sir Oswald Cheung's Chambers, Boston University, Boston, MA, USA
e-mail: margaretng@siroswald.com

N. Bar-Am, S. Gattei (eds.), *Encouraging Openness*, Boston Studies in the Philosophy and History of Science 325, DOI 10.1007/978-3-319-57669-5_20

approximation for the time being, and unless we take it seriously we cannot test it. The basic requirements of "law", that it must be clear and certain and accessible to those who live under it, of course, has its pedigree in the Magna Carta, that no one shall be deprived of his liberty except by law; but for my purpose, these requirements are necessary if the law is to be "falsifiable". The rule of law does not rest on an authoritarian foundation of judges who are fit arbiters of justice because of the guarantee of judicial independence – recalling radicalism which says that the truth manifests itself once we rid ourselves of all prejudice. The rule of law rests on its "testability", and a built-in system of self correction. Hence, at the heart of the rule of law is the reasoned judgment, in which the court states the evidence it considered and relied on or rejected and why, and the legal principle it applied, explaining how it reached its conclusion, in language accessible to the common men and women, so that all can examine for themselves the cogency or flaw, and the justice or injustice of the court's judgment, so that mistakes may be discovered and corrected so far as possible. The authority of the rule of law is, therefore, only as good as the reasoning of the latest judgment. Where an "unjust" judgment stands uncorrected, the rule of law is compromised. The fact that the judgment is nevertheless binding does not make it just. It simply makes injustice more visible. What is justice may be searched for in all sorts of ways. It may be metaphysical, analytical or intuitive. In the context of critical rationalism, diversity is good, while traditionalism might be distressed by diversity as a sign of disagreement which undermines authority. Indeed, the strength of the rule of law under the common law system in which I have my practice, depends on vigorous arguments from opposing parties.

The idea of justice is relevant to the legislator because he/she has the responsibility of approving and effecting law reform. The bulk of law reform is incremental improvement, and may not involve the idea of justice, but the dedication to the continuing review and reform of he law has its basis in the aspiration to a just system of law that is also workable. Likewise for the courts: the "landmark judgment" which overturns an old authority – these things do not happen without a strong sense of critical rationalism as applied to the law. It is correct and not ironical, that the law's authority comes from its open refutability and survival under constant and rigorous tests and criticism. "Justice," says a renowned jurist, "is not a cloistered virtue."

This is what I recommend to every legal practitioner and every jurist, and every legislator who believes in the rule of law, which is to be distinguished from the rule *by* law: the ruler using the law as a political instrument to control the people. This is all very contrary to the historical and current authoritarian regime of China under the supremacy of the Communist Party. Since the survival of the rule of law as we know it and as I have described is not possible under China's authoritarian regime, Hong Kong has no choice but to fight for a sufficient form of democracy as an autonomous region. Democracy is a necessary condition of the rule of law, though not a sufficient one. Popper stated his theory of democracy in an article which was recently re-published[3] as a principle that the government can be dismissed by a

[3]The *Guardian*, 31 January, 2016; the article was originally published on 23 April, 1998.

majority vote. This does not address the day to day business of parliamentary democracy. My theory of day to day democracy is also based on the idea of approximation: the approximation to the common good through criticism. The democratic debate is notionally an ongoing search for consensus – assuming the "common good" to be self-evident and therefore commands immediate agreement – and every resolution by majority vote is an acknowledgement of a temporary failure to reach consensus, a necessary expediency because of limitation of time and information and the list of business requiring resolution. Hence, democracy celebrates diversity, openness, rational debate, humility, and respect for others. Parliamentary procedure, which enforces as well as restricts the exercise of the freedom of speech of each member, exists for that purpose.

Critical rationalism is a theory of knowledge which is equally applicable to politics, ethics and society. Agassi most directly discussed this in his *Siblinghood of Humanity*[4] in which he criticized the banishment of politics from science and the refusal of philosophers to take positions of responsibility in intellectual leadership.[5] I started to translate *Siblinghood* into Chinese more than a decade ago in my spare time, and only recently completed it.[6] The last phase of my translation coincided with the Umbrella Movement,[7] the biggest ever civil disobedience action for democracy in Hong Kong under China. In that context, the last chapters of *Siblinghood* were blindingly relevant, and will remain so as the Movement has metamorphosed into a living dialogue on the future of Hong Kong.

In *Siblinghood*, Agassi opens with the question of rationality which he poses as "How should I choose my lifestyle?" and, further, "How do I decide? How do I decide how to decide?" which takes the reader into the field of the greatest generality, a methodology which applies to science and society, ethics and politics. The unfolding of the question and the search for answers which have been attempted and countered, takes us through the history of philosophy from the ancient Greeks to Popper. Every time we are brought back to the starting point: how should I choose my lifestyle? By only what is universal and unchanging and therefore true, i.e. living only by science? By that which is above customs and convention but applies equally to all mankind regardless of race or origin and all forms of differentiation, i.e., by the notion of all humanity as siblings? But the pursuit of science, particularly science as proof, falls into all kinds of difficulties. Proof, as Agassi demonstrates summarily, is impossible. Then there is the opposing view that there are no universals and no "reality"; only particulars and appearances: the community we live in, the traditions we live by, our faith and loyalty. But forms of parochialism also fall into difficulties: how then does on account for change? How does one account for

[4] See the article note.

[5] *Ibid.* p.266.

[6] Unpublished; the text is available on Agassi's website. The translation was a joint effort of myself and my former assistants Mr. Chan Wai Hung and Mr. Wong Man Lai to whom grateful acknowledgement is due.

[7] 28 September-16 December 2014.

adhering to "mere" customs and convention when they result in great human suffering?

It is necessary to see the arguments, as Agassi presents them, in their minimalist terms to draw what we lawyers call "the bright lines". It is because anyone with some diligence can fill in the details; what we should focus on is the debate. Debate provides criticism as well as options and alternatives. Debate tells us which are the details that matter. But the benefits of the exercise would be lost to us if we equate success with correctness. Agassi has untiringly said elsewhere and repeats here as well, that an idea which turns out to be flawed may be far more important because it opens new avenues for development some of which may lead to the truth. It is the error, not the idea, that we should discard. Examples of such flawed but fruitful ideas abound in the history of science and philosophy, and he gives many in this book. The error of classical rationalism does not deprive it of its great merits, and what is needed is not to discard but to improve upon it.

The narrative line of *Siblinghood* is simple and recognizable, but digressions into fascinating and instructive examples of great ideas and their great flaws, discovered by the rigorous criticism of their challengers, gives it great richness. Effective criticism is an art and a skill that grows with practice and knowledge. More difficult is the change of mind-set from the search for certainty and ultimate authority which is infallible, to a commitment to perpetual test and refutation in the quest for ever better solutions. The history of philosophy assists because all sorts of mistakes have been made, and all sorts of ideas have been expounded. We do not have to make the same mistake again ourselves. We can stand on the shoulders of giants, and take off from the point where their mistakes have been discovered and lasting merits have emerged.

Such an approach is much needed to-day to be applied to our search for solutions in the field of politics, when world law and order threatens to disintegrate under the shadow of terrorism, when people in developed economies are disillusioned about democracy and political leadership, even as smaller nations are struggling for democracy, when the rise of populism contributes to the problem and not the solution. When it seems that peaceful means are exhausted, people inevitably turn to look at means other than peaceful. It is time we returned to rationality. Philosophers should be promoting it by their own examples. Yet, as Agassi complains, there is instead deliberate disengagement. Science and politics do not mix, it is said. Philosophers do not interest themselves in politics.[8] Indeed, it is almost as if politics is synonymous with prejudice association with which taints the credibility of the pursuit of truth, if not the pursuit itself. Agassi thinks otherwise. The question of morality is part of philosophy, and the most important contexts for moral decisions are political and religious.[9] A question in the political context is, should we obey the law if it conflicts with morality? The classic democratic answer is, one should obey the law and urge the legislature to change the immoral law. But, what if efforts to move the legislature fails, or the legislature is itself undemocratic and in need of

[8] Agassi, *Ibid*, p.266. Lawyers say the same of law.

[9] *Ibid*. p. 249.

reform? Is civil disobedience open to those who maintain we should respect the rule of law as a solution? Mahatma Gandhi's campaign succeeded, in Agassi's view, because his adversary was the British Government, not Nazi Germany. Martin Luther King's solution was that civil disobedience upholds, not defies, the rule of law since the participants make the sacrifice of accepting the punishment of the law consequent upon the offence.[10] But is it still a valid argument if one is faced with a court injunction? Is disobeying a court order, even if the order is wrong in law and unjust, qualitatively different from disobeying a statute? Can King's argument still stand if by disobeying a court order he was guilty of contempt of court?[11] More than that, for different reasons, both Gandhi and King succeeded in their actions. What if their actions, noble as they were, failed because of an indifferent and undemocratic government – such as in the case of the Umbrella Movement? When all peaceful means fail to change an unjust law, is it moral to resort to means which are not peaceful, as some participants in the Umbrella Movement felt compelled to do in the end? In that situation, is Nietzsche to be preferred, when he spoke of going beyond good and evil,[12] and when he recommended taking risk?[13] Agassi did not carry the critique that far, but the real argument has to go even farther.

Siblinghood offers not just a methodology but also an appealing proposal for the way forward. The siblinghood of humanity as encapsulated by Agassi is the idea that everyone deserves to be treated as an autonomous individual, as one who has the right to make their own decisions and not to be enslaved to the decisions of others; everyone has the right to decide about one's own affair. It is an ideal which should be held by everyone. But it is not easy to practice, and Agassi points out that there is no way to implement it fully. Trying to do so is utopian, and both unrealistic and irresponsible.[14] It is irresponsible because to be responsible to the whole of humanity is very easy. It is harder to be responsible to one's country, friends and relations, and such responsibility in practical life is seldom discussed in propounding the ideal.

So it is irresponsible just to hold up the ideal of the siblinghood of humanity with which to beat parochialism, the idea of dividing people into "them" and "us" and treating them differently. Such divisions are there and the fact has to be recognized.

---

[10] Agassi, *Ibid.* p.251–252.

[11] Though not mentioned by Agassi, this was in fact the situation King found himself in. That was also the situation faced by the occupiers in the Umbrella Movement. In the latter case, the authoritative Hong Kong Bar Association publicly stated that defying an junction, be it rightly or wrongly issued, put the occupiers outside the justification of civil disobedience, however noble their objective. However, in the opinion of Yale law professor Owen Fiss, writing on the case of King, the differentiation between disobeying a statute and disobeying a court order had no valid basis: "Statutes are commands issued by a legislature and injunctions are noting more or less than commands issued by a court. So it is hard to understand why people owe a greater duty of obedience to an injunction." See Fiss, "The Civilizing Hand of the Law? Birmingham, 1963" published in the *Yale Law Review*, January, 2001.

[12] Agassi, *Ibid.* p.245.

[13] *Ibid.* p.269.

[14] Agassi, *Ibid.* p.253.

Rejecting parochialism because it is not universal is sterile, and sharpens the conflict between those who stand by parochialism and those who adhere to universalism without advancing the universalist ideal itself. Attempts to accept ever more demanding obligations to all humanity are getting to be regarded as irresponsibility towards people in one's own society, while the rejection of the obligation towards wider humanity in order to protect local self-interest is being regarded as discriminatory and also irresponsibility. We can see this happening in the world around us more and more, long after Agassi's book was published. In a way, the dramatic events of Brexit and the election of Donald Trump as president of the United States exposed the depth of the sense of resentment of those who believed their interests are sacrificed towards people who advocate global views and universal values. When the resentment asserted itself in popular votes, the world was shocked by the outcome, and trust in democracy was shaken. On a much smaller scale but no less poignant, this was what happened in the Umbrella Movement. The cry for self-determination free of China's domination, for at least half of the younger generation of democrats, was the demand for recognition of local identity and local interest of Hong Kong people as a people, against the old democrats who advocated universal human rights, and responsibility towards asylum seekers, torture claimants and immigrants from China.

Siblinghood of humanity is unattainable and parochialism is unpalatable. Instead of sharpening the conflict, Agassi proposes that, instead of sharpening the conflict, we should first work at creating the conditions that would better accommodate both our duty to kith and kin as well as the demand of the abstract siblinghood of humanity. Adopting this attitude, siblinghood of humanity ceases to be utopian and becomes operative.[15] This is the exploration that we should embark on, philosophers and participants of politics, and engage each other in the process. We should seriously consider taking on the responsibility as leaders – of opinion at least, if not also in government.[16] The retreat of philosophy from political debate does harm both to philosophy and politics. It makes philosophy academic and irrelevant to real life, from the individual's choice of his lifestyle to the suffering of mankind. It also makes thinking people shun politics, thereby leaving the welfare of the nation to chance and ambition.

As Agassi accepts, political responsibility is voluntary. He rejects seeing responsibility as to the whole of humanity and sees responsibility as of different kinds: first, to oneself; second, to the weak in one's immediate neighbourhood, especially to the child and the elderly; and third, to society at large, depending on one's circumstance, such as the taking on of public office. This third kind of responsibility brings us to the problem of leadership: how leadership can be defended if every individual is autonomous and equal with every other individual? Agassi has no defence of leadership, just the observation that leadership together with its

[15] Agassi, *ibid.* p.255–256.

[16] In October 2000, Agassi visited Hong Kong and gave a paper "On Civil Disobedience". His discussant was Denis Chang, Senior Counsel, a public figure and former adviser of Hong Kong's last Governor Chris Patter. The occasion was, sadly, the only one of its kind.

responsibility is universal in that it is found in every society and in nature. But the rules of leadership are not universal. Since leadership is autonomous at the expense of the autonomy of others,[17] the question is what rules of leadership are most favorable to the implementation of the siblinghood of humanity. In traditional societies where leadership is obligatory and founded on birthright, but democracy leaves the choice of leadership open. The responsibility that goes with leadership, like leadership, is a matter of contract. Choices are two-way arrangements: one has to both agree to lead and to be elected, and the same goes for every social status which must be by consent. And if consent is not to be passive but vigilant, then democracy must be participatory, and all parties of the contract participate in leadership, each to his or her chosen degree. Agassi quotes Bertrand Russell's beautiful saying that democracy is aristocracy without the snobbery.[18] I agree. I would say and have on many occasions said, that we can all choose whether or not to be aristocrats, but once we have chosen must behave like one, including all the responsibility that goes with it.

Is this too utopian? Of course it can be. But applying the new attitude Agassi has proposed, we can make it operative, by finding ways towards it instead of imposing it on all as measure of validity. Hong Kong is still being denied the democracy that China has once promised in the Basic Law it promulgated but is now reneging upon by re-interpreting its articles; Hong Kong people still cannot choose their Chief Executive[19]; half of Hong Kong's legislators are still elected by organizations favored by the Government; the people cannot dismiss the government by majority vote. But the other half of the legislature is returned by popular vote, and LegCo has the power of impeachment against the Chief Executive under the Basic Law. The Judiciary is still independent and the rule of law is still working, and by any reasonable view these are good conditions with which to seek further improvement. After the clearance of the occupy sites under court orders, most of the participants considered the Movement an utter failure, because their civil disobedience action had not resulted in compelling China to change its stance. But in truth the Movement was an unprecedented success, in that it unleashed a territory-wide inter-generation debate on the future of their city and what is to be done and legitimately can be done to achieve it. Young idealist democrats are prepared to pay the sacrifice of their personal liberty, security and career in their commitment to fight for the future for all. It's not only their future but democracy and the rule of law in Hong Kong matters to the world. The debate is on-going and is serious. Critical rationalism, not Nietzsche, hopefully will be the greater influence.

---

[17] Agassi *ibid.* p. 255.

[18] *Ibid.* p.255.

[19] The head of the government and head of the Hong Kong Special Administrative Region.

Chapter 21
En Route to *The Siblinghood of Humanity*?

Miriam Farhi Rodrig

> *Finally, historians of science should attend very critically to the marvellous but vague theory of the siblinghood of humanity and the humane but questionable claim that faith in it is the foundation of science.*
>
> (Agassi 2016)

We are in the making of an information revolution in which big data is transformed into knowledge for policy recommendations by national and local governments, global institutions and other large organizations. Against this backdrop, a growing migration crisis pokes the western world threatening its past achievements and adopted values. My aim is to discuss one grand solution and three more modest solutions to the migration crisis which unsettles the European Union in particular. Moreover, my aim is, also, to extend invitations to decision makers to critically move towards the adoption of, at least, one of the solutions discussed. In addition to discussing both the grand solution proposed by Rosentein-Rodan, and the three more modest solutions, namely the reactionary, the reformist, and the radical, I explore the way decision makers can facilitate the conditions to formulate initiatives within the new status quo regarding the migration crisis in the European Union. My bias is for the humanitarian model and I consider myself a traditional liberal for *laissez passer* and *laissez faire*. My methodology is that of critical rationalism, which critically examines competing solutions to problems. I opt for the adoption of the grand solution in conjunction with the radical solution and for facilitating the conditions within the status quo in the European Union for a more critically assessed, dignified and fruitful migration.

The Siblinghood of Humanity is the title of a book by Joseph Agassi.

I would like to thank my friend and running mate, Tina Esther Pesah, for discussing with me every detail of the article and editing it accordingly.

M. Farhi Rodrig (✉)
Tel Aviv University, Tel Aviv, Israel
e-mail: miriam.farhi@gmail.com

N. Bar-Am, S. Gattei (eds.), *Encouraging Openness*, Boston Studies in the Philosophy and History of Science 325, DOI 10.1007/978-3-319-57669-5_21

21.1 Introduction

It is more than a quarter of a century since I encountered my friend and academic advisor, Prof. Joseph Agassi. A brief 5 min were sufficient for me to realise that I was taking part in a lecture by an outstanding and very special person. I immediately corrected my posture so that I could savour the lecture. Since that day more than a quarter of a century ago, in a classroom at Tel Aviv University, the only way that I can converse or discuss a problem with Joske or ask him a question is by sharpening my focus first. Most of the time when he shares with me his examination of a problem, a small light lightens inside my mind as if its dimensions grow. There is no way I can thank Joske for brightening my mind and giving me a relentless practical morality within a democratic framework. In other words, thanks to Joske, I consider myself a 'sophisticated' liberal democrat. By sophisticated, I mean, 'not naïve". I am deeply moved and feel privileged, to contribute to this volume in his honour.

In the second half of the twentieth century, after two world wars, many regions in Europe were naturally, forcefully, and even aggressively divided into states. Thus, Europe entered a new phase in its development. The citizens of different states, as well, entered a new phase in their expectations: They expected democracy and liberalism to provide the political and financial anchors of their respective governments so that their demands would be met. The citizens of the different states expected as well, *grosso modo*, that there will be no more wars, that the states will take responsibility for their citizens, that international institutions will resolve problems among states, that advancement in science will discover the secret of the world, and that technological advancement will facilitate the different chores of life. Within this framework, states delineated borders and delegated duties and rights. It was hoped that these may eventually change for the better in the quality of life within the given states where laws protect the citizens. As to the interactions among the citizens, both at an economic level and at a social level, they were supposed to take place almost by an individual hand, a la Francis Bacon and Adam Smith. This model is the Westphalian model according to which, the sovereignty of the states is unlimited and cannot be subject to any higher or different locus of power.[1]

According to those defending humanitarian principles, however, states must be made accountable for their behavior. Humanitarian principles, as opposed to those of the Westphalian system, include the principles of humanity, neutrality, impartiality, and operational independence. They are rooted in international humanitarian law and are devised to guide the work of humanitarian actors.[2] Humanists, as opposed to nationalists, may accept a transnational or global government to which

[1]The Westphalian system of sovereign states was established in 1648 as part of the Peace of Westphalia. There were three core principles to the treaty: The principle of sovereignty, the principle of (legal) equality of states, the principle of non-intervention of one state in the international affairs of the other.

[2]These principles are given by the UN Office for the Coordination of Humanitarian Affairs (OCHA).

nation states would be accountable. The question is, what if some power gains control over the global government? Agassi answers, "There is reasonable fear that if such a central organisation were instituted, some power may usurp and control it and thus increase the risk rather than reduce it. Hence, a world constitution assembly is urgently needed." (Agassi 2014. p.8).

Until the last quarter of the previous century, the mainstream understanding was that of the Westphalian model while the humanitarian model remained marginal as if it was a script written for a fringe play off Broadway. Although the humanitarian model remained marginal, Einstein and Russell cooperated to present the highlights of this marginal model and Rosenstein-Rodan presented an outline, a working plan for this model. My bias is for the humanitarian model and I consider myself to be a traditional liberal for *laissez passer* and *laissez faire.*

Here, I critically examine the possible solutions to the migration crisis, which pokes the European Union in particular. The essay is divided into five parts: after the introduction, in the second part, I give a short description of the problems that arise at the turn of the millennium. In the third part, I discuss the solutions to the migration crisis: first the grand solution based on the outline presented by Rosenstein Rodan and then, the three more modest solutions; namely, the reactionary, the reformist, and the radical. In the fourth part, I discuss the way decision makers may facilitate conditions within the present new status quo regardless of the solutions proposed. Last, I make my concluding remarks. None of the solutions discussed are absolute or final. I invite decision makers to critically move towards the adoption of one or more of them.

21.2 Problems at the Turn of the Twenty-First Century

The twenty-first century in the western world looks different from its predecessor, as if something went wrong. What went wrong? Where did we err? Or, maybe, nothing went wrong and we did not err. Those who say that something went wrong, blame the decision makers who, according to them, did not put reigns on the advancement of technology. Those who say that we erred, blame the decision makers who, according to them, did not put reigns on the process of globalization and on the threats to the sovereignty of the states. Those who put no blame on decision makers find, in the current fluctuations, an opportunity to dare revise and improve what there is by adopting the methodology of critical rationalism; that is, the methodology of daring to move towards new solutions, examine their outcomes and revise the suggested solutions according to the situational logic and revised aims. Those who blame technological advancement may be described as closed communities; for example religious communities. Those who blame globalization may be described as illiberal or undemocratic communities, who would like to put limits to openness, democracy, and liberalism. Those who put no blame on decision-makers may be described as critical rationalists or democrats with a progressive and social twist; for example, thinkers such as Joseph Agassi (Agassi 1985) and his peers.

They dare to revise and improve upon what there is, aim at expanding the limits of openness, democracy and liberalism in a way that is progressive and socially oriented. Their emphasis is on revision according to the situational logic of what there is and of the decision makers' movements towards policy recommendations. The broad aim of the application of this methodology is to move towards an extendable and revisable liberalism within a system of evolving democracy. It is the liberal methodology of continual revision of movements towards solutions and policies.

In my search for proposing a tentative solution toward which we may move among all the solutions that I discuss here, I find myself in a team with critical rationalists, even though I am, to a great extent, disillusioned even disenchanted with the perceived extent of the problems we experience. I am, in particular, disillusioned with the gap between the pressing global problems and global politics, which has not seriously developed institutions to meet these problems. This gap is particularly notable in the ineffective and shallow policies against the expanding migration crisis in the twenty-first century. (Farhi Rodrig 2015).

The idea of making global politics more effective and vigorous by creating a global government for better global coordination is not new. Even though several institutions such as the UN, NATO, and the International Criminal Court exist since the end of World War II, the creation of a global government has not been seriously discussed. Yet, global problems such as Agassi's 4Ps (Pollution, Nuclear Proliferation, Poverty and Population Explosion), global terrorism, global migration, and a return to non- rational choices became more acute. (Agassi 1985, 2014). Of course, there could be a failure in global coordination and the emerging state of affairs may be worse for all parties involved. If no blame is put, then decision makers may have a chance for seriously reshaping coordination by daring to revise and improve upon what there is.

Is it possible that an immediate mega crisis with extensive damage be a trigger for leading decision makers to take such a decision?[3] Can escalating conflicts among desolate nations with the outcome of global migration to the West in general and to the European Union in particular qualify for such a mega crisis? After all, according to the UNHCR, there are about 65 million displaced persons moving around the planet. (UNHCR 2015a, b).

Mega crisis or not, similar to most residents of the western world, I am puzzled by the effects of the major technological and geo-political transformations around me including the migration crisis. Against this backdrop, I find myself more and more at a loss in expressing and assessing the moral dimensions of these changes.[4]

We can think of at least three revolutions that occurred lately. The first is the internet revolution of the 1990s with the creation of infinite data and the second is the birth of social media at the turn of the century. In my opinion, we are in the making of another (third) revolution: the transformation of big data into knowledge that

---

[3] This point has been suggested to me by my colleague and friend Kobi Tadmor while we were discussing the article at its dawn.

[4] Even though moral problems exist only on situations that decisions and measures are taken seriously and not randomly, when anything goes, the assessment of morals become redundant.

is gradually being used by global, national and local governments and other non-governmental institutions in their policy recommendations; for example, in the management of smart cities. It will be up to us, as residents and decision makers of Western countries and consumers of this new knowledge, if we want to make the world more intransigent/uncompromising or more social.

The western countries I am referring to are economically and politically advanced countries. An economically and politically advanced country, at least in our case, is a liberal democratic country. The decision makers of liberal democratic countries put their trust in the future by basing themselves on their past achievements as benefactors of humanity (nation building, creation of democratic procedures, technological advancement) and on their adopted values (liberalism, individualism, civil and human dignity). Today, decision makers of the Western world feel threatened and confused regarding their achievements and adopted moral values. Yet, in spite of their confusion, they have to devise policies to confront the threats to their achievements and to their adopted values such as the threats that the migration crisis poses. Their decisions and policies reflect their confusion by being less rational and more neurotic, and simply fearful of what will come next.

Nevertheless, I do not examine here the psychology of the decision makers of the West. I have a more modest aim, which actually is not so modest… As I mentioned earlier, I explore viable solutions to moderate the migration crisis in the West; in particular, in the European Union. This crisis affects decision makers of national and local governments and non-governmental institutions. A movement towards a solution to this crisis would not only give the confused decision makers a light to hold on to, but may also assist us in their critical march towards the "Siblinghood of Humanity". (Agassi and Meidan 2016). As Bauman (2013) says, "I attach myself to the third category of people today – people of hope who believe the world can be made more hospitable than it is right now." The other two categories are, according to Batman, the optimists who claim we are living in the best of world and the pessimists who suspect that the optimists may be right.

21.3 The Solutions

In general, there is basically one grand solution and three more modest solutions to moderate this growing and intensifying migration crisis in the West. The grand solution requires wide global coordination; e.g. some kind of global governance, which is not easy to achieve nor is it immediate. The three more immediate solutions, which are the reactionary, the reformist and the radical, are more modest than the grand solution, only because they require no global governance, coordination, and even cooperation.

The Grand Solution The grand solution is to return of hope to the most desolate regions of the world by investing in their infra-structure through some global authority. Though this solution may be considered to be revolutionary, it has already been proposed by Rosenstein-Rodan (1943), Austrian economist and to some extent, philosopher, in the first half of the twentieth century and discussed since then in different forums. For example, Agassi (1985) discusses this solution in his book *Technology* in 1985 and Oxford economist Collier (2008) discusses a similar solution in his book *the Bottom Billion* in 2007. Both Agassi and Collier still continue to discuss this solution in different forums. (Agassi 2014).

According to Rosenstein-Rodan, "a massive investment in the infrastructure of the most desolate regions in the world may, in the first place, trigger hope for a better future in the desolate areas and then act as a booster to accelerate their economies". In a nutshell, Rosentein-Rodan's outline for a development model for desolate regions suggests that aid to these regions must be given, massively and upfront, in terms of investment in their infrastructure in order to invite foreign capital and accommodate local plants. Along these lines, this developmental program and the ensuing spillover effect that this policy produces in nearby desolate areas may immediately result in containing and delaying migration to the West.[5] Moreover, this program may, to some extent, even invert the direction of migration provided that migrants receive practical education and training in the countries to which they migrated.

In addition, the spillover effect may still be beneficial for the conflict-ridden areas, where different interests are involved, even though there is not much that can be proposed for such regions. Conflict-ridden areas, as well, are also plagued with despair. Investment in the infrastructure of nearby desolate areas may arouse hope and help contain and even resolve the conflict there. (Syria, Sudan) The main problem with this recommendation is that it requires wide global coordination which takes time to achieve, at least for now.

Nevertheless, even though global coordination is lacking, the time may have come to invest heavily instead of sporadically not only in the infrastructure of conflict-stricken desolate areas but also in countries hosting migration. George Soros, Chairman of the Open Society Foundations and of Soros Fund Management, (2016) proposes to engage in spending a large amount of money up front in recipient countries. Accordingly, he says: "[this] would be far more effective than spending the same amount over several years. Front loading the spending would allow decision makers to address the most dangerous consequences of the crisis such as anti-immigrant sentiment in receiving countries and despondency and marginalization of refugees. Making large initial investments would help tip the economic, political, and social dynamics away from xenophobia and disaffection, and toward constructive outcomes that benefit refugees and the recipient countries alike.".[6]

[5] Spillovers of activity increases the number of others who undertake that activity.

[6] In the same article, to emphasise the impact of heavy investment, Soros adds that, The International Finance Facility for Immunization (IFFIm), which borrows against future government contributions to immunization programs, has raised billions of dollars over the past several years to ensure that vaccination campaigns are successful as soon as possible. In the long run, this is more effective than spending the same amount of money in yearly installments. IFFIm provides a convincing precedent for the current crisis.

The Reactionary Solution The reactionary solution is the fruit of decision makers who, against all odds, prefer to adopt a policy of isolating the migration of displaced people in their territory in order to restore the previous status quo of a more homogeneous society. The decision makers, who suggest adopting this reactionary solution, regardless of the damage inhered to their liberal democracy, prefer to strongly repudiate the present new status quo that includes a higher percentage of migrants in their territories. In other words, reactionary decision-makers propose to find ways to make migrants go away even if this policy disregards the values of liberalism, democracy, and human dignity and favors populist nationalism with shades of protectionism.

The Reformist Solution The reformist solution is the fruit of decision makers who prefer to adopt a restricted policy of weakly repudiating the present new status quo, that of an increased percentage of migrants in their territories. To weakly repudiate an increased percentage of migrants is to evasively reconcile with the changed demographics without looking for long-term social solutions that would significantly reduce the marginalization of the migrants. By not looking for long-term social solutions such as providing housing, education, welfare, and employment for migrants, these decision makers eternalize both the crisis and the marginalization. As such, they set the scene for a nightmare for the next generation. These decision makers offer some lip service to solving the migration crisis on the one hand, while on the other, they propose to isolate migration by implementing policies of selective inclusion only for those migrants that made it to metropolitan hubs. The aim, here, is to keep the complaints and hardships of the weak residents at a manageable level.

The Radical Solution The radical solution is the fruit of decision makers who prefer to adopt a policy of openly looking for ways to cope with the present new status quo. To cope with this status quo, decision makers have to dare to devise and implement novel policies of inclusion of the migrant community and revise them as mistakes and failures are discovered. This is the trial and error process in line with liberal democratic values and is in a bid to reach a new *modus vivend* with which all sides will be comfortable. In other words, decision makers who side with the radical solution agree to pursue new principles for coexistence.[7] I also opt for this solution.

Since the idea, even for those in favor of the radical solution, is not to maximize migration, but to keep it under control, decision makers alternate in their preferences. Sometimes, they are in favor of the reformist solution, at others they may be in favor of the radical solution. The grand solution of investing in the infrastructure of desolate areas, however, remains a viable solution all along.[8] Next, notwithstanding

[7] At times, the current Chancellor of Germany, Angela Merkel, seems to side, to some extent, with this solution: By accepting large numbers of migrants into Germany in 2016, Merkel showed that she is willing to make far reaching concessions related to the homogeneity of the German society and that she is willing to change the present new status quo.

[8] This may be the reason that, in 2016, Merkel visited factories in Ethiopia and discussed the possibility of erecting industries there that would hire locals for the benefit of only the local popula-

the grand solution, I discuss the way decision-makers may facilitate conditions within the present new status quo for a more critically assessed, dignified and fruitful migration.

21.4 Facilitating Conditions Within the Present New *status quo*

Is the status quo about numbers and ratios only? Or, is it first and above all about displaced people looking for a new shelter? I suggest that the status quo is not only about numbers and ratios. First and above all, the status quo is about displaced people looking for a shelter and my question is, in what ways can the conditions within the status quo be facilitated to this end of providing a shelter?

The conditions within the status quo may be facilitated by replacing the migration crisis with an already deeply-rooted housing crisis in the metropolitan hubs. This suggestion is also made by Eleanor Perry, writer and associate editor at the Open Democracy UK think tank, which serves as an independent global media platform covering world affairs, ideas and culture. (Penny 2016). The impact of the replacement of the refugee crisis with an overall housing crisis, in my opinion, will be, immediately, felt by both migrant residents and weak and disenfranchised residents.

In addition to that, the responsibility of migrants will be transferred from national governments to municipalities and local governments when the migrant crisis is replaced with a housing crisis. By assuming responsibility for sheltering the migrant residents, the municipalities will become the major decision makers in moderating the migration crisis as well. In other words, the hard work moves to the municipalities and local governments of metropolitan hubs where migrants are concentrated. Municipalities and local governments, then, would be responsible to resolve the deeply-rooted housing crisis in their territories, not only for the migrant residents or the locals, but for all their weak residents.

Needless to say that municipalities and local governments are more flexible in fine-tuning their policies in general, and in handling of a large number of weak residents looking for shelter within the new status quo, in particular. In fact, the more vividly the migration crisis is replaced with a deeply-rooted housing crisis, the more city municipalities and local governments will not ask themselves what to do with migrants entering their gates, but in what ways can they alleviate the housing crisis in their territory for all those suffering from the crisis. As I mentioned earlier, I think that municipalities can use their newly acquired expertise with the exploration of big data in order to formulate a viable housing model for all weak residents.

The catch, however, is that providing housing to migrant residents who are, in fact, part of the weak population may seem to jeopardise the chances of the weak

tion. This is not about a monopole firm hiring workers in desolate areas to profit from cheap work force and bring back revenues estimated in billion dollars.

local residents to be eligible for affordable housing; in this way, municipalities may find themselves playing into the hands of the right-wing populist politicians. Both groups, weak residents and right-wing politicians, accuse their governments – both local and national – for caring for migrants at the expense of weak locals.

Notwithstanding this catch, however, municipalities and local governments can assume responsibility for the housing crisis for all needy residents within their boundaries by asking themselves and the weak residents what to do with the crisis. Assuming responsibility is in line with the values of liberal democracy of being accountable for one's decisions and values, while declining responsibility over part of their weak residents is tantamount to neglecting to resolve the practical and moral conflicts arising within their territory. The question is what can municipalities do more specifically?

Once again, I have one grand, general solution and a few other more specific solutions. Actually, I already discussed some of these specific solutions to relieve the housing crisis such as the exploration of big data to extract information about demand and supply of affordable housing.

The Grand Solution The grand solution is to start a dialogue, immediately or as soon as possible, among all the parties involved – decision makers of local governments and municipalities, weak residents and migrants. This dialogue may grow into a bold grassroots public discussion or simply an open debate. Either way, it may guide responsible decision makers in the formulation of their agendas and initiatives to relieve the housing problem. I also suggest the following specific practical solutions to alleviate the housing crisis.

Finding Funds Competition to find funding for migrants and weak residents seems to be a major setback in containing the housing crisis. However, while funds for housing for weak locals come from national and regional governments, funds for migrants' housing come from charity and international organizations such as AMIF (Asylum and Migration Integration Funding of the European Union). In other words, the resources of the funds for the migrants are different from the locals. Thus, the available funds are independent of each other and do not favor one group at the expense of the other.[9]

Finding Partners for Cooperation and Reducing Housing Standards The partnership between local governments and housing associations set up to contain the problem of rising rent and homelessness may generate a higher supply of habitation.[10] In addition, a temporary reduction of housing standards or their revision in view of the recent housing crisis may, to some extent, alleviate the strained housing crisis by lowering the cost of housing both in building new accommodation and converting empty buildings into habitable shelters.

[9] The late Israeli experience of the local population in the north of Tel Aviv taking upon itself the extra care for the refugee children being integrated in the kindergartens in their privileged areas is an example of funding that is independent of municipal funding for the education of pre school children.

[10] In Amsterdam, this partnership lead to 1000 extra homes. (Penny 2016)

Increasing the Awareness of Cost and Benefit Calculations of Migrants Migrants have potential human capital that is an asset to the economy of the city and metropolis. The acknowledgement of the hidden or latent human capital in the migrants may provide the locals with the incentive to cope better with the housing crisis and adopt a more business-like attitude. In other words, profit and loss calculations may contribute to expand the enlightened self-interest of the locals and alleviate their concerns over migrants posing a threat to their future.[11]

Inclusion of Migrants and Weak Residents in Decision Making on Housing Local governments may recruit local and migrant residents for the housing project. To some extent, this may increase the dialogue among the two groups working to solve one common crisis. In other words, the voice of migrants can be heard loud and clear next to that of the weak locals and not one voice that favours one group at the expense of the other.

Exploration of Big Data and Its Sharing The exploration of big data (its disaggregation or breaking down) in order to extract knowledge may assist decision makers to have a better insight into the preferences and priorities of all the residents in need of affordable housing. The results may be used in policy recommendations as soon as possible, and mistakes detected and revised promptly. As such, decision making procedures may be more critical, open, transparent, and inclusive; that is, more democratic. Democratic decision making and the ensuing process can be shared immediately by other local governments eventually laying the grounds for some sort of global cooperation for cheap housing and moderation of the migration crisis. This is, in fact, what we have been looking for all along and it brings me to my concluding remarks.

21.5 Concluding Remarks

By somewhat alleviating the chronic housing crisis, municipalities may pave the way for all residents (locals and migrants) to consider the possibility of a common future in a more social metropolis. More specifically, this new perspective relieves the migrants from the responsibility of dealing with finding some sort of permanent shelter on their own. In this way, they may be more available to join the workforce in their new environment. This transition from the state of homelessness into a state of participation in the local work force reduces their marginalization, enables them to contribute to the local economy, and, eventually serves the weak locals in regard to their enlightened self-interest.

Ironically, regardless of these specific suggestions to alleviate the housing crisis, municipalities of metropolitan hubs, where migrants are concentrated in large

[11] The booming of the Munich economy where migrants are integrated in the workforce of Munich is an example. Many areas in Munich are gentrified and Munich is on its way to becoming a Metropolitan hub with a continuous housing crisis but not a migration crisis. (Penny, 2016)

numbers, meet the challenge presented by the influx of migrants by allocating space to informal economies and unfamiliar lifestyles in their citadels. In fact, municipalities extract significant revenues from such policies and the marginalization of the migrants is reduced as they contribute to the labor market, human capital development of receiving countries, and the public revenues. As a result, a feasible *modus vivendi* is created naturally for all residents, both migrants and locals, and the antagonism felt by the weak local residents against the migrants is, to some extent, dissipated.

Moreover, municipalities and local governments in metropolitan hubs manoeuvre between respecting national legislation and agreements concerning migrants within their area of jurisdiction (Benhabib 2006; Baubock 2005). Although this maneuvering of municipalities touches sensitivities regarding sovereignty, it alleviates the crises of migration and to some extent serves global social integration. This, however, is the subject matter of another paper. Instead, let me sum up my paper with the following two sentences: In my opinion, by investing in the infrastructure of the most desolate regions and by adopting the radical solution to manage the new status quo in the recipient countries, it is possible both to arouse hope in the most desolate areas of the world and to dismantle the chronic housing problem in a way that will meet the demands of all the weak residents in cities with a dense migrant population. Consequently, the migration crisis will be alleviated, a nightmare for the next generation will be, to some extent, avoided, and humanity will have taken a tiny critical step towards "the Siblinghood of Humanity".

References

Agassi, J. 1985. *Technology*. Dordrecht: D. Reidel Publishing Company.

———. 2014. To care for the future of the human race. *Eruditio e-Journal of the World Academy of Art & Science* 1(5), September-October. http://www.tau.ac.il/~agass/joseph-papers/to-care-for-the-future-of-human-race-jagassi-reprint.pdf. Accessed 27 Dec 2016.

———. 2016. Interview: Joseph Agassi: Interviewed by Mauro L. Condé, Raffaele Pisano and Michael Segre in September 2016. *Transversal: International Journal for the Historiography of Science 1* (December 2016) http://www.historiographyofscience.org/index.php/transversal/article/view/23. Accessed 5 Jan 2017.

Agassi, J., and A. Meidan. 2016. *Beg to differ the logic of disputes and argumentation*. Switzerland: Springer International Publishing AG.

Bamboock, R. 2005. Citizenship policies: International, state, migrant and democratic perspectives. *Global Migration Perspectives* 19 January. https://www.iom.int/jahia/webdav/site/myjahiasite/shared/shared/mainsite/policy_and_research/gcim/gmp/gmp19.pdf. Accessed 27 Dec 2016.

Bauman, Z. 2013. 150 Years of German social democracy. *Social Europe Journal*, June. https://www.socialeurope.eu/wp-content/uploads/2013/10/OccPap2.pdf. Accessed 27 Dec 2016.

Benhabib, S. 2006. *Liquid modernity*. Cambridge: Polity Press.

Collier, P. 2008. *The bottom billion*. Oxford: Oxford University Press.

Farhi Rodrig, M. 2015. Global migration: A transnational problem. *Journal of Globalisation Studies* 6(1): 3–13. http://www.sociostudies.org/authors/farhirodrig_m/. Accessed 30 Dec 2016.

Penny, E. 2016. We don't have a refugee crisis. We have a housing crisis, 5, November. https://www.opendemocracy.net/can-europe-make-it/eleanor-penny/we-don-t-have-refugee-crisis-we-have-housing-crisis. Accessed 27 Dec 2016.

Rosenstein - Rodan, P. N. 1943. Problems of industrialisation of eastern and south eastern europe. *Economic Journal* 53 (210/211):202–211. https://www.econ.nyu.edu/user/debraj/Courses/Readings/RosensteinRodan.pdf. Accessed 27 Dec 2016.

Soros, G. 2016. Bringing europe's migration crisis under control. *Social Europe* 12 April. https://www.socialeurope.eu/2016/04/bringing-europes-migration-crisis-control/. Accessed 27 Dec 2016.

UNHCR. 2015a. Figures at a glance. http://www.unhcr.org/figures-at-a-glance.html. Accessed 27 Dec 2016.

———. 2015b. Global trends forced displacement in 2015. http://www.unhcr.org/576408cd7. Accessed 27 Dec 2016.

Chapter 22
The Perfect Mediator and His Limitations: Toward a Multifaceted Approach to Peacemaking in the Israeli-Palestinian Struggle

Sapir Handelman

Who is the most suitable mediator to create a peacemaking change in difficult situations of intractable conflict? Realists will argue that an effective mediator is an intermediary with leverage who can impose settlement on the opposing sides. Pluralists and human needs theorists will claim that a suitable mediator is a problem solving facilitator who could help political elites of opposing parties find solutions to fundamental needs, fears and concerns that constantly fuel the conflict. Contractualists will suggest a consensus-building mediator who could help in establishing an effective peacemaking coalition from major elements in the opposing parties. This paper takes the Israeli-Palestinian conflict as a case study to examine the advantages and disadvantages of each proposal. It presents the different mediators and analyzes the negotiating processes that each of them suggests. It shows that no mediator can create the foundations of an effective peacemaking process by himself alone. The perfect mediator is built of a sophisticated mixture of prominent qualities of different types of intermediaries, which enables him to lead various peacemaking interactions simultaneously. The paper concludes that a multifaceted approach to peacemaking, which utilizes various types of mediation simultaneously, has the greatest potential to approach the ideal of the perfect mediator and create the conditions for an effective peace process.

S. Handelman (✉)
Achva Academic College, Arugot, Israel

The Minds of Peace Organization, St. Louis, MO, USA

The Minds of Peace Organization, Tel-Aviv, Israel
e-mail: sapir.handelman@gmail.com

N. Bar-Am, S. Gattei (eds.), *Encouraging Openness*, Boston Studies in the Philosophy and History of Science 325, DOI 10.1007/978-3-319-57669-5_22

22.1 Introduction

The Israeli-Palestinian struggle is an archetypical type of intractable conflict where ordinary citizens, and not standing armies, are at the center. It is a long-standing time struggle where violence is a dominant language in the relationship between the opposing societies. History shows that these types of conflicts cannot be resolved by force and militaristic means. Indeed, a long socio-political process was required to create a revolutionary transformation[1] in similar types of conflicts, such as the 'troubles' in Northern Ireland and the struggle against Apartheid in South Africa.[2] This paper focuses upon the question: Who is the most suitable mediator to lead an effective, and even a revolutionary, negotiating process that could help the opposing societies to discover the road to peace and stability in the Israeli-Palestinian situation?

Competitive schools of thoughts – which emphasize different aspects of the conflict and provide different tools to analyze it – suggest different answers to the question: Realists will offer an intermediary with leverage that could use militaristic, political and economic power to push opposing leaders to reach a settlement (cf. Kriesberg 2001: 378); Pluralists and human needs theorists will suggest a problem-solving facilitator that could help political-elites of opposing parties to find solutions to fundamental needs, fears and concerns that constantly fuel the conflict and make it intractable (Kelman 1996); Contractualists will vote for a consensus building mediator that could establish a broad peacemaking coalition of major societal elements from the opposing parties (cf. Curran and Sebenius 2003; Handelman and Pearson 2014).

The paper presents the different mediators and analyzes the advantages and disadvantages of the negotiating processes that each of them suggests. It shows that no mediator can create the foundations of an effective peacemaking process by himself alone. The perfect mediator is built of a sophisticated mixture of prominent qualities of different types of intermediaries, which enables him to lead various negotiating interactions simultaneously. For example, the super mediator should be able to create active public support in the peacemaking struggle, lead political-elites to construct a negotiated vision, and maneuver the opposing leaderships to conclude agreements. Is it possible to find such a super mediator?

The paper shows that even a super mediator – who probably does not exist in reality – can apply his special skills effectively and efficiently only in an adequate framework of rules and institutions. Moreover, he will need socio-political support of major societal forces to apply his special talents for the benefit of the opposing parties. In other words, the complexity of the situation indicates that even a perfect mediator – who can lead different negotiating tracks simultaneously and can always provide the right answer at the right time and place, similar to a legendary

[1] "'Revolution' as Eugen Rosenstock-Hussy has written, "brings on the speaking of new, unheard of language, another logic, a revaluation of all values."" (Friedrich 1966: 4)

[2] For a comprehensive account of the South African case, see Mandela (1994) and Sparks (1994). For a comprehensive account of the Northern Ireland case, see Mitchell (1999) and McKittrick and McVea (2002).

Machiavellian prince – is limited and cannot bring an effective change by himself alone (cf. Hayek 1994). The challenge is to construct a multifaceted approach to peacemaking that attacks the conflict from various sides, directions, and dimensions simultaneously.

A multi-faceted approach to peacemaking creates the conditions for different mediators to operate competitive negotiating processes (such as bargaining, conflict analysis and resolution, and consensus building) in a complimentary fashion. Thus, it provides the socio-political framework to approximate the ideal of the perfect mediator operating in a supportive environment.

22.2 Realism, Pluralism and Contractualism

Realism, Pluralism and Contractualism are three competitive schools of thought in Peace and Conflict Studies (cf. Kelman 1981; Handelman and Pearson 2014). Realists interpret conflict situations in terms of power struggle between competing interests; Pluralists and human needs theorists point out that conflicting parties tend to believe that the conflict is about their most fundamental needs and concerns (Banks 1984: 19–20; Fisher 1977: 3; Nye 2009: 120). Contractualists emphasize that the absence of a peace contract – accepted by the majority in the opposing sides – and effective mechanisms to enforce and keep it (social, political and militaristic means) is a classical symptom of intractable conflict (cf. Murshed 2002; Handelman and Pearson 2014).

Negotiation plays a key role in the three paradigms. But each approach emphasizes and concentrates on different aspects of the process: Realists emphasize the competitive aspects of negotiation – negotiation as bargaining; Pluralists and human needs theorists emphasize the cooperative aspects of negotiation – a joint search for solutions to the major issues that fuel the conflict – negotiation as a process of Conflict Analysis and Resolution (cf. Hopmann 1995); Contractualists look at negotiation as an attempt to build a broad active support among disputing parties for a peaceful settlement – negotiation as a consensus building process.[3]

Naturally, each school will suggest a different kind of mediator to lead an effective negotiating process: Realists will suggest a mediator with leverage that could push leaders to reach a settlement and would be able to provide guarantees to enforce and keep it (Kriesberg 2001: 378); Pluralists and human needs theorists will offer a problem solving facilitator that could help political elites reach an agreement that could address the fears and needs of the opposing parties "on a basis of reciprocity" (Kelman 1996: 100); Contractualists will suggest a coalition builder mediator who could establish a broad peacemaking coalition that is able to marginalize the impact of extremists and motivate the leaderships to conclude agreements. Table 22.1 presents the three different suggestions.

[3] Scanlon (2000: 5), who suggests a somewhat narrow perspective of moral contractualism, points that the contractualist project is "finding principles that others similarly motivated, could not reasonably reject."

Table 22.1 Realism, pluralism and contractualism

Paradigm	Negotiation	Mediator
Realism	Bargaining	Intermediary with leverage
Pluralism	Conflict analysis and resolution	Problem solving facilitator
Contractualism	Consensus building	Coalition builder intermediary

Each of these approaches can have a valuable impact on the efforts to create a constructive change in situations of intractable conflict in general and in the Israeli-Palestinian struggle in particular. However, reality shows that none of these approaches can sustain an effective peacemaking process by itself: the US – a mediator with leverage who operated in a Realist fashion from the Madrid conference of the 1990s to John Kerry's closed door diplomacy of 2013/2014 – succeeded in bringing Israeli and Palestinian leaders to the negotiating table but failed to broker a peace agreement (cf. Kriesberg 2001: 384–386); The Norwegians – who functioned as a pluralist problem solving facilitator in the Oslo accord of the 1990s – succeeded in providing the optimal conditions for political elites to initiate an innovative peace process that eventually collapsed into the violence and despair of the second Intifada in 2000 (Kelman 2007b). The idea of negotiation as a consensus building process never got serious attention and consideration in the Israeli-Palestinian situation.[4] Political-elite diplomacy – negotiations between leaders and political elites in realist and pluralist fashions – have been the main peacemaking channels in this case. The idea to involve the people in the peacemaking efforts – in order to create a broad active support base for a negotiated settlement – has not been explored seriously by leaders, policy makers and the international community.[5] Indeed, Senator Mitchell, who led a consensus building process in Northern Ireland during the 1990s – known as the 'all-party talks' – was the US envoy to the Middle East for a very short time and hardly left any impact (Deane 2009).

The complexity of intractable conflicts indicates that the prefect mediator is a superman who could lead different kinds of peacemaking interactions simultaneously. He should be able to function as an intermediary with leverage, a problem solving facilitator, a coalition-building mediator and an intermediary of any other useful peacemaking engagement. It is quite clear that such super mediator does not exist in reality.

The central argument of this paper is that a multi-faceted approach to peacemaking is needed in order to approach the ideal of the perfect mediator. A multi-dimensional peacemaking approach provides socio-political framework (rules and

[4] The idea of a mediator as a coalition builder intermediary is inspired by the peacemaking work of Senator Mitchell in Northern Ireland as described in Curran and Sebenius (2003). However, I do not know if the authors will agree to categorize their description and analysis as contractualist.

[5] For example, the idea to establish a consensus building institution – such as a major Israeli-Palestinian public negotiating congress – was never examined seriously by the relevant decision makers. This peacemaking institution could be a Middle Eastern version of the multi-party talks that prepared the ground for a revolutionary transformation of entrenched conflicts in Northern Ireland and South Africa during the 1990s (Handelman and Pearson 2014; Handelman 2016).

institutions) to operate different negotiating processes – such as leadership's bargaining, political elite's problem solving, and public consensus building – by different mediators simultaneously.

Let me present the advantages and disadvantages of each mediator in order to build and support the central argument of this paper.

22.3 Realism – An Intermediary with Leverage

22.3.1 Intractable Conflict and Peacemaking

Realists interpret a conflict situation in terms of power struggle between competing interests. For example, Edward Azar et al. (1978) has pointed out that global forces, international trends and regional factors – such as nationalism, religious fundamentalism, and clash of cultures –are motivating vehicles of intractable conflicts. They fuel the struggle and complicate the situation beyond imagination.[6] The practical meaning is that building and keeping peace, order and stability requires reliance on hard-nosed cost and benefit analyses and political maneuvers associated with realpolitik (cf. Allison 1971; Schelling 2005). Implied tactics include: sanctions, threats, rewards, political manipulations, and the final determination that power determines outcomes (cf. Bueno de Mesquita and Lalman 1992).

Indeed, there is a long tradition of political thought advocating the idea that any solution to civil war and intractable conflict depends on the drastic initiative of a strong political force (Wantchekon 2004). This tradition, began with Niccolò Machiavelli (Machiavelli 1979), continued with Thomas Hobbes (Hobbes 1985) and encompasses contemporary thinkers such as Samuel Huntington (Huntington 2006) and protagonists of the free market system, such as Friedrich Hayek.[7] Following this logic, it is quite acceptable to believe that Egyptian President Anwar Sadat's astonishing visit to the Israeli parliament in Sadat 1977 created a momentum for a major transformation in the geo-political structure of the Arab-Israeli conflict. However, this drastic move, which can be viewed as a diplomatic offense on a hardline Israeli government, was not enough to create a change. An active intervention of a mediator with leverage – that could put pressure on the Israeli and Egyptian leaderships – was required to use this momentum and broker an agreement.

[6] See, for example, Azar et al. (1978: 45): "[…] to understand the present and potential futures of the conflict in the Middle East (and elsewhere in the Third World), one must understand the 'rules' of the international political game, as established by the superpowers who dominate the theory and practice of international relations." It is quite acceptable that Azar is not a classical realist thinker. However, he provided a realist analysis of the phenomenon of intractable conflict. This paper presents a modified and actual version of Azar's view which was formulated during the cold war.

[7] Hayek F.A., interviews in El Mercurio (Santiago, Chile, 1981) from http://coreyrobin.com/tag/el-mercurio/

In contrast, mediators with leverage have constantly failed to broker a peace agreement in the Israeli-Palestinian situation. The questions are: Why? What are the differences between the two conflicts, which are interwoven? What lesson can be drawn from a critical comparison between the two cases?

22.3.2 Case Study – The Israeli-Egyptian Peace Process

In 1977, Anwar Sadat, the leader of the strongest Arab country and most rigidly entrenched of Israel's enemies, declared his intention to come to Jerusalem to talk peace in the Israeli parliament, the Knesset. Menachem Begin, the Israeli Prime Minister, invited him (or, more precisely, did not have much choice but to invite him). No doubt that Sadat's astonishing visit to Jerusalem in 1977 was the beginning of a turning point in Middle Eastern politics in general and in the Arab-Israeli conflict in particular (cf. Kelman 2005).

However, after the visit the parties were far from reaching an agreement. Perhaps, in order to breach the gaps they needed a mediator with leverage to motivate them to accept tough decisions which were necessary for a peace treaty. Indeed, the climax of Sadat's initiative was the Camp David accords of 1978. The two leaders, Begin and Sadat, met for thirteen days of closed door diplomatic interaction under the mediation of the president of the US, Jimmy Carter – a mediator with leverage.

Carter describes the summit as one of the most frustrating experience in his life (Bickerton and Klausner 2007: 190). It was very difficult for him to broker an agreement. He used his persuasive skills and the political and economic power of the US – threats and generous offers of financial support – to push the two leaders to reach an agreement. Had Carter actually controlled and manipulated the situation? Or, was he a victim of manipulative behavior of two sophisticated leaders, Begin and Sadat, who wished to increase US financial support?[8]

Sadat, who challenged the very existence of Israel in the 1973 war, did not abandon his desire to return the Sinai desert to Egyptian control and was looking for an opportunity to establish strong relationships with the United States (Hirst and Beeson 1981: 252–54). Moreover, after his bold peacemaking visit to Jerusalem, returning to Egypt without a deal would have seriously damaged his prestige at home and in the Arab world (Gilboa 1998, 220). His drastic initiative had such an impact on the international community in general and the Israeli public in particular that Begin – a longtime advocate of greater Israel – also could not afford the summit to end without a deal. There were good chances that coming back to Israel without an agreement would have led Begin and his government to lose credibility among the Israeli people and to face a serious confrontation with the USA (Kelman 2005: 130–31).

In short, the two leaders understood that they could not allow the negotiations to fail. The tough bargaining helped the two leaders to "motivate" Carter – who was caught in the same trap because failure would have meant damage to his political career and

[8] Handelman (2009) argues that not every manipulation is indecent. There are manipulations that could be desirable, even from a liberal perspective.

damage to the prestige of the USA as the leading superpower- to increase his generous financial offers. Following the agreement both sides (Israel and Egypt) were to receive massive economic and military aid from the United States: Israel was to receive $3 billion in military and financial assistance; Egypt was to receive $2 billion in military equipment. This was in addition to the existing 1979 foreign aid allocation of $1.8 billion to Israel and $1 billion to Egypt (Bickerton and Klausner 2007: 194).

It is hard to know if our assumption of Sadat and Begin manipulating Carter is true, half true, or completely imaginary. However, it is quite clear that the Israeli-Palestinian situation is a quite different story. It is a struggle were ordinary citizens are at the center of the conflict and not standing armies facing each other. Perhaps, Sadat and Carter were fully aware of the differences. Sadat, on the one hand, fought as much as he could for the Palestinian cause: "As for the Palestinians cause, nobody could deny that it is the crux of the entire problem. Nobody in the world could accept, today, slogans propagated here in Israel, ignoring the existence of the Palestinian People, and questioning their whereabouts. The cause of the Palestinian People and their legitimate rights are no longer ignored or denied today by anybody. Rather, nobody who has the ability of judgment can deny or ignore it" (Sadat 1977). On the other hand, neither Sadat nor Carter invited the Palestinians to participate in the negotiating process (Kriesberg 2001: 379). Was it a coincidence, absent-mindless, or deliberate intention?

The Realist tradition leaves no room for inviting Palestinians to take an active role in such initiative. As we will see later, pluralists and human needs theorists argue that this is one of the limitations of the Realist paradigm.

22.3.3 Limitations and Obstacles

Israel and Egypt were two functioning states with quite stable regimes and established institutions. It has been relatively easy for both sets of authorities to keep agreements and maintain a cold peace. In contrast, the Palestinian people, back then, did not even make their first steps toward the establishment of an independent state. And states are the key actors in the international arena according to the Realist paradigm (e.g. Nye 2009: 220). States are the only entities that could provide the conditions for keeping peace agreements and maintaining order and stability. Following the Realist logic, the initiative of the US secretary of state John Kerry, which involved Israeli and Palestinian officials in direct negotiations about 35 years later, was doomed to fail for very similar reasons.

Israeli Realists point out that the Palestinian Authority – in contrast to Israel – can be branded as a "failed state"[9] that hardly made its first steps toward independence. The Palestinian authority is facing major domestic socio-political problems that

[9] A "failed state" is a term used to describe a failure of the central authority to maintain law and order and meet basic needs of the population. Inbar (2006) claimed that the Palestinian Authority met the criteria for a failed entity. The disfunction of the Palestinian Authority led to a civil war between Hamas and Fatah in 2007 and a political division between Gaza and the West Bank for about 7 years.

constantly appear in modernizing entities during transitional periods. These symptoms include dysfunction of the political system and major difficulties in maintaining law and order. According to realism, it is impossible to achieve peace, security and stability in developing entities that lack the institutional framework of well-established states (cf. Huntington 2006).

A Realist looks at negotiation as bargaining. And bargaining is a form of competition. Constructive and effective competition – which means, in this context, negotiation by peaceful means – can emerge only in an adequate framework of rules and institutions of well-established states (cf. Buchanan 2001; Vanberg 2005). As stated previously, the Palestinians have hardly begun to build a unified governmental system that can maintain law and order, is able to represent the interests of its people, and is capable of coping with strategic and existential problems (cf. Rubin 1999). Therefore, any attempt to engage the two parties in a constructive competition (negotiation by peaceful means) can quite easily slide into a destructive contest, which means the renewal of the violent struggle. Indeed, the 2000 Camp David Accord – a closed door diplomatic interaction between Israeli prime minister, Ehud Barak, and Palestinian Authority chairman, Yasser Arafat, under the mediation of the US president, Bill Clinton – collapsed into another cycle of violence, known as the Second Intifada (Bickerton and Klausner 2007: 327–334).

Israeli Realists argue that in this chaotic situation the conflict cannot be resolved and has to be managed (Inbar 2006). Any serious negotiations on solutions to the conflict with "representatives" of a failed Palestinian entity ('conflict resolution approach') under the pressure of a mediator with leverage only endanger the very existence of the state of Israel. Israeli Realists provide policy recommendations for improving the domestic and foreign position of each society in order to reduce the intensity of the irresolvable struggle ('conflict management approach'). Hopefully, conditions for negotiation of a peace agreement will ripen in the future (cf. Bar-Siman-Tov 2007; Inbar 2006; Landau 2006). Should the opposing people sit and wait for the ripe moment? Or, perhaps, they should be engaged in a multi-dimensional peace process that might create the socio-political conditions for the ripe moment?

22.3.4 Concluding Remarks

One of the unintended consequences of the peace treaty between Israel and Egypt was the Palestinization of the Arab-Israeli conflict, a process that started after the 1967 war (Kelman 1997: 184–85). No Arab state would try to go to war with Israel and the Palestinians became a central player in the confrontation. The results of the failures of the 2000 Camp David Accord and the 2014 John Kerry's peace initiative were desperation on both sides which, probably, contributed to bloody confrontations: the Second Intifada in 2000 and Operation Protective Edge in 2014.

The Palestinization of the Arab-Israeli conflict is an unintended consequence of a peacemaking initiative while the desperation from peace and escalation in the conflict following the failed negotiations in 2000 and in 2014 were predicted by

Israeli Realists. A mediator with leverage is not a Machiavellian Prince that can make miracles and always provide the right answer at the right time and in the right place. A mediator with leverage can function well and contribute to negotiations only in a supportive environment (Handelman 2011).

Generally speaking, the Realist paradigm gave some satisfactory answers to major security problems during the Cold War when two superpowers were engaged in intractable competition to dictate and control the world order.[10] Realists point out that their sophisticated doctrine provides a good explanation for one of the most astonishing miracles of the twentieth century – 50 years without a nuclear war (Schelling 2005). Moreover, the Realist paradigm afforded insights related to power and interests and in some cases provided forced endings to ethnic and religious-based conflicts such as the most recent war in Bosnia. Richard Holbrooke, the American mediator, led an aggressive peace process to end the bloody ethnic strife in Bosnia and Herzegovina, in 1995. He used American and NATO military and economic power to shape and manipulate the situation in a way that resembles a Realist textbook for "peacemaking" diplomacy (Curran et al. 2004).[11] However, the initiative, known as the Dayton accords, is an exceptional example. Mediators with leverage, even if they are official representatives of the US government, cannot simply impose an agreement, especially in difficult situations of intractable conflict where it is not clear if the local leadership is able to control their constituents and implement agreements (Kriesberg 2001: 386).

Realism often failed to provide satisfactory strategies and tactics to cope with protracted intrastate conflicts, civil wars and widespread carnage and loss of innocent lives. To better cope with the challenge, scholars like John Burton (1969) and Herbert Kelman (1981), have begun to develop a new pluralist approach to the study of peace and conflict (Fisher 1977). The key assumptions of this fresh perspective contradict cornerstones in the traditional Realist paradigm (Nye 2009: 220; Fisher 1977: 6; Banks 1984: 19–20).

22.4 Pluralists and Human Needs Theorists – Mediator as a Problem Solving Facilitator

22.4.1 Intractable Conflict and Peacemaking

Protracted conflicts often begin with "real problems" (not everything is only psychology). "Real problems" are problems that in principle can be quantified, managed, and in an ideal world could be, probably, resolved by a reasonable compromise. Examples of 'real problems' could be: territorial claims, dispute over allocation of

[10] One pf the highlights of the Realist paradigm was Kennedy's peaceful resolution of the Cuban Missile Crisis, which threatened to destroy the world in 1962 (Allison 1971).

[11] http://www.foreignaffairs.com/articles/65352/patrice-c-mcmahon-and-jon-western/the-death-of-dayton

resources and demographic complications. However, in our real human world the subjective dimension (the mental sphere) complicates the situation beyond any possible imagination (cf. Kelman 2007a: 63).

Pluralists and Human Needs theorists, such as John Burton (1988) and Herbert Kelman (2007a), have emphasized that opposing parties, involved in intractable conflict, tend to believe that the struggle is about basic needs. The basic needs at stake include material needs, "such as food, shelter, physical safety, and physical wellbeing" and psychological needs, "such as identity, security, recognition, autonomy, self-esteem and a sense of justice." The negotiating process is a joint research of and exploration for an agreement that addresses the fundamental needs and fears of both parties "on a basis of reciprocity" (Kelman 1996: 100).

The mediator becomes a problem solving facilitator that leads a process of Conflict Analysis and Resolution. The analysis focuses upon discovering basic needs, fears and concerns that have led different groups to be engaged in intractable conflict. The resolution is an agreement that can satisfy those basic needs and give guaranties that can help opposing factions to overcome their fears.[12] The idea of negotiation as a process of Conflict Analysis and Resolution led to the development of unconventional mode of diplomacy, known as – track II diplomacy.

Track-II diplomacy involves informal discussion, research and exploration between unofficial political elites from both sides. Its aims are to understand the difficulties of the other side, to discuss unpopular and creative ideas, to construct a vision for a peaceful social order and to prepare the formal negotiations.[13]

To achieve the maximum results from the interaction, the participants in track-II diplomacy are usually "mid-level" elites. This means that they have to obtain a social status of political-elites but they cannot have an official position in the government. Examples of "mid-level" political-elites can be retired politicians, senior journalists, scholars, businessmen, and former military generals (Agha et al. 2003; Kelman 1996).

The mid-level status of the participants in track II diplomacy (unofficial political elites) has advantages on two levels. On the one hand, their unofficial status gives them more space to maneuver and a great degree of freedom to explore creative and sometimes unpopular ideas. On the other hand, their elite status and connections are supposed to help them in transferring knowledge gained from the interaction to current leaders and policy makers. In an ideal track II interaction, the participants are able to: think "outside of the box" and come up with innovative insights for making changes in a desperate situation of intractable conflict; to share the new ideas with

[12] Scholars, who have endorsed a pluralistic approach to the study of destructive social conflicts and their resolution, developed various methodological approaches to peacemaking and named them differently. For example, Burton (1969) named his version "control communication"; Fisher (1977) called his understanding of the peacemaking process "interactive conflict resolution"; Kelman (1996) labeled his approach "interactive problem solving workshop."

[13] Compare to Agha et al. (2003), who distinguish between "soft" and "hard" track-II interactions. The first is "aimed at an exchange of views, perceptions, and information between the parties to improve each side's understanding of the other's positions and policies." The second is intended to "help negotiate political agreements between governments."

influential policy makers; and to shape the public opinion by civil activities, such as: giving lectures, writing newspaper articles, and publishing books.

The Oslo accord of the 1990s is a classic example of political-elite peacemaking diplomacy. The accord progressed as an impressive experimental exercise in a peacemaking laboratory. The initiative progressed gradually through various diplomatic channels as track-II diplomacy (unconventional mode of diplomacy) turned into secret diplomacy (secret negotiations between officials) that, eventually, led to track-I interactions (formal negotiations). The unofficial part involved a problem solving facilitator while the official part involved a mediator with leverage (Agha et al. 2003; Kelman 1997; Handelman 2011: 83). Thus, a peacemaking engagement that started in a pluralist manner developed into a peacemaking interaction in a realist fashion. Contractualists would argue that a consensus building process – that could involve the opposing people in the struggle for change – was missing from this combination (Handelman and Pearson 2014). It could have prepared the opposing publics for an innovative peacemaking process, supported the accord and, perhaps, led to different results.

22.4.2 *Case Study – The Oslo Accords of the 1990s*

The accord was initiated through secret track-II meetings in Oslo, Norway in 1993. It was an unofficial exploration of possibilities for reaching a peace agreement. The primary participants were two Israeli academics and several low ranking officials from the PLO.[14] The Norwegians – who did not have leverage over the peace talks and did not have geo-political interests in the region – acted as an impartial third party. They helped to keep the process secret and provided optimal conditions to the participants to develop trust and working relationships, to understand the fears, needs and concerns of each side, and to explore possibilities to end the conflict (cf. Kelman 1997: 189–93; Kriesberg 2001: 378).

Although the initiative was unofficial, it was quite clear that the Norwegian government was behind the project. It was an official third party intermediary that could offer an optimal framework for an unofficial problem solving engagement. They provided logistical support and some forms of facilitation in a secret and delicate process and, at the same time, helped contacting high level officials in the opposing societies (Kelman 1997: 191).

Perhaps setting up a facilitative working environment was the best way to help Israeli and Palestinian political elites – who came from different cultures and held

[14] In peacemaking literature, classical track II diplomacy is supposed to include unofficial elites from both sides. Since the Palestinian delegation included officials, some scholars suggested labeling the Oslo talks as track 1.5 diplomacy. The important issue is that that the first stage of the Oslo process (the unofficial talks) had almost all the important characteristics of track II meetings. They were secret, not obligatory, and mediated by a third party (Norwegians) that did not have any direct interest in the region. Compare to Agha et al. (2003, 2–3).

different perceptions of reality – to form a problem-solving team. However, almost from the beginning the participants were not alone in the peacemaking game. Any substantial progress was reported back to the political leaders from both sides. The feedback that was received indicated to the process participants the limits of any possible solution to the main controversial matters.

As soon as it became clear that the Oslo talks might result in a formal agreement, the Israeli team was expanded to include official negotiators. Track-II diplomacy (unofficial problem-solving contact) turned into secret diplomacy (secret negotiations between officials) (cf. Agha et al. 2003).

The Oslo accord was formalized with the Declaration of Principles (DOP) in Oslo Norway, in August 1993. It was officially signed in Washington on September 13, 1993, by Israeli Prime Minister Yitzhak Rabin and PLO leader Yasser Arafat. The agreement was a framework for the future relations between Israel and the anticipated Palestinian state (although there was not any specific commitment to establish an independent Palestinian state). After the agreement was signed, Track-I talks – formal negotiations under the leadership of the US, a mediator with leverage with clear interests in the region – became the main channel for negotiations between the two sides, while back-channel diplomacy continued to support the process (Savir 1998). The intentions were to prepare the ground for the negotiations of a permanent agreement, which was planned to begin no later than May 1996.

The Oslo peace process looks like an impressive exercise taken from an academic textbook for the peacemaking diplomat. The communication channels of political-elite diplomacy were used effectively and efficiently (Handelman 2011: 86–89). The secret and unofficial problem-solving interaction at the beginning of the process enabled the two leaders (Rabin and Arafat) to explore different ideas to promote peace and stability in a way that they could not do in the official negotiations between Israelis and Palestinians that were conducted at the same time in Washington (following the Madrid conference in 1991). To put it differently, a problem-solving facilitator (Norwegians) created optimal conditions to engage political-elites in an effective process of Conflict Analysis and Resolution, while the US – a mediator with leverage operating in a realist fashion in Washington – led a frustrated and unsuccessful official negotiating interaction. The subsequent secret negotiations between officials who joined the Oslo talks enabled the formalizing of agreements and prepared the ground for making the process official. The signing ceremony in Washington and the beginning of implementing agreements made the peace process concrete to Israelis, Palestinians, and the international community.

It is important to emphasize that the delicate nature of track II diplomacy, under the hospitality of a mediator without leverage that did not have a direct interest in the region, enabled Rabin and Arafat to explore peacemaking opportunities in a way that could not have been done in any official negotiation (Kelman 1997). The secret, unofficial, and not obligatory characteristic of the track II interaction – which resembles a Conflict Analysis and Resolution workshop – made it possible to explore creative and unpopular ideas that even contradicted the official position of both sides. Indeed, the breakthrough which eventually led to an historic agreement – that included mutual recognition between Israel and the PLO, and the establishment of the Palestinian Authority – was achieved at the beginning of the process at the track II engagement.[15]

The sophistication and the effectiveness of the track-II interactions enabled Rabin and Arafat to appear as visionary leaders who were ready to challenge entrenched conventions. However, the new negotiated vision was composed by a group of Israelis and Palestinians in unofficial voluntary backchannel diplomacy. It is not clear if Rabin and Arafat fully endorsed the vision and committed fully to its implementation. For Example, as Kelman (2007b) noted, both leaders continued to be suspicious of each other and to prepare the ground for the process not to progress as anticipated. Thus, both sides continued to keep the struggle alive: the Palestinians built up military capacity beyond the level stipulated in the agreement and did not stop the incitement against Israel, while the Israelis continued to expand and strengthen the settlement project which is perceived by Palestinians as a violent strategy that can put an end to any peacemaking process.

The Oslo process did not involve the public in any substantial way. It was almost entirely dependent on political leaders (Rabin and Arafat) who were ambivalent toward the peace process that they approved and blessed. Perhaps, secrecy, unconventional methods of back channeled diplomacy, efficient functioning of a problem solving third party facilitator, and the central part that top Israeli and Palestinian leaders played in the accord – were the only way to reach a historic agreement. However, it was not enough to give the peace process a life of its own and create the conditions to end the Israeli-Palestinian conflict, to say the very least.

The public from both sides were not prepared to endorse innovative ideas and accept necessary compromises that a peace process required and to trust the other side in times of expected and unexpected crises. The lack of social foundations for an effective and reasonable peace process enabled extremists, radicals, and professional spoilers to continue dictating conditions for the rest of the societies. These actors perpetuated the struggle via violent tactics, such as ongoing suicide bombing attacks inside Israel; and the expansion of Israeli settlements in the Palestinian territories (Kelman 2007b: 292–293). These tragic episodes were discovered by extremists to be effective tools of destructive public diplomacy.[16] They clearly demonstrated to the people of each society that there are no partners for peace on the other side, resulting in a massive pressure on the leadership to stop the process. Indeed, Israelis, supporters of the process, began to associate Palestinians with suicide bombers and Palestinians, who were initially supporters of peaceful methods of diplomacy, started to correlate Israelis with Jewish settlers who, they believe, wish to take their land (cf. Kelman 2007b: 292).

[15] Agha et al. (2003) claim that "the Oslo talks can be considered "the mother of all track II talks" in the Middle East."

[16] For a comprehensive discussion on the meaning of public diplomacy, see Gilboa (2008). Handelman (2012a) suggested an expanded perception of the concept 'public diplomacy', which was initially developed during the cold war.

In 1995 Rabin was assassinated by a radical Israeli who probably opposed the Oslo accord. The ongoing suicide bombing within Israel, during this time, gave the impression that the new Israeli government, under the leadership of Shimon Peres, Rabin's successor, was caught off-guard and burdened with an imaginary peace process. The dissonance between the grand peace vision of Peres – "New Middle East" (a prosperous peaceful region) – and the tragic reality on the ground paved a way for a new right wing Israeli government. The new government, under the leadership of Benjamin Netanyahu, advocated a different line of thinking than the Oslo path and the grand vision of Peres remained a fantasy (Handelman 2009: 83–99).

Unfortunately, one crisis led to another until the collapse of the Oslo peace process into the second Intifada in 2000. Ironically, every side blamed the other one for the failure of an innovative process that was intended to put an end to the Israeli-Palestinian struggle (Kelman 2007b: 294–295).[17]

22.4.3 Limitations and Obstacles

Political-elite diplomacy can be very effective in initiating a peace process and helping political leaders and policy makers conclude agreements. Its back channel mode of communication (such as: track II talks) can be very efficient in creating a negotiated vision and dealing with "sensitive" critical problems that require careful treatment, delicate process of analysis, and complete secrecy. The price is that the public is not involved in the peacemaking efforts. Consequently, the people are not prepared to accept necessary compromises, to support new innovative solutions to old problems and to cope with expected and unexpected crises that a revolutionary peace process entails. Kelman (1996) has already pointed out the inherent tension between the ability to engage political-elites of the opposing sides in an efficient and innovative process of conflict analysis and resolution (pluralist perception of negotiation) and the possibility (or more precisely lack of possibility) of transferring the results to the general public.

The Oslo Accords, like many other peacemaking efforts in similar cases of intractable conflict, indicate that as the peace process begins to accelerate, violence tends to increase.[18] One of the reasons is that extremists are making all possible efforts to wreck the process and continue dictating conditions to the rest of the society. People, on opposing sides, who are prepared to open a new page in

[17] There is a lot of literature focused on the question of why the Oslo peace process failed. I present only one hypothesis that certainly does not give a complete picture. My main point is that political-elite diplomacy, diplomatic interactions between elites involving a problem solving facilitator, cannot bring a sustainable peace by itself alone in the complicated situation of the Israeli-Palestinian struggle.

[18] Indeed, violence in different forms, shapes and variations carried out by the enemies of the peace process appeared during the all-party negotiations in Northern Ireland (Mitchell 1999: 117), the multi-party talks in South Africa (Sparks 1994; Sisk 1994:52), and the Oslo accords in the Israeli-Palestinian case (cf. Kelman 2007b: 292).

their relationships, understand that the only answer to such provocative episodes is to continue promoting the culture of peace by all means. However, they need to see hope, instruments to cope with expected and unexpected complications, and a strategy to end the conflict. This is a major difference between the Oslo accord and the revolutionary peace process in Apartheid South Africa that happened around this time.

The violence during the beginning of the peace process in South Africa increased to a level that could not be tolerated. Any mechanism that was employed to cope with the phenomenon, such as the National Peace Accords in South Africa, could not stop the aggression. The level of violence in South Africa began to decrease only when the people saw that the peace process began to have a life of its own and could not be stopped (Sisk 1994: 53–66).

It is hard to imagine a problem solving facilitator engaging the mass public in a process of conflict analysis and resolution. Bar-Tal (2001) has argued that it is much easier to mobilize the masses by approaching basic impulses such as fear than to create hope by appealing to their cognitive skills. The involvement of the public in the peacemaking efforts and preparing them for a change requires a different strategy.

22.4.4 Concluding Remarks

The failure of the Oslo accord to bring peace and stability touches a fundamental question concerning the whole idea of negotiation as a process of 'conflict analysis and resolution' – who should negotiate?

Elitists argue that political-elites are the perfect participants in such a problem solving negotiating process. They have the ability to create a vision, reach a settlement and draft a practical program that can address the needs and fears of the opposing societies (Kelman 1996). No doubt, populists will disagree. Populists will argue that leaders and political-elites are part of the problem (Ionescu and Gellner 1969): they certainly cannot determine the needs and fears of ordinary people and, often enough, they are interested in preserving a conflicting situation which is beneficial for them. Contractualists will accept the populist criticism of elitism.[19] As liberals, they will point out that there is an urgent necessity to involve the people – those who suffer from the continuation of the conflict on a daily basis – in the negotiating process (Handelman and Pearson 2014). The questions are: Who are the people? Is it possible to involve all of them in the peacemaking efforts?

A coalition builder mediator intends to provide answers by using a sophisticated and powerful tool – multi-party negotiating forum (or public negotiating congress).

---

[19] The endless debate between populism, elitism and liberalism is beyond the scope of this paper.

22.5 Contractualism – Mediator as a Coalition Builder

22.5.1 *Intractable Conflict and Peacemaking*

One of the symptoms of intractable conflict is that extremists, who have a clear agenda, dictate conditions to everyone. The moderate majority – those who wish to end the conflict by peaceful means – are desperate.[20] They do not see any viable solution to the violent struggle and they let radicals shape the socio-political reality.[21]

The contractualist mediator's major task is to create an effective peacemaking coalition that could establish active public support for a negotiated settlement, marginalize the impact of extremists and motivate the leaderships to conclude agreements. One of the methods to achieve this goal is to offer a political platform to involve major elements of the opposing societies – leaders, political-elites and ordinary people – in the efforts to reach a comprehensive peace contract. The multiparty negotiations in Northern Ireland and South Africa during the 1990s were designed to create such a peacemaking effect.[22] Curran and Sebenius (2003) argue that Senator Mitchell, the independent chairman of the all-party talks in Northern Ireland during the 1990s, is an archetypical example of coalition builder mediator. The senator succeeded in building working relationships between entrenched enemies, formed an effective coalition of major centrist parties, and paved the way to the Good Friday agreement in 1998.

22.5.2 *Case Study: The All-Party Talks in Northern-Ireland*

The all-party talks invited representatives of at least eight major political parties active in Northern Ireland to discuss, debate and negotiate solutions to the struggle. Each representative had to commit to general rules of non-violent dialogue, known as 'The Mitchell Principles.' Political parties with paramilitary arms, such as Sinn Fein, which refused to accept Mitchell Principles, were excluded (Mitchell 1999: 35–36). The intention was to transform the aggression in the streets into a peaceful discussion around the negotiation table.

The all-party talks were designed to facilitate negotiation between local political leaders and create public and international support for reaching a reasonable

[20] Public surveys show that most Israelis and Palestinians wish to end the conflict by peaceful means. However, the majority on both sides believe that it is not possible. See for example: http://www.timesofisrael.com/survey-most-israelis-palestinians-support-2-states/

[21] For example, the settlement project, in the Israeli-Palestinian case, was born in a political vacuum (cf. Gorenbeg 2006).

[22] In both cases there were various forms of multi-party negotiations. The South African assemblies were: Convention of Democratic South Africa (CODESA) and Multiparty Negotiating Forum (MPNF) (Sparks 1994). The Northern Ireland's multi-party negotiations are known as the 'all-parties talks' (Mitchell 1999).

contract. To be more specific, the all-party talks had three main functions: (1) to provide a political platform for negotiations between local political leaders; (2) to operate as a public diplomacy device that could create public support and public involvement in the struggle for change; (3) to create pressure on, or at list to prepare the ground for, the leaders of Britain and the Irish Republic – the two countries that were directly involved in the conflict – to conclude agreements (Mitchell 1999; Handelman 2014). This was a complicated consensus building process that created tensions that had to be resolved.

The involvement of the major political parties in the negotiations was intended to give the process credibility and gain the confidence of the people. However, it was almost impossible to reach a peace contract – which addresses the most fundamental problems and sensitive concerns – in an assembly that included so many political parties with different socio-political agendas. To cope with the difficulty, the senator used a sophisticated (and even manipulative) tool – sufficient consensus. Sufficient consensus is a compromised decision-making process between simple majority (more than 50% vote yes), that could not get the confidence and support of the minority, and full consensus (unanimous agreement), that would be impossible to reach. However, the sufficient consensus mechanism, which Senator Mitchell constructed, provided the actual decision making power to the two major moderate political parties while giving the impression that every participant has a major impact on the procedure (Mnookin 2003: 199–215). For example, according to the rules of sufficient consensus, one of the tests that any proposal had to meet was gaining "support from parties representing a majority of both Catholic and Protestant communities" (Curran and Sebenius 2003). This test enabled Mitchell to form a coalition of the two major moderate parties and marginalize the impact of the more radical ones. Nevertheless, it was still very difficult to reach a peace agreement by sufficient consensus.

Unfortunately, the all-party talks were teetering from one crisis to another and could not finalize an agreement. Moreover, paramilitary groups continued to perform violent actions on a daily basis (Mitchell 1999: 126; McKittrick and McVea 2002). The senator's main focus was on building trust and working relationships between leaders of the major moderate political parties. This strategy helped to give the peace process a life of its own and motivated dissident political parties, such as Sinn Fein, to accept Mitchell Principles and to join the assembly (Mitchell 1999; Mnookin 2003).

The all-party talks had an important contribution in creating a public atmosphere of change and to prepare ordinary people to accept necessary compromises for reaching a settlement. They succeeded in involving the mass public in the peacemaking process through reports back to the community by the leadership cadres, press conferences, demonstrations, scandals and through provoking public debate in the opposing communities. Extensive and exhaustive media coverage of the peace talks slowly began to concretize in the minds of the people that a peaceful pathway could resolve long standing political disputes (Mitchell 1999). The persistence, determination and commitment of major public players to the peace process significantly diminished public support for the political violence pathway. Consequently, the preference for collective violence among dissident elites fell from grace as non-violent principles of negotiation gained legitimacy. Multi-party talks were successful in establishing an effective peacemaking coalition of the moderates who could marginalize the impact of radicals and extremists who, previously, dominated the political discourse (cf. Curran and Sebenius 2003). This consensus building process created the conditions for the leaders of the two governments – Britain and the Irish Republic – to reach negotiated solutions to the most difficult and sensitive problems (Mitchell 1999, 143–183).

The idea of mediation as a process of consensus building did not get any serious attention in the Israeli-Palestinian situation. The dominant peacemaking experience in this case is interactions between leaders and political elites through various diplomatic channels almost without any public involvement. This makes any peace process fragile and vulnerable to aggressive actions of extremists (Handelman 2012b).

History shows that in situations like the Israeli-Palestinian case, any effective peacemaking initiative will cause an increase in the level of violence. Extremists on both sides will deliver their maximum efforts to crash the initiative. A broad peacemaking coalition of the moderates, which is able to push the peace process forward, can be an appropriate answer. Indeed, in South Africa and Northern Ireland multidimensional peacemaking efforts eventually succeeded in gaining broad support and the people's confidence in the process. The results were that violence became ineffective politically and started to decrease (Mitchell 1999; Sparks 1994). Moreover, political parties, which were associated with the violent struggle, began searching for ways to join the process. From a contractualist perspective, the questions in the Israeli-Palestinian context are: How can a consensus building process be created in this difficult situation? Who can be a suitable mediator that could create a major Israeli-Palestinian peacemaking coalition that will include leaders, political elites and ordinary people, and get the support of the international community?

22.5.3 Limitations and Obstacles

Senator Mitchell, who led the all-party talks in Northern Ireland, was a third party coalition builder mediator who operated without leverage. His main instruments were his persuasive ability and his experience as a politician (Curran et al. 2004). Curran and Sebenius (2003: 16–28) describe Senator Mitchell as an exceptional coalition builder mediator who knows how to play the right game with the right people at the right time and place. However, as talented and sophisticated as the Senator is being presented by these authors, he did not act in a vacuum. The establishment and operation of the all-party talks was supported by exogenous forces, such as the UK-Irish rapprochement and US urging; and they were inspired by the EU example of peaceful relationship-building, the peaceful revolutions of 1989 in central and eastern Europe, the unification of Germany in 1990, and the beginning of a negotiated revolutionary transition in South Africa around the same time (Mansergh 1999). No such conditions exist in the Israeli-Palestinian situation.

The international community and incumbent local leadership focus on traditional forms of diplomacy and do not show any signs of considering the importance of involving the general public in the peacemaking efforts and establishing broad peacemaking coalitions (Handelman 2011: 99–100). Moreover, the geo-political situation in the Middle East is teetering between the hope for a democratic region that the Arab spring of 2010 created and the despair of ongoing civil wars and the brutal spread of radical Islam in countries like Iraq and Syria which began to have a momentum about 2 years later. Therefore, it is not clear what an Israeli-Palestinian coalition building process will look like.

There is a bold grassroots attempt to build a major Israeli-Palestinian public negotiating congress with political power, named the Minds of Peace project. The initiative focuses on conducting short-term Israeli-Palestinian public negotiating congresses in different locations. These short-term congresses – named the Minds of Peace Experiment – are designed to demonstrate the peacemaking power of a major congress, evaluate its potential outcomes and create domestic and international support for its establishment (e.g. Handelman 2014). However, it is not clear if the socio-political conditions in Israel and Palestine are ripening for such initiative to grow and have an impact on the political reality. Nevertheless, it is worth remembering that the multiparty talks in South Africa during the 1990s (CODESA) – a major consensus and coalition building process – was inspired by the Congress of the People, a grassroots initiative which set up people's vision of a new South Africa, about 40 years before, in 1955 (Thompson 2001: 209). Perhaps, the Minds of Peace Experiment – short-term Israeli Palestinian public negotiating congresses, which are being conducted in different formats, sizes and locations – is the first cornerstone in the establishment of a major congress (Handelman 2014).

It seems that the road to establishing a coalition building process in the Israeli-Palestinian crisis is paved with almost infinite impediments. The cultural barrier is an example of a major obstacle for peacemaking in the Israeli-Palestinian case that did not play a significant role in Northern Ireland. It helps to demonstrate that the two conflicts, which share many symptoms and characteristics of intractable conflict, are also very different.

It is quite clear that the process of building broad peacemaking coalitions requires a deep understanding of the culture and mentality of the opposing parties. Broadly speaking, Senator Mitchell and the people in Northern Ireland share western traditions. Israelis and Palestinians do not share the same tradition and mentality. Israel is a modern Western Democracy while the Palestinian society is part of the Arab world. There are major differences between the Western approach and the Arab approach to critical issues, such as mediation, negotiation, trust and conflict resolution (Hamdan and Pearson 2014). It is likely to assume that change in general and a coalition building process in particular will develop differently in the two societies.[23]

The cultural gap between the two societies is a well-known barrier for almost any peacemaking endeavor (cf. Alon and Brett 2007). It is an obstacle to the negotiating

[23] In a quite similar way, Huntington (Huntington 2006) noted that revolutions in the East and revolutions in the West tend to grow and develop differently.

process itself and the ability to transfer progress and results to the people. A coalition-builder mediator would need to be able to understand the culture and mentality of the opposing sides, gain their trust and legitimacy, and explain the benefit of a reasonable settlement to the people on both sides in their own language. It is almost impossible to find a third party mediator that could fit the challenge.[24] In contrast, a team of Israeli and Palestinian mediators, perhaps, can do a better job in leading a process of coalition building.

The interaction in a major Israeli-Palestinian Public Negotiating Congress – which can be viewed as a Middle-Eastern version of the multi-party talks in Northern Ireland and South Africa (Handelman 2014) – will probably be, at least, a three dimensional process: negotiations between the delegations of the opposing societies (Israelis and Palestinians), negotiations within each delegation (Israelis and Palestinians have internal disagreements), and negotiations between the delegations and the general public (cf. Handelman 2014; Putnam 1988). It seems that Israeli and Palestinian skilled mediators – who could build trust, operate in harmony, and develop working relationships –have the potential to contribute to such a coalition building process. The results of the Minds of Peace Experiment support this view.

The negotiating assemblies in the Minds of Peace Experiment – small, mid and mass-scale Israeli-Palestinian public negotiating congresses that have been conducted in different locations and with delegations of Israelis and Palestinians from all walks of life reflecting almost the entire political spectrum[25] – were co-chaired by Israeli and Palestinian mediators. The 32 rounds of the assemblies demonstrated that such a team could lead a major Israeli-Palestinian Public Negotiating Congress quite efficiently and help in establishing peacemaking coalitions that could motivate the opposing leaderships to conclude agreements (Handelman and Pearson 2014; Handelman 2014).

22.5.4 Concluding Remarks

In the 1970s, the 'troubles' in Northern Ireland, the struggle against the Apartheid regime in South Africa, and the Israeli-Palestinian situation were considered to be the most entrenched conflicts in the world. In the 1990s the first two conflicts experienced a major transformation while the Israeli-Palestinian conflict is still teetering between hope and despair (Courtney et al. 2005: 277).

---

[24] Senator Mitchell was an envoy to the Middle East for a very short time. In 2001, he tried to initiate a peace process by suggesting general principles for negotiation similar to the Mitchell Principles in Northern Ireland (Deane 2009).

[25] Mini-scale congresses included 14–20 negotiators and audience; mid-scale congresses included 40–100 negotiators and audience; mass-scale congresses included, at least, 500 negotiators and audience.

The multi-party talks in South Africa and in Northern Ireland played a major role in a revolutionary transformation of the conflicts.[26] In both cases these talks helped to involve the people in the struggle for change and establish a major peacemaking coalition of the moderate against the extremes (cf. Mnookin 2003). In the South African case, the cooperation between exceptional leaders – Nelson Mandela and William De-Klerk – was the main force behind the operation of the multi-party negotiations (Sparks 1994; Handelman and Pearson 2014). In the Northern Ireland case, Senator Mitchell, the independent chairman of the talks, was discovered as an effective and talented coalition builder mediator, at least according to the analysis of Curran and Sebenius (2003). However, the Senator did not operate in a vacuum. The support of major political forces – such as the British and the Irish governments – and drastic actions of brave political leaders – such as the dramatic visit of the British secretary of state for Northern Ireland, Marjorie Mowlam, to Maze prison to meet personally with paramilitary prisoners to ensure their support for the negotiations – enabled the talks to survive and bypass major obstacles and crises (Mitchell 1999, 129–133).

In the Israeli-Palestinian case, the idea of establishing a Middle-Eastern version of multi-party negotiations did not receive serious consideration by the central players in the confrontation. And, it is hard to imagine how to build such a major peacemaking institution. Moreover, as the Northern Ireland case indicates, it is quite clear that even the most skillful mediator – who can bridge the cultural gap between Israeli and Palestinians – cannot lead an Israeli-Palestinian coalition building process by himself alone. A coalition-builder mediator – who wishes to create an effective change – needs to work in a supportive environment that would provide adequate conditions for building consensus. These conditions include: the cooperation of influential local leaders, the support of different socio-political groups and organizations, and, even, the backing of the international community.

22.6 Summary

Who is the most suitable mediator to lead an effective, and even a revolutionary, peacemaking process in the endless Israeli-Palestinian case? The complexity of the Israeli-Palestinian conflict indicates that the most suitable mediator should be a superman: he should be able to enforce leaders to reach a settlement in a Realist fashion; He should be capable of involving political-elites in a process of conflict analysis and resolution (problem-solving workshop) in a Pluralist manner; And he should have the talents to build a major peacemaking coalition in a Contractualist mode. In short, the perfect mediator should have capabilities similar to Adam Smith's invisible hand, a magical entity that can lead conflicting agents to work for the benefit of society in an ideal free market (Smith 1976).

[26] Zartman (1995) labeled the transition of South Africa – 'negotiated revolution.'

No doubt that such a super-mediator does not exist in reality. Approaching the vision of the perfect mediator requires a multi-dimensional peace process that creates the conditions to operate different types of negotiating processes in a complementary fashion. For example, a consensus building process that involves Israeli and Palestinian people in the peacemaking efforts could have supported the Oslo accord of the 1990s and Kerry's closed-door diplomacy of 2013–2014 and led them to different results. To be more specific, a major Israeli-Palestinian public negotiating congress – a contractualist consensus building institution – could have supported different tracks of negotiation between leaders and political-elites operating in realist and pluralist fashions.

Johan Galtung (1976) has taught us that peace has to be made, built and kept. An imaginary super-mediator can navigate negotiations effectively and efficiently in situations of intractable conflict, like the Israeli-Palestinian struggle, only when the foundations for peace-building and peace-keeping are being built. In a similar mode, Adam Smith's invisible hand, the market forces operating in a free society, can bring progress and prosperity only in an adequate framework of rules and institutions (Buchanan 2001; Vanberg 2005, 25–26).

Adam Smith, who noted that economics is the science of the legislator, created the foundations for the development of Constitutional Economics (Vanberg 2005). Constitutional Peace Studies could be a new research field focusing on the study of rules and institutions necessary to create and support a multidimensional peacemaking process. This approach can help us to better cope with the question, what are the socio-political mechanisms that could help mediators and peacemakers to build a momentum for a revolutionary transformation in difficult situations of intractable conflict.

References

Agha, Hussein, Shai Feldman, Ahmad Khalidi, and Zeev Schiff. 2003. *Track-II diplomacy: Lessons from the Middle East.* Cambridge, MA: MIT Press.

Allison, Graham. 1971. *Essence of decision; explaining the Cuban Missile crisis.* Boston: Little Brown.

Alon, Ilay, and Jeanne Brett. 2007. Perceptions of time and their impact on negotiations in the Arabic-Speaking Islamic World. *Negotiation Journal* 23: 55–73.

Azar, Edward, Paul Jureidini, and Ronald McLaurin. 1978. Protracted social conflict: Theory and practice in the Middle East. *Journal of Palestinian Studies* 8 (1): 41–60.

Banks, Michael. 1984. In *The evolution of international relations theory in conflict in world society: A new perspective in international relations*, ed. Michael Banks. Sussex: Wheatsheaf books.

Bar-Siman-Tov, Yaacov. 2007. Dialectic between conflict management and conflict resolution. In *The Israeli-Palestinian conflict: From conflict resolution to conflict management*, ed. Yaacov Bar-Siman-Tov. New York: Palgrave Macmillan.

Bar-Tal, Daniel. 2001. Why does fear override hope in societies engulfed by intractable conflict, as it does in the Israeli society? *Political Psychology* 22: 601–627.

Bickerton, Ian and Carla L. Klausner. (2007) A Concise History of the Arab Israeli conflict, 5th. edition. Englewood Cliffs: Prentice Hall.

Buchanan, James. 2001. *Moral science and moral order*, Vol. 17 of The collected works of James M. Buchanan. Indianapolis: Liberty fund.

Bueno de Mesquita, Bruce, and David Lalman. 1992. *War and reason: Domestic and international imperatives*. New Haven: Yale University Press.

Burton, John W. 1969. *Conflict and communication: The use of controlled communication in international relations*. London: Macmillan.

———. 1988. Conflict resolution as a function of human needs. In *The power of human needs in world society*, ed. Roger A. Coate and Jerel A. Rosati. Boulder: Lynne Rienner.

Courtney, Jung, Ellen Lust-Okar, and Ian Shapiro. 2005. Problems and prospects for democratic settlements: South Africa as a model for the Middle East and Northern Ireland? *Politics and Society* 33 (2): 277–326.

Curran, Daniel F., and James K. Sebenius. 2003. The mediator as coalition-builder: George Mitchell in Northern Ireland. *International Negotiation* 8 (1): 111–147.

Curran, Daniel, James Sebenius, and Michael Watkins. 2004. Two paths to peace: Contrasting George Mitchell in Northern Ireland with Richard Holbrooke in Bosnia–Herzegovina. *Negotiation Journal* 20: 513–537.

Deane, Shelley. 2009. Instituting peace: Third party principles and Mitchell Effect. *Journal of Intervention and Statebuilding* 3 (1): 65–91.

Fisher, Ronald. 1977. *Interactive conflict resolution*. New York: Syracuse University Press.

Friedrich, Carl, ed. 1966. *Revolution*. New York: Atherton Press.

Galtung, Johan. 1976. *Three approaches to peace: Peacekeeping, peacemaking, and peacebuilding in peace. War and defense: Essays in peace research*. Vol. II. Copenhagen: Christian Ejlers.

Gilboa, Eytan. 1998. Secret diplomacy in the television age. *Gazette: The International Journal of Communication Studies* 60 (3): 211–225.

———. 2008. Searching for a theory of public diplomacy. *The Annals of the American Academy of Political and Social Science* 616 (March): 55–77.

Gorenbeg, Gershom. 2006. *The accidental empire: Israel and the birth of the settlements, 1967–1977*. New York: Henry Holt and Company, LCC.

Hamdan, Nahlah, and Frederic Pearson. 2014. *Arab approaches to conflict resolution*. London/New York: Routledge.

Handelman, Sapir. 2009. *Thought manipulation: The use and abuse of psychological trickery*. Santa Barbara: ABC-CLIO.

———. 2011. *Conflict and peacemaking in Israel-Palestine: Theory and application*. London/New York: Routledge.

———. 2012a. The minds of peace experiment: A laboratory for people-to-people diplomacy. *Israel Affairs* 18 (1): 1–11.

———. 2012b. Between the Israeli-Palestinian conflict and the East-west Pakistan struggle: a challenge to the conventional wisdom. *Israel Affairs* 18 (1): 25–26.

———. 2014. *People-to-people diplomacy in Israel and Palestine*. London/New York: Routledge.

———. 2016. Peacemaking contractualism: A peacemaking approach to cope with difficult situations of intractable conflict. *Global Change, Peace & Security* 28 (1): 123–144.

Handelman, Sapir, and Fredric Pearson. 2014. Peacemaking in intractable conflict – A contractualist approach. *International Negotiation* 19: 1–34.

Hayek, Friedrich. 1994. *The road to Serfdom*. Chicago: University of Chicago Press.

Hirst, D., and I. Beeson. 1981. *Sadat*. London: Faber and Faber.

Hobbes, Thomas. [1660] 1985. Leviathan. ed. Tuck Richard. Cambridge: Cambridge University Press.

Hopmann, Terrence. 1995. Two paradigms of negotiation: Bargaining and problem solving. *The Annals of the American Academy of Political and Social Science* 542 (1): 24–47.

Huntington, Samuel. 1968 [2006]. Political order in changing societies. New Haven: Yale University Press.

Inbar, Efraim. 2006. Israel's Palestinian challenge. *Israel Affairs* 12 (4): 823–842.

Ionescu, Ghiţa, and Ernest Gellner, eds. 1969. *Populism: Its meanings and national characteristics*. London: Weidenfeld and Nicolson.

Kelman, Herbert. 1981. Reflections on the history and status of peace research. *Conflict Management and Peace Science* 5 (2): 95–96.

———. 1996. Negotiation as interactive problem solving. *International Negotiation: A Journal of Theory and Practice* 1 (1): 99–123.
———. 1997. Some determinants of the Oslo breakthrough. *International Negotiation* 2 (2): 183–194.
———. 2005. The psychological impact of the Sadat visit on Israeli society. *Peace and Conflict: Journal of Peace Psychology* 11 (2): 111–136.
———. 2007a. Social-psychological dimensions of international conflict in Peacemaking in international conflict: Methods & techniques (rev. ed.), ed. William Zartman. Washington, DC: U.S. Institute of Peace.
———. 2007b. The Israeli-Palestinian peace process and its Vicissitudes: Insights from attitude theory. *American Psychologist* 63 (4): 287–303.
Kriesberg, Louis. 2001. Mediation and transformation of the Israeli-Palestinian conflict. *Journal of Peace Research* 38 (3): 373–392.
Landau, Simha. 2006. Settings, factors and phenomena of conflict in the Israeli society. In *Conflicts and conflict resolution in Middle Eastern societies – Between tradition and modernity*, ed. H. Albrecht, J. Simon, H. Rezaei, H. Rohne, and E. Kiza. Berlin: Duncker and Humblot.
Machiavelli, Niccolo. [1532] 1979. The Prince. In *The portable Machiavelli*, ed. Bondanella Peter and Musa Mark. New York: Penguin Books.
Mandela, Nelson. 1994. *Long walk to freedom: The autobiography of Nelson Mandela*. Boston: Little, Brown and Company.
Mansergh, Martin. 1999. The early stages of the Irish peace process. Conciliation Resources, http://www.c-r.org/accord8/early.htm.
McKittrick, David, and David McVea. 2002. *Making sense of the troubles: The story of the conflict in Northern Ireland*. Chicago: New Amsterdam Books.
Mitchell, George. 1999. *Making peace*. New York: Alfred A. Knopf.
Mnookin, Robert. 2003. Strategic barriers to dispute resolution: A comparison of bilateral and multilateral negotiations. *Journal of Institutional and Theoretical Economics* 159 (1): 199–220.
Murshed, Mansoob. 2002. Conflict, civil war and underdevelopment. *Journal of Peace Research* 39: 387–393.
Nye, Joseph. 2009. *Understanding international conflicts: An introduction to theory and history*. 7th ed. New York: Pearson Longman.
Putnam, Robert D. 1988. Diplomacy and domestic politics: The logic of two-level games. *International Organization* 42: 432.
Rubin, Barry. 1999. *The transformation of Palestinian politics: From revolution to state-building*. Cambridge, MA: Harvard University Press.
Sadat, Anwar. 1977. Statement to the Knesset by President Sadat. Special Meeting of the Knesset: The Forty-Third meeting of the Ninth Knesset (November 20, 1977), Jerusalem.
Savir, Uri. 1998. *The process: 1,100 days that changed the Middle East*. New York: Random House.
Scanlon, Thomas. 2000. *What we owe to each other*. Cambridge, MA: Harvard University Press.
Schelling, Thomas. 2005. Nobel Prize Lecture. Available at: http://nobelprize.org/nobel_prizes/economics/laureates/2005/schelling-lecture.pdf.
Sisk, Timothy. 1994. South Africa's national peace accord. *Peace and Change* 19 (1): 50–70.
Smith, Adam. 1976. A*n inquiry into the nature and causes of the wealth of nations*. Chicago: University of Chicago Press.
Sparks, Allister. 1994. *Tomorrow is another country: The inside story of South Africa's negotiated revolution*. Sandton: Struik Book Distributors.
Thompson, Leonard. 2001. *A history of South Africa*. New Haven: Yale University Press.
Vanberg, Viktor. 2005. Market and state: The perspective of constitutional political economy. *Journal of Institutional Economics* 1 (1): 23–49.
Wantchekon, Leonard. 2004. The paradox of "Warlord" democracy: A theoretical investigation. *American Political Science Review* 98 (1): 17–32.
Zartman, William. 1995. In *Negotiating the South African conflict in Elusive peace: Negotiating an end to internal wars*, ed. William Zartman. Washington: Brookings.

Chapter 23
Joseph Agassi's Educational Thoughts in *Interchange* (1970–1987)

Ian S. Winchester

For over 50 years Joseph Agassi has thought and written about education and published some of his original thinking in the journal *Interchange*, originally the official journal of the Ontario Institute for Studies in Education in the University of Toronto and presently a journal in the Springer collection of journals edited out of the Werklund School of Education in the University of Calgary. Joseph joined the editorial board of *Interchange* in the 1980s where he also made a vigorous contribution to the improvement of the work of others. In the next few pages I would like to comment on some of his own written contributions as they appeared to me as the editor of that journal from 1982 and throughout the 1980s. Although he has published recently in the same journal I shall say nothing about these contributions here.

Perhaps what struck me most about Joseph's educational thinking was its being direct and to the point. I once recalled asking Joseph what he thought about the much befuddled contemporary debate between the teacher-centred view of education and the child-centred view which has tended to dominate discussion from roughly Locke and Rousseau onward. He replied simply: "A child learns by itself, with the indispensable aid of a teacher." This remark has often haunted me for its clarity, its simplicity and its truth. In the next day few pages I hope to comment partly on that debate and his remark and to extend the discussion by directly discussing the implications of his insight.

From Rousseau onward there has been a continuing debate in educational circles over whether education ought to be centrally teacher driven (children should learn the wisdom of the past and the necessary skills for the present from competent teachers) or child centred (children should learn on their own what their natural powers and interests dictate in relation to the world as they now find it). Contributions from the Greeks onwards had tended to emphasize the importance of the teacher in

I.S. Winchester (✉)
Werklund School of Education, University of Calgary, Calgary, AB T2N 1N4, Canada
e-mail: winchest@ucalgary.ca

N. Bar-Am, S. Gattei (eds.), *Encouraging Openness*, Boston Studies in the Philosophy and History of Science 325, DOI 10.1007/978-3-319-57669-5_23

educational contexts. Thus the Platonic Socrates in Plato's dialogue the *Meno* brought out the importance of the teacher in the education of a slave boy who, through Socrates' questioning, was able to produce the Pythagorean Theorem correctly though previously untutored in geometry. This sort of picture, while not accepting the Platonic/Socratic view that all human knowledge was already internal to every human and only had to be elicited by a competent teacher, was enhanced by John Locke in his educational writings and in his *Essay Concerning Human Understanding*. In particular Locke's view that a human mind is initially like a "blank slate" onto which the teacher has only to write the knowledge tended to emphasize the importance of the teacher and of right teaching methods and of a necessary set curriculum for all.

On the other side of the debate Rousseau had an entirely different picture of the educational circumstances facing a child that became central to European thought. His concern was to emphasize the importance of human freedom of thought and action that he saw as under many social constraints including those of contemporary educational pratices. As a consequence he criticized contemporary education from the vantage point of its impact on that freedom. A teacher who either fills a blank slate with a set curriculum or who elicits those aspects of human knowledge from a pupil that only relate to that prior set curriculum is just imposing on the child's freedom and ultimately on his or her whole life. Thus, for Rousseau, a child's education should emphasize the importance of the child-centredness of education, not of its teacher-centeredness.

Joseph Agassi's insight that "a child learns by her or himself with the indispensable aid of a teacher" seems to me to paralyze the dispute between the parties to this debate and to give one a tool to cast out many absurdities in educational thinking.

What I explore in the next few pages is how this insight worked its way through some of Agassi's own writings on education that appeared in the journal *Interchange* from 1970 to the present day and why it is important in many contexts.

Agassi's first contribution to *Interchange* appears in the first Volume of *Interchange* in 1970 in a special issue on "Education for the Future" in which such luminaries as Marshall McLuhan, Carl Rogers and R.S. Peters among others appeared. Agassi writes as a reviewer of a book by John Holt called "The Underachieving School" a popular and best selling critique of American education. In this book Holt had suggested, as Rousseau had before him, the notion that if children were presented with interesting reading material and left to their own devices they would all learn to read on his own. The suggestion is that if a child has interesting material and can read then a curriculum is in principle unnecessary as such a child would find interesting things to read. Agassi, for his part, thinks that it is "quite possible that even in a world full of interesting printed material children will grow illiterate and inarticulate". This is because "the young are such that under some favourable conditions they will ask for help, under some conditions they will not". Agassi thinks that "we should create positively the minimal favourable conditions enabling them to ask for help". Whereas Holt had drawn the conclusion that schools were no longer necessary, Agassi thinks that it may well be the "best arrangement that the community at large should pay a few of its members to administer places where the young will feel free to ask for that help that they desire, with some measure of the possibility that they will find something there that they may appreciate". So for Agassi, but not for Holt, "the abolition of the curriculum and the classroom should take precedence over the abolition of compulsory attendance at school". It seems to me that all of this depends on and presupposes the truth of Agassi's remark that "a child learns by itself but with the indispensable aid of a teacher".

In another special issue of *Interchange*, some 15 years later, in Volume 16, Number 1, 1985 and with the editor no longer Andy Effrat but myself, Agassi produced another educational piece "The Myth of the Young Genius". In this piece Agassi looks at the destructiveness of the notion of the young genius and at the impossibility of identifying geniuses adequately, of determining through some educational process their life's career course or of protecting them from their envious peers. He concludes "We should help these young people, and everyone else possible, to develop their abilities and to enjoy developing them, and we should put no pressure on them to excel." Again the truth of this refers back to Agassi's insight that a child learns by itself but with the indispensable aid of a teacher". But the child's wanting and needing a teacher is one thing. The imposition of the teacher's discipline and desires for the child on the child is quite another. For sometimes at least "quite a few brilliant individuals suffered from pressure so much that as adults they were resolved not to use the special skills and talents they had developed under that pressure".

In a slightly earlier piece, also in *Interchange* (Volume 15, Number 4, 1984) Agassi tackles what he refers to as "Training to Survive the Hazard Called Education". This paper was originally read at the Ontario Institute for Studies in Education in 1970 but had been rejected earlier by a variety of educational journals he tells us. In this paper he says that he considers "most of the curriculum not so much a great evil as an utterly redundant one and as a hindrance and impediment for educational experimentation". He does not think that schools as such should be abolished but only the silly and redundant activities that they engage in. Even if schools abolished the curriculum the law of the land would undoubtedly impose some part of it. And those who wish to practice certain skills such as masonry or medicine will have to pass examinations to qualify. So some required curriculum will always remain. But perhaps the time will come when students need not stay in the classroom because" forcing them to do so is criminal folly". But such pressures from society can be met by students and teachers together in co-operation. "It is the teachers volunteering to be society's instrument that must be abolished, and by the teacher himself".

Agassi's final appearance in *Interchange* in the 1980s, while it was still the official journal of the Ontario Institute for Studies in Education and wholly owned before becoming part of the Kluwer and then Springer journal empires was "The Autonomous Student". In this piece he is responding to paper by John Wettersten writing "On the Unification of Psychology, Methodology and Pedagogy". Agassi's stalking horse or horses in this piece are the educational absurdities of The Age of Reason with Francis Bacon, Rousseau and especially Immanual Kant as exemplars.

Agassi's brief characterization of the main absurdity of the educational thinking of the Age of Reason is the apparent truism that "education is inconsistent with autonomy".

Agassi's characterization of Kant in this regard is that "we need self-discipline, and lots of it, to be autonomous; that we need disciplining in order to acquire self-discipline; and that schooling has as a major aim the acquisition of discipline. Each of these ideas is absurd". Why so? "Were self-discipline the high road to autonomy" Agassi tells us, "then flagellants and yogis would be highly autonomous and asceticism without autonomy would be unthinkable". Much of the remainder of this article is devoted to how Karl Popper and ultimately Agassi himself actually taught classes to students, particularly in the context of mathematics. The deep disagreement between Popper and Agassi appears to be that Popper thought that students are not, generally speaking autonomous, and as a consequence they have to be taught initially rather dogmatically in the hope that they will ultimately rebel and gain autonomy. Agassi for his part thought that dogmatism should have no place in education at all. Ultimately his model for this was the early Lakatos before he became a dogmatic teacher himself. Again how are we to understand all this? I suggest that it becomes clear that Agassi continues to presuppose that a student learns by her or himself, but with the indispensable aid of a teacher. Consequently there is no room for dogmatism in education. A teacher is an aid to a student's learning, not an authority. And the teacher's help may always be autonomously accepted or rejected by the learner.

Are there any obvious objections to Agassi's educational presupposition? Well one might suppose that some like, say, Abraham Lincoln, learn by themselves and apparently need no teacher. He seems to have taught himself to read and went on to an astonishing achievement of education and accomplishment without any teachers. Perhaps that was true of Gauss as well, at least in the context of mathematics and languages. And I suppose that there are some extreme cases in which no learning would occur because of personal difficulties or biological defects that make autonomous learning practically impossible such as the case of the blind and deaf Helen Keller who would likely have remained trapped in her limited inner world without the direct and persistent teaching help of Anne Sullivan. But for most of us, while we learn by ourselves with our normal equipment of senses and mental powers, we certainly need the aid of a teacher from time to time. So I think we might well embrace Agassi's dictum as a general truth with almost no exceptions on the educational front.

References

Agassi, Joseph. 1970. The preaching of John Holt: John Holt, "The underachieving school". *Interchange* 1 (4): 115–118.

———. 1984. Training to survive the hazard called education. *Interchange* 15 (4): 1–14.

———. 1985. The myth of the young genius. *Interchange* 16 (1): 51–61.

———. 1987. The autonomous student. *Interchange* 18 (4): 14–20.

Chapter 24
Can the Israeli Academia Replace Mandatory Conscription?

Amir Meital

Certain conditions have to obtain before serious attempts can start to implement great ideas; in the case of autonomy these conditions include relative peace and relative prosperity.

(Joseph Agassi, *The Siblinghood of Humanity*, p. 253)

In 2015, Israeli President Reuven Rivlin launched a new civic initiative: "Israeli Hope". The main thrust of this initiative was to establish a new Israeli civic order based on mutual recognition and cooperation between four different "tribes," in Rivlin's words, that are in an ever-increasing state of mutual hostility. These are: the secular tribe, the Arab tribe, the ultra-orthodox tribe, and the national-religious tribe. President Rivlin presented a dire picture[1] of a conflicted Israeli society but stated his belief that it is possible to generate consensus around four pillars: "each sector's feeling of security, a shared responsibility for the fate and future of the state of Israel, fairness and equality, and the creation of a common Israeli identity". This brings to mind Joseph Agassi's view that any Israeli political initiative is tantamount to crossing the Rubicon, since it challenges the Jewish "ghetto mentality" and the passive tradition of diaspora Jewry (Agassi 1993, 208, fn. 56). In his speech, President Rivlin emphasized the fact that two out of four of these "tribes" do not serve in the Israeli military, but that all of them can find themselves side by side in the academy and in the workplace.[2] Needless to say, the Israeli army, which is com-

[1] In his 2015 "Four Tribes" speech, given at the Herzliya Conference at the Interdisciplinary Center (IDC), Herzliya, Rivlin spoke about "a boy from Beit El, a boy from Rahat, a girl from Herzliya, a girl from Beitar Ilit – not only do they never meet, but they are also raised on completely different perceptions of the basic values and desired nature of the state of Israel. Will this be a secular and liberal, Jewish and democratic state? Will it be a Jewish theocracy? Or a halachic democracy? Will it be a state of its citizens or of its nations? Tribe, tribe, tribe, tribe – and each with its own media campfire." See: https://israeli-hope.gov.il.

[2] "Academia is the first meeting place [for all sectors] of Israeli society, and therefore it is also the most effective space for the creation of a shared language and goals. Higher education is the ticket to economic influence and the gate to the fulfilment of the Israeli dream. Academia is the most significant force driving the realization of Israeli talent wherever and whatever it may be, and it is

A. Meital (✉)
Sapir Academic College, Tel-Aviv, Israel
e-mail: amirmeital@gmail.com

N. Bar-Am, S. Gattei (eds.), *Encouraging Openness*, Boston Studies in the Philosophy and History of Science 325, DOI 10.1007/978-3-319-57669-5_24

monly referred to as "the people's army" cannot live up to this title if only two of four of its tribes are eligible for service. In the absence of a substitute for this function in Israeli society, can the academy become the site for the formation of an Israeli national identity?

In his masterful two-part work, *Between Faith and Nationality: Towards an Israeli National Identity* (1993), Joseph Agassi foresaw all of the problems that Rivlin's program seeks to address today and proposed a new practical plan for the Israeli nation. While Rivlin's plan has a long shot of succeeding, due to its denial of the core of Israel's inner conflict – the lack of an Israeli national identity – Agassi's proposal is much more interesting and far-reaching. Moreover, Agassi's framework for a solution to the Arab-Israeli conflict remains, to this day, the only practical, balanced plan that is brave enough to deal boldly with the root of the problem in the Middle East. The biggest difference between Rivlin's initiative and Agassi's proposal is that Rivlin clings to the definition of the state of Israel as "Jewish and democratic" (without discussing the inherent contradiction of these two terms), while "conditioning national identity with Jewish or religious terms" (Barzilai 1992, 39). Agassi, on the other hand, offers a plan for a normal western nation-state, based on democratic, liberal, and pluralistic principles drawn from the western philosophy and experience.

This paper deals briefly with two aspects of Agassi's program: the first of these is mandatory conscription to the IDF, a central tenet of Agassi's thought; the second explores the role of higher education as an expression of normalcy in a liberal society. In the first part of this paper I will address the subject of mandatory conscription in Israel and ask: Can institutions of higher education offer a viable alternative to the army's role in the formation of national identity in Israel? In the second part of this paper, I will present the Bedouin Academic Program at Sapir Academic College, whose academic principles and values are drawn from Joseph Agassi's philosophy, as a case study for a liberal education.

24.1 Part I – Mandatory Conscription

Mandatory conscription to the Israel Defense Forces (IDF) is anchored in the Israeli Security Service Law (Integrated Version) (1986)[3] and applies to only three groups of Israeli citizens: secular Jews, national-religious Jews, and Druze men. A sweeping exemption from military service is granted to ultra-orthodox Jews,[4] Arabs, religious-Jewish women, and those who are not deemed suitable for military service for health or other reasons of unsuitability. As a result, since 2007, less than half of those who reach the age of military conscription actually enlist. Over the decades,

also the space and the key for the creation of a diverse leadership and elite." See: https://israeli-hope.gov.il/Academy-Launch

[3] https://en.wikipedia.org/wiki/Exemption_from_military_service_in_Israel

[4] See Israeli Security Service Law (amendment no. 21) 2015.

we have seen a decline in the status of the IDF and an increasing skepticism as to its role as a "melting pot" for Israeli society. Joseph Agassi proposes the inclusion of all Israeli citizens under the law of mandatory conscription to the IDF as part of their civic duties. According to Agassi, the conscription of Arabs to the Israeli Defense Forces is not only a moral thing to do but will also hasten the coming of a Middle East peace agreement. The exemption from military service granted to some of the Arabs and to ultra-orthodox Jews[5] prevents Israel from becoming a normal state and turns its Arab citizens "more and more into Palestinians, because they have not been given the right to define themselves[6] [...] and those who think that such a national minority can exist in Israel for the long term, are fooling themselves" (Agassi 1993, 194, 228). In his novel, *The Secret Life of Saeed: The Pessoptimist*, Palestinian-Israeli author Emil Habibi brilliantly and sarcastically describes the life of the Arab minority in the state of Israel as one of constant collaboration. Saeed's origin, past, and humanity have been taken away from him and he is forced to spy after himself in his own country. An extreme example of the impossibility of his situation can be seen in the absurdity of celebrating Independence Day of oneself and of one's operators on the same day (Habibi 1974/1995, 100). According to Agassi, in Israel's Jewish theocracy, selective conscription to the army is a by-product of the lack of separation between religious establishment and state (Agassi 2016, 32). Unfortunately, conscription, or the selectivity of it, perpetuates the rule of the religious establishment by means of "divide and rule," splitting Israeli society into different and separate groups: religious and secular Jews, Jews and Arabs, and different groups of Arabs, since Druze, Bedouin, and Christian Arabs can volunteer if they wish to serve in the IDF.[7] According to Agassi, the consensual inclusion of the state's Jewish and non-Jewish citizens into a unified Israeli national identity is a precondition for the creation of an Israeli national identity (Agassi 2000, 194). Any of the individuals currently living in Israel will be free to choose whether to live in the country as citizens with full and equal rights and duties, or as residents (Agassi 1991, 58). By giving his or her personal agreement to the political change, the Arab or ultra-orthodox citizen will become by his or her own volition an inseparable part of the new nation. The Muslim citizen will live in a state where there is a separation between state and religious establishment; in which all citizens have the same rights and duties (such as the obligation of military service), and in which belonging to the Muslim faith is not seen as inferior. As part of the Israeli nation, the Arab-Palestinian citizen can feel pride in his origin.

---

[5] In 1948–9, when David Ben-Gurion was plotting to banish the Israeli Communist party and right-wing "Herut" party from the Israeli consensus, he "bought" the ultra-orthodox parties' parliamentary support by granting an exemption from the army to a few hundred yeshiva students. Later on, this exemption was given to all ultra-orthodox Jews.

[6] In July 1954, there was a short-lived changed in the Israeli Security Service Law, allowing Muslim Arabs to enlist into the security services. No less than 4000 young Arab men signed up for service, however, the Israeli military preferred not to enlist these young men. (Kais 2011, 59)

[7] http://english.alarabiya.net/en/perspective/profiles/2013/04/24/Bedouin-army-trackers-scale-Israel-social-ladder-.html

Is mandatory conscription incompatible with the values of an open society? In the modern era, mandatory conscription appeared for the first time during the French Revolution (September 5, 1798) and it was one of the outstanding characteristics of the French *Grande Armée* under Napoleon. Throughout the nineteenth century, mandatory conscription was instituted in the Prussian and Russian armies as well, and by the outbreak of WWI in 1914, conscription to most armies in the West was no longer done on a volunteer basis (the USA and UK being two exceptions to this rule).

Ancient Athenian society imposed mandatory military conscription of 2 years on all 18-year-old males. Besides its contribution to the security of the Polis, service as a horseman, hoplite, or seaman was considered to be an important phase in a young man's education, preparing him for future political activity. 18-year-old sons of Athenian citizens took the Ephebic Oath (quoted by Lycurgus in his fourth-century B.C.E. work *Against Leocrates*) upon induction into the military, swearing to obey those in authority and to submit to the laws of the land. On the other hand, oligarchical arguments against mandatory conscription appear as early as the late fifth century B.C.E. In the pamphlet *Athenaion Politeia*, Athenian critic Pseudo-Xenophon, also known as the Old Oligarch, argues that because it is the people who man the navy's ships and give the democratic Polis its strength, the masses are given jobs, the right to vote, and the right to speak ([Xen]1.2–3). However, this does not answer the question of whether mandatory conscription grants legitimacy to the ruler, whether it expresses the will of the people, and whether it necessarily leads to the formation of a national identity.

Karl Popper (1991 [1945]) briefly mentions the subject of conscription in *The Open Society and its Enemies*, in his chapter on Hegel and the new tribalism. According to Popper, before Hegel, modern nationalism was a possible liberal revolutionary belief despite the fact that nobody could actually define what a nation is. He wrote on the Hegelian national state: "It is a myth. It is an irrational, a romantic and Utopian dream, a dream of naturalism and of tribal collectivism" (Popper 1945, 12 III, 264). On military conscription, he writes: "[…] and a great practical step in the nationalistic direction was made when the French Revolution inaugurated a people's army, based on national conscription." He reiterates this claim (Popper 1945, 12 V, 280) when he explains Hegel's view on true courage as being the willingness of the individual to give himself entirely to the state and to be "as one among many […] in self subordination to the universal cause". From here, the road to total war is a short one. Popper than gives us a partial answer to the questions above: mandatory conscription does indeed grant a false legitimacy to the ruler; it can dangerously express the will of the people but it can also lead to the formation of a national identity.

Conscription to most armies in the West today is done on a voluntary basis. Issues of national budgets, surplus manpower, military technology, the advancement of weapons, and a trend towards professionalization, have reduced the size of the armies. Mandatory conscription is seen as militaristic and anti-democratic (as against freedom of employment and freedom of conscience), and as coming at the expense of progress in important societal issues such as education, welfare, health, economy, defeating crime, etc.

In the absence of a stable national identity, the military in Israel plays more than just its primary role of ensuring national security. From its very outset, it was clear that the army would be responsible not only for the security of the state but it would also be expected "to serve as a melting pot for the Jews coming to the homeland from the diaspora [...] and a factory for a cohesive nation. The IDF was meant to turn into a greenhouse for pioneers, nurturing people who would build the nation and make the wilderness bloom" (Ben Gurion 1953; Drory 2009). Slogans such as "The people are the army," "Every good young man to arms," and "A conscripted society" only reinforced this lie. In fact, the exemption of large parts of the citizenry from mandatory conscription, on the one hand justified the second-class status of non-Jewish citizens and stymied any political initiative; and on the other hand gave a sweeping consent for the ultra-orthodox (approximately 10% of the population in Israel today) not to enlist into the army on religious grounds and thus established the principle of selective justice, giving part of the public rights without duties and turning Israel into a theocracy.

Nowadays, military service is no longer a ticket to success in later civilian life in Israel. Only a minority of citizens is willing to pay with their lives for the protection of the state. The army is manned mostly by young people from the national-religious sector (many of them settlers from the West Bank) or from the periphery, who do not accept the foundations of democracy, liberalism, and human rights as taken for granted (Barzilai 1992, 40). For the national-religious sector, the army is a manifestation of religious values, and for the young soldiers from the periphery it is an opportunity to improve their social standing, to learn a profession, and to gain occupational security (Levy 2007). Because rates of enlistment remain relatively high, however, there have been calls to shorten the length of service, to invest in the professionalization of the soldiers, and to choose only those who are suitable for conscription.

This reality undermines even more the army's role as a "melting pot" as was seen in Israel's early days. It underscores the need to find an alternative place for the consolidation of Israeli identity. Oddly enough, President Rivlin chose higher education as the new "melting pot" for civil society.[8] In this context, some 50 years ago, Agassi claimed in an interview with Abraham Peleg of the *Ma'ariv* newspaper, that "the IDF is more democratic than academia," because of the strict entrance requirements to the various faculties and the anti-democratic organizational behavior commonly found in institutions of higher education in Israel.[9]

[8] https://israeli-hope.gov.il/sites/default/files/%D7%AA%D7%A7%D7%95%D7%95%D7%94%20%D7%99%D7%A9%D7%A8%D7%90%D7%9C%D7%99%D7%AA%20%D7%91%D7%90%D7%A7%D7%93%D7%9E%D7%99%D7%94%20-%20%D7%9E%D7%AA%D7%97%D7%99%D7%9C%D7%99%D7%9D.PDF

[9] http://jpress.org.il/Olive/APA/NLI_Heb/SharedView.Article.aspx?parm=kP1qfA43aVZaW89m45My8BqJGLXVezHgpErbkOj7iBdtyEYrH0lGAMA7kHgAMoQsYw%3D%3D&mode=image&href=MAR%2f1972%2f03%2f29&page=28&rtl=true

For this reason, it is worth adding that any civic initiative that is not based on the separation of state and religious establishment, on political progress toward a peace agreement (which will also give a solution to the abnormal situation in Jordan as a state without a nation, much like Israel), on the possibility for minorities to live with dignity as part of the Israeli nation, and on mandatory conscription for all citizens of the state, is destined to fail.

24.2 Part II – A Gateway to Academia

The 225,000 Bedouin Arabs living in the Negev Desert in the south of Israel[10] enjoy neither relative peace nor relative prosperity, and their future as impoverished Palestinians in the south of Israel is certainly not a bright one. They are ostensibly citizens of the state of Israel; *de jure* they are entitled to equal rights but *de facto* it seems that the state of Israel is tired of them and they are tired of the state. Joseph Agassi writes on the situation of the non-Jewish citizens of Israel: "[…] their status is built on logical contradictions and a murky and untenable situation of citizenship-that-is-not-citizenship and equality under a law-that-is-not-a-law, and that is not in effect recognized by the people or by the government." (Agassi 1993, 207). Unlike in the case of most of the Muslim citizens of the country, the non-assimilation of the Bedouin into Israeli society occurs despite the fact that they are "loyal citizens by all measures and they serve in the IDF […] and that they are ethnically speaking more Arab than other minorities who are exempted from military service just because they are Arab" (Buber et al. 1991, 44). It has often been said that the Bedouin in the Negev are undergoing a process of "Palestinianization," assimilating national sentiments that make them a part of the Palestinian-Israeli conflict. Aside from Joseph Agassi (1993, 225), very little has been said about the resemblance of the narrative of non-Jews in Israel to the narrative of Jews in Israel, i.e. that the Jewish nation is comprised of the Jews living in Israel and those living in other countries. The theocratic reality in Israel has cast the Israeli Bedouin into this mold, forcing upon them a narrative that is analogous to that of diaspora Jewry. It points to them and says, with typical insensitivity, they are like us – a religious and tribal minority living away from its nation's geographical center (In the case of the Bedouin, it is not clear who this nation is and where it is located). On this, Agassi writes that, because the state of Israel does not recognize itself as a nation, the Bedouin consider their citizenship to be fictitious, and they are, like the rest of the minorities, a horde of tribes whose national identity was stolen by Israel and by

[10] The statistics about Bedouins in the Negev are unclear at best and suspicious at worse. Many strange numbers appear in the national statistics, such as a 38% rate of polygamy in Bedouin families, a birth rate in the unrecognized villages of 11%, that the Bedouins double their population every 15 years, and more. The numbers that can be partially trusted are those that present the minimum and maximum figures. The above numbers, even if only an estimation, pull the rug out from under the age-old Zionist demographic principle that "time is on our side" (Agassi 1993, 230).

Jordan. It is important to emphasize that Agassi's prescription for the creation of an Israeli identity does not ignore the fact that the contradiction-filled Jordanian Kingdom (with its 35% Bedouin citizens) is a poor substitute for a Palestinian state. A solution to the Jordanian paradox (of a state without a nation), which is only growing as a result of the Syrian civil war, will allow the Bedouin to decide in which nation, if at all, they wish to live in as citizens: Israel or Palestine (former Jordan).

By every economic, academic, and social measure, the Bedouin in the south of Israel are at the bottom of the food chain of Israeli society – a position that in Israel of 2017 is not so easy to achieve. These citizens traditionally live in tribes or large extended families, in recognized and unrecognized settlements. A full history of the primary, high school, and higher education of the Bedouin in Israel has yet to be written. When historical facts about education are needed, the Bedouin in Israel will turn to the "old person who remembers everything" to get information. But even though there is no official data, it is clear that the story of Bedouins in higher education in Israel is a story of ongoing failure that has become a trauma. Some statistics that we do have: As of 2015, the Negev has 37 high schools that produce 5500 graduates per year. Of these, 500 go on to higher education – the men in Jordan, the Palestinian Authority, or Europe, and the women in Israel. Of the latter, 50% drop out for economic, social, or academic reasons. Until 2014, Only 4% of Bedouin in the south of Israel who began studying at Sapir Academic College succeeded in completing a bachelor's degree (without attaining an exemption from English, which is required of all other students in Israeli academia) over the course of an average of 5.6 years of study. 90% of those who completed their academic studies did not succeed in finding work in their field of study.

The Bedouin Academic Program at Sapir College, known also as the "Gateway to Academy", has evolved greatly since it was founded in 2011 under the name Taqadum, with the goal of changing this grim reality. Today, 350 Bedouin students are enrolled to its various frameworks and the expectation is to reach 500–600 students by 2019. From its first day, this special program for Bedouin students has yielded impressive results. In the BA program in the Multidisciplinary Studies Department, there is a 75% success rate (of students earning a BA, including English); in the rest of the departments (10 academic departments: law, social work, communications, cinema, engineering, culture, computers, economics, business and public policy, and human resource management), beginning in 2016 the percentage of students going onto their second year is approximately 65%. Likewise, 60–70 percent of the Bedouin students successfully pass courses with a mathematical orientation or in language. These remarkable results were attained due to a new curriculum that was developed especially for Bedouin students, the good will of the staff and faculty at Sapir Academic College, and a generous government budget that was earmarked specifically for this goal. As the academic director of the program, I would like to list a number of the philosophical principles I have learned from Joseph Agassi and which guided me in the writing of the academic guidelines of the program[11]:

[11] In this paper, I focus on guidelines for academic liberalism.

24.2.1 The Siblinghood of Humanity

"Any discriminations by 'us' against 'them' are in the wrong" (Agassi 1991, 281). This principle holds true not only for being admitted into higher education institutions but also for being able to understand and critically process diverse academic material. Success on exams is not a testimony to the fact that "they" have understood and internalized "our" knowledge and that "we" have succeeded in teaching "them," but a recognition on the part of everyone that the new enlightenment is shared by everyone to the same degree. It is very important to keep in mind that knowledge is a universal and not a particular thing. Given the political reality in Israel, convincing people of the universality of education is not an easy task. The teachers as well as the Bedouin students come from a world in which the us/them dichotomy is a matter of religious, political, economic, and cultural survival. And nevertheless, the universal academic language is apparently the only language with which enemies can speak to one another in the Negev. Here, it is important to emphasize that Muslim and Christian Arabs learn Hebrew and Jews learn Arabic out of a variety of interests, primarily economic, and only rarely for the creation of a shared language and culture. The academy, as a universal institute, is not a substitute for the "melting pot" or for any other political conspiracy, and outside of the halls of academia, only the separation between religious establishment and state will be a meaningful practical step towards a siblinghood of humanity.

24.3 Critical Rationalism

The academic language used in the program and taught in the proficiency courses and, as much as possible, in the rest of the academic courses at the college, does not encourage the justification of one's deeply held views. The students are asked to raise active problems and give them reasonable solutions. This sounds simple and surprisingly it is so, with the help of dialogue and methodical problem-solving based on trial and error, conjectures, and criticism (Agassi 1991, 274). The intellectual discussions take place with the aspiration toward autonomy (decision without coercion), self-criticism, good will, and mutual respect (as well as humor). In any problem that is discussed, questions, definitions, assumptions and counter-assumptions are presented, after which a methodical attempt is made to check a number of alternative solutions and to find opposing or refuting arguments for all of them.[12]

[12] "[…] Rationality does not need defense; it needs improvement," writes Agassi. And we may try to improve it piecemeal. We are all rational to some degree and are all interested from time to time in using reason more effectively than we now do. We cannot help but be rational, since thinking is, like seeing, innate to some extent. No one is always rational or perfectly rational any time. Our best hope, then, is to use rationality to improve the partial and limited rationality which we all use to one degree or another. We use a bootstrap process in that we use the rational methods we now have at hand to develop better methods, whereby the methods we use may very well be corrected or even discarded." Accessed at http://www.iep.utm.edu/cr-ratio/

24.4 Humanism

The principle that all human beings have the right to live in dignity is at the core of the "Gateway to Academy". This is by no means a purely academic statement, but also a social and economic statement having to do with the unique situation of the students in the program. Many of the Bedouin students who enroll in the Sapir Academic College have never set foot before in an institution of higher education. Many have never met Jews or spoken Hebrew or English. Their knowledge about the West is limited. Furthermore, differences in the cultural norms and fear of the different (64% of the Bedouin students are women and most of the students are aged 18–22) prevent them from speaking in public or getting to know the Jewish students at the college. Most of them come from a closed, tribal, impoverished society in which bigamy is permitted and the authority of the father or husband is not questioned. Some of them come from unrecognized villages that have no connection to water or electricity. Because public transportation does not reach all Bedouin areas, some of the students can spend four or more hours a day commuting to and from the college. But these students are no inferior to any other student at Sapir. They have a rich culture, and their contribution to campus life is felt throughout the college. Academically, their studies are divided into two stages: the first stage focuses on language study, enriching general knowledge, and internalizing the principles of critical rationalism. In the second stage, they enter academic courses in the various departments. Throughout their studies in the "Gateway to Academy" program, some of the students stand out for their social initiatives, others for their natural leadership, and all, almost without exception, in their attempts during their years at the college to engage in dialogue with the Jewish students at the college.

In lieu of a conclusion to this paper, I would like to make a wish. 25 years have passed since Joseph Agassi's books about religion and nationalism were published in Hebrew in Israel, and many events have changed the face of reality in Israel and in the world since then. Therefore, it is my wish that a third edition of *Between Faith and Nationality: Towards and Israeli National Identity* be published, and for this book to take its rightful place in Israeli public life.

References

Agassi, Joseph. 1991. *The siblinghood of the humanity: An introduction to philosophy*. New York: Caravan Books.

———. [1983] 1993. [Hebrew] *Between faith and nationality: Towards an Israeli national identity*. Tel Aviv: Papirus.

———. 2000. *Letters to my sister concerning contemporary philosophy*. Tel Aviv: Yedioth Aharonoth Books and Chemed Books.

———. 2016. [Hebrew] "The next national philosopher", *Katharsis: 25*. Kfar-Sava: Mezaref.

Barzilai, Gad. 1992. [Hebrew] *A democracy in wartime: Conflict and consensus in Israel*. Tel Aviv: Sifriat Poalim Publishing.

Ben Gurion, David. 1953. In: Gal, Reuven. [Hebrew]. 1999. The motivation to serve in the IDF. INSS 2(3): 14.

Buber, Judith, Agassi Joseph, and Moshe Berent. 1991. [Hebrew] *Who is Israeli?* Rehovot: Kivunim.

Drory, Zeev. 2009. [Hebrew] "Ben Gurion – People's army for the people". In *The IDF – the people's army or professional army*, ed. P. Yehezkeli, Holon: Ministry of Defence Press.

Habibi, Emile. [Hebrew]. 1974/1995. *The secret life of Saeed: The Pessoptimist*. Tel Aviv: Hakibutz-Hameuchad.

Kais, Firro. 2011. [Hebrew] "The Israeli Palestinians: Mandatory conscription of the Druze into the Israeli military: Historical background". In *The Palestinians in Israel – readings in history¨ politics and society*, ed. Nadim N. and Areej Sabbagh–Khoury Haifa: Mada al Carmel Press.

Levy, Yagil. 2007. [Hebrew] From *the 'people's army' to the 'army of the peripheries*. Jerusalem: Carmel Publishing House.

Popper, R. Karl. [1945] 1971. *The open society and its enemies*. Princeton: Princeton University Press.

Chapter 25
Education for Democracy

Ronald Swartz

> *Authority, if it is to govern education, must rest upon [...] the State, the Church, the schoolmaster and the parent [...] no one of them can be trusted to care adequately for the child's welfare, since each wishes the child to minister to some end which has nothing to do with its own wellbeing [...]. But in view of the fact that no authority can be wholly trusted, we must aim at having as little authority as possible, and try to think out ways by which young people's natural desires and impulses can be utilized in education.*
>
> (Russell 1935, 1977, pp. 144–145)

In relationship to education for democracy, we can ask at least the following two questions: (1) What kinds of educational programs should be tolerated, and perhaps even encouraged, in democratic societies? and, (2) Is it reasonable to create educational programs that are organized as democratic learning communities? The first question can be referred to as the problem of creating educational programs in democratic societies. The second question can be called the educational problem of creating democratic learning communities. In this brief essay my goal is to say more about the second question, but I will also say a few words about the first.

Views related to creating democratically organized schools have received little attention in the professional literature related to philosophy of education. However, in relationship to creating democratic schools Joseph Agassi has noted the following:

> The literature on education and its aims was traditionally directed at socialization [...] in modern democratic society, socializing is at best democratizing, and so the best school is the democratically run school [...] criticism of this view is the argument that in democratic schools pupils will make study noncompulsory and then will not study at all [...] humans learn at any given opportunity. But, as Daniel A. Greenberg of the famed Sudbury Valley School observes, in democratic schools pupils learn what they want, not what their elders and betters think is good for them. (Agassi 1977, pp. 314–315)

R. Swartz (✉)
Professor Emeritus of Education, Oakland University, Rochester, MI, USA
e-mail: swartz@oakland.edu

N. Bar-Am, S. Gattei (eds.), *Encouraging Openness*, Boston Studies in the Philosophy and History of Science 325, DOI 10.1007/978-3-319-57669-5_25

In one of his numerous attempts to explain the sense in which the Sudbury Valley School is a democratic learning community Daniel Greenberg once wrote the following:

> [...] the ideal environment for children [...] subject only to constraints imposed by safety, is totally open for exploration, free of restraints, free of external imposition, a place where each individual child would be granted the freedom to reach out everywhere and anywhere they wished so that they could follow through in all of their curious probing [...]. Both society and the individual in modern, post-industrial America require that schools be an environment in which children are free, and in which children can learn to use freedom, how to be self-governing, how to live together as free people in peace and harmony and mutual respect. [...]. Every child at Sudbury Valley has a vote in every matter that pertains to the school. The school is governed by a School Meeting in which four year olds have the same vote as adults. Every decision in the school is made by that School Meeting [...]. Our community is a model of democratic governance, much like the New England communities that we serve. (Greenberg 1993, pp. 71–73)

Agassi and Greenberg belong to small group of recent writers who have made such claims as, in democratic societies "the best school is the democratically run school." Needless to say, this claim should at best be viewed as a conjecture that needs to be explored in detail. Moreover, given that Agassi endorses a fallibilistic intellectual perspective that is an outgrowth of the one developed by his famous teacher Karl Popper, the claim about democratic schools being the best, may, indeed, be seen as inadequate or unsatisfactory at some future time. Nevertheless, in this essay I do not plan to criticize or attempt to argue that there is much merit in the very unconventional claim that in democratic societies the best school is a democratically run learning community. What I would like to suggest here is that in answering the problem of creating educational programs in democratic societies it is worthwhile to note that among the educational alternatives that need to be tolerated in democratic societies we should do our best to make sure that some schools are democratic learning communities. That is, it is indeed a good sign of the democratic society in the United States that it has allowed the Sudbury Valley School to remain an educational alternative for over 50 years. In addition, in order to understand whether the Sudbury Valley School, or any other existing or suggested innovative school should be tolerated, it is important to have standards for acceptable and unacceptable educational alternatives. What these standards are at any moment in time may of course change; at any moment in time standards may become more or less restrictive. But the standards for acceptable programs should never be tolerant of "anything goes."

In the history of Western education the Sudbury Valley School should not be viewed as totally unique. Although all democratic schools will no doubt have some unique aspects, the fact that they are democratic means that some aspects of all these schools will have a common thread. And in relationship to how the Sudbury Valley School (SVS) is similar to schools such as A.S. Neill's famous Summerhill School, it is worthwhile to note that Donald Arnstine once wrote the following:

> The organization of SVS is about as open and flexible as one could imagine, and its governance is democratic in the most straight forward sense of the term. Its main difference from A. S. Neill's Summerhill is the fact that the latter was a boarding school. (Arnstine 1995, p. 150)

Since it was founded nearly 100 years ago A.S. Neill's Summerhill School in many ways has been a beacon for democratic learning communities. Whenever one speaks about Summerhill it is important to note that it is not an anarchistic community. On the contrary, Neill did his best to explain that democratic self- governance at Summerhill had limits on individual freedom. In his book *Freedom – Not License!* Neill summarized his views about the need to have constraints on freedom with the following:

> Freedom, over-extended, turns into license [...] in my school a child is free to go to lessons or stay away from lessons because that is his own affair, but he is not free to play a trumpet when others want to study or sleep. (Neill 1966, p. 7)

At Summerhill and all other reasonable democratically run schools that become viable educational alternatives the liberal ideal of having laws, rules, or regulations that in some ways restrict an individual's freedom to act is a must. The distinction between freedom of action and freedom of thought is a liberal ideal that is wonderfully explained in Chap. 3 of John Stuart Mill's book *On Liberty*. (Mill 1974, 119) If a democratic school ignores the need for some reasonable limits on individual freedom of action, then it can be viewed as an unsatisfactory educational alternative. Thus, to bring democracy to a school does not necessarily mean that a person or a group has suggested a reasonable or satisfactory educational alternative. What I am suggesting here is that some democratic educational alternatives may be degenerate and not acceptable for implementation; a democratic educational alternative is unacceptable when it becomes an unsafe and dangerous learning community which can do significant harm to individuals. An ongoing problem for all democratic learning communities is the continuous challenge to learn the limits of democracy, self-governance, liberalism, and freedom.

Education for democracy is in no way a stable or easily defined entity. And learning that all the values we cherish and wish to live by should not be excessive is, indeed, a difficult task to learn both for any individual, or for any group that one belongs to. The standards we wish to live by need to be critically evaluated, and at times, limited. Democracy, liberalism, freedom, and self-governance without some restraints may create unlivable situations that subvert the ideals we cherish. In relationship to the need to not have excessive standards in education it is worthwhile to note that Agassi once wrote the following:

> The excess of standards becomes a disaster in education. The Enlightenment idea of an Emile learning only from Nature herself is too naïve to practice. Even Emile himself had a mentor. And so schools simply remained, to date, in paternalistic hands [...]. It is a sad reflection on our culture that until recently the liberal schools were almost all reform schools or orphanages – Homer Lane, Mgr. E. J. Flanagan, Janusz Korczak, Anton Makarenko – or else schools for difficult children – A. S. Neill, and even Bertrand Russell. (Agassi 1977, p. 279)

Paternalistic educational programs have indeed been the dominant educational alternative in Western societies. These programs come in many varieties and, at times, have been excessively authoritarian options which were extremely cruel and abusive towards children. Fortunately, in our modern societies many authoritarian paternalistic programs are somewhat benevolent. Yet, in spite of this, Agassi has made a sweeping suggestion that authoritarian paternalistic programs continue to do a great deal of harm. For Agassi

> Students from elementary school to graduate school all over the world suffer from two very painful complaints. One is boredom. The other is bewilderment, or a sense of being lost, or a sense of inadequacy, or even, an incomprehensible but profound sense of failure, often tainted with a sense of guilt, and, in severe cases, of despair. (Agassi 2014, p. 11)

Throughout his long career as a philosopher Joseph Agassi's writings have consistently criticized authoritarian paternalistic educational programs beginning with those articulated in Plato's *Republic* to Karl Popper's suggestion in *The Open Society and Its Enemies* that Socratic teachers need to include "one element of authoritarianism" in the methods they use to teach students. [Popper 1962, 130] Agassi's views on democratic educational communities should be seen as an attempt to improve upon the educational programs that are an outgrowth of philosophical works from Plato to Popper and beyond. And Agassi seems to suggest that modern educators have often too hastily dismissed the viability of the educational innovations such as the Sudbury Valley School, Summerhill, Lane's Little Commonwealth and Korczak's Orphans Home.

In conclusion, the potential dialogue between those who wish to create democratic educational communities and those who continue to support paternalistic authoritarian educational communities can be seen as a lost dialogue that has yet to take place. Joseph Agassi's educational writings can be seen as providing one place to begin this very important dialogue. (Swartz 2016, p. 373).

References

Agassi, Joseph. 1977. *Towards a rational philosophical anthropology*. Dordrecht: Springer.

———. 2014. *The hazard called education: Essays, reviews, and dialogues on education from forty-five years*, eds. Ronald Swartz and Sheldon Richmond, Rotterdam/Boston: Sense Publishers

Arnstine, Donald. 1995. *Democracy and the arts of schooling*. Albany: State University of New York Press.

Greenberg, Dan. 1993. School for a post-industrial society. In *The exhausted school*, ed. John Taylor Gatto, 65–75. Berkeley: Berkeley Hills Books.

Mill, John Stuart. 1974 [1859]. *On liberty*. In *John Stuart Mill: Utilitarianism, on liberty, essay on Bentham*, ed. Mary Warnock, New York: New American Library.

Neill, A.S. 1966. *Freedom-not license!* New York: Hart.

Popper, Karl. 1962. *The open society and its enemies, volume 1*. London: Routledge.

Russell, Bertrand. 1977 [1935]. *Sceptical essays*. London: Allen and Unwin.

Swartz, Ronald. 2016. *From socrates to summer hill and beyond: Towards a philosophy of education for personal responsibility*. Charlotte: IAP, INFORMATION AGE Publishing, Inc..

Chapter 26
How Can We Attain Both Democracy and Constitutionalism?

Kiichi Tachibana

> *Its [The state's] ultimate purpose is not to exercise dominion nor to restrain men by fear and deprive them of independence, but on the contrary to free every man from fear so that he may live in security as far as is possible, that is, so that he may best preserve his own natural right to exist and to act, without harm to himself and others. It is not, [...], the purpose of the state to transform men from rational beings into beasts or puppets, but rather to enable them to develop their mental and physical faculties in safety [...]. Thus the purpose of the state is, in reality, freedom.*
>
> (Spinoza, *Theological-Political Treatise*, Chapter 20, 567)

Democracy and constitutionalism are usually regarded as indispensable elements of modern constitutions. Some constitutional scholars and political scientists, however, argue that constitutionalism and democracy are mutually incompatible, and that one or the other must be preferred. Particularly in Japan, both the constitutionalist and the most influential of the democratic positions presuppose that democracy is nothing but majority rule. Thus these scholars observe that in a parliamentary cabinet system a political opinion of the Prime Minister, who is the head both of the Cabinet and of the ruling (*majority*) party, clashes with jurists' or lawyers' ideas or interpretations regarding that country's constitution[1], thus showing that constitutionalism and democracy are incompatible. On the other hand, the layperson

[1] The second Abe Cabinet reinterpreted the current Constitution, and allowed the use of the right to collective self-defense for the first time and then approved security bills. Against the Abe Cabinet decision, almost all constitutional scholars have been opposed to the reinterpretation as unconstitutional and as against constitutionalism. Besides, the Cabinet Legislation Bureau in Japan has conflict of interests, playing the role of *judicial review* while being an office of the *executive* branch. These problems will lead to a reflection on the relationship between democracy and constitutionalism.

K. Tachibana (✉)
Faculty of Education and Human Studies, Akita University, Akita, Japan
e-mail: tachiba@ed.akita-u.ac.jp

N. Bar-Am, S. Gattei (eds.), *Encouraging Openness*, Boston Studies in the Philosophy and History of Science 325, DOI 10.1007/978-3-319-57669-5_26

unfamiliar with political sciences or constitutional studies takes for granted that democracy and constitutionalism are compatible with each other[2]; but there are few powerful arguments why democracy and constitutionalism are compatible with each other[3]. I shall attempt to rectify the omission in this paper. Above all, despite the *persistent* idea of democracy as majority rule, majority rule is not a characteristic of democracy, as Popper has argued. Popper's theory of democracy, therefore, may provide a solution allowing for both democracy and constitutionalism as compatible elements of a constitution.

26.1 What Is Constitutionalism?

26.1.1 Constitution Without Constitutionalism

To the layperson on constitutional studies, constitutionalism seems to apply to any constitution. Then who should observe the constitution? Everybody? There are indeed some provisions that everybody should observe, such as tax obligations. But it is commonsense among constitutional scholars that those who should observe the constitution are *governors*[4]. This is an important aspect of constitutionalism. It should be noted that the mere existence of a constitution does not automatically imply the existence of constitutionalism. As Okoth-Ogendo in his paper 'Constitutions Without Constitutionalism'[5] says, "While clinging to some notion of the constitution, they [the military] have advanced the issue of constitutionalism no further than civilian regimes have," and, "The most fundamental of the functions of a constitution, at least in liberal democratic theory, to regulate the use of executive power, is clearly not one that African constitutions that have survived military intervention now perform"[6]. In Africa there are *constitutions without constitutionalism.* This insight reminds of Julio Faundez, a western legal scholar, that constitutionalism is not simply in compliance with the rule of the constitution[7]. I shall therefore investigate whether the Meiji Constitution is with constitutionalism or not.

[2]The phrase, "constitutional democracy" is often used as the title of various books. See my references.

[3]Kis (2003) is exceptional, but regrettably he does not use Popper's theory of democracy.

[4]In the current Japanese Constitution governors imply the members of government such as Ministers of State, members of the Diet, judges, and all other public officials, and the Emperor or the Regent as well (Article 99). They are *public officials*.

[5]Okoth-Ogendo (1993), 65–82.

[6]Ibid., 78–79.

[7]Faundez (1993), 358.

26.1.2 Is the Meiji Constitution with Constitutionalism?

What are the requisites for a constitution to be properly called "constitution"? In order to answer this question, the Declaration of the Rights of Man and the Citizen (1789) is very useful.

Article 16:

> Any society in which the guarantee of the rights is not secured, or the separation of powers not determined, has no constitution at all.

In short, constitutionalism is to limit political powers by separating them in order to guarantee the human rights.

According to this criterion, it is commonly judged that the Constitution of Prussia, for example, is based on apparent-constitutionalism (Scheinkonstitutionalismus). Then how about the Meiji Constitution, which was drawn up after the model of this Constitution? Russell closely compares the two and answers to this question. He says[8]:

> Foreigners unacquainted with Japan, knowing that there is a Diet in which the Lower House is elected, imagine that Japan is at least as democratic as prewar Germany. *This is a delusion.* It is true that Marquis Ito, who framed the Constitution, which was promulgated in 1889, took Germany for his model [...]. But there were many points in which the Japanese Constitution differed from that of the German Empire. To begin with, the Reichstag was elected by manhood suffrage, whereas in Japan there is a property qualification which restricts the franchise to about 25 per cent of the adult males. This, however, is a small matter compared to the fact that *the Mikado's power is far less limited than that of the Kaiser was* [...] *in Germany the Reichstag had the power of the purse, whereas in Japan, if the Diet refuses to pass the Budget, the Budget of the previous year can be applied, and when the Diet is not sitting, laws can be enacted temporarily by Imperial decree – a provision which had no analogue in the German Constitution.*
>
> *The Constitution having been granted by the Emperor of his free grace, it is considered impious to criticize it or to suggest any change in it, since this would imply that His Majesty's work was not wholly perfect.*

If Russell's observations are correct, the Meiji Constitution is inferior to the German Constitution concerning the limitations of *the supreme power*[9], which was Tenno in prewar Japan. The Meiji Constitution, therefore, may be regarded as a constitution with even less constitutionalism than the Constitution of Prussia. It would be safe to say that while the latter is with apparent-constitutionalism, the former is with *pseudo*-constitutionalism.

On the other hand, Yoichi Higuchi, one of Japan's leading constitutional scholars, who *advocates constitutionalism instead of democracy*[10], also refers to Hirobumi

---

[8] Russell (1922), 63–64. Italics are mine.

[9] In constitutionalism *some limitations of the supreme power* is vitally important. In the past the supreme power was a king (Tenno in prewar Japan). Now in Japan the Diet is the supreme organ of the state power. From a constitutionalist point of view, therefore, it is important to limit *the Diet's power by any means*.

[10] Higuchi prefers constitutionalism to democracy. For him democracy is a vague and empty concept, because modern dictatorship has not denied democracy. Stalin purged people's enemies in the

Ito but depicts him as a figure who knew what constitutionalism was, and cites his remark on constitution as follows:

> The spirit of establishing constitution is in the first place to limit the monarchical power, and in the second place to protect the rights of His subjects.

And Higuchi deplores the poor understandings[11] of constitution by *contemporary* Japanese politicians, who repeatedly say that in the Constitution *there are few articles that prescribe the duties of the people* in comparison with the prescriptions of their rights. Then Higuchi concludes that their sense of constitution is anachronistic, compared to Ito[12]. And he claims that constitutionalism was commonsense in prewar Japan. Thus according to the Meiji Constitution a so-called constitutional government had been accomplished in the "Taishō demokurashī"[13] era, with the names of the political parties such as Rikken-Seiyūkai (Constitutional Association of Political Friendship) and Rikken-Kaishinto (Constitutional Progressive Party) clearly indicating the existence of constitutionalism. The Meiji constitutional system, according to Higuchi, is constitutional monarchy, and the politics of prewar Japan had been ruled according to constitutionalism *until the rise of militarism*[14]. In short, there existed constitutionalism in prewar Japan.

Which, then, is right: Russell or Higuchi? I shall take up a speech by Ito, whom both Russell and Higuchi mention. In his opening address at the first meeting for the draft for the Great Japanese Imperial Constitution in 1888, Ito, the chairman of the Privy Council, said as follows[15]:

> As the axis of our nation, there exists Imperial Family exclusively. Therefore, in the draft for this constitution we concentrated on this point and *respected the imperial power and tried not to limit the power*. On one hand, some claim that it is not without fear of the danger of abuse when the imperial power is too strong. This claim has a slight truth, but when such

name of people. Hitler seized power by democratic means of popular election. Higuchi (2000), 84–86.This is a bad argument. In order to overcome some equivocal ambiguities of the concept of democracy, all you have to do is to distinguish the views of democracy into democracy$_1$, democracy$_2$, and so on, and critically examine them. See Tachibana (2015), 40–47.

[11] Such poor understandings are shown from Prime Minister Abe's following utterances: He said in his response at the Budget Committee of the lower house of the Diet, February 3, 2016, that constitutionalism is to limit king's power and it is unnecessary in Japan which is a country of democracy; he said in his response to the same Committee, May 16, 2016, that the Prime Minister is the head of the *legislative* organ (actually the *executive* organ). His remark shows the problem of the parliamentary cabinet system and malfunctioning of the Diet.

[12] Higuchi (2000), 74–75.

[13] The technical term, "Taishō demokurashī" is automatically translated into "Taishō democracy", but this translation is misleading. See Tachibana (2015), 37–51.

[14] Higuchi (2000), 75–76. In his earlier book, however, he develops an argument against the view of the Meiji Constitution with constitutionalism, by appealing to the significance of Article 16 of the Declaration, and concludes that the Meiji Constitution is "Scheinkonstitutionalismus", a term which Higuchi does not use in his book of 2000, and makes the criticism that so-called "Rikken-shugi [constitutionalism]", or "Rikken-seiji [constitutional government]" are not the same as the corresponding ideas after World War II. Higuchi (1989), 23–43. On this point there is a great disparity between his book of 1989 and that of 2000.

[15] OMP (1888), p. 157. Italics are mine.

> things happen, the prime minister takes the responsibility, or rather, it is not that there are no other ways of prevention. *It is unreasonable to try to narrow the realm of the imperial power by unnecessarily fearing the abuses.* Thus this draft sets the imperial power as the axis and attempts not to destroy the power one-sidedly, and dares *not to rely on the spirit of separating powers*, the spirit of which is adopted in Europe. Japan is originally different from the way of conjoint body of monarch and its subjects in the system of the several European countries. These are the fundamental principles in drafting.

It is evident that the italicized passages are against constitutionalism. For he tried not to rely on the spirit of separating powers *so as not to limit the imperial power*[16]. He rejects even the conjoint body of monarch and its subjects. Besides, he never mentions the right of the subjects in his address. His intentions are not based on constitutionalism, or rather, they are anti-constitutional. For the direction of the vector is completely opposite to constitutionalism. I am convinced that this reason alone is decisive to conclude that the Meiji Constitution is a constitution without constitutionalism, and therefore, it is not a constitution but a pseudo-constitution. Further, I claim that what Ito did in his drafting is much worse. For, he *designedly* tried to draw up the constitution without constitutionalism, *although he certainly knew what constitutionalism was.*

26.1.3 What Is Constitutionalism?

Then what is constitutionalism? As I stated, Higuchi mentioned Ito's remark on constitutionalism, which was a paraphrase of Article 16. But Higuchi's view of constitutionalism is exclusively concerned with the first half of his remark and says: Constitutionalism is, in a word, to limit the power in order to prevent abuse of the power[17]. And he ignores the second half of Ito's remark, that is, to protect the rights of His subjects. But he does not give the reason for his omission.

I do not offer Article 16 as the *final* answer of what constitutionalism is, but I take up the Article as a springboard for discussion.

As Ito adequately paraphrased, this Article is divided into two clauses, to secure the guarantee of rights, and to determine the separation of powers. According to the criteria of the Article, a constitution, which prescribes the two conditions, can be regarded as a constitution with constitutionalism. Let us call such a constitutionalism, prescribed in the Article, "basic constitutionalism". Then which of the two is the aim and which its means? Let us take note that in this case both elements are

---

[16] There is a metaphor that Tenno is the king of "shogi". All shogi pieces excepting the king piece are just the means of the king's survival in order to win the game. This metaphor is clearly against constitutionalism. Besides there is a penetrating remark that Ito and other political leaders really wanted to be professional shogi players who can freely move the emperor like the piece of king and other pieces as if they were puppets. They wanted to utilize the Tenno system as the tool of their political ambitions. Tanaka (2000), 55–58.

[17] Higuchi (2000), 84.

regarded as *good*[18]. I regard the former as the aim and the latter its means. The reason is simple. Powers should be separated in order to secure the guarantee of rights, but not vice versa. Why, then, try to determine the separation of powers? In order to limit powers. Finally, which is the aim, to limit powers or to secure the guarantee of rights? There is no doubt that to secure the guarantee of rights is the aim and to limit powers is its means. Then it turns out that Higuchi's view of constitutionalism as limitation of powers includes only a means and *lacks the aim*[19].

In Article 16, however, *there is no expression on democracy at all.* I shall therefore try to integrate democracy with the Article. In order to do this, let us view democracy using the end-means relation, as I have already done in my examination of Article 16. Is democracy an aim which is at least equal, or superior to the guarantee of the rights? It may be proper to say that you try to make the government more democratic in order to secure the guarantee of the rights, but not vice versa. If so, democracy is a means to attain the higher aim of securing the guarantee of the rights[20]. Then what is democracy as a means?

26.2 What Is Democracy?

Because Popper is hardly familiar to the Japanese, his theory of democracy is almost unknown. On the other hand, Popper's theory of democracy is well known today in the West. Even in the West, however, it is only in recent years that Popper's theory of democracy has become well known and accepted. In his paper of 1988 Popper himself could still note that his theory of democracy was scarcely understood; but the situation has improved. On democracy, Bobbio, for example, says the same thing as Popper: democracy is "a set of rules [...] for the solution of conflicts without bloodshed"[21]. Democracy-Dictatorship (DD) index is often used in empirical researches on democracies in the world. This index was invented by Adam

[18] For Popper closely examines various cases of end and means from his moral point of view. Popper (1945), 161, 623–24.

[19] If you favorably interpret Higuchi's constitutionalism, the limitation of power is simply a means to attain the aim of securing the guarantee of the rights, the latter not mentioned as a matter of course.

[20] In the current Constitution the rights are explicitly prescribed as *fundamental* (Article 11). The provision on fundamental rights is also stated as *one of the Supreme Laws* (Article 97). I do not insist that there is no room for what is included as the fundamental right, but it seems evident that freedom of thought, which naturally includes freedom of speech (freedom to *criticize*), is at the heart of democracy.

[21] Bobbio (1987), 156. Incidentally, Dahl, a prominent theorist of democracy, admits that Popper is right and excuses himself that in his book, "Who Governs?" he considers not only 'who' but also 'how', in response to Bosetti's sharp and frank question: according to Popper, Dahl's question who governs is a wrong one and should be replaced by the good question of how to control the ruler, and whether Dahl is ready to correct his view of democracy. Dahl (2002), Japanese translation, 45–47.

Przeworski[22] by utilizing Popper's distinction between democracy and tyranny[23]. Thus Popper's theory of democracy has provided an effective instrument for social sciences. Additionally, it is of great significance that not only the supporters of liberal democracy in a wide sense but also socialists such as Bobbio and Przeworski accepted Popper's ideas of democracy.

The two figures were socialists, who saw democracy as nothing but a means of realization of their socialist ideals, but they found Popper's theory of democracy as a peaceful replacement of political power indispensable and thus changed their position concerning the status of democracy[24].

Another example. On January 31, 2016 *the Economist* republished a paper, Karl Popper on democracy, whose content is the same as that of his paper of 1988, which I mentioned earlier, one day beforeAmerica's presidential race began with the Iowa caucus.

Most Japanese, however, do not know Popper's theory of democracy and still regard democracy as majority rule. Here I shall present Popper's theory of democracy.

26.2.1 Democracy Is Not Majority Rule

Popper's paradox of democracy can be used as a powerful argument against democracy as majority rule. There is an argument by Rijpkema, however, that Popper regards democracy as majority rule, by using Popper's paradox of democracy[25]. So I shall examine his argument, and try to make it clear that Popper does not regard democracy as majority rule. Rijpkema paraphrases Popper's paradox of democracy as follows:

P1 Democracy is majority rule
P2 Anti-democratic parties can seize a majority (the truth of this premise seems obvious)
C A democracy can abolish itself

Rijpkema interprets that Popper *tacitly* accepts that the essence of democracy is in majority-rule and the conclusion is inescapable when one accepts majority-rule as the defining aspect of democracy. As an evidence for Popper's acceptance of

[22] Cheibub et al. (2010), 68, note2.

[23] Przeworski (2003), 12–13, 16.

[24] Mentioning Popper, Przeworski says: I came to this conception of democracy as a result of the 1973 coup against Allende in Chile. I realized how important democracy is and that any policy that may undermine democracy is irresponsible, because it may lead to mass murder. My minimalist view of democracy dates from that experience. [...] That was a major transformation in my thinking. Przeworski (2007), 475. Bobbio says: I have never forgotten Karl Popper's dictum according to which the essential distinction between a democratic state and a non-democratic one is that only in the former can citizens get rid of their government without bloodshed. Bobbio (1987), 41–42.

[25] Rijpkema (2012), 93–96.

democracy as majority-rule, Rijpkema cites Popper that he recommends a system with some form of *majority rule* as the best, but not infallible, form of *government control*[26].

According to Rijpkema, in order not to draw this unpleasant conclusion, Popper says: there can be no democracy for the anti-democrats, just as there can be no tolerance for the intolerant. And he interprets Popper's solution to the paradox of democracy as that *we simply need to deny anti-democrats the right to democracy for democracy to survive*. And he accuses Popper of repudiating, out of desirability, a logically compelling conclusion *in retrospect*, despite his attempted solution being *logically untenable*[27].

Rijpkema's interpretation is misleading. Popper, I suggest, appeals to the paradox of democracy in order to refute P1 that democracy is majority rule by using *modus tollens*. For Popper does not recommend *a system with some form of majority rule as the best form of government* though Rijpkema claims so. Rijpkema seems to mistake government *control* for government altogether though he uses the phrase, "government control." Exactly speaking, Popper says[28]: We demand *a government* [...] *that is controlled by*, and accountable to, *the public*. [...] that *some form of majority vote*, together with institutions for keeping the public well informed, is *the best*, though not infallible, *means of controlling such a government*. You have to pay attention that Popper does not use the phrase, majority *rule*", but "majority *vote*." It is clear that Popper does not regard a system with some form of majority rule as the best form of government at all. Popper's concern here is not with the problem of the best form of government but the problem of democratic control, that is, how to control government. Popper definitely denies majority rule, which is traditionally mistakenly regarded as democracy, as one of the answers to the question, "Who should rule?" Instead, he tries to answer the new question, "how is the state to be constituted so that a bad ruler can be got rid of without bloodshed, without violence?", and his answer is that the modern so-called democracies adopt *the principle that the government can be dismissed by a majority vote*. Popper calls his theory—his problem and its solution—a theory of "democracy", even though it is emphatically not a theory of the "rule of the people."[29] The rule of the people, as a matter of fact, corresponds to majority rule. Far from accepting the first premise (P 1), Popper *explicitly* denies the first premise (P 1). Popper says[30]:

> [I]t[31] has put in a hopeless intellectual position *all those democrats who adopt*, as the ultimate basis of their political creed, *the principle of the majority rule* [...]. On the one hand, the principle they have adopted demands from them that *they should oppose any but the majority rule*, and therefore the new tyranny; on the other hand, the same principle demands

[26] Ibid., 93. Italics are mine.

[27] Ibid., 94–95.

[28] Popper (1945), 602–603. Italics are mine.

[29] Popper (2016).

[30] Popper (1945), 123. Italics are mine.

[31] "It" here designates that the majority of the people clamours for a tyrant.

> from them that *they should accept any decision reached by the majority*, and thus the rule of the new tyrant. *The inconsistency of their theory* must, of course, paralyse their actions.

Popper is a convinced democrat, but he does not subscribe to the principle of majority rule as the democratic decision-making process, since this leads to inconsistency. Instead, Popper proposes his theory of democracy as follows[32]:

> *[T]he theory of democracy is not based upon the principle that the majority should rule*; rather, the various equalitarian methods of democratic control, such as *general elections* and representative government, are to be considered as no more than well-tried and, in the presence of a widespread traditional distrust of tyranny, reasonably effective institutional safeguards against tyranny, always open to improvement, and even providing methods for their own improvement.

I hope that Popper's reflections on the paradox of democracy may help to remove the (commonsense) prejudice that democracy is nothing but majority rule. In short, according to Popper, democracy is a form of government which enables the ruled effectively to criticize their rulers and to change them without bloodshed[33]. Any government lacking this device cannot be called a "democracy", except as a pretence. This is a minimum and indispensable criterion to distinguish democracy from non-democracy such as tyranny or dictatorship. In Popper's theory of democracy, the keyword is "without bloodshed", in other words, by peaceful means. Then why peace? The answer is simple. Recall social life at the beginning of the Iraq war under President George W. Bush in 2003, or the series of coercion and purge by the Turkish government after the failure of the attempted coup d'état in 2016. In wartime, be it revolution, civil war, a coup d'état, or an international conflict, it is as plain as day that far from being guaranteed, human rights are oppressed. Or consider the smooth and efficient performance of the emergency services: in wartime or in peacetime? Democracy in Popper's sense is a very useful means to secure the guarantee of rights. In Popper's theory of democracy, an indispensable element of election is that it be peaceful, not a majority vote. Even election by coin-toss or lottery would be perfectly acceptable as a peaceful means. The loser, at all events, must accept the result of the election in order to avoid violence and maintain peace[34]. Why? This answer is also simple. Suppose the consequences of the case that the loser rejects the result. While the winner, as a matter of course, will keep holding on to the seat of power, the loser, who does not accept the result, will attempt to take the seat of power. Then it may be natural that the people are divided into the supporters of the winner and those of the loser. And as a worse consequence, by appealing to the name of justice or to the support of the people both parties will eventually

---

32 Popper (1945), 125. Italics are mine.

33 This precise formulation of Popper's theory of democracy is Magee's. Magee (1994), 80.

34 Murphy says: It is far easier for losers to accept the results of elections if they know that they will soon have other chances to gain governmental power and in the interim will retain their basic rights to life, liberty, and property. *If democratic politics were played by the rules of the ancient Mayan version of basketball, which allowed the victors to eat the losers, each election would be followed by civil war.* Murphy (2007), 11. Italics are mine. His remarks really resemble Popper's view of democracy.

resort to force. Therefore, it should be a duty for losers to accept the results of elections. And winners should be ready to accept the possibility of dismissal by the people in the *next* election. Otherwise, a power struggle may shift or degenerate from peaceful to violent.

Plato classified forms of government (monarchy, aristocracy, democracy) according to who should rule. Spinoza classified them according to *who could choose the ruler*. According to Spinoza's criterion, monarchy is a form of government where the ruler's succession is hereditary[35]. Next, comparing aristocracy to democracy, Spinoza says as follows[36]:

> The difference between this state [democracy] and the aristocratic state consists mainly in this, that in an aristocracy *it depends solely on the will and the free choice of the supreme council that any particular person be made a patrician*. Thus *no one has a hereditary right to vote and undertake offices of state*, and *no one can demand that right for himself by law*, as is the case with the state now under discussion [democracy]. For in this state *all* who are born of citizen parents, or on native soil, or have done service to the commonwealth, or are qualified on any other grounds on which the right of citizenship is granted by law, *all*, I say, *can lawfully demand for themselves the right to vote in the supreme council and to undertake offices of state*; *nor can they be refused* except for crime or dishonour.

In short, in aristocracy *the rulers alone* can choose the rulers, and in democracy *the ruled* can choose the rulers. The sharp difference between them is clearly evident[37]. According to Spinoza's criterion, an institution of election, which can be made by all the people having the right to vote, can properly be called as democracy.

According to Spinoza, therefore, you cannot call election by coin-toss or lottery as democracy, even though the election is performed with peaceful means. Of course, it is easy to integrate Spinoza's theory of democracy as popular election[38] with Popper's theory. For, though Popper seems to pay attention only to the dismissal of the ruler by the ruled, a *general election*[39] is, as a matter of course, included in his theory. There is no doubt that the most important element of democracy as popular election is that the right to choose (and to dismiss[40]) the ruler or the governors is in the hands of the people.

---

[35] Spinoza (1677), 718.

[36] Ibid., 752. Italics are mine.

[37] In Spinoza's theory of democracy you find there is a seminal idea of the democratic control of the ruler *by the ruled*.

[38] Though I use the term "popular election", women and slaves had no right to vote or stand for election at that time. "Popular election" is so ambiguous that we commonly use "*universal* suffrage". The transition from limited suffrage to universal suffrage is clearly the process of democratization *after* democracy in Popper's sense. As Agassi points out, even the present world suffrage is not yet "universal" suffrage. Universal suffrage is the antonym of *limited* suffrage, but so-called universal suffrage has some *age limit*, and therefore, the suffrage cannot be regarded as "universal". I think Agassi's proposal for the abolition of age restriction is worth considering. Agassi (1977), 350.

[39] See text to note 32.

[40] In the current Constitution there is an article on the right to choose and to dismiss *the public officials* (Article 15). See also note 4.

From the viewpoint of democratic control of the ruler by the ruled, it is evident that majority rule differs from majority vote. Popper claims that majority rule, or the rule of the people, is practically impossible. Popper says as follows[41]:

> [A]lthough 'the people' may *influence* the actions of their rulers by the threat of dismissal, *they never rule themselves* in any concrete, practical sense.

It is an undeniable fact that in any form of government, which includes democracy as well, one cannot eliminate the relationship between the ruler or the governors and the ruled or the governed. Therefore, it is important for the ruled to control the government, and democracy is the sole form of government where some peaceful controls of the ruler by the ruled are possible.

Concerning political opinions, which is the minority or the majority view changes as a result of deliberative discussions or critical examinations: Malala's UN speech reminds us that even *one child's opinion may influence the opinion of many*. This is a democracy.

Though Popper often mentions majority vote, he is not a majoritarian who claims the majority vote is most desirable as a principle of the decision-making process for democracy. He prefers a two-thirds or even a three quarters (qualified) majority to unqualified majority *in some cases*. Besides, Popper's attitude toward majority vote is cool and detached. He says[42]:

> When we say that the best solution known to us is a constitution that allows a majority vote to *dismiss* the government, then *we do not say the majority vote will always be right. We do not even say that it will usually be right*. We say only that this very imperfect procedure is the best so far invented.

Popper, therefore, can consistently conclude that no majority, however large, ought to be qualified to abandon this rule of law [a rule that enables us to get rid of the government][43].

For Popper, majority vote is a useful peaceful means to control the government by the ruled.

26.2.2 *Decision by Vote as a Compromise*

When there is a time limit, and a collective decision should be made, and further the agreement is not reached by deliberative and critical discussions, we may appeal to a vote, usually to a majority vote. This is a *compromise*[44] in order to solve the collective decision problem with a *peaceful* means.

---

[41] Popper (1945), 125. Italics are mine.

[42] Popper (2016). Italics are mine.

[43] Ibid.

[44] Though Popper concedes that compromise has a pejorative meaning, he values the usefulness of a compromise in democracy, by saying that there will be a possibility of *reaching a reasonable compromise* and therefore of *achieving the improvement by democratic methods*, and says that it is important for us to learn *its proper use*. Popper (1945), 159. Italics are mine.

A family trip may exemplify what is meant by a compromise here. Suppose the family is composed of a father, mother, son, and two daughters; they do not agree over their destination. There are several means for arriving at a decision. Father as a patriarch always decides; father allows his son as a prospective patriarch to decide; or all members accept the result of a majority vote. Though the decision by majority vote is not necessarily right or good, it is better than the decision by other means from the viewpoint of fairness or equality, and is easier to reach the *compromise*, even if the minority may feel sorry for the result. This collective decision may have a kind of *legitimacy*, which is completely different from legitimized or justified legitimacy[45]. This may be called *compromise legitimacy*. This example of a compromise may apply to a majority vote *in politics*.

According to my understanding of Popper's theory of democracy, a government composed of a party or representatives who are ready to accept the possibility of dismissal by the people in the next election has a *tentative legitimacy* while a government composed of a party or representatives who keep on rejecting dismissal by the people has no legitimacy at all. Democracy may be based on institutions which are composed of various results of good and reasonable compromises.

Adding democracy in Popper-Spinoza's sense to the "basic constitutionalism" leads to the attainment of a (consistent) inclusive constitutionalism, in which there are three elements: to secure the guarantee of rights; to limit powers; and democracy as peaceful means of controlling the government by the ruled or the governed by vote (election and dismissal). Democracy is not majority rule but an institution which enables peaceful attempts at problem-solving, in which majority vote is included as a peaceful means. A form of government, which is based on a constitution with such a constitutionalism, may properly be called a "constitutional democracy". According to Popper's theory of democracy, it is evident that both constitutionalism and democracy are useful peaceful means of controlling political power. Both of them are inseparable partners, probably, with friendly-hostile co-operation.

Finally, let us consider the so-called problem of incompatibility between democracy and constitutionalism. Let us take up one example[46]. Yasuo Hasebe, another of Japan's leading constitutional scholars, claims as follows[47]:

> Though *from a democratic point of view* it may be reasonable for the majority as popular sovereignty to deprive the people belonging to a particular race or class of some of their

[45] This remark on legitimacy has nothing to do with the so-called problem of the legitimation of legitimacy of government. For I regard this as a problem produced from justificationism. Social contract theory is one of its attempts. On criticism of social contract theory, see Agassi (1999), Chap. 2.

[46] Similar arguments (pro and con) are developed in the West. For example, Christiano (2009), Chaps. 15–16, and Bellamy (1996), Chaps. 1–5.

[47] Hasebe (2011), 12–13. Italics are mine. Murphy succinctly summarises such a conflict or clash: In constitutionalism every exercise of governmental power should be subject to important *substantive limitations* and obligations; representative democracy would impose *few substantive limitations* on the people's freely chosen representatives. Murphy (2007), 6–7. Italics are mine. This is also not an incompatibility thesis because "few" does not mean "nothing" but "something".

rights, it is unjust *from a constitutionalist point of view*, in light of the ideal of limiting the state power based on the respect of the human rights. Then there is a possibility of conflict or clash between the governmental (legislative and executive) organ and the judicial organ.

This is a schematic argument but it is understandable if one accepts *his concepts* both of democracy and constitutionalism. His concept of democracy is popular sovereignty[48] and representative democracy based on popular sovereignty, and his concept of constitutionalism is "basic constitutionalism," which does not include democracy in Popper's sense.

Concerning this argument at least four points can be stated. Firstly, logically speaking, conflict or clash is not the contradiction, so these difficulties do not lead to the logical incompatibility of democracy and constitutionalism, but can be solved by reasonable compromise. Second, Popper's concept of democracy is completely overlooked, so it is better to examine his concept and Popper's. Third, the problem of incompatibility between constitutionalism and democracy is proposed *after* the separation of powers is performed and democracy in Popper's sense is introduced, so his argument is already in the *framework* of "constitutional democracy". Lastly, the incompatibility problem, which is one of the contemporary political and legal problems[49], is the task of democratization *after* democracy[50] and may be solved by various attempts at peaceful problem-solving. In order to solve these problems or difficulties it is important not only for the professional but also for the ordinary citizen to *watch and check* the governors of the organs in the government (not only the executive and the legislature *but also the judiciary*[51]), and not only to vote *but also voice their opinions* in their own right. These peaceful approaches can be done solely in "constitutional democracy". People in such a country having no experience of civil war or coup d'état, may have forgotten the irreplaceable value of democracy. The transition from bloody power struggle to bloodless power struggle is of immeasurable significance in human political history.

References

Agassi, Joseph. 1977. *Towards a rational philosophical anthropology*. The Hague: Martinus Nijhoff.

———. 1999. *Liberal nationalism for Israel: Towards an Israeli national identity*. Jerusalem: Gefen Publishing House.

[48] Hasebe seems to regard that in democracy the representative government can be *justified* by popular sovereignty. But according to Popper's theory of democracy this view is beside the point. Popular sovereignty may be seen as a *regulative* idea, whose standard no *extant* government yet satisfies. It is better to use the concept of popular sovereignty *negatively* to eliminate even *the majority* as not true sovereign. See Tachibana (2015), 46–47. I propose a way to integrate the theory of popular sovereignty with Popper's theory of democracy.

[49] Concerning authoritative interpretation of constitution or constitutional amendment there are serious problems, but they are out of the extent of this paper.

[50] On the differences of democratization *before* and *after* democracy, see Tachibana (2015), 45.

[51] In the current Constitution *popular review of the Supreme Court justices* is prescribed (Article 79–2).

Bellamy, R., ed. 1996. *Constitutionalism, democracy and sovereignty: American and European perspectives*. Aldershot: Avebury.

Bobbio, Norberto. 1987. *The future of democracy*. Trans. Roger Griffin. Cambridge: Polity Press.

Cheibub, José Antonio, Jennifer Gandhi, and James Raymond Vreeland. 2010. Democracy and dictatorship revisited. *Public Choice* 143 (1): 67–101.

Christiano, T., ed. 2009. *Contemporary debates in political philosophy*. Chichester: Wiley-Blackwell.

Dahl, Robert A. 2002. *Intervista Sul Pluralismo* by Robert A. Dahl, ed. Giancarlo Bosetti, Roma-Bari: Gius. Laterza & Figli S.p.a. *Dahl Demokurashī wo Kataru* [*Dahl Talks on Democracy*]. Trans. Takeshi Ito. Tokyo: Iwanami-shoten, 2006.

Faundez, J. 1993. Constitutionalism: A timely revival. In *Constitutionalism and democracy: Transitions in the contemporary world*, ed. Douglas Greenberg, Stanley N. Katz, Melanie Beth Oliviero, and Steven C. Wheatley, 354–360. New York: Oxford University Press.

Hasebe, Yasuo. 2011. *Kenpo to Heiwa wo Toinaosu [Constitution and peace: A reconsideration]*. Tokyo: Chikuma-shobo.

Higuchi, Yoichi. 1989. *Jiyū to Kokka [Liberty and the state]*. Tokyo: Iwanami-shoten.

———. 2000. *Kojin to Kokka: Ima Naze Rikken-shugi Ka [Individuals and the state: Why is constitutionalism now?]*. Tokyo: Shūeisha.

Kis, János. 2003. *Constitutional democracy*. Trans. Zoltán Miklósi. Budapest: Central European University Press.

Kokuritsu-Kobunsho-kan [National Archives of Japan]. 1888. *Sūmitsuin-Kaigi-Gijiroku [Official records of the meetings in the Privy Council]*. Vol. 1. Tokyo: Tokyo-Daigaku-shuppankai, 1984. OMP (1888).

Magee, Bryan. 1994. *Philosophy and the real world: An introduction to Karl Popper*. La Salle: Open Court.

Murphy, Walter F. 2007. *Constitutional democracy: Creating and maintaining a just political order*. Baltimore: The John Hopkins University Press.

Okoth-Ogendo, H.W.O. 1993. Constitutions without constitutionalism: Reflections on an African political paradox. In *Constitutionalism and democracy: Transitions in the contemporary world*, ed. Douglas Greenberg, Stanley N. Katz, Melanie Beth Oliviero, and Steven C. Wheatley, 65–82. New York: Oxford University Press.

Popper, Karl. 1945. *The open society and its enemies*. London: Routledge, Golden Jubilee edition 1995.

———. 2016. Karl Popper on democracy, from the archives: The open society and its enemies revisited. *The Economist*, January 31. http://www.economist.com/node/21689663/print. Accessed 8 Nov 2016.

Przeworski, Adam. 2003. Minimalist conception of democracy: A defense. In *The democracy sourcebook*, ed. Robert Dahl, Ian Shapiro, and José Antonio Cheibub, 12–17. London: MIT Press.

———. 2007. Capitalism, democracy, and science. In *Passion, craft, and method in comparative politics*, ed. Gerardo L. Munck and Richard Snyder, 456–503. Baltimore: Johns Hopkins University Press.

Rijpkema, Bastiaan. 2012. Popper's paradox of democracy. *Think: Philosophy for Everyone* 11 (32): 93–96.

Russell, Bertrand. 1922. *The problem of China*. Teddington: The Echo Library, 2007.

Spinoza, B. 1670. Theological-political treatise. In *Spinoza: Complete Works*. Trans. Samuel Shirley, 387–583. Indianapolis: Hacket Publishing Company, 2002.

———. 1677. Political treatise. In *Spinoza: Complete works*. Trans. Samuel Shirley, 676–754. Indianapolis: Hacket Publishing Company, 2002.

Tachibana, Kiichi. 2015. Is Taishō demokurashī the same as Taishō democracy? In *Japan's multilayered democracy*, ed. Sigal Ben-Rafael Galanti, Nissim Otmazgin, and Alon Levkowitz, 37–51. Lanham: Lexington Books.

Tanaka, Akira. 2000. *Meiji-Ishin [The Meiji restoration]*. Tokyo: Iwanami-shoten.

Chapter 27
Civil Nationalism: The Only Buffer Against Fascism and Fundamentalism

Chen Yehezkely

27.1 The Two Questions: Presentation

There are two questions that form the backbone of both realms – that of political theory and that of political practice – in all of history: the question of the source of the legitimacy of the state, and the question of political collective identity.[1] The same inner logic determines the centrality of said questions in both said realms: the answer that is assumed to either one of them has an immediate effect upon all other questions in the field. In what comes I hope to offer the reader a summarized overview of the discussions of these two questions.

27.2 The Question of the Source of the Legitimacy of the State

Let us start with the question of the source of the legitimacy of the state: what is the source of this legitimacy? This question is necessitated, to state the obvious, by the combination of the fact that the state forces its will upon its citizens, on the one hand, with the general consensus that forcing people to do things against their will is wrong unless there is a sound justification for it, on the other. This reflects the universal default, or guideline, or ideal of anarchism. Anarchists should rejoice at this and cash it in. Sadly, they don't, possibly because anarchy as default, or guideline, or ideal, is not good enough for them: they want anarchy, period. This has

[1]The question of political collective identity is mostly referred to as the question of nationality. Significantly, the first major scholarly survey was published after WWI (Hayes 1928).

C. Yehezkely (✉)
Western Galilee College, Acko, Israel
e-mail: chenyehe@gmail.com

N. Bar-Am, S. Gattei (eds.), *Encouraging Openness*, Boston Studies in the Philosophy and History of Science 325, DOI 10.1007/978-3-319-57669-5_27

been the main source of criticism against them.[2] Their obliviousness to said criticism is tantamount to intellectual suicide.

The question of the source of the legitimacy of the state is further necessitated by the need for social and political stability and by the desire to avoid excessive cruelty. An illegitimate state invites resistance thus leading to social and political upheaval. Alternatively, it is forced to take extreme measures all of which are exceptionally oppressive and cruel.

The answers to this question boil down to two: the theory from divine appointment,[3] and the theory of the social contract. Each of these two theories has two versions: the traditional and the modified.

The traditional version of the theory of divine appointment is the idea that the source of the legitimacy of the state is the divine appointment of its rulers. This theory rules out rationality, checks and balances and accountability.[4] These considerations, coupled with the loss of grace of said theory due to the French and the American revolutions forced the advocates of the theory of divine appointment to take the modified version, namely, the idea that the source of the legitimacy of the state is the divine appointment not of the state's rulers but of the state itself. This idea has two variants. (1). The state is divine in and of itself and, thus, is itself the source of all legitimacy. This is fascism. (2). The state is divine in as far as it fulfills the role of materializing a pre-given, determinist, divine plan. This is fundamentalism.

A few words seem warranted. It is commonly held that fascism is intrinsically secular and fundamentalism is intrinsically religious. This is a mistake. Both fascism and fundamentalism can be secular or religious as the case may be. What is intrinsic to both is not any attitude vis-à-vis religion, but their attitude vis-à-vis democracy and human choice: they are averse to both. This aversion on their part is their chief feature, as can be seen, for example, in the ease with which their adherents shift between the two, the dramatic difference between them notwithstanding.

So much for the theory of divine appointment. The traditional version of the theory of the social contract is divided into two complementary parts – one genealogical and one ethical. The genealogical is the famous description given by Hobbes, Rousseau, Locke, Paine and others, of the transition from the real or hypothetical state of nature to the state of statehood.[5] The transition was necessitated because the state of nature was "war of every man against every man," (Hobbes), or because it was every man for himself (Paine). Thus, people had to become organized. This involved the mutual consensual renunciation of the hypothetical freedoms that were afforded by the state of nature, in exchange for the real ones afforded by the state. The description of said transition carries no pretense of historical truth – but the

---

[2](Nozick 1974).

[3]Burgess 1992. I replace "right" with "appointment" since I deem this fitting.

[4]The most famous and illuminating defense of the theory of the divine appointment is that Sir Robert Filmer's. (Filmer 1680). The most important attack thereupon is Locke's. (Locke 1689, 2003).

[5]Munro 2016; Hampton 1986.

inner logic of the situation.[6] Yet even as such it is flawed, due to the following three implicit assumptions: firstly, that legitimization is either conferred upon all states or denied to them all; secondly, that the question of legitimization is binary, thus not allowing for any shades of grey;[7] and thirdly, that the process of arriving at consent is necessarily free of coercion. All three assumptions are obviously false, thus rendering the theory of the social contract indefensible.

Even if the genealogical part of the social contract theory was completely sound, it still does not answer the question, why does it oblige us, those who did not participate in the original process of arriving at the alleged consent? This is where the supplementary ethical part of the theory comes in: we re-validate the social contract in that we all give it our consent by the very act of having been born or having grown up into the state.[8]

The traditional theory of the social contract is a step forward, to be sure, compared to the traditional theory from divine appointment, both because it is more humane and because it appeals to rational reasoning. Yet it is unsustainable just the same. The idea that one gives the state one's consent by the very act of having been born or having grown up into it is ludicrous. The answer to this is, of course, that the idea must not be taken at face value, i.e., as a statement of fact but as… indeed, as what? A metaphor? A metaphor to what?

The social contract theory enjoys a special status, shared by an exclusive family of theories: no matter how many times they are discarded, someone always comes to their rescue.[9] Of course this always comes at the price of inserting modifications. Question: whence the motivation to always change the theory to meet the criticism? Why not admit defeat and be done with it? Answer: because we feel that we cannot – may not – forgo the demand for consent, in some form or another, as a condition for legitimacy.[10]

---

[6] The most popular such account goes to Rousseau (Rousseau 1761, 1987), probably because of his apeal to deep rooted and popular ideas of his times.

[7] Says Agassi: the theory of the contract either fully justifies a regime as liberal or it fully condemns it as illiberal. This conflicts with the common recognition that liberalism is a matter of degree. Agassi 2016, abstract.

[8] Locke met this difficulty by suggesting that people join the body politic only by explicit consent. (Friend, *The Internet Encyclopedia of Philosophy*). This raises problems all of its own, of course. For a concise and illuminating summary of the social contract theory I refer the reader to the excellent review of Celeste Friend in *The Internet Encyclopedia of Philosophy*. Friend traces the philosophical roots of the theory to Plato's Crito where "Socrates uses something quite like a social contract argument to explain to Crito why he must remain in prison and accept the death penalty".

[9] For a summarized account of the problem of the social contract theory, the attempts to overcome it, and their insufficiency, see Braybrooke 1976.

[10] This is a dilemma within the liberal camp: what is more important, consent, or minimal intervention? The social contract theory opts for the former, Nozick and others for the latter. Nozick argues that as long as the state does not exceed its role as a night-watchman, it requires no consent in order to be legitimate (Nozick 1974). Beyond that, Nozick stipulates, nothing is legitimate, except with the consent of all, no less. This is populism, often viewed as ultra-democratic, though in truth it is anti-democratic. Perhaps with democracy, ultra is anti.

The only way for the traditional theory of the social contract to make any sense is if we read it to say that the source of the legitimacy of the state is not *that* we give it our consent but *if* we do, and to *the extent* that we do. This is the modified theory of the social contract. It comes at a price: consent is only meaningful when freedom of dissent is assumed. Freedom of dissent is assumed in democracy alone. It thus transpires that the modified social contract theory is that the only thing that confers legitimacy upon a state is democracy and that any given state is legitimate only in so far as it is democratic. Another way to present the same argument is that only the democratic state is legitimate because democracy is the best approximation to anarchy. More precisely: democracy is the only platform upon which collaborative efforts towards ever improving our approximation to anarchy can take place.[11]

To sum: the state of the discussion of the question of the legitimacy of the state, up-to-date, is that it has two answers: the view that the state is legitimate by power of divine appointment, i.e., fascism and/or fundamentalism (which is the modern modification of the historic theory of divine appointment), on the one hand; and democracy (which is the modern modification of the theory of the social contract) on the other.

27.3 The Question of Political Collective Identity

The question of political collective identity is this: what is the "we", the body of people, the collective, to which the state belongs? This necessitates two distinctions: one, that the question is not about a mass of people but a collective. Thus the answer that the state belongs to all its citizens is not acceptable, whereas the answer that the state belongs to the body of all its citizens is. The reason for this technical: a mass of unorganized, un-unified, completely individualized people cannot own things. The distinction is subtle: it is between "you and I and he and she, etc." on the one hand, and "we" or "us" on the other.

The second distinction is between collective identity and *political* collective identity. The question at hand pertains to the latter. Every person is a member in countless collectives and, thus, shares in many collective identities. I myself am Jewish, Israeli, a kibbutznik, a father, an academic, a husband, a philosopher, a jazz lover, middle-aged, shortsighted, and many other things besides. Each of these traits about me renders me a member of the group of all those who share it. Some of these groups identify themselves as groups and some do not, and so on.[12] The questions pertaining to all this are countless, and many of them are intriguing. The literature

[11] This idea is sometimes attributed to Giovanni Sartori: "If understood to the letter a democracy must be a stateless society [...]" (Sartori 1987).

[12] This trivial idea actually had to be discovered. The discovery goes to Georg Simmel. [Simmell [1908] 2009.]. A clear and simple articulation thereof goes to Lakshmi Ramarajan. (Ramarajan 2014). Also see Chayko (2015).

about them is vast. The question at hand pertains only to one specific type of collectives: nations that own – or aspire to own – a state; in a word: political.

A small digression seems to be in order. The question of political collective identity is the modern variant of the classical question of sovereignty, namely the question, who does the state belong to. The answers to that question include not only collectives, but also God. The idea that a state belongs to God is theocracy. This idea is marginal enough to not warrant discussion. Even Israel, which is allegedly constituted upon a Godly decree, is defined not as the state of God, but as the state of God's flock. Thus Israel is more of a clericalism than a theocracy. Oren Yiftachel described it as an *ethnocracy* (together with Northern Ireland, Estonia, Latvia, Serbia, Croatia, Lebanon, Cyprus, Sri Lanka and Malaysia).[13] Gershon Shafir, Yoav Peled and Sammy Smooha prefer the term *ethnic democracy*.[14] Having ruled out God as the sovereign of any state, causes the classical question of sovereignty to completely unify with the question of political collective identity, to which I now return.

Interestingly, a significant part of the vast literature about nationalism deals with the question, what is a nation. This question is new, historically. It was first raised by the Romantics in the nineteenth century as a reaction to the French and the American revolutions: the French and the Americans wanted to include, the Romantics to exclude. More recent political scientists seem to have an additional motivation: they need to know what constitutes a nation because of the principle of the right of all nations to self-determination, i.e., the ethos that all nations are entitled to states.[15] This idea is liberal yet in obvious collision with civil nationalism which too is liberal. This, to state the obvious, is a dilemma facing all liberals as such. Sadly, most choose to ignore it. The principle of the right of all nations to self-determination is true only in an ideal world or – which is the same thing – in a world in which entitlement does not entail any obligation on the part of others.[16] Exclusive definitions of nationhood tend to bestow nationhood upon non-political collectives (the Jewish nation, the Arab nation and the Catholic nation) or to fictitious ones (the Arian nation, the Islamic nation).

The question of what constitutes a nation is not necessitated – at least not initially – by the question of political collective identity. This question, in turn, is

---

[13] Yiftachel 2006.

[14] Ram 2010, pp. 63–67.

[15] The principle of the right of nations to self-determination is often referred to as a *jus cogens* (compelling law) in modern international law. It's first source is possibly the Atlantic Charter, signed on 14 August 1941, by U.S. president Franklin D. Roosevelt, and British Prime minister Winston Churchill. An earlier version was articulated already by US President Woodrow Wilson at the wake of WWI.

[16] Entitlement, i.e. right, is usually taken to imply at least the minimal obligation on the side of others of refraining from impingement. But what beyond this? This question has haunted natural-right advocates throughout history, starting with the Stoics, through the Protestant Reformation and up to humanists and liberals of present day. The dilemma can be seen in the description of natural right as unalienable; does this also mean unimpingeable? If yes: why, then, do we need to bother? If no: why, then, do we need to bother? (For more see Tuck 1979).

necessitated not merely by our wish to do justice to all nations, but also by a far more pressing concern. The state is the most powerful instrument ever to have been in human hands, and today more than ever. An instrument is dangerous in proportion to how powerful it is, and thus also in need of responsible handling. This obviously calls for ownership. It is thus in the highest interest of humanity that we answer the question of political collective identity correctly or, at least, that we do our best, especially in regards countries that have large and powerful armies and or weapons of mass destruction at their disposal.

The answers to the question of political collective identity fall into either one of these two categories: organic / deterministic / essentialist / ethnic / tribal nation, on the one hand; and artificial / nondeterministic / liberal / civil nation, on the other. Let me refer to them as ethnic nations (or nationhood, or nationalism), and civil nations (or nationhood, or nationalism), respectively. Political ethnic nationalism is the idea that the collective to which our state belongs is an ethnic nation. The concept of ethnic nationality is commonly used to denote a group of people – all sharing in or somehow relating to a common culture and/or religion and/or ancestry and/or geographic origin. Taken to the extreme it is described as an extended tribe, which is an extended clan or family, i.e., an organic body of people, all of whom are the descendants of a single ancestry, united by blood, and sharing in a single culture, history, religion and set of values and beliefs. This brand of nationalism is often referred to as *tribal nationalism*, or even simply *tribalism*, thus confused with traditional or primitive tribalism. The difference can hardly be stressed enough: primitive, traditional tribalism is pre-rational. Tribal nationalism is anti-rational. Perhaps it can be described as tribalism with a vengeance.[17] The prominent spokesperson for this brand of nationalism is Friedrich Hegel, but the pioneer, I suppose, is Johann Gottlieb Fichte. The school of thought that both thinkers represent is Romanticism.[18] Significantly, Fichte argued that this brand of nationalism follows directly – no less – from universalism.[19] These people are also the pioneers of the grand and extensive discussions surrounding the question of the definition of nation and nationhood. According to Agassi, though the Romantics differed widely among themselves on a hundred and one issues regarding nationalism, they all agreed upon this one criterion: a definition of nationhood is acceptable only if it excludes the Jews. Jews. In Israel this criterion was replaced with the demand to include the Jews and to exclude the Arabs. This was problematic due to the fact that many Jews are Arabs.To circumvent the problem, Arab Jews are not referred to in Israel as Arab Jews but as Easterner, North African or Sepharadim Jews. They mostly also agreed upon the best means by which to attain the desired goal – the rite of roots and of authenticity. The Jew, needless to say, is rootless (the word is: alienated, i.e., crazy)

[17] For more on the distinction between different types of tribalism see James 2006.

[18] "Romantic nationalism (also national romanticism, organic nationalism, identity nationalism) is the form of nationalism in which the state derives its political legitimacy as an organic consequence of the unity of those it governs." Wikipedia. The key words being *organic* and *unity in a word: Nazism.*

[19] Anderson 1983, p. 34.

and unauthentic. Hence, also, the rite of unity – which is a euphemism for homogeneity – so characteristic of said ideology. This explains, firstly, why political tribal nationalism is so often (always?) monistic, and, also, why it so often leads to racism.

Civil nationalism, in contrast, is the idea that the collective to which our state belongs is the body of all its citizens, with all their ethnic, tribal, cultural and religious diversity. Perhaps the prominent value of this brand of nationalism is diversity, or, at least, tolerance thereof, mostly known as multiculturalism.[20] Civil nationalism in its modern form was invented by the fathers of the French and the American revolutions, but earlier forms are found already in the great empires of old, all of whom aspired to include in their nationhood all the ethnic nations they conquered. Still, their concept of nationhood was hierarchic, which is contrasted with the modern concept of multiculturalism.

To be sure: the distinction between ethnic and civil nationhood or nationalism is not black and white. Features typical of each can be found in the other. For example, some degree of homogeneity is required by all civil nations. The paradigm here is language: the very advent of civil nationalism was manifest in the creation of one national language. Also, by the same token, some degree of tolerance of diversity is found in all ethnic nations.

27.4 Civil Nationalism

In most parts of the world civil nationalism is unheard of to this day. Israel, sadly, is a case in point. This has a powerful reason: from the point of view of ethnic nationalism, civil nationalism is invisible. This is why, for example, in high school textbooks in Israel, the United States of America is sometimes described as a nationless state, while at other times it is described as a multi-national state. Some writers acknowledge civil nationhood as a "mere formality", being defined, as it is, by citizenship, which is a formal status.

More striking is the observation that even citizens of civil-nation-states, say the United States, often refer to their own civil nationality as a "mere formality." Such people are utterly surprised when you point out to them the words "one nation under God" which appear on the dollar bill or in the Pledge of Allegiance, "indivisible, with liberty and justice for all"; or national grief (say, after the JFK assassination), pride (say, after Neil Armstrong's first steps on the moon); or heroic acts of solidarity (such as in the civil rights movement).

[20]The claim that civil nationalism is also (possibly? Inherently? I do not know) multicultural goes to British political scientist Tariq Modood. This idea is recurrent in both his writings and oral addresses. (Modood 2011). Strangely, and sadly, the literature is almost entirely void of reference to this idea, sympathetic or critical. I have heard Agassi say in many of his lectures, tolerating diversity is good but not good enough: we need to celebrate it (also see Agassi 1985, Chapter 1).

It thus transpires that civil nationalism is invisible not only to the adherents of ethnic nationalism, but also to its own adherents. This is one explanation for the scarcity of its mention in the vast literature about nationalism and nationhood. Another is that it is tabooed. Either way, one riddle is replaced by another; and either way, the damage is stupendous. A glimpse at said damage can be attained by considering that:

1. Civil nationalism is intrinsic to democracy; democracy is the only alternative to – thus also buffer against – despotism[21]; it thus follows that weakening civil nationalism is strengthening despotism.
2. Civil nationalism is the only alternative to – thus also buffer against – ethnic nationalism; ethnic nationalism is at the root of most – possibly all – atrocities committed in modern history; it thus follows that weakening civil nationalism is enhancing the likelihood of future atrocities.

Upon one thing there seems to be full agreement between all who write on these topics: political ethnic nationalism is, indeed, at the root of most atrocities, crimes against humanity, uncontainable political upheavals and intractable conflicts worldwide in modern history, including in the Middle East. Indeed, political ethnic nationalism is the active ingredient in all political ideologies that inspire all the above, including racism, fascism, chauvinistic nationalism, Nazism and fundamentalism. This calls for the question of the alternatives to ethnic nationalism. It so happens that there are two: Universalism, or cosmopolitism on the one hand, and Civil nationalism, on the other. Universalism ignores the universal need for non-universal national alignment, leaving the field to the monopoly of ethnic nationalism. It is thus utopist in principle and defeatist in practice. This leaves civil nationalism as the only viable alternative to ethnic nationalism. Civil nationalism is the only platform extant upon which it is possible to combat with any degree of effectiveness all the above-mentioned ideologies, to attempt diffusing racial and ethnic hatred, and strive to end intractable political conflicts. This is not to deny that civil nationalism is problematic, of course. It is problematic especially from the point of view of the political philosophy that begot it – i.e., liberalism. It is liberal in that it is an approximation of sorts to the liberal ideal of the siblinghood of humanity.[22] i.e., universalism or cosmopolitism, and in that it allows for all other collective identities to thrive under its umbrella. Yet it is illiberal in that it decrees that the state must define the nation and not vice-versa, and, consequently, in that it opposes national secession thus also national-self-determination.

The fact that almost all Western democracies are civil-nation-states suggests a wide consensus that civil nationhood is intrinsic to democracy, and that civil nationalism is intrinsic to the democratic outlook. It is strange to observe that this is neither ever stated nor ever denied.

[21] De Tocqueville 1988, Book Four, Chapter VII.

[22] The traditional title of this idea is *the brotherhood of man*; the present one is a suggestion for an upgrade by Agassi (Agassi 1991).

27.5 The Role of Determinism

Possibly the most obvious difference between ethnic and civil nations, is that the latter is openly artificial; that is to say: it is a conscious, deliberate human construct. The former, too, is largely a human construct: but not openly or consciously so, and possibly also not deliberately so. Many political theorists claim that all nations are human constructs, i.e., artificial. At least to some extent they are right. On the most part they completely ignore the question of the import of deliberateness in the context of nation building, thus implying that it carries no import. This is a crude mistake. Many infer that all nations are illusionary. According to this logic cars, too, are illusionary.

Alternatively, scholars acknowledge the import of deliberateness in the context of nation building but only negatively: nation building is but the business of securing the hegemony of one ethnic group or social class over others. Truth be told: this has been and still is the case way too often. But to say that this is necessarily the case is an exaggeration. For the record, then, let it be said: human beings can and do construct nations. Sometimes they do it tacitly, in hiding, possibly undeliberately and even unconsciously. At other times they do it openly, deliberately and consciously. At those times they often declare that their goal is that of creating a free and an equalitarian political collective such as in the United States Declaration of Independence, quoted so eloquently by Abe Lincoln in his famous Gettysburg Address: *Four score and seven years ago our fathers brought forth on this continent, a new nation, conceived in Liberty, and dedicated to the proposition that all men are created equal.* The words that elicit the greatest attention here are, *all men are created equal*, naturally, but the operative words in our present discussion are, *brought forth, new,* and *conceived.*

A common response to the above is to suggest that the founding fathers were less than sincere as is demonstrated in the fact that they all were slave owners and married to women deprived of civil rights. This indeed taints their intentions. Yet they gave their commitment to later be cashed by sons and daughters of later generations – including Abe Lincoln, The suffragists, Martin Luther King and others.

27.6 Nation Building

The very possibility of deliberate nation building invites an entire array of questions, all of which are passed on by literature with deafening silence, which only in recent years is beginning to be broken. These questions, are all worthy, pressing, beautiful, and potentially fruitful beyond words. These include – among many others – questions such as: How are nations constructed? What are the minimum desiderata for nationhood? How is national solidarity fostered? What is the minimal degree of solidarity required for a healthy national feeling? How much – or what kind of, or in what areas – homogeneity is required in order to allow for national solidarity?[23] How much – or what kind of, or in what areas – homogeneity is too much? To what extent, if at all, must collective rights within the nation be acknowledged?[24] What model of nationhood is most consistent with universal human values, such as tolerance, the love of diversity, equality, democracy, the rule of law, human freedom, human rights and the strive for peace? And so on and so forth. Possibly they are all variants on this one:

Q.1: What model of nationhood is most consistent with the ideal of the siblinghood of humanity?

Civil nationhood is not the answer to this question – but the platform upon which this question can be asked and discussed. The natural answer to this question is universalism, or cosmopolitism, i.e., the idea that all of humanity is one nation. Cosmopolitism has only one flaw: it is completely unlikely, since it demands of people to forego all other national affiliations, which the vast majority of us are not ready to do. Thus we need to forfeit cosmopolitism as a demand and keep it as an aspiration. How do we aspire to the universalist, the cosmopolite ideal? How do we strive to get ever closer to that ideal? Indeed, it seems that our question aims at just that. We can rephrase it, then, in a way that better reflects this:

Q.2. What model of nationhood is the best available approximation to cosmopolitism?

Alas, the answer to this question is that we do not know. Yet this does not stop us from keeping looking for an answer. So we might want to rephrase our question yet again:

Q.3. What model of nationhood can serve as a platform upon which we can collaboratively endeavor to ever improve our approximations at cosmopolitism?

To this there is only one viable answer: civil nationalism.

The silence over nation building constitutes a taboo, in effect if not by design. Whence this taboo? Possibly determinism. It seems odd to attribute determinism to an entire milieu, especially given that most of the writers therein seem avowed enemies of determinism. This is a mistake: most writers are avowed enemies not of determinism, but only of one brand thereof, namely, primordialism or perennialism. Primordialism is, historically, the most important brand of national determinism, so it seems natural to expect that anyone who rejects it rejects national determinism as such. Natural though it may be, the expectation is wrong. Question: if we reject primordialism but hang on to determinism, what concept of nationhood, then, do we endorse or even merely acknowledge? Answer: probably none. This depicts all such expert writers as experts on illusion. Many of them declare just this, in the wake of Benedict Anderson. He called this illusion an *imagined community*.[25] Does "imag-

[23] On questions pertaining to national solidarity see Malesevic 2013.

[24] On this issue see Kymlicka 1995.

[25] Anderson 1983.

ined" imply non-real? Anderson never says, but leads his reader to reach this conclusion on their own.[26]

The magnitude of the damage of a taboo over an idea, is in direct proportion with the importance of said idea; which, in turn, is in direct proportion with the importance of the question to which it is the purported answer, minus the number of viable competing answers. The question to which civil nationalism is the only viable purported answer, is of the highest import imaginable, and viable competing answers there are none. Let me repeat the question: What model of nationhood can serve as a platform upon which to collaboratively strive to ever improve our approximations at cosmopolitism? The importance of this question can be seen in that striving to improve our approximations at cosmopolitism is the logical equivalent of fighting against the evils of ethnic nationalism. It thus follows that our question is, in its turn, the logical equivalent of this one:

Q.4: What model of nationhood can serve as a platform upon which to collaboratively seek an alternative to – thus also buffer against – ethnic nationalism?

To this, again, there is only one viable answer: civil nationalism.

27.7 Between Civil Nationalism and Democracy

Perhaps the claim that civil nationalism is intrinsic to democracy is not obvious and warrants some explanation. Ethnic nationhood is monist. Civil nationhood, in contrast, is monist in one sense and pluralist in another. It is monist in the sense that it monopolizes the political arena. All Western democracies are civil nation states with the exception of Belgium and Great Britain. Even Switzerland is not an exception for it is, contrary to popular belief, a nation state: true, it is confederation of 26 cantons, with almost as many languages and cultures, yet all share in one national affiliation. You may say that this amounts to saying that their nationality is merely formal. Indeed, that is exactly what it *merely* is – to ethnic nationalist eyes.

It is by power of its political monism that civil nationalism can be pluralistic in the sense of allowing for all other collective identities to co-exist peacefully under its umbrella. This is possibly due to the distinction between the political and the communal realms, also known as the distinction between nationhood and ethnicity, or community. To the extent that ethnic or communal pluralism – i.e., multiculturalism – is intrinsic to democracy, it thus follows that said distinction, too, is intrinsic to democracy. Since the distinction is only made possible under the umbrella of civil nationalism, it thus follows that civil nationalism, as such, is intrinsic to democracy.

[26] One of these is Omar Dahbour who argues that since nationhood is intrinsically illusory, the principle of the right of all nations for self-determination is false. (Dahbour 2003). A counter argument to both is brought forth by James: Nations, he says, are not imagined but Abstract Communities. (James 1996).

Civil nationalism is intrinsic to democracy for yet another, closely related reason. Due to the distinction that it reinforces between the political and the communal, it allows for the separation between the political and the communal powers. This separation is known as the principle of separating state from Church: the word Church is used here to refer to any religious establishment and, by extension, to all communal authorities or powers as such. Democracy consists not in one but in two principles of separation of powers – one internal and one external. The internal principle of separation of powers is said to have first been conceived in ancient Greece for separating the different branches of the state from each other. The external principle of separation of powers separates the state as such from other powers that bear upon the lives of the citizens. This is meant not as a measure for the protection of the state against these other powers (as is the prevailing view in Israel) so much as for their protection against it. But mostly it is felt to be needed to protect the citizens from religious persecution and discrimination, without which protection religious freedom, which is intrinsic to democracy, cannot even be envisaged. This renders civil nationalism intrinsic to democracy once again. End of explanation.

The above logic is trivial, but its ramifications are far-reaching. You recall – democracy is one of the only two answers to the question of the source of the legitimacy of the state. Civil nationalism is one of the only two answers to the question of political collective identity. If you add to this the realization that civil nationalism is intrinsic to democracy, the result is that the two questions collapse into one another: they are really not two questions but one.[27] This question can be referred to as the question of sovereignty, since sovereignty speaks of both the legitimacy of the state and the political collective to whom the state belongs. The mutual collapse of both questions into one another is eminently demonstrated in how the answers to one dictate those to the other and *vice versa*. People who reject civil nationalism are forced to also reject democracy and vice versa. This explains why those who reject civil nationalism are forced to endorse fascism and/or fundamentalism all good intentions notwithstanding. Israel, sadly, is a case in point: by Israeli standards the fascists and the fundamentalists in Israel form a minority – a big and a growing one, to be sure, as well as a loud one, but still, a minority. By Western standards, in contrast – though most Western friends of Israel shy from saying so – almost all Israeli political parties are fascist and/or fundamentalist. By this I mean no condemnation of all members of said parties: on the most part they are nice, gentle people, who sincerely wish nobody harm. Yet they are all fascists and/or fundamentalists in the sense that they adhere to the theory of the divine appointment of the state. Rejecting civil nationalism leaves them with no other choice.

Civil nationalism is not only intrinsic to democracy, but it is also similar to democracy in two chief features: firstly, in the vein of Churchill's famous aphorism, civil nationalism is the worst form of political collective identity, except for all the others. And, secondly, in the vein of the famous characterization of democracy,

---

[27] The realization that the question of legitimacy and the question of sovereignty collapse into one another goes to Gellner. More specifically, Gellner suggested that nationalism is an answer not only to the latter but also, and primarily, to the former. (Gellner 2009).

attributed to Alexis de Tocqueville, as the only buffer against despotism, so is civil nationalism the only buffer against fascism and fundamentalism. Israel, sadly, is not a case in point. The good news, however, is that we still have choice.[28]

References

Agassi, Joseph. 1985. *Technology: Philosophical and social aspects*. Episteme/Dordrecht: Kluwer. Online revised edition: http://www.tau.ac.il/~agass/joseph-papers/technoln.pdf.

———. 1991. *The siblinghood of humanity: An introduction to philosophy*. Delmar: Caravan Books.

———. 2013. *Liberal nationalism for Israel: Towards an Israeli National Identity*. Jerusalem/New York, Gefen.:Translation from the Hebrew book of 1984.

———. 2016. Toward a fictionless liberalism. *Philosophy of the Social Sciences* 46 (1): 77–91. Sage Publications.

Anderson, Benedict. 1983. *Imagined communities: Reflections on the origin and spread of nationalism*. London: Verso.

Braybrooke, David. 1976. The insoluble problem of the social contract. *Dialogue* XV (1): 3–37.

Burgess, Glenn. 1992, October. The divine right of kings reconsidered. *The English Historical Review* 107 (425): 837–861.

Chayko, Mary. 2015. The first web theorist? Georg Simmel and the legacy of 'The web of group-affiliations'. *Information, Communication & Society* 18 (12): 1419–1422.

Dahbour, Omar. 2003. *Illusion of the peoples: A critique of national self-determination*. Oxford/New York: Lexington Books.

De Tocqueville, Alexis. 1988. *Democracy in America*. Trans. Henry Reeve. New Brunswick: Rutgers University Press. (Originally published: 1840)

Filmer, Sir Robert, Bart. [1680]. *Patriarcha or the natural power Of kings*. E-version: http://www.constitution.org/eng/patriarcha.htm.

Friend, Celeste. *Internet encyclopedia of philosophy*, a peer-reviewed academic resource. Social Contract Theory: http://www.iep.utm.edu/soc-cont/.

Gellner, Ernest. 2009. *Nations and Nationalism* (2nd ed.).

Hampton, Jean. 1986. *Hobbes and the social contract tradition*. Cambridge: Cambridge University Press.

Hayes, Carlton J.. [1928] 1931. The historical evolution of modern nationalism. New York: R. R. Smith.

James, Paul. 1996. *Nation formation: Towards a theory of abstract community*. London: Sage Publications.

———. 2006. *Globalism, nationalism, tribalism: Bringing theory back in*. London: Sage Publications.

Kymlicka, Will. 1995. *Multicultural citizenship: A liberal theory of minority rights*. Oxford: Clarendon Press.

Locke, John. [1689] 2003. *Two treatises of government and a letter concerning toleration*. Yale University Press.

Malesevic, Sinisa. 2013. *Nation-states and nationalisms: Organization, ideology and solidarity*. Cambridge: Polity Press.

Modood, Tariq. 2011, January 27. Multiculturalism, Britishness, and Muslims. In *Open democracy: Free thinking for the World*.

[28] The choice is described, explained and argued for by Agassi in his Liberal Nationalism for Israel, together with the history of its original conception by Herzl, its rejection by Ben Gurion and the consequences of it all (Agass, 2013).

Munro, André. 2016. State of nature. In *Encyclopedia Britannica*. Online https://www.britannica.com/topic/state-of-nature-political-theory.

Nozick, Robert. 1974. *Anarchy, state and Utopia*. New York: Basic Books.

Plato. [360 B.C.E]. *Crito*, E-version: trans. Translated by Benjamin Jowett: http://classics.mit.edu/Plato/crito.html.

———. [360 B.C.E]. *Republic*. E-version: trans. Translated by Benjamin Jowett: http://classics.mit.edu/Plato/republic.html.

Ram, Uri. 2010. *Nationalism: Social conflicts and the politics of knowledge*. Abingdon: Taylor & Francis.

Ramarajan, Lakshmi. 2014. Past, present and future research on multiple identities: Toward an intrapersonal network approach. *The Academy of Management Annals* 8 (1): 589–659. http://www.hbs.edu/faculty/Publication%20Files/Ramarajan-2014-Annals-Multiple%20identities-intrapersonal%20network%20approach_b2358fa9-050f-46c2-b57a-d855b4a14d22.pdf.

Rousseau, Jean-Jacques. [1761]. Editor: G.D.H. Cole. *The social contract and discourses*. E-version: http://oll.libertyfund.org/titles/rousseau-the-social-contract-and-discourses.

Sartori, Giovanni. 1987. *The theory of democracy revisited*. Thousand Oaks: SAGE Publications.

Simmel, Georg. [1908] 2009. *Sociology: Inquiries into the construction of social forms*. Leipzig: Duncker & Humblot.

Tuck, Richard. [1979] 1998. *Natural rights theories: Their origin and development*. Cambridge University Press. Online: http://assets.cambridge.org/97805212/25120/sample/9780521225120ws.pdf.

Yiftachel, Oren. 2006. *Ethnocracy: Land and identity politics in Israel/Palestine*. University of Pennsylvania Press.

Chapter 28
Obstructing the Realization of Civil Responsibility: Political, Sociological, Historical, and Psychological Aspects of Civil Responsibility

Chemi Ben-Noon

This article connects two of my research fields: obeying the law (understanding the essence of civil disobedience and civil responsibility in democratic states), and flaws in reasoning (understanding the role of cognitive biases in ordinary and fateful decisions). Prof. Joseph Agassi made a decisive contribution to these research directions and accompanied me over many productive years.

I met Prof. Agassi in 1979, the first year of my Master's degree studies. I participated in his seminar *Philosophy of Science*, and immediately knew that I would choose him as my dissertation supervisor. A short time later, we talked about this and Agassi asked me to phrase my research question.[1] I chose the question "What conditions or circumstances justify civil disobedience?", and we started a dialogue that concluded with my submission of my dissertation for evaluation. This Master's dissertation later became the book *Civil Disobedience*. Agassi agreed to supervise my doctoral dissertation, entitled *Civil Responsibility*. One of the chapters in the dissertation, which was the opening shot in this field, was entitled "Obstructing the Realization of Civil Responsibility", which is also the title of this paper. I have processed and reshaped this chapter, and it found its place in my book published by Paragon, *Civil Disobedience – The Israeli Experience*. A few years after this book was published, upon receiving the request to write for this festschrift, I decided to rework the chapter, to change it and adapt it to the present day, and to turn it into this paper.[2] My writing has undoubtedly been greatly shaped and influenced by Prof. Agassi's thought.[3]

[1] On Agassi's way of phrasing research questions and of teaching, see Ben-Noon 1997, 31–36.

[2] This is also how Agassi operates, in a spiral motion starting with renewed, broader examination of what has been written, rephrasing it to provide deeper meaning, and unceasingly striving for greater rationality.

[3] To mention at least these three books: Ben-Noon 1997, 1980; Agassi 2000.

C. Ben-Noon (✉)
The Academic Center of Law and Science, Hod HaSharon, Israel
e-mail: benoon@netvision.net.il

N. Bar-Am, S. Gattei (eds.), *Encouraging Openness*, Boston Studies in the Philosophy and History of Science 325, DOI 10.1007/978-3-319-57669-5_28

This paper deals with the ways in which our cognitive structure, the state, its organs, and those who cooperate with it consciously or unconsciously, obstruct the functioning of the responsible citizen in a democratic state. I shall present the ways actions the state employs for this purpose, along with insights from research into cognitive biases. The paper represents a search for a more accurate presentation of the idea of civil responsibility in a democratic state, along with the state's ways of obstructing responsible citizens' attempts to realize their civil responsibility.[4]

Being a responsible citizen is not easy, nor is becoming one. Democracies are rarely blessed with a proliferation of responsible citizens, but real and solid democracies where the government does not hold absolute power even when based on a majority require more such citizens.[5] In solid democracies, not all laws must be obeyed and the majority may not restrict minority rights. The sovereign must accept that some of its laws may be wrong, and therefore it should tolerate civil disobedience and conscientious objection, if only as a necessary and contributing evil.

In states where "that government is best which governs least",[6] responsible citizens are the pillars of democracy and carry out Emerson's maxim that "good men must not obey the laws too well".[7]

Emerson's maxim characterizes responsible citizens. Such citizens ask themselves, at important democratic junctures, where the state's authority reaches. In other words, where is the limit of the duty to obey the orders and instructions of a legal system *that individuals, as citizens of the state, willingly recognize and accept.* Aristides de Sousa Mendes, the Portuguese Ambassador in Bordeaux, acted against the instructions of his superiors, granting visas to thousands of Jews and saving their lives. This is an explicit case of civic responsibility of a senior official in his country's foreign service who acted illegally in contrast to his government's policy for a higher cause, saving human lives.[8] This is how other diplomats acted when they

---

[4]The paper focuses on the obstruction methods that are employed with sophistication and malice rather than random actions or direct law enforcement. The paper does not deal, for example, with the arrest of responsible citizens during law breaking (the state's use of the security forces) or direct orders to employees not to break the law (the case of Aristides Mendes described below).

[5]In such a democracy, it is not usually necessary to refuse to obey the law, because a law that people refuse to obey is probably a law that contravenes its principles anyway. On this, Marcuse wrote: "As existing democracies become directed and controlled democracies that legally (rather than illegally) restrict democratic rights and freedoms, so it becomes more necessary to add to these forms of opposition action also extra-parliamentary means of struggle" (Marcuse 1961, 72).

[6]H. D. Thoreau https://www.brainyquote.com/quotes/quotes/h/henrydavid124849.html. One of the most important principles is the *principle of the regime's self-limitation.* Such a limitation appears, for instance, at the base of the United States Constitution. Congress knows it is not entitled to legislate laws that contravene the Constitution. Furthermore, even if Congress represents the majority of citizens, there are laws it cannot legislate, even if the vast majority of the people support them.

[7]R. W. Emerson, https://www.brainyquote.com/quotes/quotes/r/ralphwaldo134991.html/.

[8]The full story of Mendes and others is told by Avshalom Elizur, "The Power Drunkenness of General Consul Aristides de Sousa Mendes", http://a-c-elitzur.co.il/site/siteArticle.asp?ar=140&stext=%F1%E5%E0%F8%F9&page=1.

disobeyed their governments' instructions, and some of them paid a heavy price for this.[9]

Responsible citizens form a key foundation of democracies, because they are willing to face a hostile consensus, break laws, and pay personally to make their protests heard (Ben-Noon 2015, 24). Why, then, are only few of us responsible citizens? A series of personal, political, psychological, historical and social reasons can explain this. The following intends to lay the infrastructure for the discussion of possible explanations. These numerous reasons may be classified into two major categories: individual-personal- and public-state-generated (Ben-Noon 2015, 78).

An examination of this division into two levels of reference may miss another level of explanation that does not contradict these two levels, and even cuts across them and complements them: the level of our cognitive flaws. While these faults apparently exist on the individual level, their outcome pervades the public policy explanation level. My main claim in this paper is that the cognitive level should be prominent within and alongside the two other levels, and we must act to reveal it as an additional level of reference, perhaps no less important than them. Bringing the discussion regarding the cognitive level into the spotlight could lead to an important and fascinating public discourse.

28.1 The Individual Level

Individuals generally fear public opinion, being an exception, making decisions based on insufficient data, the personal hardships that might follow disobedience, and so on. Furthermore, individuals wish to promote their personal interests, which might clash with public interests, and may not fully comprehend why they disapprove of certain laws. How do you explain that you violate the law to protect the law?.In view of the above, individuals may form a set of beliefs that actually prevents them from assuming responsibility, or may fail to distinguish between acting for their communities and promoting state interests.

Thinkers over the years have disagreed over the source of the reasons why individuals do not assume responsibility, mostly citing personality traits or the environment,[10] or a combination of the two. Erich Fromm represents the environmentalist approach, while Martin Buber, Raymond Aron (1968), and others argued

[9] These include: Raoul Wallenberg, the Swedish ambassador who paid with his life for his intervention in support of Hungary's Jews; Chiune Sugihara, Japan's Consul in Lithuania, who lost his job due to helping local Jews; the Chinese Consul General in Vienna, Feng-Shan Ho, who was reprimanded for granting visas to Jews; Georg Duckwitz, the Nazi trade attaché in Copenhagen, who risked his life by warning the Jews of Denmark about the intention to send them to concentration camps. These and others together saved hundreds of thousands of Jews.

[10] For a shocking account of how environments shape humans, see Packard 1977.

that all humans share a similar response structure that makes them shed responsibilities for the sake of conformity.[11]

Fromm argued that the modern industrialist system, primarily its monopolistic version, created insecure, helpless, anxious, and lonely individuals. This state of meaninglessness is dangerous, mainly because such individuals are easily influenced by fascist messages, which allow the development of antidemocratic regimes. He believes that modern democracy has failed to attain its goals: Though we are presently free to express ourselves, we are incapable of producing independent thoughts. Free Man, according to Fromm, is merely a conformist and an automaton.

Fromm noted sadly that fake liberty is actually a new form of enslavement in which people fail to understand the situation and actively change it, and sacrifice their personal integrity. Fromm argued that what matters is not whether social democracies promote human liberties, but whether civilians are actively involved in shaping their personal goals and the public agenda. Thus, citizens should not settle for casting their votes on Election Day, but should be actively and daily involved in their workplace and social, communal ties.

Aron is more pessimistic on this issue, claiming that the chances of reaching an anomaly are the same as the chances of establishing orderly democracies. Failing to find a difference between conformism and consensus, he states that anomia and conformism threaten every modern society. Aron argues that sociologists and psychologists should help people gain more self-awareness, believing this could change the behavior of nations and individuals.[12]

Martin Buber concurred with Fromm, saying that technological development created a situation where peace is minimized to the periods between wars, and decreased human cooperation makes individuals lose their autonomy and surrender to collective and centralist regimes (Buber 1958).

Buber and Fromm, each in his own way, speak about states' brutality that crushes individuals, and they try to make people aware of this. Current history has proved them right, which is why I believe that cultivating responsible citizens is a possible solution for the modern social ailments. Being more original and spontaneous, such citizens could cooperate with their peers and change that modern trend.

Leading Israeli psychoanalyst Yehoyakim Stein discussed the banality of good and evil, asserting that no distinctive characteristics could foretell who might commit morally deviant acts. When that banality is coupled with conformism (satisfying popular norms) and opportunism (quick, self-promoting adaptation), certain citizens might become totally irresponsible. They would be passive at best, but in the worst case they might undermine the work of responsible citizens or anyone the

[11] The main finding indicated by Tversky and Kahneman, the pioneers of research into cognitive biases, is the uniform response of most humans to situations in which our cognition is biased. Our mistakes are systematic. Recent studies imply that intuitively operated probabilistic or statistical judgment is liable to systematic failings and is guided by thinking principles that deviate from the criteria that mathematics and formal logic determine for rationality. See Menashe 2008, 282–283.

[12] Ibid., p. 302.

government may instruct them to harm. Stein believes that under given circumstances, evil might be found in everyone. In a press interview, Stein said: "Conformists and opportunists are the vast majority and their motivations are eternal." He added: "The banality of evil enjoys hiding behind the broad shoulders of the law, order, and common civil decency." The concept of the banality of evil (which was angrily rejected in Israel) was popularized by Hannah Arendt's book on Eichmann's trial where she described the Nazi officer as a law-abiding and mentally sound family man: "The trouble with Eichmann was precisely that so many were like him, and that the many were neither perverted nor sadistic, that they were, and still are, terribly and terrifyingly normal." (Arendt 2006, p. 12).

28.2 The Public-State Level

Dazzled at the Top, a book published shortly after the Yom Kippur war, raised the question of how to suppress protest "without violence, dying without pain, good spirited self-elimination [...]" (Ofer 1974, 172). The recipe is as follows: First, ignore the matter; if an outcry still arises, and the protest movement grows and increases, the Prime Minister summons the heads and organizers of the protest movement for a face-to-face meeting, in order to study them and examine the possibility of coopting them into the establishment, or else he threatens some of them. Since this conversation constitutes a sort of cure, the chance increases that the protest will be eradicated already at this stage. But the protest leaders who were not shifted from their position by this conversation are considered "dangerous" and the regime applies against them means such as spreading rumors about them and instructing the relevant bodies to keep an eye on them. Moreover, the regime may plant people within the protest movement, and these people, in addition to supervising and reporting, will direct the activity to safer places for the government (Ofer 1974, 174–175). The result:

> In Israel, it operates exceptionally. The revolution immediately scatters (and is divided) into ideological splinter groups. For long days and nights, they discuss and phrase each clause and clause in the format of youth movements, until the purpose for which the protest movement was established is pushed aside in actually ceases to exist [...] So in the end the whole affair ends as a youthful experience full of goodwill [...] This is the uniqueness of Israel, its power and hope, fortunately for its leaders (ibid., 172 ff).

Alon Harel wrote:

> [...] During a crisis, the jurist who translates into words the political wishes of his masters is obliged to rebel against his juridical-technocratic calling. This rebellion is both possible and necessary when the political values that guide society clash with the internal values of the legal system. Israel is the place to demonstrate the way the political leadership promotes, with unhindered militancy, values that do not accord with the legal values such as the rule of law, equality before the law, and freedom from the regime's ruthlessness.
>
> [...] In the struggle between the jurist and the politician, the politician is certain to win. But the duty of every Israeli jurist – state employee, academic, or in the private sector – is to understand that the role of the jurist in Israel has changed beyond recognition. He must

> not serve the values of Israeli politics. The role of law in Israel today is subversive: to delay and scupper the brutalization of Israeli politics (Harel 2011).

Sometimes, law acts in a way that is similar to an autoimmune disease, and harms both the popular protest activity and the chances of the regime to benefit from it: social activists believe that the law has already “swallowed” the protest because protest organizations have changed their character and undergone a process of “legalization”, meaning that the main activity of the protest organizations is legal. Thus, the protest energy has been redirected from its traditional place of activity (the street) into the legal channels: instead of ground roots activists, there are black-robed professionals (Elbashan 2011, 92 ff). I consider this as a victory of the state and the legal system over the protest, without a battle or bloodshed. This also redirects the aggressive energy of protest to words rather than actions. In my opinion, this is one of the factors for recent protests not existing in their traditional forms, and future protests will be weakened.

Two elements feature on the public level that could deliberately create situations and requirements that hinder civil responsibility: government systems and interest groups. They employ tactics such as imposing law obedience, creating false consensus regarding the criminal liability of offenders, or naming them irresponsible. They can exploit the law and declare states of emergency using propaganda and disinformation, cooperate with the legal system, and so on. Here are some of the key tactics:

28.2.1 Imposing Law Obedience

Directly or indirectly imposing law obedience is usually a positive element of state management, but the indiscriminate or even violent imposition of laws or policies – particularly when they conflict with democratic values – is inappropriate. Such an imposition is based on the common view that obeying the law is a supreme democratic value because the laws express the wishes of the majority.

28.2.2 Discussion-Diverting Methods

Diverting public, political, and even legal discussions away from relevant issues might also impede individuals’ desire to become actively involved. Regimes provide the public with partial or false information, which might motivate people to do nothing or to act wrongly. They may further encourage citizens to exercise individual censorship before they act politically, placing obedience before responsibility. Citizens may also choose not to respond because they feed on information from the government that makes them defensive and sets their agenda for them. Alternatively, the government may incorporate political activists in the establishment (cooptation); or it may dominate the meta-debates, determining which is politically relevant and which is not.

28.2.3 Dominating Debate Subjects and Boundaries

A well-known rhetorical technique, known as *ad hominem*, is addressing speakers, rather than the issues they raise. Governments focus on activists personally, referring to them as "enemies of the people," which forces them to defend themselves and ignore the issues. This helps the rulers reestablish the boundaries of discussion. Responsible citizens can develop some sensitivity to this technique and formalize strategic or tactical reactions.

28.2.4 Preemptive Action – Rationalization

A clever and cheap way of stopping political protest before it becomes active is known from the philosophy of decision-making and combines knowledge of the nature of mankind with introducing skilled agents and using the media. A decision-making theory maintains that before we make a decision, we suffer from cognitive dissonance and inner disharmony because each choice we may make is accompanied by regret for not choosing others, and discomfort because the consequences of our choice are unknown to a large extent. To minimize this unease, people tend to rationalize the reality that follows the decision, making the world appear emotionally harmonious and cognitively stable in their minds. State agents (including the media) step into that gap and deliver messages that intensify the protesters' discomfort before they make a decision, or act upon one. Some states use psychological experts to provide the undecided with data that would weaken their resolve and nullify their potential protest act.

28.2.5 Cooptation

This is a paradoxical way in which governments incorporate protesters into the establishment or "flood" them with state resources (or promises of funds, status, etc.) to make protesters cooperate with or even join the establishment (state jobs), or deny their access to national resources and neutralize them. States may even seek out weaker members of the opposition to influence this way, using the old and trusted "divide and conquer" technique.

28.2.6 Grading Political Issues

Controlling political debates and issues on the national agenda (by dominating the media, for example), enables states to brand issues that protesters promote as irrelevant, which could make citizens disengage with protest movements. Regimes can also argue that political issues that the opposition promotes are actually moral, or judicial. This supposedly makes them matters for experts, turning common citizens into passive participants.

28.2.7 Creating False Consensus

Institutionalizing protests or making them seem redundant further prevents citizens from realizing their responsibility in real time. In Israel, the Supreme Court under President Aharon Barak introduced the norm that everything is "justiciable", which forced protest organizations to change their focal points, from socially-oriented groups to legal-minded ones. Though perfectly legitimate, reasonable, and even positive in various contexts, this new legalistic approach weakened protest movements from within and naturally minimized their impact. Another method that I classify with the false consensus family is institutionalizing protests. Protest movements are required to strictly follow the rules of law and democracy, which makes them appear superficial and irrelevant, or face the justice system to deflect charges of illegality. When protesters follow that formula, their protests are stripped of meaning and sense, and becomes ineffective.

Civil disobedience is inherently illegal. When the regime permits only legal protests, in effect it shuts down protest movements, making them lose their uniqueness and unexpected moves, and makes political debates predictable and, therefore, controllable. Regimes further employ the media and propaganda techniques to institutionalize protests by, for example, presenting all political felonies as negative and harmful.

Propaganda, indoctrination, agitation, and the various other names this phenomenon has been given all imply that they deal with shaping public opinion and motivating people to act in ways that the propaganda initiators intended. Propaganda campaigns present only certain carefully-selected facts, conceal others, take things out of context, explain certain events but ignore others, direct public attention to specific issues, present their views on disputed issues as the final word, or create disputes, promote or undermine loyalties, and instigate or curb the population. For example, governments may promote certain laws, presenting them as democratic even if they are not. Then, they insist that citizens obey those laws, stating that even if they are questionable, they should be discussed only in the parliament. This type of institutionalized disinformation helps to subdue protesters and to deter potential objectors.

28.2.8 *Employing State Establishments, Particularly the Justice System*

This serves the politicians when fending off civil protest. They exploit legal formalities and codes while possessing an unrestrained ability to enact laws as they see fit. This method has served parliamentarians when passing laws that harmed human and civil rights and liberties, justifying such laws as emergency or security-related necessities, and suspending democracy while providing the ruling majority with the power to act at will.

In effect, the legal system provides support for the political system by trivializing political offences in this way. Ideological protesters are then viewed as ordinary criminals and therefore their acts are investigated in isolation, while ignoring the motives behind them. Furthermore, when illegal protests are examined only using standard legal tools, the government crimes that prompted these protests are seen as legal norms. Legitimizing the regimes' acts, the legal system errs primarily because it fails to look into the future and does not distinguish between rebellion and civil disobedience, for example. Thus, paradoxically, the legal system hurts civilians who wish to uphold the law.

What is more, the legal system naturally cooperates with the government and demands that law violators explain and justify their deeds, which upsettingly reverses the normal order of things in which the system must prove guilt. Thus, while politicians are not required to justify the undemocratic laws they introduce, the wrongful policies they promote, or the intimidation and propaganda methods they use, the confused and humiliated protesters must face the court. Legal systems should have placed the real culprits on the defendants' bench. The greatest threat this poses in democracies is that civilians might lose faith in the legal system. I, therefore, believe that the two state systems must be politically and economically independent of each other.

28.2.9 *The Betrayal of the Elites (and Academic Irresponsibility)*

The elites create governing cultures and serve as guardians of public order. The importance of their democratic role far exceeds their proportion in the population. While this is a vague term, "the elites" are composed of groups of individuals who hold the majority of state resources, appear in the media, control and create public debates, shape the public agenda, advise politicians, and often serve as a redeeming factor on political and social missions. These social, economic, and academic elites betray the population through dereliction of their duties when they help politicians harm democracy, neutralize protests, mislead the public, and justify wrongful government moves.

Academics must draw a line between receiving state grants for their studies and placing the results of these researchers at the negative disposal of the government. Intellectuals must beware and stop their colleagues from being tempted by the powers of darkness and forsaking the people they really work for.

Leading intellectuals are responsible for preventing politicians from introducing wrongful laws and institutionalizing protests in order to neutralize them; for pointing out what is relevant when governments try to change public priorities; for severing the lethal ties between politicians and leading economic forces that prevent protest; for stopping the politicization of the judiciary; and they must encourage free and creative thought. The academic elites, particularly in law schools, are required to assume *academic responsibility*, step out of their ivory towers, and protest when protest is due. They need to stand up when human rights are harmed, warn against wrongful laws and suggest better ones, update legal standards and values to reflect changing realities, intensify criticism, and minimize the influence of the systems that feed them economically - even if that could harm the academia (Freund 1960).

28.3 The Cognitive Level – Cognitive Biases in the Regime's Service

Considering the sub-section about the betrayal of academia, we could assume that academics and experts on decision-making would cooperate with the government, given the possibilities raised by the results of studies on cognitive biases. Several experts have indeed exceeded the boundaries of academia and become civil servants.[13] I have no direct proof that experts in cognitive psychology, decision-making, or game theory have collaborated to harm the protests, but, as I shall demonstrate, there is circumstantial evidence for this.

Earlier we hinted that the means of eradicating protest originate in cognitive biases. It is reasonable to assume that the government applies tactics in its war against protest and protesters that rely on the study of cognitive biases, in order to divert the protesters to places that do not endanger the regime and public order. The tactics of anchoring, representativeness, and availability may be a sophisticated instrument of government to prevent protests or to thwart their possibility through labeling, creating a sense of guilt, and confusing the protesters. Several cognitive biases are presented below, followed by discussion of how they can hinder responsible citizens from actualizing their citizenship.

[13] For example, Guy Rotkopf, a researcher from Tel Aviv University, became the Director General of the Justice Ministry, and David Haan moved from Bar Ilan University to head the Administrator General's Department. Overseas, Cass Sunstein served as the Administrator of the White House Office of Information and Regulatory Affairs in the Obama administration, and Richard Thaler headed the British Behavioral Insights Team under the UK Prime Minister, David Cameron.

28.4 Cognitive Biases

Studies in the field of cognitive psychology have shown that when we are required to decide (usually rapidly and intuitively) in information-dense scenarios, we use various rules of thumb and mental shortcuts. As a result, we do not gather complete information and we do not systematically internalize or process it. The use of these shortcuts and rules of thumb constitutes, in effect, a substitute for gathering and processing data, even the available data. These rules of thumbs, known as heuristics, are guidelines that usually function reasonably well, but we do not expect them to apply to every event or circumstance. Due to their simplicity and inability to cover the event from all sides, the use of these heuristics and shortcuts lead to cognitive biases, which are, among other things, simple statistical errors and thinking errors that characterize our thinking. They were first identified as systematic by Amos Tversky and Daniel Kahneman (Menashe 2008, 282–283).[14] In their classic article, "Judgment under Uncertainty: Heuristics and Biases" (Tversky and Kahneman 1974), Tversky and Kahneman argue that human beings rely on a limited number of heuristic principles that reduce the complex tasks of estimating probabilities and predicting values into simple judgment actions. A coincidental mistake is not interesting to research, but a systematic error is particularly interesting, as it enables us to predict how people will err in particular evaluations and judgments.

The studies by Tversky and Kahneman did not attack the rational model, but offered empirical evidence for cognitive flaws, and left the conclusions regarding human rationality to be drawn by the readers themselves. Some readers concluded, incorrectly, that Tversky and Kahneman were arguing that in general people are incapable of thinking logically without biases. However, this was not their argument. Kahneman believes that his study does not demonstrate human irrationality, and that this perception of his research is mistaken. He claims that research into heuristics and biases just disproves an unrealistic perception regarding rationality that considers it completely consistent.

The model accepted by most scholars today is Kahneman and Tversky's hypothesis that our cognition contains two systems: System 1, the intuitive one, is heuristical, automatic, effortless, associative, rapid, insufficiently self-aware, insufficiently critical, unreliable, vague, and it has high emotional value. System 2, the rational one, is critical, effortful, deductive, slow, systematic, conscious, rule-operated, and flexible.[15]

---

[14] Tversky and Kahneman found that most people respond uniformly when taking intuitive decisions. They found that our mistakes are systematic.

[15] On the two-system theory, see: *Kahneman* 2003. In my opinion, even though the professional literature presents these two systems as distinct, there may be some "leakage" of System 1 into System 2, and vice versa. In other words, the two systems operate in parallel, but there is a local "overpowering" in any given decision of one of the systems over the other. Support for this theory can be found in the model by Shane Frederick (and Daniel Kahneman) whereby we make an intuitive judgment (System 1) that can be changed by consideration from System 2: "System 2 monitors the quality of these proposals, which it may endorse, correct, or override". Shane 2005, 40.

I will present here a few biases, and then demonstrate how they can be used to harm potential protesters and even active protesters.

Representativeness Bias Representativeness is a common cognitive bias that can create attribution or a causal relation ex nihilo. This bias leads to evaluating the probability of attributing A to group B according to the similarity of A to members of group B. The greater the similarity between A and members of group B, the higher the tendency of many to determine that the probability of A belonging to group B, often without any justification. This similarity can be manipulated: A can be called a criminal and attributed to the group of criminals, or called an "enemy of the people" or "enemy of the state" and attributed to the group of those who threaten the people or the state (Ben-Noon 2009).

The Availability Heuristic We tend to rely on our memory and experience when retrieving "available" information from our memory, instead of analyzing the facts before us. We rely on the first examples that occur to us in order to assess the probability of an event, and as a result we may assess the size or influence of a group in an exaggerated manner. We tend to attribute great weight to recent events, even if rare, and ignore the high frequency of events in the past. Thus, for example, we may fear a marginal protest while forgetting that most protests, even the dramatic ones, cause relatively little damage (ibid., 207 ff).

The Framing Effect We are influenced by the way information is presented. The verbal presentation of the alternatives in a decision (even if they are identical) may lead to a different decision. Thus, we can err in assessing the risk resulting from a certain protest due to its verbal presentation (ibid., 211 ff).

The Anchoring Heuristic This leads to bias in our assessment of a point created at random or deliberately, which is not really relevant (ibid., 215 ff). When performing a numerical assessment, people tend to rely on a preliminary value that is available. This number provides a starting point that "anchors" the rest of the assessment process. We can become fixated on some point that sticks in our mind without realizing, and continue to stick to it without employing our judgment. Moreover, even the boundaries of protest can be formed through the influence of anchoring.

Cognitive Dissonance This originates in the sense of discomfort and regret resulting from our difficulty to accept the implications of our decisions (ibid., 222 ff). Our decisions have clear emotional implications, some of which constitute a source of anxiety and guilt. In this way, protesters can be associated with guilt, and their struggle weakened.

28.5 How Can Cognitive Biases Deter Responsible Citizens from Fulfilling Their Civic Responsibility?

Cognitive biases can explain the failure of citizens to protest against unjust laws or policies. The lack of knowledge about the rules, foundations, and boundaries of democracy can lead to a range of cognitive biases. An additional function is the regime's rhetoric and the use of the media, which also diverts individuals from realizing their civic responsibility. I will attempt to present here the classification of possible biases.

Diverting the public and political (and even the legal) debate and distancing it from dealing with political acts that are considered responsible may deter citizens wishing to realize their political involvement. Partial information can create a system of unfounded beliefs and damage the citizens' decisions. While wondering about the possible results of the political acts at their disposal, they become convinced that the law takes precedence over responsibility.[16] Another way is when citizens convince themselves not to respond politically (or alternatively, to obey the law) because the information they receive from the regime (including value labeling of the types of protests) leads them to defensiveness or sometimes determine the agenda and boundaries of the discussion. Another way is to control the meta-debate, determining what is relevant to the political discussion and what is not.

Labeling the protesters as enemies of the public fits in well with the representativeness bias. The deliberate attribution of people to dangerous groups can lead to inaction, and this directly harms the protesters and indirectly harms their state. The representativeness bias is apparent when the regime finds fault with the protesters by identifying them with people who protested violently. In this way, they can classify actions as "good" and "bad", label the protesters as "anarchists" and "criminals", and directly damage their protests, which were often correct and positive. Creating internal censorship or self-persuasion that makes potential protesters defensive anchors them outside the limited realm of action, and thus harms the potential results of their actions. Anchoring is a cognitive bias that slants our assessment to a random or deliberately created point that is not actually relevant. In this way, the control of the meta-debate – what is relevant to the discussion and what is not – anchors the citizens to a fruitless discussion.

One of the techniques know from the field of rhetoric is diverting the discussion to the person, what is called *ad hominem*, instead of discussing the matter itself. The use of this technique reduces the weight of the original topic of discussion by focusing on the protester himself. The regime or one of its conscious or unconscious collaborators calls the protesters "enemies of the people", "anarchists", or "enemies of the public. The protesters then have to defend against this label, rather than

[16] This diversion is described as a group diversion and termed "groupthink" by Janis and Mann. This type of thinking occurs when the influence of a "senior" group member over the rest makes the group support the "senior" member's opinions. For our purposes, the senior member is the state or its organs. See Janis and Mann 1977.

discussing the reasons for their actions. The main purpose of the *ad hominem* technique is to reshape the boundaries of the discussion around the protesters rather than the topic on the agenda.

A sophisticated and inexpensive way is stopping the political action before it takes place. This way can be realized by inserting skilled agents, exploiting the media, and good familiarity with human motivation and bias mechanisms. The most prominent of these biases is cognitive dissonance to deter the protester from realizing the action he wished to take. The state may use expert knowledge in these areas in order to "provide" the individual with "materials" and "contents" that would eventually weaken or eliminate his possible protest action.

Controlling the contents of the political debate in the country and dividing the discussion into relevant and irrelevant subjects constitutes real anchoring. When citizens "understand" (or are helped to "understand") that they are protesting against something irrelevant, they may redirect their efforts to another channel or cease completely to be active. This is how the anchoring bias operates. When the state diverts the discussion of the political act from the political realm – which is the relevant one – to the moral or legal realm (or proposes setting up a committee, which is usually no help whatsoever) – which are irrelevant, the chances decrease that the protest will grow, because the discussion is focused on what will not endanger the state. Institutionalizing political protest constitutes anchoring it to the realm of the establishment's interest, thus making the protest fruitless.[17] The institutionalization of protest or ignoring it are the "sins" of a regime that prefers to act without control and without checks and balances. This institutionalization occurs in many ways. One of the familiar ways is the demand that the protester act in an institutional, legal, and democratic way. Dividing protests into "permitted" and "forbidden" (representativeness bias) and diverting the protester to justify the protest under the "law" – which was obtained democratically and therefore the only way to change the law is legally and democratically – close the door to any illegal protests. In this respect, the availability bias has an effect: since the "law" is the first thing that comes to mind in cases of law-breaking, rather than the protest's possible benefit. It is clear that in such circumstance civil disobedience is neutralized, since it is necessarily illegal. Legal and institutionalized protest may not be sufficiently effective in cases where effectiveness is required, so the protest becomes fruitless. Employing permitted forms of protests means that citizens are unable to influence the regime in matters of legislation and policy. Furthermore: institutionalizing protest makes the accuser (the protester) into the defendant (legal offence or moral guilt), and the just (politically) into a justifier. This inversion constitutes serious damage to the protesters and to the state.

[17] Great danger arises from the Supreme Court, under President Justice Aharon Barak, creating the "revolution of judicializing protest". This revolution leads many organizations to turn from having social dominance to having a legalistic orientation. See Elbashan 2009, 146. This revolution is important in itself when the protest is legal, but may be an obstacle and a burden when it is not legal. The institutionalization and judicialization of protest is very important, as long as it does not contain civil disobedience.

The regime also uses propaganda and similar techniques to institutionalize protest and harm the protesters (On propaganda, see Confino 2002, 9–28; Volman 1990). Propaganda can neutralize citizens by presenting, for example, all political law-breakings (responsible and irresponsible alike) as negative and reprehensible. Propaganda deals with shaping public opinion and motivating the public to particular positions or actions. When the propaganda's origin is a regime body, it seeks to motivate citizens and political groups in accordance with purposes it considers important[18] and therefore directs the public's interest to certain issues while diverting it from others, nurtures support or opposition to political or social goals, tips the scale on public disputes, creates or silences disputes, supports or undermines loyalties, incites,[19] or restrains the population, and so on.

28.6 Conclusion

The basic assumption in this paper is that the government uses various tactics in order to eliminate protest and to deter responsible citizens from realizing their civil involvement. The two classical levels of reference regarding disrupting protest are the individual level and the public-state level. This division into two levels may miss another level that does not contradict these two levels of explanation and even complements them: the cognitive flaws level. The central argument in this paper is that the cognitive level possesses important and independent validity alongside the other levels. As far as I know, the discussion of the cognitive level in the research literature is lacking compared to the other two levels.

We can assume that academics and experts on decision-making could collaborate with the regime relying on the possibilities entailed in the results of studies regarding cognitive biases. If cognitive biases are a means of eradicating protest, it is reasonable that the government applies tactics in its struggle against protests and protesters that are based on the study of cognitive biases. Using them diverts the protesters to places that do not endanger the regime and public order. The tactics of anchoring, representativeness, and availability may constitute a sophisticated tool for the government to prevent or disrupt protests through labeling; creating a sense of guilt and confusion among the protesters, and the framing effect and cognitive dissonance may have similar applications. I briefly presented these cognitive biases and described how they could prevent responsible citizens from realizing their citizenship.

---

[18] Breaking the law for ideological reasons, for example, could be named by the regime "an undemocratic act that harms state security"; total obedience to the law will be called "the state's interest"; acts of civil disobedience can be described as "it is unthinkable that a citizen of the state, who is supposed to be loyal to it, could do such a thing", and so on. See Marcuse 1961, 11–13; George Orwell 1949.

[19] Two serious examples that probably originated in incitement that eventually led to murder – even if it is difficult to find a direct line from the incitement to murder – are the murders of Emil Greenzweig and Yitzhak Rabin. See Ben-Noon 1993.

Naturally, being a responsible citizen is not easy, and is becoming one is quite hard. It means that citizens must face themselves and their surroundings daily. Considering the public opinion, responsible citizens must not surrender to conformity. These cognitive biases may cause protesters to surrender and thus damage civil responsibility.

References

Agassi, J. 2000. Letters to my Sister. Tel Aviv, Yediot Achronot. [Hebrew]

Arendt, H, and A. Elon. 2006; first published 1963. *Eichmann in Jerusalem: A report on the Banality of Evil*. New York: Penguin Classics.

Aron, R. 1968. *Progress and disillusion: The dialectics of modern society*. London: Frederick A Prager.

Ben-Noon, C. 1980. *Conversations on the philosophy of science*. Tel Aviv, Defense Ministry Publishing. [Hebrew]

———. 1993. On the despicable margins of freedom of expression. *International Problems, Society and State* 32: 3–4 [Hebrew].

———. 1997. *The limitations of reason*. Jerusalem, Keter. [Hebrew].

———. 2009. Intuitive decisions and judgment decisions. *Shaarey Mishpat* 5: 195 [Hebrew].

———. 2015. *Civil disobedience: The Israeli experience*. Paragon: St. Paul.

Buber, M. 1958. *Paths in Utopia*. Boston: Beacon Press.

Confino, M, ed. 2002. *The power of words and the weakness of reason: Propaganda, incitement, and freedom of speech*. Tel Aviv, Yitzhak Rabin Center for the Study of Israel/Am Oved.

Elbashan, Y. 2009. Aharon Barak: Between law and protest. In *Barak Book*, ed. E. Zamir, B. Medina and C. Pessberg. http://www.yedid.org.il/?id=3633. p. 139 [Hebrew].

———. 2011. *"The Absence of Law", Reading the Protest*. Tel Aviv. [Hebrew]

Freund, L. 1960. Responsibility: Definitions, distinctions, and applications in various contexts. In *Responsibility*, ed. C.J. Friedrich. New York: Liberal Art Press.

Harel, A. 2011. *A time for judicial subversion*. Haaretz. July 15.

Janis, I., and L. Mann. 1977. *Decision making: A psychological analysis of conflict, choice, and commitment*. New York: Free Press.

Kahneman, D. 2003. A perspective on judgment and choice: Mapping bounded rationality. *American Psychologist* 58 (9): 697–720.

Marcuse, H. 1961. *The end of Utopia*. Tel Aviv, Am Oved [Hebrew].

Menashe, D. 2008. *The logic of evidence validity*. Jerusalem, Magnes. [Hebrew]

Ofer, S. 1974. *Dazzled at the top*. Tel Aviv, Bazaq. [Hebrew]

Orwell, G. 1949. *1984*. London: Secker & Warburg.

Packard, V. 1977. The people shapers. Boston, Little, Brown.

Shane, F. 2005. Cognitive reflection and decision making. *Journal of Economic Perspectives* 19 (4): 25–42.

Tversky, A., and D. Kahneman. 1974. Judgment under uncertainty: Heuristics and biases. *Science* 185 (4157): 1124–1131.

Volman, M. 1990. *Demagogy and rhetoric*. Tel Aviv, Papyrus. [Hebrew]

Chapter 29
The Secular Problem of Evil

John Kekes

It is an honor and a pleasure to join in the celebration of Agassi's 90th birthday. He is a philosopher who in the work of a lifetime has continued in the long, noble, and always unpopular tradition in which philosophers question the shibboleths of their age. Reading and conversing with him is a bracing experience that is never boring and always challenging. He goes to the core of whatever problem he discusses, articulates the assumptions on which it rests, considers the alternatives to them, and ever so innocently inquires why we should accept one of these assumptions rather than contrary ones. May he long continue his subversive work. I offer the discussion of evil that follows in the same spirit.

29.1 What Is Evil?

Evil is the most serious and succinct moral condemnation in English, much worse than moral badness. It involves depraved, malignant, monstrous actions that violate universal values and conventions in a way that breeches the bulwark between civilized life and barbarism. Facing evil makes us realize that if human beings can be that callous, cruel, fanatical, or ruthless, then there are no limits to man's inhumanity to man. An abyss opens up and a glimpse into it is horrifying. No wonder we are reluctant to face evil. But if we do not keep it at arm's length and think, feel, or imagine ourselves into the positions of the evildoers and their victims, we come to deeply disturbing malignant motives and immense suffering.

Clarity about evil calls for concrete examples. They are unfortunately readily available and familiar. The ones that follow involve murder, deliberate homicide committed, as the law says, with malice aforethought. The examples are from

J. Kekes (✉)
Research Professor, Union College, Schenectady, NY, USA
e-mail: jonkekes@nycap.rr.com

N. Bar-Am, S. Gattei (eds.), *Encouraging Openness*, Boston Studies in the Philosophy and History of Science 325, DOI 10.1007/978-3-319-57669-5_29

the last hundred years or so, but countless others could be drawn from other historical periods.

Between 1914 and 1918, the Turks massacred about a million and a half Armenians. In 1931, Stalin ordered the murder of prosperous peasants, called kulaks, and about two million of them were executed or deported to concentration camps where they died slowly as a result of forced labor in extreme cold and on starvation diet. During the great terror of 1937–1938, two million more were murdered at Stalin's orders. In 1937–1938, the Japanese murdered about half million Chinese in Nanking. During WWII, about six million Jews, two million prisoners of war, and half million gypsies, mental defectives, and homosexuals were murdered in Nazi Germany. After India's independence in 1947, over a million Muslims and Hindus were murdered in religious massacres. In the 1950–1951 campaign against so-called counter-revolutionaries in Mao's China about one million people were murdered, and the so-called Great Leap Forward of 1959–1963 caused the death of an estimated sixteen to thirty million people from starvation. Pol Pot in Cambodia presided over the murder of about two million people. In 1992–1995, about two hundred thousand Muslims were murdered in Bosnia by Serb nationalists. In 1994, almost one million people were murdered in Rwanda. To this list of mass murders many more could be added from Afghanistan, Argentina, Chile, the Congo, Iran, Iraq, Sudan, Uganda, and numerous other places. These examples, however, only illustrate, but do not define, what evil is.[1]

The definition I propose focuses, in the first instance, on evil actions, and identifies five features that jointly make actions evil. First, the actions cause grievous harm. In the examples, the grievous harm is murder, but it may be dismemberment, torture, enslavement, blinding, and so forth. Second, the victims are innocent. They were not murdered because of something they have done, but because they were Armenians, kulaks, Chinese, Jews, gypsies, homosexuals, Muslims, Hindus, city dwellers, Tutsis, or just those who had the misfortune to be anathematized by some ideologically motivated fanatic. These victims have done nothing to deserve the grievous harm inflicted on them. What is done to them, therefore, cannot be explained as punishment, revenge, or self-defense. Third, if the agents are normally intelligent and not incapacitated, for instance, by mental illness, drugs, or alcohol, then they must know that their actions cause grievous harm. Normally intelligent people cannot fail to know that murder, dismemberment, torture, enslavement, and blinding do cause grievous harm. If they know it and nevertheless do it, then their actions are deliberate. Fourth, the deliberate infliction of grievous harm on innocent victims may or may not reflect the agents' ill will toward them. If it does, then their ill will, in turn, reflects their beliefs and emotions (a point to which I will return). For the moment, it will suffice to say only that if their deliberate infliction of

[1] For further examples, see e.g. Jonathan Glover, *Humanity: A Moral History of the Twentieth Century*, (New Haven: Yale University Press, 1999); Stephanie Courteous, et. all. *The Black Book of Communism*, (Cambridge: Harvard University Press, 1999); Martin Gilbert, *The Holocaust*, (New York: Henry Holt, 1985), and my own The *Roots of Evil*, (Ithaca: Cornell University Press, 2005).

grievous harm on innocent victims reflects their ill will, then the agents' motives are malevolent. Cruelty, envy, fanaticism, hatred, jealousy, prejudice, rage, ruthlessness, or a combination of some of them are examples of malevolent motives. Fifth, actions that have the first four features may or may not be morally unjustifiable. It is hard to conceive how a malevolently motivated action that deliberately causes grievous harm to innocent victims could be morally justified. But let us suppose that there are some extreme situations in which such an action would help to avoid even more grievous harm. If, however, no moral justification is available, then the action is evil. Those who think that an action that has the four features is nevertheless morally justified must provide the justification, since there is an overwhelming presumption against its possibility.

An action is evil, then, if (1) it causes grievous harm to (2) innocent victims, and it is (3) deliberate, (4) malevolently motivated, and (5) morally unjustifiable. Actions that have these features may still be more or less evil, because they may differ in the quantity and quality of the harm they cause, and in the degree of deliberation and malevolence involved in them.

Evil actions are, by definition, morally unjustifiable, but they may be excusable if there are exempting or extenuating circumstances, such as irresistible coercion, great provocation, momentary insanity, or ignorance of relevant facts. The crucial difference between justified and excused actions is that justified actions are not evil, whereas excused actions are. Excused actions do not reflect adversely on their agents, because there are circumstances that acquit the agents or lessen their responsibility for doing them. Perhaps the most frequently proposed excuse is that the agents sincerely but mistakenly believe that their victims are guilty or that the grievous harm they cause is morally justified. If such beliefs were true, the actions would not be evil. But the beliefs are false, although the agents believe that they are true. Whether such beliefs qualify as excuses depends on how reasonable they are. If they are self-serving rationalizations, they fail as excuses. If others in the agents' position would also be likely to hold them, then they might serve as excuses. It must be remembered, however, that the excuse is sought for mass murder that includes children and obviously harmless adults. It is extremely unlikely that anyone could reasonably believe that such victims are guilty or that murdering them could be excused. If, nevertheless, a legitimate excuse is available, then the agents' responsibility is lessened, although, of course, the actions remain evil. In this paper, I aim to provide an explanation of why actions, people, or social conditions evil.

We may say, then, that people are evil if their habitual actions are evil and there is no pattern of good actions that may counterbalance the evil ones. Social conditions are evil if they regularly lead people to do evil. The primary subjects of which evil can be predicated are actions; the secondary subjects are people; and tertiary subjects are social conditions. This is not merely a grammatical observation. It indicates the direction in which understanding evil must proceed. We must explain first what makes actions evil; then we can go on to ask why people do evil habitually; and lastly inquire why people create and maintain social conditions that lead to habitual evildoing. Social conditions would not be evil if they were not created and maintained by evil people. And people would not be evil if their actions were not habitually evil.

Actions, people, and social conditions can be more or less evil, and evil actions may shade into morally bad actions, because the grievousness of the harm, the innocence of the victims, the malevolence of motivation, the deliberateness and moral justifiability of their performance are often matters of degree about which reasonable and morally committed people can disagree. Although there is often a clear difference between evil and bad actions, there are likely to be cases in which the difference becomes unclear. I do not think that this has a substantive bearing on the account I will provide. Given the numerous mass murders, there can be no reasonable doubt that evil is prevalent. The question is why it is prevalent.

29.2 The Secular Problem

It is well known that there is a theological problem of evil. If God is perfect in goodness, knowledge, and power, then why has He created our world in which evil is prevalent? The favored answer, that evil is created by human beings, not by God, is unsatisfactory. God is said to have created human beings, as well as everything else, and if human beings cause evil, then God has created them in a way that they would cause it. Why would a perfect God create a very imperfect world? Christians, of course, have struggled with and proposed various solutions to this problem. My present concern, however, is not with their putative solutions, but with pointing out the less well known fact that there is also a secular problem of evil that is surprisingly similar to the theological problem.

The secular problem arises in the following way. Normal human beings in normal circumstances prefer well-being to its opposite. Reason requires that we should do what we can to protect the conditions of well-being. Yet the prevalence of evil shows that we often violate these conditions. If we accept the optimistic view inspired by one version of the Enlightenment that human beings are basically reasonable and good, then how could evil be prevalent? The favored answer is that evil is prevalent, because human beings are corrupted by social conditions. But this is as unsatisfactory as the theological answer that attributes evil to human beings rather than to God. For just as a perfect God would not have created imperfect human beings, so reasonable and good human beings would not have created corrupting social conditions. Just as evil seems to be incompatible with the nature of God as conceived by many Christian theologians, so evil seems to be incompatible with human nature as conceived by many secular thinkers.

That many influential thinkers in the Enlightenment tradition do hold this view of human nature is borne out by what they say. Here is Rousseau: "Man is naturally good; I believe I have demonstrated it" and "the fundamental principle of all morality, about which I have reasoned in all my works [...] is that man is a naturally good creature, who loves justice and order; that there is no original perversity in the human heart."[2]

Kant says virtually the same: man is "not basically corrupt (even as regards his original predisposition to good), but rather [...] still capable of improvement" and "man (even the most wicked) does not, under any maxim whatsoever, repudiate the moral law. [...] The law, rather, forces itself upon him irresistibly by virtue of his moral predisposition."[3]

John Stuart Mill speaks of the "leading department of our nature [...] this powerful natural sentiment [...] the social feeling of mankind – the desire to be in unity with our fellow creatures, which is already a powerful principle in human nature, and happily one of those which tend to become stronger, even without inculcation."[4]

John Rawls claims that "the capacity for moral personality [...] no race or recognized group of human beings [...] lacks" and that "moral personality is characterized by two capacities: one for a conception of good, the other for a sense of justice." A conception of the good gives "a moral person [...] a fundamental preference for [...] a mode of life that expresses his nature as a free and equal rational being." And "the sense of justice is continuous with the love of mankind." He claims that "men's propensity to injustice is not a permanent aspect of community life, it is greater or less depending in large part on social institutions."[5]

It is not unfair, I think, to ascribe a secular faith to the numerous thinkers who hold that human beings are basically reasonable and good.[6] It is a faith, because no evidence is allowed to count against it. When human beings act reasonably and morally, it is counted as evidence that confirms the faith; when they act unreasonably and immorally, it is counted as evidence that they have been corrupted by social conditions, and that also confirms the faith. Defenders of the secular faith should pause to ask: how is it possible that supposedly reasonable and good human beings create and maintain social conditions that are contrary to the well-being of people in their own society? For these defenders of the secular faith, the problem of evil is as serious as it is for their Christian brethren.

Instead of facing the problem, however, they nurture a pleasing illusion and do not ask embarrassing questions about the prevalence of evil. They go to absurd lengths to find excuses for those who violate conditions of wellbeing. Rawls, for instance, says that "the effort a person is willing to make is influenced by his natural abilities and skills and the alternatives open to him. The better endowed are more

[2] Jean-Jacques Rousseau, *Discourses on the Origin and Foundation of Inequality Among Man*, trans. Donald A. Cress, (Indianapolis: Hackett, 1988/1754), 89 and "Letter to Beaumont" in *Oeuvres completes*, 5 vols. (Paris: Galliard, 1959–95), 935; translated by Timothy O'Hagan in Rousseau, (London: Rutledge, 1999), 15.

[3] Immanuel Kant, *Religion within the Bounds of Reason Alone*, trans. Theodore M. Greene and Hoyt H. Hudson, (New York: Harper & Row, 1960/1794), 39, 31.

[4] John Stuart Mill, *Utilitarianism*, (Indianapolis: Hackett, 1979/1861), 3, 31–2.

[5] John Rawls, *A Theory of Justice*, (Cambridge: Harvard University Press, 1971), 506, 561, 476, 245.

[6] Annette Bayer explicitly refers to it as such in her "Secular Faith" in *Postures of the Mind*, (Minneapolis: University of Minnesota Press, 1985).

likely, other things equal, to strive conscientiously, and there seems to be no way to discount for their greater good fortune." Efforts to contribute to rather than violate well-being, therefore, are "arbitrary from the moral point of view."[7] Enemies of well-being are just as reasonable and good as its friends. Only different social conditions lead them to act differently. But defenders of the secular faith do not ask the obvious question of why some social conditions are corrupting.

The secular view, however, can be held without this faith. There is another tradition whose defenders may or may not be secular thinkers, but they share realism about the prevalence of evil and unwillingness to accept optimistic illusions about human reason and goodness. Some who belong to this tradition are Thucydides, Euripides, Machiavelli, Hobbes, Montaigne, Burke, James Fitzjames Stephen, and Freud. In this paper I continue this realistic tradition. Realism, however, does not remove the need to explain why evil is prevalent. To be sure, human beings cause it. But why do they cause it?

29.3 Inadequate Explanations

I will consider four inadequate explanations. They attribute evil to natural causes, moral monsters, uncharacteristic actions, and corrupting social conditions. Each accounts for some cases of evil, but each fails to account for many others. In the next section, I will move toward a more general explanation. The first inadequate explanation is that evil has natural causes. This may be interpreted as the claim that evil is the outcome of natural processes, not of human actions. The grievous harm that befalls innocent victims is the result of earthquakes, draught, tidal waves, and similar natural disasters, or of the non-culpable scarcity of food, money, knowledge, expertise, and other resources. These natural processes cause epidemics, poverty, starvation, and agony.

Such calamities undoubtedly cause much suffering to innocent victims, but they are not evil. For they are neither malevolently motivated nor deliberate. Of course, if the calamities were foreseeable and preventable, then the failure to avert them may be evil. In that case, however, it is not the natural processes that are evil, as this explanation claims, but the actions or possibly the people who failed to avert them. And we still need an explanation of the failure to avert them. Furthermore, none of the murders I have listed can be blamed on natural disasters. They were caused by malevolently motivated deliberate human actions.

Natural processes, however, may be interpreted in a much wider sense to include psychological processes within human beings. To say that evil actions are the outcome of natural processes, then, becomes a truism that is no more informative than the banality that what will be will be or that everything has a cause. What we want from an explanation is to tell us why people act on malevolent rather than on benevolent motives, why they deliberately murder innocent victims, why they violate

[7] Rawls, *Theory of Justice*, 311–12.

conditions of wellbeing. To say that they are led to do so by natural processes does not even begin to provide the needed explanation.

The second inadequate explanation attributes evil to moral monsters, among which we may distinguish two classes. One contains those who do evil for its own sake. They deliberately and malevolently cause grievous harm to their innocent victims. They know what evil is, but they do not recognize that it limits what they may do. Their evil actions are not means, but ends in themselves. They may be misanthropes, cynics, or sadists, who believe that everyone, or only those who belong to a particular group, are incorrigibly depraved; or they may be deep ecologists who think that human beings despoil the world, which would be a better place if the species became extinct; or they may believe that their victims are loathsome, because of some characteristic or belief shared by all members of the group to which they belong, as the Nazis believed of Jews and inquisitors of heretics; or they may be ruled by savage, primitive urges, as do the instigators of tribal murders in Africa and ethnic cleansers in Bosnia.

Another class of moral monsters do evil as a means to some end. They know quite well the requirements of morality, but they think that some other consideration is vastly more important. They talk about the teleological suspension of the ethical, where the telos may be religious, as do Muslims terrorists, Christian crusaders, or Hindu fanatics. Or the telos may be political, as it was for the Jacobins during the terror phase of the French Revolution, or for Lenin as he was consolidating Bolshevik rule in Russia. Or it may be sexual, as exemplified by the Marquis de Sade.

Moral monsters may be born as such, or they may be made. They may start out with dispositions for both good and evil actions, but become monstrous, because they are brutalized by their experiences; or rewarded for evil actions; or embittered by disappointed expectations; or scorned, humiliated, or ridiculed time after time; or because they have accepted an evil creed. Regardless of what made them moral monsters, however, they are alike in wanting power in order to transform the world, or some segment of it, in accordance with their ghastly vision. They become mass murderers in order to destroy whoever stands in their way.

Moral monsters undoubtedly exist, and they account for some evil actions, but the vast majority of evil we are familiar with cannot be attributed to them. First of all, moral monsters are bound to be rare, because it is very difficult to be one. It requires great strength of character, self-reliance, and the capacity to live without love, friendship, and trust; the willingness to systematically squelch the little compassion or sympathy they may feel; and the capacity to sustain a perfervid, all-consuming misanthropy. Moral monsters, however, not only have to have these exceptional dispositions, they must also hide their true nature from others. If they did not, they would not be at large long enough in a civilized society to pursue their evil designs.

Secondly, the large-scale evil that led to the murder of tens of millions of people in the examples I have given would not have been possible without the cooperation of a large number of ordinary people. Perhaps Stalin, Hitler, Mao, and others were moral monsters, but they could not by themselves have arrested, transported, sorted, and guarded the innocent victims, organized the details of their murders, found and supplied the necessary resources, disguised from the rest of their society what was going on, pulled the triggers, and got rid of the corpses. Only a fraction of the evil that has been done to the multitude of innocent victims during the last century can be attributed to the actions of moral monsters. The rest had to be done by accomplices who lacked the strength, detrmination, and bitterness to become moral monsters. It needs to be explained why they did it.[8]

The third inadequate explanation is that evil actions are uncharacteristic responses of normally reasonable and decent people to extreme situations, sudden emergencies, great provocations, or serious threats. In such cases, people are called upon to deal with predicaments that are far removed from the customary context of their life. It is hard to know what anyone, including oneself, might do when faced with unusual situations. This is partly why many examples of the what-would-you-do-if sort, such as presented by the Milgram experiments, have an air of unreality. What people may or may not do in unusual situations is an unreliable indicator of what they may or may not do in the normal course of events. Uncharacteristic actions, by their very nature, are rare. Most people most of the time act characteristically. Life would be in an intolerable state of permanent chaos if it were otherwise. Some evil actions are no doubt uncharacteristic, but the prevalence of evil must be explained by the characteristic actions of habitual evildoers.

Cases in point are the mass murders committed by the moral monsters and their accomplices in the examples I have given. They were done over several years by the same cadre of murderers. They were the routine habitual patterns of action of the secret police, the concentration camp guards, the executioners, the military, and their vast supporting cast. They together made standard operating procedure the religious, ethnic, and tribal massacres, the death marches, and the starvation and forced labor of people suspected of being less than fully enthusiastic supporters of vicious regimes. An explanation of evil must account for the actions of ordinary people who become the accomplices of moral monsters, but who lack the monsters exceptional traits.

The fourth inadequate explanation proceeds from the optimistic secular faith I have discussed earlier. According to it, most evil actions are caused by corrupting social conditions, such as poverty, discrimination, persecution, and injustice. Evil actions are the enraged responses of basically reasonable and good people to being victimized by these conditions. Their actions are evil, but they reflect the social conditions that corrupted them, not their evil nature. The prevalence of evil, therefore, is not to be explained by the cruel, envious, fanatical, hateful, jealous, prejudice-ridden, or ruthless dispositions of murderers, but by the social conditions that cause them to have these dispositions. The agents of evil actions are held to be

[8] This is one of the themes of Hannah Arendt's work on evil. See especially *Eichmann in Jerusalem: A Report of the Banality of Evil* (New York: Viking, 1963), and also *The Origins of Totalitarianism* (New York: Harcourt, 1941). Arendt would have avoided a great deal of controversy and vituperation directed at her if instead of the banality of evil, she talked about the banality of most evildoers.

as innocent as those on whom they vent their rage. Of the four defective explanations, this is the most widely held one and the least plausible. It is, in fact, obviously untenable, and cries out for explanation how normally intelligent people could take it seriously.

To begin with, there are ways and ways again to respond to corrupting social conditions. We may treat them as adversities and strive to overcome them; become politically active in order to reform them; decide to leave the society where the conditions hold for another we believe is better; resign ourselves to them and seek consolation in religion, family life, or solitude in nature; or become mass murderers. Since the possible responses are many, the response of mass murder cannot be explained merely in terms of corrupting social conditions. It must be explained why the response of moral monsters and their accomplices was mass murder, rather than any of the other possible responses. An adequate explanation, therefore, must take into account the psychological dispositions that inclined some people, but not others, to become mass murderers.

Next, suppose that rage at corrupting social conditions blinds evildoers to possible responses other than violent ones. Why does the violence take the form of mass murder, rather than, say, the destruction of property, sabotage of public works, or wholesale theft? If the violence must take a murderous form, why is it directed indiscriminately at everyone, such as children, young men and women, and other victims, who cannot possibly be blamed for the social conditions? And if the violence is murderous and indiscriminate, why is it often combined with the humiliation, rape, torture, and mutilation of the victims? Once again, the malevolence of evil actions must be attributed to psychological causes, in addition to social ones.

Furthermore, many of the Nazi, Communist, religious, ethnic, and tribal mass murderers came from relatively prosperous families and were certainly not the victims of poverty, discrimination, persecution, or injustice. If anything, they belonged to a class whose members were exempt or actually benefited from the corrupting social conditions. Their evil actions, therefore, cannot be the result of corruption by social conditions.

Lastly, this often-repeated explanation simply takes for granted the existence of corrupting social conditions. But social conditions do not fall ready-made from the heavens. They are created and maintained by people. If they are corrupting, it is because those who created and maintain them are corrupt. It may be said that they were also corrupted by social conditions, but that merely postpones facing the question of why they created and maintained those corrupting social conditions? Sooner or later the evasion must stop, and reason compels the recognition that it is corrupting social conditions that must be explained by the psychological dispositions of people who create and maintain them, rather than the reverse, which is what this putative explanation attempts to do. Its basic defect is to explain the cause by its own effect: evil actions by corrupting social conditions, when, in fact, evil actions are not the effects, but the causes of corrupting social conditions.

Why, then, do so many normally intelligent people accept this farrago of an explanation? Because they are unwilling to question their optimistic faith that human beings are basically reasonable and good. They refuse to face the prevalence of evil, because they would have to give up their comforting faith. But their faith is an enemy of well-being. It leads those who hold it to ignore the most serious threat to well-being: the psychological dispositions of human beings to inflict grievous harm on innocent victims, malevolently, deliberately, and without moral justification. The secular problem of evil is to explain why people act on these dispositions rather than on ones that protect the conditions on which well-being, including their own, depends.

29.4 Toward a Better Explanation

The prevalence of evil is a good reason for rejecting the secular faith that human beings are basically reasonable and good. But it is not even a bad reason for accepting the equally implausible view that human beings are basically unreasonable and evil. The facts point to the conclusion that human beings are basically ambivalent. Basic dispositions are mixed. Some incline us to act reasonably and morally, some to act unreasonably and immorally, and some have little to do with either reason or morality. The prevalence of evil shows that in some circumstances unreasonable and evil dispositions come to dominate contrary dispositions. To explain evil is to explain why and how this happens. The explanation must specify both the external circumstances and the internal dispositions that jointly motivate evildoers to do evil.

29.5 The Explanation I Propose Proceeds in Five Steps

The first step is to recognize that, unlike moral monsters, evildoers typically do not see their actions as evil. In cases of mass murder, for instance, they may acknowledge that what they have done, and are doing over and over, is killing, but deny that it is evil. They deny that their victims are innocent, that their own motives are malevolent, and that the killings are morally unjustified. They may sincerely say to themselves, or to others if challenged, that their victims are dangerous enemies of their society, that their motive for killing them is the obligation to protect general well-being, and that the killings are morally justified, because the protection of well-being depends on the destruction of its enemies. Killing them is no different from killing in war or in self-defense. Evildoers may readily grant, indeed might insist, that murder is evil, and indignantly deny that their killings are murders. Their denials are grotesquely mistaken, of course, because their victims are innocent, their motives are malevolent, and their actions are morally unjustifiable, but evildoers may sincerely believe the opposite.

The ordinary people who became the accomplices of moral monsters may be Nazi or Communist party members, devout defenders of various true faiths, ardent nationalists, dedicated terrorists, or merciless revolutionaries, who may be genuinely convinced of the justice of their mass murders and the viciousness of their victims. If they have doubts about treating their imagined enemies in ways that are normally evil, they can appeal to like-minded political, religious, or historical authorities who reassure them that they are doing the right thing, and praise them for their selfless service to a fine ideal. We know from the testimony of mass murderers and from their histories and biographies that they see themselves as righteous champions of God, history, justice, racial purity, national glory, or whatever. They see evil all right, but ascribe it to their victims, not to themselves.

The second step of the explanation is to make clear that moral monsters and their accomplices see the evil they do very differently. Moral monsters see it as evil, and they do it either for its own sake or for the sake of something they regard as far more important than morality. Their accomplices do not believe that their actions are evil; they think that they are morally justified. In this, they are flagrantly, absurdly, obviously mistaken. Their mistakes are not ordinary errors we are all prone to make from time to time, but perverse interpretations of plain facts that rest on no credible reason. They are akin to blaming mental illness on possession by the Devil, infertility on the evil eye, defeat in battle on misreading signs contained in the entrails of freshly slaughtered chicken, or crop failures on not having sacrificed fat enough cattle to the god of plenty.

If we focus on the egregiously false beliefs of ordinary people who become accomplices to mass murders, we can only marvel at them. Communists say that their victims are plotting a counterrevolution to reverse the laws of history that move us toward a classless egalitarian society in which all manner of things will be well. Nazis claim that Jews are part of a conspiracy to rob the superior Aryan race of its right to dominate the world. Nationalists charge their supposed enemies, often an incompletely assimilated ethnic group in their society, of treacherously obstructing the nation's return to the past glory that rightfully belongs to it. Defenders of what they regard as the true religion condemn defenders of other religions for worshiping false gods, and hold that the true God demands ridding the world of them. Terrorists accuse arbitrarily selected targets of supporting a corrupt system that opposes the rule of those who know where goodness and truth lies.

These mass murderers appeal to ideals to justify the evil they do, but their ideals cannot withstand even superficial examination. Ineradicable differences in capacities, moral conduct, experiences, and reasoned judgment make an egalitarian society that ignores them blatantly unjust and detrimental to human well-being. Racial superiority is untenable, because of decisive historical and genetic evidence for the intermingling of races, because of normative biases involved in selecting for statistical comparisons supposedly important inherited racial characteristics, and because it is impossible to determine the relative importance of inheritance and environment for subsequent performance. The distinction between true and false religions simply ignores that all religions ultimately rest on faith, that all of them equally lack evidential basis, and that it is sacrilegious to suppose that God, if there were one, would demand mass murder. The terrorists' dream of rule by authorities with special access to goodness and truth ignores the overwhelming historical evidence of the incomparably greater suffering caused by such authorities than whatever it is the terrorists blame democratic regimes for causing.

The plots of which these mass murderers accuse their victims of hatching are, if possible, even less plausible than the ideals to which they appeal. Counterrevolution, worldwide Jewish conspiracy, worshiping false gods, and democratic conspiracies to perpetuate injustice are, one and all, absurd inventions used by evildoers to justify mass murder. But even if their justifications were accepted, they would not explain why the victims of mass murders include children, barely intelligent adults, and the decrepit old who could not possibly be plotting anything. Evildoers add to the false ideal and the absurd plot the grotesque claim that all their victims are essentially corrupted by an ineradicable component of their character – capitalist class consciousness, Jewish parentage, the values of a false religion, or Western attitudes – that makes them enemies regardless of what they actually do. That is why their victims cannot be innocent; why they have to be killed; why evildoers can claim to act on motives that have the highest moral credentials; and why their mass murders are in fact the morally justified killings of the most dangerous enemies of human well-being.[9]

The third step is to explain how these grotesquely mistaken beliefs could be held by normally intelligent people, who are the accomplices of moral monsters, and who believe themselves to be committed to morality. The explanation I propose is that they have been led to these beliefs by their acceptance of an ideology. I hasten to stress that I do not claim that all ideologies lead to mass murder or that all evil is ideological. I claim only that ideologies readily lend themselves to the justification of mass murder, and this makes them dangerous, regardless of whether they are secular or religious, political or moral, nationalistic or racial. The evidence for this is the historical record of the evil done in the name of Communism, Nazism, religion, nationalism, and terrorism.

The essential feature of ideologies is an evaluative attitude that combines a set of beliefs about the nature of the world, an ideal of how human life ought to be, an explanation of why there is a yawning gap between the ideal and the miserable actual conditions, and a program for closing the gap between how life ought to be and how it is. Ideologies are dangerous, because they tend to breed true believers with dogmatic attitudes who hold that theirs is the one and only ideal of how human beings ought to live. The ideal may be set down in a sacred text, exemplified in the life of an exceptional individual, dictated by the laws of history, sociology, or psychology, located in a past idyllic condition uncorrupted by civilization, or represented in a future Utopia of perfected human nature, and so on.

All ideologies hold that human well-being requires living in conformity to whatever the ideal happens to be. The failure to conform to it obstructs the pursuit of well-being, and reason and human well-being make it obligatory for ideologues to remove the obstruction from the way. This may be done by educating those who do not understand the ideal, or by preventing those who understand and reject the ideal from jeopardizing its pursuit. Mild coercion by education or forceful coercion by whatever means are necessary are natural concomitants of the pursuit of the ideal.

[9] For the detailed study of this mentality, see Norman Cohn's *The Pursuit of the Millennium* (London: Secker & Warburg, 1957), and *Europe's Inner Demons* (London: Heinemann, 1975).

According to ideologues, coercion for these reasons is not only justified but required by reason and goodness, because human well-being depends on it.

The widespread appeal of ideologies is undeniable. They cut through the complexities, moral ambiguities, compromises, and malaise that beset so many people in modern life, and fill the gap left by the waning of traditional religious belief. They provide clarity about good and evil, and offer simple and consistent judgments to those who yearn for guidance about how they should live. Countless people survey their comfortable or miserable existence, the routines of everyday life, uninteresting or soul-destroying jobs, dispirited relationships with similarly disenchanted souls, and say to themselves, or perhaps to each other, that there must be more to life. Ideologies provide what they yearn for. By committing themselves to an ideology, people gain a sense of identity, acquire membership in a community of like-minded people, find an ideal they could follow, receive an explanation of why the existing conditions are wretched, and are told what needs to be done to improve them. One might be tempted to say that all this is to the good. But the temptation should be resisted, because ideologies encourage their accomplices to justify the evil they do.

The perniciousness of the justification the accomplices derive from ideologies becomes apparent if we do not allow abstractions to obscure the concreteness of what is going on in mass murders. On the one hand, there are the accomplices who time after time pull the triggers, hang, bury alive, bayonet, or burn hundreds of helpless victims, including children, old people, pregnant women. The immediate, undeniable experience of the accomplices is the face-to-face massacring of people who bleed, beg, pray, hold on to each other as they helplessly face imminent death. On the other hand, there is the grotesque story the accomplices tell themselves of why they are murdering these people. They reassure themselves with ideological catchphrases that they are working for the classless society, saving the world from a conspiracy, doing God's will, making the world just, and so forth. What could possess them to believe that these absurd abstractions, whose theoretical background they understand, at best, only faintly, could justify the suffering they cause and witness?

The answer, of course, is that the catchphrases are not the reasons for what they are doing. The accomplices use the ideological slogans to justify to themselves the horrendous actions that they know are normally evil. They can, then, say to themselves that their actions are justified, because they are saving the world from the enemies of human wellbeing. The ideologies whose slogans they parrot allow them to believe obvious falsehoods about their victims and thereby deceive themselves about the true nature of their actions. But what the accomplices care about is not human well-being, but murdering their imagined enemies. They are not engaged in the ruthless pursuit of an ideological ideal. The ideal is only for show. That is why they can shrug off its utter implausibility. Their rhetoric of dedication to human well-being is phony, but the rhetoric excoriating imagined enemies is heartfelt. Why, then, do the accomplices join moral monsters in excoriating and murdering their imagined enemies? Why do they bother with the ideology and with swallowing a highly implausible case involving false ideals, ridiculous plots, and the ineradicably corrupt character traits of their imagined enemies? Why do they invent imagined enemies?

These questions lead to the final, fifth, step of my proposed explanation. I have stressed the importance of ideologies for the explanation of mass murders. Their importance, however, is not that they provide motives for the murders, but that they direct and justify the evil actions prompted by motives evildoers have independently of ideologies. These motives are the various forms of malevolence – cruelty, envy, fanaticism, hatred, jealousy, prejudice, rage, and ruthlessness – that are one of the defining characteristics of evil actions. In civilized societies and normal circumstances, most of us try not to act on such motives. We are constrained by moral limits, by commitment to the well-being of others, and by benevolent motives, which we have alongside malevolent ones. If these constraints are effective, they stop us from acting on malevolent motives, but they do not stop us from having them. They remain as possibilities, on which we may act, if we face serious hardships, provocations, or temptations.

We all have malevolent motives of various kinds and strength. If we did not have them, self-control, conscience, moral education, prudence, discipline, legal prohibitions, and law enforcement would be largely superfluous. But, of course, they are needed, as we all know directly from our own case, indirectly from the observation of our children and the testimonies of those close to us, and, a step further removed, from witnessing occasions on which people's self-control has slipped and their malevolence was momentarily revealed, as the venom may be in a clever witticism.

Malevolent motives clamor for expression when for good or bad reasons we become dissatisfied with our lives. Ideologies are dangerous, because they encourage the expression of malevolent motives by providing scapegoats – counterrevolutionaries, conspirators, idolaters, traitors, enemies of the good and the true – and justifying their persecution. Ideologies supply abstract claptrap whose slogans the accomplices can parrot to explain to themselves and others why they are right to vent their malevolence on their victims.

Ideologies provide tangible benefits for the accomplices: they make their evil actions acceptable; they legitimate the relief they feel when they express their malevolent motives; they transform their unsatisfying lives by providing them with activities said to be meaningful and important; they allow them to lay to rest their lingering doubts by camaraderie with fellow accomplices; and they receive authoritative praise for the evil they do. Accomplices shrug off the implausibility of the ideology, the absurdity of its ideal, the grotesque plots and character traits it attributes to the victims, because the benefit the ideology provides for them is a justification for venting their malevolence.

The foregoing are steps toward an explanation of evil, but they do not amount to a full explanation. I have discussed only mass murders committed by ordinary people who became accomplices of moral monsters. But there are many kinds of evil about which I have said nothing. I doubt that their explanation would be very different from what I have given of mass murder, but that, of course, needs to be shown. Nor have I claimed that this explanation of mass murder is the only one needed. Evil actions may be explained as the results of culpable negligence, or as the characteristic actions of moral monsters, or as uncharacteristic aberrations of ordinary people, or as responses to corrupting social conditions. I do think, however, that these explanations are inadequate, because they account only for a few evil actions, not for the prevalence of evil. The explanation I have given seems to me to go much further toward understanding why evil is prevalent than these other explanations do.

29.6 Conclusion

The just-completed explanation of why many normally intelligent people, who are not moral monsters, become accomplices to mass murder has two essential components: one external, the other internal to these evildoers. The internal explanation is their malevolent motivation. The external one is the ideology they opportunistically adopt to explain and justify the evil they do. Ideology enables them to express their malevolence and to disguise from themselves and others the true nature of their actions. They do evil, but they do not see it as evil.

Evildoers often give reasons for their evil actions, and, as we have seen, reasons can be given to explain why they disguise from themselves their real reasons. Some of these reasons are rationalizations, but evildoers may have reasons for their rationalizations as well, even if they are not aware of them. Evildoing, therefore, serves a purpose. It need not be unreasonable, even though it is certainly immoral. The significance of the possibility of giving reasons for evildoing is considerable. It shows that a central assumption of Western moral thought is mistaken: the requirements of reason and morality do not coincide. Immorality need not be unreasonable. The secular problem of evil is to cope with evil in light of the failure of this hoary assumption. Evil is an enemy of human wellbeing, but malevolent motives in combination with ideologies supply reasons for doing evil, just as benevolent motives in combination with moral commitments supply reasons for not doing evil. Our motives are divided, and we both seek and hinder our well-being. This gives rise to the secular problem of evil and forms part of the human condition.

The first requirement of responding to the secular problem of evil is to acknowledge its existence. A great obstacle to it is the optimistic faith of Rousseau, Kant, Mill, and their numerous followers. Their faith rests on the assumption that the requirements of reason and morality coincide. But the assumption is mistaken and their faith is just as unsubstantial a dream as the religious faith in the existence of a moral order that permeates the scheme of things. The assumption and the faith lead to their failure to face evil, and they ought to be abandoned. The second requirement is to hold evildoers responsible for what they do, even if they do not believe that what they do is evil. The third requirement is to weaken the reasons for doing evil and to strengthen attractive and morally acceptable alternatives to it. There is, however, a great obstacle to meeting these requirements. If we are indeed divided in our motivation, then our efforts to cope with the secular problem of evil will be handicapped by the same problem with which we try to cope. The proposed treatment is infected with the disease it is meant to cure. This is also part of the human condition and a reason for abandoning the optimistic faith. It has been well said that out of the crooked timber of humanity nothing straight has ever come.

Part V
The Human Study of Human Society

Chapter 30
Towards A General Sociology of Science

Ian C. Jarvie

30.1 My Philosophical Caveats

Despite the sweeping title of this essay, which suggests a descriptive, even ethnographic survey of science, its aim is more modest. I offer at most a sketch that is free from three faults, three philosophical assumptions and conclusions that much received sociology of science takes for granted. The first is the assumption that science stands in some special place of authority in the contemporary social order and, authority being a social construction, that both explains it and obliges us to accept it. Although the sociology of science should explain the authority of science, uncritical endorsement of authority is a recipe for stagnation. The second assumption is that there is no social reason to distinguish science from technology. Muddled as the two are in the public and the official mind, the sociology of science should present them as differing in aim and in social structure. The third assumption is that socioeconomic factors are sufficient to explain the growth of scientific knowledge. Suffice it to say, I concede that research is not as immune to the influence of these factors as the classical thinkers of the Enlightenment Movement assumed.

Before I examine further each of these assumptions, I need to indicate the nature of my objections, which come from a particular philosophical understanding of science. Before that, I want to link this fault-finding of mine to the title.

If we are to dislodge these pernicious philosophical assumptions about science from its sociology, making a philosophical case against them is obviously not enough. After all, adherents of the received sociology of science could say, they are describing the social order, however reasonable or perverse the philosophy behind it may be. I accept this in part: I concede that, at least in the case of so-called scientific authority and the mixing of science with technology, confusion occupies the public mind, confusion that reflects features of society that depend on (or at least

I.C. Jarvie (✉)
York University, Toronto, ON, Canada
e-mail: jarvie@yorku.ca

N. Bar-Am, S. Gattei (eds.), *Encouraging Openness*, Boston Studies in the Philosophy and History of Science 325, DOI 10.1007/978-3-319-57669-5_30

originate from) the philosophical views of science that I shall challenge here. Yet I also claim that even descriptively the picture is mixed: evidence indicates that the public sometimes allows that science and scientists are fallible, that the public sometimes allows that the latest scientific news is not the last word on its subject matter. Perhaps the public sometimes allows that science and technology are different kettles of fish in the very preference for funding technology and applied science as opposed to pure science.

To make my argument stick, then, I have to sketch the possibilities for a sociology of science that plays up the conjectural side of science and thereby emphasizes its difference from technology. To that end, I seek a general sociology of science that makes allowance for skepticism towards the authority of science and which attributes some authority to science-based technology. As to explaining the progress of science, advocates of the view known as the Strong Programme claim that they can explain it by reference to socio-economic factors. I reject this claim: I find it reductionist and at odds with widely diffused ideas about creativity. That is not a knock-down objection: but it is a good starting point, and so I will adopt it here.

Let me now expatiate a little on the three philosophical faults discussed above that are popular within the sociology of science. The problem with the ascription of authority to science is simple: it is a social fact, it is true, but it nonetheless rests on a cardinal mistake. I hold that science is no authority and need not command belief. It is an intellectual adventure that throws up some astounding, fertile ideas. These ideas do not thereby carry any authority and if authority is ascribed to them, or, worse, if it ascribes authority to individual scientists (as "experts") then this exposes a serious muddle.[1] The prestigious status that the public ascribes to science is, obviously, a social construct. But then, I assert, the authority ascribed to science is similarly a social construct, just like political, religious, bureaucratic, and other authority sub-systems. None need be endorsed. Adherents of the received sociology of science may fear that withholding endorsement undermines the status of science.[2] Or they may say that its authority is among the social facts with which they deal. (Calling something a fact is tacitly to endorse it.) I would question all this and argue to the contrary that deconstructing the so-called authority of science would be better employment for sociologists of science, and that this way they will increase, not undermine, its social status. Their claim that the authority of science is a social

[1] I constantly argue against fusing science with belief, and counsel against "believing science". Belief is a private matter; a free society does not require that one believe this or that. One can endorse the scientific project without believing any of its findings. The focus on belief invites the accusation that science is a religion-substitute. True, some people use it that way. They simply miss the very different sort of cognitive effort science is, including its different sociology. From the days of the scientific revolution, its advocates repeatedly denied that it imposes faith in any doctrine. It is only because of that different sociology that some scientists can both conduct successful research and hold some antiquated religious beliefs.

[2] To the extent that the careers of sociologists of science in the academy depends on ambitious members of appointment committees who are professional scientists who view their task as to maintain the authority of science, to that extent the work of sociologists of science is as inherently suspect as any other public-relations effort.

construct, that science is also a social construct, may be transcendentally debunking but, because it brooks no appeal beyond the social it amounts to sleight of hand: debunking that is not debunking at all. The authority traditionally claimed for science is special: it is allegedly intellectual (*viz.*, "epistemic"). Science has no such authority because there is none to be had. Adherents of the received sociology of science would do well to view the ascription of authority to science as a myth to be explained or as a social institution to be described, or both, but if both then separately – without intellectual submissiveness. It is unwise to rule out *a priori* the obvious possibility, not to mention the obvious fact, that it is in the interests of some advocates of science, including some scientists, to pretend to (a non-existent) authority. Particularly on the assumptions that I wish to challenge, but also without them, the very possibility of this kind of pretense suffices to make room for my investigation, and efforts to deny this *a priori* are not serious.

Adherents of the received sociology of science could no doubt reply that they do deconstruct scientific myths and pretensions, particularly in their studies of laboratory life and ongoing research practices. Perhaps; but reducing science to routine practices of work, much like studies of the factory floor, is to miss another philosophical point. I may distinguish between two kinds of the output of scientific work: efforts to refute old ideas and efforts to create and test new ones. Ideas are abstract and intangible products, whose value is only settled by putting them on the intellectual market for others to seek, find, and examine. There are three intellectual sub-markets where this seeking, finding, and examining take place. There is the *initial market*, in which ideas circulate as brainstorms, research proposals, preprints. (Think of Watson's account of how he and Crick sought the permission of Bragg at the Cavendish to expend some resources pursuing their ideas.) The initial market it quite small and intimate. The second market is the *futures market* where ideas that made a big impact in the initial market may continue to glow, or may fade, those that made little or no impact either continue that way or start to make a larger impact. (Think of the short communication to *Nature* that Crick and Watson used to report their DNA structure, and the tiny hints they offered about the potential ramifications of their discovery for future research. Then think of the exponential growth of their citations.) The third market is that of the *general public*. An example would be the revolution in medical thinking, the rise of germ theory, its refinement into systemic, bacterial, viral, auto-immune, and other variations that entered the public consciousness, became part of the view of disease that we teach schoolchildren, and of the rise of the popular practices regarding hygiene, nutrition, and immunization. (Think of the award of the Nobel Prize to Watson, Crick, and Wilkins; and the subsequent stellar careers of all three, which included popularizing science.)

None of these three markets turns abstract and intangible products into authority, the way markets in religious or legal ideas turn them into authority by the use of familiar social mechanisms. Indeed, the only authority it makes philosophical sense to claim for science is the view that considered rational inquiry into problems should receive more serious attention than opinions that are not thought-out and therefore untested – but such caution is too timid if it rules out wild ideas just because they are wild. (Think of Niels Bohr's famous disposition to dismiss ideas that were not sufficiently wild.)

Some of the confusion over authority, I suggest, comes from the fusion of science and technology in the public mind – and in the sociological mind as well. I am not a pedant: for some purposes of discussion, the difference between science and technology is not important and may be overlooked. But when we talk of authority, the difference is very important. Technology does possess authority – not intellectual but institutional. Society invests technology with its authority, and in doing so society stipulates the standards that technology has to meet to claim authority. Mostly, these standards are legal and quasi-legal, that is, they are enforced by the power of the state and its organs. (Think of local building codes; of best practice in the engineering and design of buildings, bridges, tunnels; think of the régimes of testing that precede the licensing of medicines; of the licensing requirements of the professions; of airworthiness certificates.)

Unlike scientists, technologists are legally and otherwise responsible for their products.[3] Pure science knows no parallel to this. Cosmologists are not in some way socially responsible for the fact that in their standard models human life is a tiny and insignificant episode in the story of the origins and evolution of the universe. One can argue that these ideas, or similar ones, demoralize and have ill effects; even so, there is no one to blame here. Individual scientists can of course warn, be compassionate, try to muffle the implications, and so on. But that is not their scientific responsibility. All scientists are citizens, but scientists claim the freedom to set aside the wishes of their fellow citizens if they think the pursuit of science would be inhibited – but they can do so only within the law. (Think of the Helsinki convention.) Problematic citizens can be unproblematic scientists. (Think of Kurt Gödel and of John Nash.)

The science/technology boundary is worth studying. The aims, surrounding social structures, and ethics of technologists are very different from those of the rational community of free inquiry that is science. Confusion of the two may be deliberate, as when the atomic bomb or the cure for cancer are dangled around in order to extract money for research, and when the military authorities dangle money carrots in order to obtain the services of scientists. Society as a whole should take responsibility for the creation of such an incentive system; adherents of the received sociology of science may of course construct models that foist these incentives on science as though there was no help for it; but social technologists should examine these models before applying them, and seek the best way to test them all, and make attempts to improve upon them in manners that may have some likelihood of success.

The profound cooperation between applied science and pure science explains some of the confusion between them. Applied science comprises efforts to put extant pure science to practical use, or even to develop pure science to serve that end.

[3] Sometimes sociologists refer to science as a profession. I take a stricter view: a profession is an organized and legally defined monopoly. Scientists, like any workers, can be spoken of as acting professionally without being members of a profession.

This last activity is known as basic science. Medicine is an applied science that draws on the pure sciences of biology, biochemistry, chemistry, quantum mechanics, and even classical mechanics (when diagnosing skeletal injuries for example). The technology of medicine is the production of drugs, surgical procedures, and diet regimens, perhaps also the design of clinics and computer-expert-systems, as these put the applied science to work on the patient. It is important to stress – with Gellner – that certain general social forms and certain specific kinds of social arrangements precede both the possibility to apply science and the possibility to devise new technologies. Medicine is first and foremost a social technology: a set of social arrangements that trains specialists in human disorders along with inculcating into the population at large that the best way to deal with their suffering is to set up clinical encounters with these specialists and to "present" themselves accordingly. Social technologists devise ways to apply medical technology, and they do so within that social framework, including all sorts of factors, including modern bureaucracy and industrialization, that enable them to make available to doctors wide ranges of tested and affordable medical technology. How all this interacts with the insurance system at large is extremely complex, requires much care and planning, and continuous monitoring and revision. In liberal societies the social technology of organized medicine is optional, not compulsory.

Classical sociology of knowledge, out of which social studies of science grew, did not presume to explain the growth of scientific knowledge by invoking socio-economic variables. To see why, apply to itself the theory that says, scientific knowledge is the result of socio-economic factors. We have then the claim that the theory in question is the result of some socioeconomic factors. Is the claim true? If yes, then some socio-economic factors lead to the truth! Suppose then that we may discover what these factors are. Then, producing them will generate science. This is scarcely credible. Note: this is not to deny that *reception* of scientific ideas in the *futures market* and *general public market* are susceptible to socio-economic examination. Any biologist could see that the Crick-Watson DNA breakthrough created immense possibilities for further research in several cognate fields. That some financial organizations were receptive and threw money at the projects are matters of the reception of innovations, not to be confused with the innovations themselves.

Innovation, it seems to me, is an intellectual achievement explained best, but still only partially, by the intellectual pressures that thinkers face, by their problems and their problem-situations. Understanding these problems and problem-situations throws light on ensuing breakthroughs and on breakthroughs that were not seen as such in the day when they became available. Furthermore, this renders it obvious that all purported Strong Programme explanations of changes of scientific ideas explain not the ideas and their emergence but the *reception of these ideas*. Scientific progress is equated with scientific reception. This makes sense, as ideas ignored by the learned public are forgotten. Yet it is erroneous. Counterexamples to it come from both directions. The successful reception of ideas is no proof that they are not duds, and the dismissal of ideas that at first look dud is no proof that they will not eventually gain recognition as explosive. A case of the first kind would be the long-held psychosomatic theory of gastric ulcers. A case of the second kind would be

Wegener's theory of continental drift.[4] To these sorts of cases sociologists of science could reply by stressing that consensus changes: it clusters for a while, dissolves, and then re-clusters elsewhere. This response is quite in order. Such clustering, dispersal, and re-clustering, being social processes, are indeed susceptible to socio-economic explanation – at least in some measure. Geographers and geologists explain why Wegener was pooh-poohed: they usually say that there was a big problem with this theory when it first appeared, namely mechanism.[5] What forces could possibly be shifting these massive continental plates? Only after new ideas about the heat and convection currents beneath the earth's surface were devised did it seem possible to account for continental drift and to turn to the independent ecological data to test it, and it turned out to offer a powerful tool for the explanation of the diffusion of plant and animal species of all sorts. This way Wegener's ideas were open to serious discussion, and, indeed, they were then discussed widely and seriously. Socio-economic factors are at play here, of course, especially communication; but the intellectual aspects are the best part of the story, so the response about the vagaries of the consensus, though true, is no reply to criticism: they cannot parry the attack that exposes the received sociology of science as tending to miss the meat while centering on the sauce.

The meat of the debate is intellectual: deficiencies of a theory explain its neglect, and supplements and modifications of it explain its change over to a positive reception. Adherents to the received sociology of science will respond further. What they observe, what they hear scientists say, what they write, tells a story of uncertainty and dispute, a story of the marshalling of arguments and of evidence, and of the differing suggestions as to the conclusions to draw. This all sounds to adherents of the received sociology of science like Debate and Rhetoric 101. Debate and Rhetoric 101 usually teaches that disputes are settled by a vote, of judges in a debating contest, of members present in a debating society like the Oxford Union. Debates in both sorts of places are sometimes framed around issues where there is genuinely something at stake (other times they are tongue-in-cheek). Similarly, adherents of the received sociology of science opine, scientists engage in rhetoric and persuasion

[4] See the interesting study of Naomi Oreskes (1999), and the comments of Rachel Laudan in her review (2000). Oreskes also co-edited an anthology of memoirs by those taking part in the scientific revolution in earth science (Oreskes 2001). In both volumes she resorts to the language of belief (see note 1 above). My view is that this is unobjectionable and in all cases can be replaced with the language of "ideas" without loss. For example, her aphorism "*science is not about belief; it is about how belief gets formulated*" translates seamlessly into "science is not about ideas; it is about how ideas get formulated" (Oreskes 1999, p. 6). In fact this translation is helpful and one can see at once that a false dichotomy is involved: science is both about ideas and about the formulation of ideas. It is indeed not about belief in those or any other ideas. Ideas can be formulated and promoted without being endorsed, still less believed.

[5] Oreskes (1999) argues most cogently that this was hindsight; several mechanisms had been put forward. Her explanation is different: resistance was due to inductivist dislike of theory-driven science. Oreskes also argues that inductivist American geophysicists operated with multiple "working hypotheses". It is unclear whether she sees that her model of this science is a fascinating counterexample to Kuhn's claim that genuine science requires consensus on a paradigm. The beauty of it is, geophysics is a "hard science".

every time they try to forge a new consensus, and these are social constructs like any other. This leaves too much of the situation needlessly puzzling. Why do scientists not describe the issues in this way? Is it because they are not sociologists and do not know how to frame and describe sociologically? Or is it because, while fully aware of the features that resemble a debating contest, they consider the underlying issue and the content of the arguments to be utterly serious? Even highly sociologically self-conscious scientists like Michael Polanyi, Thomas S. Kuhn, and Paul K. Feyerabend do not reduce the glories of science to mundane practices of negotiation, rhetoric, debate, and their socio-economic determinants. Those are practices received in the community, to be sure, just as they are received in other social institutions. The question is, what is the scientific bottom line? Is it consensus? Is it public *éclat*? Is it research money? Is it medals and prizes? None of these is sufficient explanation. The consensus is about something or other being the case; the money is given on promise; the medals and prizes are given for achievement. What is the referent of these expressions, "the case", "promise", or "achievement"? If adherents of the received sociology of science say it is self-validating then the argument is circular. The sociologist may be inclined to grant that and then say, but what other measure is there?

The difference between science and magic need not be discussed in order to agree that received sociology of science is pro-science and anti-magic. Yet they can distinguish between genuine and fake magic – by reference to it as a social institution, of course. The situation with science is different, as the genuineness of science is much more a matter of objective test than of science being a social institution. To be precise, the testability of science is its major institutional characteristic. I contend that received sociology of science does its best to overlook this, thus pushing itself towards the view of science as effective magic.

The reason for this blunder in the received sociology of science is simple. The question on the agenda of the sociology of science is philosophical, yet the sociologists want to be sceptical towards the philosophers and the scientists who argue that there is a measure that is conjectural and uncertain and not monotonic with the social consensus of scientists. This measure is variously called the facts, the case, the truth, reality, Mother Nature.[6] Sociologists of science respond that all attempts to capture and state scientific results begin with joint endeavours and end with assertions framed in language and these are necessarily social. And, they conclude, scientific results are to be seen as social constructions: necessarily. This inference is invalid: social effort at something does not make that something social. The social efforts have a goal, the goal is socially defined, but whatever the goal is, it need not be a social entity. Democracy, museums, and universities are social entities; citizens and works of art and of science are not.

Ernest Gellner (1973, p. 17 and elsewhere following Russell) notes that the arguments that sustain all forms of metaphysical idealism usually rest on the doctrine of

[6] Oreskes is characteristically clear and blunt: "Scientists are interested in truth. They want to know how the world really is, and they want to use that knowledge to do things in the world." (Oreskes 1999, p. 3.)

internal relations. This is the doctrine that relations are "internal" to things, namely, that relations [between things] are assimilated to properties [of things]. He also suggests that thinking about society renders this doctrine very tempting. Hence, thinking about society renders idealism very tempting. The idealism of the Strong Programme in the sociology of science has two sources. One is the circular idea that all human activity is social activity and that hence social activity is all there is. Second is the linguistic idealism of the later Wittgenstein, who merely hinted that since language is a social activity anything said in language is shaped, constrained, and explained by language and form of life. (By the term "*Lebensform*", he seems to have meant social form of life, namely, society *simpliciter*: this way he managed to smuggle in much metaphysics with the aid of one obscure technical term.) We are then landed with a special version of the anthropomorphism that takes universal science to be an illusion because all of science, from its fundamental measures to its language, is shaped from the point of view of *Homo sapiens* and cannot possibly map on to non-human observers or even be intelligible to them.

The question then is can we create a general sociology of science that does not take for granted the authority of science but rather deconstructs it, a general sociology of science that unfuses science and technology, and circumscribes an autonomous domain for intellectual achievement as such? I do not wish to hide my value system: I do value ideas and I think they have autonomous existence. This includes sociological ideas by distinguished ancestors: Smith, Marx, Durkheim, Simmel, Weber, Evans-Pritchard, Popper, Gellner. Sociologists of science fuse the commonwealth of learning with the diverse institutions that house it. These institutions change faster than the commonwealth. When Popper wrote on objective knowledge as an institution, he surprised and distressed many, as he avoided this erroneous fusion of institutions with the whole they serve and constitute. Objective knowledge is a part of the commonwealth of learning. It too is not to be fused with the institutions that house it, such as museums and libraries and the internet.

30.2 My Targets

So far I have left the specification of sociology of science aside. Although I am an occasional attentive reader of it, I would not claim the status of a specialist. I can now be a tad more specific. I propose to rely on a survey article by Thomas F. Gieryn, "Science, Sociology of", published in the Elsevier *International Encyclopedia of the Social and Behavioral Sciences*, 2001, that strikes me as excellent.[7] Gieryn is a scholar at Indiana University. He divides sociology of science into three main periods (plus a pre-history at the beginning and an uncertain future at the end). The first period is Mertonian; it focuses on science as a form of social organization.

[7] I do have some criticism of the piece as scholarship, though: I lament its failure to acknowledge the immense debt that today's sociology of science owes to Kuhn, and through him to Polanyi; neither is listed in Gieryn's bibliography.

The second period focuses on the processes that make science. The third period concerns itself with science as authority, or, more precisely, with scientists as possessors of cognitive authority.

Gieryn presents the sociology of science as value-laden. He iterates the discrepancy between the claims of universalism and communism (Merton) on the one hand and on the other the hierarchical organization of science, the principle of cumulative advantage, the systemic and systematic sexism, and utter lack of democracy. These criticisms are congenial, although with some reservation as to democracy since science as a whole has never succeeded in suppressing criticism to the degree achieved by undemocratic political regimes. Indeed, if I was to sum up my complaint, it would be that the sociology of science gives far too much credence to sham and corrupt claims about science and is itself quite insufficiently critical (and thus not scientific). My aim is a sufficiently critical sociology of science that begins by sifting of the grain from the chaff in science. Science is of supreme importance as a human endeavour. It is however in the hands of people who do not always subscribe to that view. Most of the powerful and many of the lesser ranks are normal scientists Kuhn-style, i.e., plumbers, hacks, and worse, whose primary interests are a living, a career (fair enough), or, the accumulation of power, prestige, and the policing of turf (not so fair enough). On the most charitable interpretation, all are careerists. Abundant evidence for this is available directly or through the critical reading of the current, uncritical sociology of science. A critical sociology of science would be one that knew the difference between science as the disinterested search for the truth about the world, wherever that is pursued, and the various manifestations and masquerades that parade themselves as institutionalized science or the Voice of Science in the present world. Yet I do find the miracle of petty individuals producing noble harvest, a miracle that interested many an author.[8]

My impression is that Merton's studies of science, and those of his students, Barber, Zuckerman, Ben-David, etc., stemmed from a genuine respect for science as an enterprise and for its importance in the human adventure (sociology being part of it). Merton's work disclosed some interesting social novelties, not all of which were attractive, as, for example, the Matthew effect, namely, the principle of cumulative advantage. The very notion of gaining or taking advantage struck me, on first reading, as an exposure of corruption at the very heart of emerging Big Science. Merton did not treat it that way, preferring a detached functionalism: possibly, he held, it serves the interests of science that such an institutional deformity persists. True. It does serve certain functions but they may have nothing to do with science as the pursuit of knowledge. After all, the idea that those with strong track records should be showered with advantages is inductivist and philosophers, at least, know that Hume showed induction to be utterly, irremediably invalid. Is this principle then really the best that the invisible hand of society can come up with for science? Is the efficiency that the editors of periodicals enjoy by yielding to the Matthew effect not

[8] Examples: "Some have greatness thrust upon them", *Twelfth Night*, ii, 5; Rudyard Kipling, "The Man Who Would Be King", 1888; Robert Louis Stevenson, *Father Damien: An Open Letter to the Reverend Doctor Hyde of Honolulu*, 1890.

a public loss? If so, then the function that the effect serves is short-term: in the long run it is a nuisance.

Leaving Gieryn's Mertonian Science section for now, I move to his "Making Science" and "Science as Authority". These two succeeding periods or "lines of inquiry" that he isolates raise serious doubts about whether there remains any "genuine respect" for science, both of the natural sciences under study and of the sociological science being practiced. Deeply embedded in these two later lines of inquiry I find the three philosophical faults outlined at the beginning of this essay: face-value respect for authority, uncritical mixing of the sociologies of science and of technology, and the claim that intellectual progress (in science) is sociologically explicable (reducible). Lurking behind all three I find an attitude that is quite different from that of Merton: debunking or demystifying. I have the strong impression that between the lines of studies of this later strain of the sociology of science hides the aim of unmasking science and presenting it as just another form of work, not too dissimilar from factory work or plumbing or engineering, even story-telling ("narratives" is the jargon for this). Of course, studies of science might benefit greatly from contrasting it with these forms of work. The question is, do sociologists of science keep a sharp eye on where science *differs* from these sorts of practical, technological modes of work? At some level they must, otherwise they cannot claim to be doing sociology of science. What then do they allow to be special about science? Not its claims to be pursuing the truth, which they class as persuasive rhetoric. Answer: its authority, its publicly recognized claims to authority, the public ascription of authority, these features make science sociologically special. There is here a pretence to be offering value-free research; yet their stance makes them power-worshippers. If only they were a little more critical and skeptical.

Let us insert a fragment of sociology here using a different comparison. Art is a form of practice, a form of labour. Writers, musicians, visual artists, all labour at what they do. Most of them must train for long periods, apprentice to masters, and work away with materials as they put together their works. The result may be written works, cut records, and visual art works. They expose their works to peers – publisher's readers, hanging committees, gallery owners, other artists – and to the public at large. Their work is subject to the judgement of peers whenever they seek exposure to the public. Their work also enters a popular market which is rough and tumble – at least a bit more than the specialist market. They may have very high aspirations, as did Bernard Shaw, Somerset Maugham, and Rudyard Kipling. All three writers were very successful with the general public and sold many copies. Yet the judgement of their peers was mixed. Kipling's sinking reputation with the literati was more or less single-handedly reversed because T. S. Eliot, a high priest of literary modernism, had a change of heart. In the nineteenth century Mendelssohn and Mosheles rehabilitated the relatively neglected J. S. Bach by daringly moving his immense *Mass in B Minor* from the cathedral to the concert hall. In our own lifetime John Constable moved from chocolate box to towering genius. I do not deny these are examples of social processes. My claim is that the social process has a goal: the recognition of aesthetic merit. And aesthetic merit as such drastically differs from the social recognition of it, and the proof of this is exactly the cases at hand, where the two diverge. (This proof holds even on the false assumption that merit is potential recognition, since the potential and the actual still differ.)

At issue here is the difference between what the ethnographer could find out by being critical as opposed to uncritical. If a meretricious artist is treated with great reverence and showered with honors, then it is incumbent upon a critic to report that. But is it disingenuous to overlook that underlying the struggle for reputation and honor today there is a yearning for artistic standing and appreciation forever? Rhetoric is in use here as everywhere else, of course. That is the way of the world. And some artists are strongly interested in this-worldly success. None of their activities makes much sense, however, without the underlying yearning. This applies to all kinds of success, worldly success as well. There are easier ways to get worldly success than practicing the arts. The double demand for worldly success and being a great artist sometimes incorporates the idea that initial scorn for a work of art is a mark of its ultimate worth that will ultimately find its just recognition. Hence, misunderstanding and suffering are to be welcomed as promises that better is to come. And "better" may mean simply future reputation, an obvious social construct. This is true as far as it goes but it does not go nearly far enough. The artist hopes the reputation derives from ascending Parnassus.

How does art compare to science in this respect? Very closely, in my view. Working in pure science, let us say cosmology, requires arduous training, intense intellectual effort, some degree of competitiveness, and poor job prospects. What drives such actors? What do we lose if we leave out or explain away the basic one, the attempt to get it right? To find the ideas that will survive testing against the facts and move us a little closer to a true picture of the cosmos? Unlike the ability to produce an acknowledged artistic masterpiece, however, there is little market reward. There is no pure science equivalent to the hugely rewarded Picasso. On the average, a plum job in academe for a cosmologist will pay far less than the medical specialty of cosmetic surgery. Although there seems to be a struggle for power and research money, and although oldsters seem to resist new ideas, what are the material stakes? They are all within the narrow band of bourgeois levels of living. To get filthy rich you have to find something you can patent or copyright and in cosmology there is nothing like that. (True, a cosmologist may become a media idol and then win fame and fortune. This is as rare as the proverbial unknown youth who manages to climb the social ladder to become Pope or President of the United States.)

So, to continue with the comparison between art and science, the sociologist can readily view both as simply individuated kinds of work. No quarrel. They can view art and science as through and through social. The media of art are comparable to the media of science, these being natural language, scientific terminology, and mathematics. Media and languages are congeries of social institutions. By means of these institutions artifacts are produced and discussed. Both involve work, some performed in private some in social settings of all sorts. Both have been monopolized by the romantic philosophy, heroic, titanic, alienated, misunderstood, prophets without honor in their own time, and other fancy stories. And both put their products out into markets of sorts. The art market is rather different from the market into which scientific productions enter, be it the *initial market*, the *futures market*, or the *general public market*. Indeed, as soon as the market uses money we might suspect that the discussion has shifted to technology or at best to applied science. One need not be a snob to see the usefulness of this differentiation. The aims of technology, applied science, and science differ; hence, their sociologies need to differ.

Technology is social – from its stipulated problems through to its mandated licensing, safety and quality régimes. Applied science is not quite the same. Its problems are sometimes technological and sometimes not. Medicine is a prominent example of applied science (as well as of technology). There is little or no pure "medical science". (Theoretical pathology is the paradigm here, yet the very concept of theoretical pathology was introduced only in 1959 or later.[9]) What there is, is biology in various of its branches, especially molecular biology, biochemistry and chemistry, and various inputs from engineering, statistics, and so on. The structure of DNA was a discovery in pure (molecular) biology. It answered the question, what is the structure of the nuclear molecule? This piece of pure science has consequences of the greatest importance to forensics, to genealogy, to the understanding of certain kinds of disease, and of course to medicine more generally. Thinking up ways to apply pure science to the world is applied science. How could DNA be used as an identifier? How could DNA be modified for various purposes? These are questions of applied science. Technological applications catered to the need for better methods of identifying criminals. The use of identifiers would aid us to link family connections, to link the whole human race as one family, to link all life to Darwin's famous Tree of Life. In the decades since Crick and Watson's breakthrough the technology has been devised and refined to allow work on all these and other lines of inquiry. It remains possible that both the technology and the applied science can throw up results that refute pure science, or basic research.

(Notice: the applications of molecular biology to the theory of evolution belong to pure science, not to technology; the technology seen in research laboratories often belongs to pure science. And so it goes.)

It is the logical relation of contradiction that stands behind all science,[10] and that can be a side effect of applied or technological applications. As I noted, much science engages in refuting earlier, especially pre-scientific ideas and thereby creating a new objective problem. Any reader of Galileo will know how he patiently tried to show that the evidence goes against received ideas, including those that are strongly intuitive. He marshaled the evidence to put intellectual pressure on the received ideas and compel the audience to see the reasonableness of the alternative. He was a great writer and of course deployed his terrific rhetorical and dramatic skills to put over his case. Such instances led some sociologists of science to claim that science is rhetoric. This is much too vague. That science uses rhetoric is obvious, yet, rhetoric aside, Galileo had a case to put over, namely, that some received ideas are false

[9] Peter Hucklenbroich says in the opening of his "System and disease: On the fundamental problem of theoretical pathology", *Theoretical Medicine and Bioethics*, 1984, 5, pp. 307–323: "Theoretical Pathology as a purely intellectual and theoretical discipline". He gives the date of its birth as 1959 or 1981.

[10] Karl R. Popper, *The Logic of Scientific Discovery*, London, Hutchinson, §11.

and better ideas are to hand. By contrast, the rhetoric of Kurt Gödel was so poor that as soon as his classic unreadable paper appeared people better at rhetoric than he was translated his ideas to render them more accessible. Were science and rhetoric the same, his followers, not he would be the winners.

The way I have just framed the matter is of course going to be said by the sociological debunkers to be itself advocacy and rhetoric, an attempt to negotiate with the audience by manipulating facts, etc. No doubt all such lines of objection are impeccable. However, this applies to them as much as to me, and then the question is, does the decision of readers to follow my argument or theirs depends on their ability to reason or on their susceptibility to rhetoric? Or will it depend on their assessment of my relative prestige? The sociologist is treating the intellectualist account of science as the received view and trying to marshal evidence to show that its lack of sociology dooms it. "There is nothing not-social about science" (Gieryn, p. 13696). Is nature social? Yes, nature is a rhetorical construct in our natural languages. True. Better question. Is nature *nothing but* a social construct in our natural languages? This again is much too vague. By "nature" I mean the natural context for society. Yet to represent and discuss nature we have to have society and language. Does that mean nature is "nothing but" its representations and discussions? Some sociologists are strongly tempted by this *non-sequitur*. Nature is like a hologram: transparently created for social purposes. It is, but is or is not the hologram trying to show us something? More technically, in order to assert that X is nothing but Y one needs a larger context that includes both X and Y. That larger context is nature. Were nature nothing but its hologram, natural scientists would lose their task of exploring it. And no research in natural science will save it, no sociology of science, and no rhetoric. At this point I should begin offering my suggestions for the skeleton of a critical sociology of science.

30.3 My Aims

By a sociology of science I intend a sociology of scientific institutions – from the Royal Society of London, founded (at least twice) in 1660–63, to the present day. By a general sociology of science I mean one that draws on an available general sociology, its treatment of institutions, and the place it finds for science in society (explicitly or tacitly – as is the general case in the social sciences). My view is that the current general sociology gets science wrong, and that more specific sociological studies need remedy.

What is the problem? First there are the three philosophical faults I have been discussing. Second, can I compensate for the narrowness and partiality of most of the classical work in the sociology of science? Third, my most general question is, to what extent are the institutions of science typical social institutions, to what extent does their being institutions *of science* make them significantly different? (And significant for what?)

I suggest that the main function of social institutions is coordination.[11] Before language could develop there must have been institutions, or what we can call institutions (perhaps by mere extension): they had to grow hand-in-hand. All accounts of how language could have come into being see it as a collective endeavour, a means of improving coordination. The invention of language was undoubtedly a huge step forward in human ability to coordinate and became itself a major and universal social institution. As Goody and Watt argue, the later invention of written language was also a quantum leap. It affected human memory, made possible the idea of law, and also the idea of assertions having in some sense an existence of their own. Without written language there is scarcely any science. Otherwise put: the enterprise we know as science is vastly different from those cognitive enterprises that exist in societies without writing.

A general sociology of science should, I agree, begin with language. Although ordinary language is in constant use in the dealings of the natural sciences – in the conduct of research as well as in the transmission of its results, including the teaching of science – it is a commonplace that its esotericism is partly due to its having a specialized language, or languages, that usually must be mastered before making efforts to contribute to it. The growth of specialized languages was driven by various considerations, mostly quite legitimate: the more background information serves a discussion, the more terms that allude to items in it are of use; and these are technical terms, the use of which is conditioned on the acquisition of some information. The functional effect of technical terms, however, was exclusionary: those not familiar with the background knowledge assumed in a discussion are excluded from it. C. P. Snow discussed this already in his Rede Lecture of 1959.[12] He saw the culture of science as a specialized subcategory of culture in general, utilizing, for example, ordinary language some of the time but finding the need to invent and use technical languages a lot of the time. Since these technical languages were not taught to every schoolchild, and since they were not easy to master, the effect was that science placed walls around its activities. These abstract, social walls preceded the gated compounds with barbed wire, with security and other clearances that are common on scientific campuses since World War II.

Social anthropologists used to teach their students that mastery of the local language was a *sine qua non* for successful fieldwork. Not so sociologists of science. Although they often pretend to be following anthropological methods, they declare and valorize the distance that not understanding the science they study allegedly gives them. They look at the work practices and at the discourse as social interactions directed at social ends (persuasion, power, research money, professional

[11] J. Agassi, 'Methodological Individualism', *Brit. J. Soc.*, 11, 1960, 244–70, reprinted in J. Agassi and I. C. Jarvie, *Rationality: The Critical View*, Dordrecht: Kluwer, 1987.

[12] C.P. Snow, "The Two Cultures", 1959, reissued 1993. Snow suggested that the teaching of technical terminology in schools to all pupils would help close the gap between scientists and the lay public and to that end he suggested emulating the Soviet educational system. This showed the poverty of his familiarity with the Soviet school system and his lack of understanding of what the acquisition of technical terms involves.

prestige) only. The assumption seems to be that in a social interaction all that is at stake are social ends, power, money, status, all achievable by persuasion. What argument persuades? No answer. Hence, neither logic nor language supports their position. Social activities often have non-social ends, such as biological reproduction or building shelter. Reproduction and shelter have social aspects but it does not follow that they are social *tout court*. Or, if someone maintains that they are, then everything is social and we are mired in sociological solipsism. Language, similarly, is used for purposes other than linguistic – to describe, to express, to reason with regard to *something or other*. To equate any and every something or other with another linguistic object is to fall into linguistic idealism. Some philosophers and sociologists find this position obvious and correct. I find it obviously incoherent.

The walls of language conceal the first two markets discussed earlier, the *initial* specialized market and the *futures* market where the results move out to peers. The *public market* is different: there scientific ideas are offered with a gesture backward to the other markets. Since these markets are relatively closed to outsiders, the gesture is one of "trust us". When in an argument someone says, "trust me", alarm bells should go off. Why not just explain? It is too technical? Then make it simpler. Is that too hard? Is there any idea, any body of evidence that cannot be explained to people who need to know? Even so, at least one can try. Consider the changes in medical practice in the last generation. Many of us can remember when doctors were trained to behave like gods and to invite us to think of their expertise and technology as a magic wand: the patient should let them wave it and be grateful if the outcome is positive. This attitude was and to some extent still is inculcated into medical students. And yet, long ago there was a countervailing institution known as the second opinion. It presumed fallibility and was invoked often when a lot was at stake. It was also resisted.

The atmosphere has improved somewhat, the air is a bit clearer. Today insurance companies are ready by default to pay for second opinions. Today some medical schools teach that the best expert on a patient's ailments is the patient. The patient monitors minute to minute, the patient directly experiences therapy, the patient is the single factor connecting all the different doctors, nurses, lab technicians and others employed by the newly patient-oriented institutions. It is the patient with whom medical professionals should engage in a careful discussion of the case, what is known about it, what is not known, what research shows as regards various kinds of therapies. This change is coming about in a legally regulated profession, one with much to gain from a hermetic and authoritarian attitude.[13] The relatively clear organization of the profession makes it more tractable to reform. Both in the education of entrants to the profession and in the institutional machinery of practice it is possible to initiate reform and to monitor its success. It is an ongoing question which way and how much the posting of medical information on the internet will impact this process.

[13] This is not to say that cultivating the view of the physician as infallible is in the interest of physicians, even though many of them think so. See Nathaniel Laor and J. Agassi *Diagnosis: Philosophical and Medical Perspectives*, Dordrecht: Kluwer, 1990.

Gieryn refers to science as a profession (e.g. Gieryn, p. 13697), also comparing it to plumbing in undertaking "double tasks". His language is illustrative of a tendency to hyperbole, as are references to society's need for "certified knowledge" (Gieryn, p. 13693). These comparisons distort and exaggerate. The exaggeration is in the wrong philosophical direction. Science is better not treated as a profession.[14] The disorganization of science is a feature that functions to promote free inquiry. The amateurs, dabblers, dilettantes, cranks, etc., are not wholly excluded. The undertaking of tasks that are very far from "doable" is just as characteristic of science as the reverse, and closer to its philosophical better self. The problem with factory work and plumbing as analogues for science is that the kind of imagination such work requires bears little or no resemblance to the kind of imagination required to change our views of things. As to whether society has a functional need for certified knowledge: what does society do when it has no such certified knowledge, when certified knowledge fails, or when certified knowledge is declared no longer certified? The answer is, it does without. The reason is, in a deep sense all certificates of knowledge are fakes, no matter how many social endorsements they bear. "Certified knowledge" or "Certified expertise" is a myth; the need for certified knowledge is a myth. (Not so certified competence, namely, certified technical expertise.) We want sociologists to separate the fact of myth from other social facts. Yet received sociology of science ignores the difference between the two. This wins it the status of a myth.

"Science" is a rather loose concept (dating from the 1850s). Its German cognate "Wissenschaft" is even looser (allowing for the term "dogmatic science" to denote theology and the law). Hence the range of institutions and roles that need cataloguing in basic sociology of science is wide and difficult to demarcate. High school and undergraduate science instruction is normally the beginning. People without such background are, regrettably, no longer considered scientists. The scientists of the Scientific Revolution sometimes had no scientific education; they were self-educated. Later examples would be Michael Faraday and Thomas Alva Edison. Francis Upton, who was Edison's assistant, had a Master's degree. Science is more formal nowadays and a PhD is normally required. Such credentialing is recent, not necessary, and is as much a barrier as a qualification. Polanyi explicitly says barriers are needed to keep out cranks and dabblers. Given that from Kepler to Schrödinger, Pauli, Dirac, and Pauling many scientists of the highest rank have also been cranks or dabblers of one sort or another it is difficult to empathize with Polanyi's fear of the damage they might do.

[14]The scientific revolution established science as amateur and efforts at its professionalization began in 1830 and ended after World War II. See below. See also Dorothy Stimson, *Scientists and Amateurs: A History of the Royal Society*. New York: Henry Schuman, 1948. Even Michael Polanyi and Thomas S. Kuhn ignore this fact.

30.4 Elementary Forms of the Scientific Life

Let us take a step backwards. To make things simple, let us date science from the Scientific Revolution. We have then to start with individuals who were making science only in the sense that the outcome ("Science") was an unintended consequence of their activities, which might be described as tinkering, dabbling, observing, speculating, discussing (often by letter), and, particularly and significantly, having great fun. We would say that science was an unintended consequence of those informally organized inquiries and discussions and that it flourished before it was named, it flourished under its first name, "natural philosophy", and later when the preferred name became "science"; the term changed its meaning to allow for the revolution in physics of the early twentieth century. Today's sociologist of science may object to beginning with these elementary forms of the scientific life. After all, we are constantly reminded that, as Derek J. de Solla Price put it, the majority of all the people who have ever been considered scientists are alive today, and "we should have two scientists for every man, woman, child and dog in the population".[15] Even if this is not an equivocation on the word "scientist", it hardly explains why so many sociologists and historians of science do historical case studies, often in the seventeenth and eighteenth centuries[16]. What does explain this and what trumps the argument from taking the majority of those alive today as a norm is that the sociology of science is primarily concerned with institutions. Just because the earliest institutions are informal, and it is the informal institutions that create the first formal institution – the Royal Society of London and its imitators – does not mean that they are sociologically out of date. To the contrary, the staying power of institutions is well known. Institutions that just grow, being often below the radar, sometimes display greater inertia than designed institutions. In the excellent BBC dramatization of the Crick-Watson discovery the film makers go out of their way to show the great researchers as also tinkerers, eclectics, dabblers, enthusiasts, and so on.[17] To this day one of the characteristic institutions of laboratory science is the group. This may have formal aspects (all those paid to work in a given lab on a given project), but it also has informal aspects of a Simmel-style relatively loose social network, since there is interaction at the edges with non-group members and with other groups, there is the informally developed network with other researchers around the world, there are the water-cooler, tea cup and coffee mug conversations, the encounters at meetings, the tips handed over, none of which can be easily mapped. They are like

[15] Derek J. de Solla Price, *Little Science, Big Science*, New York, Columbia University Press, p. 19.

[16] Of course, quite possibly the sociologists of science are often unable to follow "hard-science" texts of later periods.

[17] I am referring to the 1987 film directed by Mick Jackson, with Jeff Goldblum as Watson, Tim Piggott-Smith as Crick, Alan Howard as Maurice Wilkins and Juliet Stevenson as Rosalind Franklin. It is variously known as "Double Helix", "The Race for the Double Helix", and "Life Story". Another excellent depiction of science in a critical manner was the 1978 BBC series "The Voyage of Charles Darwin" which attended closely to the running dialogue Darwin had with Charles Lyell, Capt. FitzRoy, the geological and fossil evidence, and himself.

the complex and shifting social arrangements among any group of people who have something in common about which they eagerly communicate. Within some groups there is a great deal of tinkering, dabbling, speculating, brainstorming. This is the sociology of science at level zero. One of the favorite representations of the future scientist in the mass media is the kid who tinkers in a kind of workshop-cum-lab young-Edison-style. Are the media so far from the truth?

To put the matter a little more formally, in its pre-history and early history science emerged within civil society and the associations that made possible. When formal institutions emerge the role of civil society does not disappear. Within scientific institutions there are informal civil society associations and these are especially important in leaping over and around the fences of the institutions to formal and informal associations beyond. Electronic communication has made civil society communication networks a world-wide phenomenon.

30.5 The Growth of Institutions

Institutional patterns of scientific activity producing science made possible formal scientific institutions and scientific education. The earliest scientific education is mathematical. That was Galileo's training. That was Descartes' training. That was where Newton's and Leibniz's innovations lie. Already at Pisa, where Galileo studied, mathematics was part of the university curriculum. The first modern[18] labs were private: at the courts of wealthy dilettantes who wanted to tinker; in the studies of those with the space and the know-how. For over two centuries The Royal Observatory at Greenwich employed one astronomer and no more than five assistants – the only professional scientists in Britain till 1800, when Britain's Royal Institution was founded, which conducted research and communicated to the public, and was and is a standalone from academe and the Royal Society. The laboratory as institution developed as an unintended consequence of tinkerers wanting to have a place to tinker and to set up their equipment. On the model of the Royal Society, even before it was professionalized, all kinds of institutions grew up to bring together mathematicians, statisticians, physicists, geographers, anthropologists in Britain alone. The first scientific laboratory in an English university was the Cavendish Laboratory in Cambridge. It is sociologically important to stress all this because in the long historical perspective the association of science with universities and big business is recent and anomalous, not natural or predestined. As for the military, we know from Merton's first monograph (1938) that the association is long-standing. This does not mean that the extraordinary intrusion of the military during the Second World War and the Cold War should be taken for granted. Merton was correct in saying that communism was fundamental to the ethos of science.

[18] Preceding these were the medical, alchemical and mineralogical laboratories that did not possess scientific status by the standards of the Royal Society of London.

Hence, the large amounts of military money tied to secrecy and isolation were and still are obviously inimical to the best practices of science.

Little modification of this is needed if we shift focus to the growth of labs in industry, especially the chemical industry all over Western Europe in the mid-nineteenth century. Tinkering with the product, studying the issue of scaling up from lab bench to industrial capacity, and so on, were the obvious incentives for in-house labs. That they should gradually add to their repertoire the exploration of possible new products is something that comes from the imperatives of industry even though the pursuit requires science, applied science, and science-based technology.

30.6 Final Resting Place: Academe

Last but not least we have academe, which gets onto the science bandwagon only in the mid-nineteenth century, and in some countries very late in the nineteenth century. By that point there is some consensus on science teaching needing laboratory work, there is some promise of scientific careers, there are organs of publication, and there is some agreement on modifying the divisions of knowledge handed down from Aristotle (an example of the longevity of institutions) so as to create Departments tasked with responsibility for particular areas of scientific inquiry. At its simplest that division was physics, chemistry, and biology, as expressed in the founding of technical universities and schools of engineering in older universities in the United States and in Europe. Readers are invited to look at the web site of their own university or some other and count the number of divisions and sub-divisions of these three they can find nowadays. They can also reflect on the variant names and on the different organizational charts of each university.

This departmental fissiparousness is sociologically under-explored. Every department that splits has usually had activities that will now be split off going on under its aegis. What drives such splits? Autonomy is the most obvious. But then this obviously has its limits, and sociologist of science might profitably explore just what forces were thought to be ranged against autonomy within the previous structure. Is it as simple as being resistance to the new, including new challenges to the boundaries? And why, when a split is engineered, do newly minted departments have a tendency to draw up their own boundaries and make efforts to police them? That is, why do they reproduce the institutional resistance to autonomy the split was designed to overcome? This too is a case of what I mean by persistence of institutional patterns over long periods of time. Why are boundaries and boundary-policing so part and parcel of academic life when they are not inherited from the tinkerers, dilettantes, dabblers, observers, and speculators out of which academic science developed? Why is the curriculum such a battleground in almost every university department? Why do universities consider it their place to promise that someone with a degree in such and such will have mastered the subject of so and so? Professions often enough administer their own tests for admission and consider it sufficient that the syllabus is made public. It is interesting how things worked in German-speaking universities in the nineteenth and the early twentieth century. Once enrolled, a student could wander from course to course in any part of the university and from one university to another, studying whatever took their attention. When they wished to graduate they needed to opt for an area and be sure they were prepared for the examinations in that area. They could emerge with degrees in the most unusual and hybrid combinations. (There were exceptions, for sure, law and medicine, especially medical specializations, being the most obvious ones.) It is unclear that any harm was done by this to the sciences or to the students.

To answer some of the questions in the previous paragraph sociologists might have to look at the intellectual stakes. More often than not, fissiparousness is promoted by disputes about the world and consequent disputes about how research into it should proceed. There are schools of thought in science over intellectual matters such as which problems are interesting, which lines of inquiry are promising, what certain research approaches show, and so on. These schools sometimes have great longevity and are in constant running battle over the nature of things. Inquiry into the nature of things being a form of friendly-hostile cooperation, to use Popper's apt expression, can naturally lead to collateral disputes about power which can be used to discourage, underfund, underpromote, and generally disparage other parties. To ignore status and power as objects of the exercise is a very naïve utopianism. To see status and power as the sole object of the exercise is worse: it is a very naïve reductionism, the sort of thing that used to be called the viewpoint of the observer from Mars. (Sociology itself is a case in point. What distinguishes the *Journal of Classical Sociology*, say, from all of the other journals? One answer is that it takes the classics seriously; another is that it encourages sociological research that denies the relativist version of social constructivism.[19])

Finally I come to the modern era, the era of Kuhn, the apologist for the new bureaucratic routinization of science so often taken as the norm in today's received sociology of science. Kuhn divided the life of a science into two cycles, the normal and the exceptional. Normal science consisted of reasonably routinized work on the foothills of an established paradigm. Exceptional science was the situation that developed when the established paradigm began to seem to be in trouble, its anomalies and patches making research difficult or fruitless. A period ensues in which there is a struggle either to renovate it, or to ditch it, seek a new paradigm. This revolutionary period he viewed as not a period of science at all, but an *interregnum*. Research is disorganized hence of unknowable status. (The dabblers and cranks are ascendant!) Science only begins again once a new paradigm is established, generally received, and plays the role of the new base for newly routinized research. I do not know how to resolve the contradiction here between there being "normal science" and "revolutionary science" and revolutionary science not being science after all. What I do know is that much received sociology of science to an astonishing

[19] One of the most extreme of relativist texts that I have come across (Edwards et al. 1995) was published in, of all places, the journal *History of the Human Sciences*, even though it was not historical but constructivist and philosophical.

degree swallowed this picture wholesale despite the fact that any observation of any historical detail should lead a follower of Kuhn to the view that it is *part of an interregnum*. It is not mentioned by Gieryn because, I conjecture, he takes it for granted as "deep background". I conjecture that it is buried deep because, notoriously, Kuhn considered the social sciences to be pre-paradigm and hence pre-science or at best proto-science. The counter-claim that they were a new species, multi-paradigm was fruitless. Multi-paradigm means many, which means none in Kuhn's system, where there must be consensus on a single model to make a discipline scientific. Kuhn's admission that his theory is sociological and that the social sciences are multi-paradigm is an admission of defeat. Hence his inspiration is also an embarrassment.

Normal science, Kuhn said, is most efficient. His theory looks attractive as it offers a view of science according to which tinkering, dabbling, and speculating in a dilettantish way are all minimized. It has science reasonably routinized so that comparisons can be made with the factory or with plumbing; this allows the assimilation of science to the sociology of work. Yet both factory work and plumbing are technological. The tendency then is to see science as a form of technological work.[20] Scientists are all like Edison, Eastman, Bell, Land working towards some technological end-product. Cosmology is of course the challenge, since it invents nothing tangible, it operates comfortably with several competing paradigms, and is very high-powered. If we look at the strange attempt of the ethnomethodologists Garfinkel et al. (1981) to reduce astronomers' activities to routinized work (on the pulsar) we find that the lab notebooks, the recorded lab chat, the discussions of what it is their instruments are registering, are all used to show that they produce (i.e., make) the pulsar on exactly the same model that actors generally produce society by and in their interactions. Hence there is no "Galilean object" called a "pulsar" discovered in the Heavens. There are certain arcane social practices of instrumentation, processing, and discussion that lead to discourse that misleadingly suggests that such is the case. The actors produce a play called science and the play tells a story.[21] The story is not real, only the acting and production are real. Since on this model restaurants produce not food but the story of food, I would say: the sociological solipsist is under illusion. Or perhaps there is no sociology of science and no ethnomethodology; it is just a make-believe that there is.

Ethnomethodology, in all fairness to sociology, is only one school or paradigm for the sociology of science. Like all the other paradigms, it ignores the paper and pencil[22] cosmologists, preferring to study those who can be found "at work" in a "workplace". But this is general in the sociology of science. The farther away they

[20] Indeed, as sociologists of work care for the quality of working life and so oppose excessive routinization, some of them reached the conclusion that if Kuhn is right then the quality of research life needs improvement too.

[21] This theory R. G. Collingwood advocated to explain magic since he was reluctant to take it at its face value, refusing to view its practitioners as irrational.

[22] A famous joke ascribes to Einstein's wife the assertion that all a researcher needs is a pencil and a used envelope on the back of which to write something.

can get from intellectual work the more confident they can be. The more they can focus on discourse and less on ideas the more startling their conclusions. Above all ethnomethodology is nakedly idealist; I have labelled it a version of solipsism, "sociological solipsism" (the world, the facts, the pulsar, are social dreams). They are like the famous solipsist Christine Ladd-Franklin, who wrote to Bertrand Russell that she could not understand why everyone doesn't adopt her position.[23] Not all sociologists of science are idealists veering towards solipsism. But I have said enough to show that there is widespread endorsement – be it serious or not – of poor philosophy and in that sense it is uncritical. This is funny. Above all they want to be critical, to be radical. They think of themselves as debunkers, demystifiers. Instead they take at face value all kinds of myths, illusions, delusions, errors, *non-sequiturs*, and refusals to face the facts. To do that is to endorse, or to re-endorse, the *status quo*. This is certainly a counter-Enlightenment project. As it is unintended, we may invite its practitioners to return to Reason.

Acknowledgements I thank Joseph Agassi for his help with, and hence input to, this paper.

References

Agassi, Joseph. 1960. Methodological individualism. *British Journal of Sociology* 11 (3): 244–270.

Edwards, Derek, Malcolm Ashmore, and Jonathan Potter. 1995. Death and furniture: The rhetoric, politics and theology of bottom line arguments against relativism. *History of the Human Sciences* 8 (2): 25–49.

Garfinkel, Harold, Michael Lynch, and Eric Livingston. 1981. The work of a discovering science constructed with materials from the optically discovered pulsar. *Philosophy of the Social Sciences* 11 (2): 131–158.

Gellner, Ernest. 1973. *Cause and meaning in the social sciences*. London: Routledge.

Gieryn, T.F. 2001. Sociology of Science. In *International encyclopedia of the social & behavioral sciences*, ed. N.J. Smelser and P.B. Baltes, 13692–13698. Amsterdam: Elsevier.

Laudan, Rachel. 2000. Review of Oreskes 1999. *Philosophy of Science* 67 (2): 343–345.

Merton, Robert K. 1938. Science, technology and society in seventeenth century England. *Osiris* 4 (2): 360–632.

Oreskes, Naomi. 1999. *The rejection of continental drift. Theory and method in American earth science*. Oxford: Oxford University Press.

———, ed. 2001. *Plate tectonics. An insider's history of the modern theory of the earth*. Boulder: Westview.

Snow, C.P. 1959. *The two cultures and the scientific revolution*. Cambridge: Cambridge University Press.

[23] "As against solipsism it is to be said, in the first place, that it is psychologically impossible to believe, and is rejected in fact even by those who mean to accept it. I once received a letter from an eminent logician, Mrs. Christine Ladd Franklin, saying that she was a solipsist, and was surprised that there were no others. Coming from a logician and a solipsist, her surprise surprised me." (Bertrand Russell, *Human Knowledge: Its Scope and Limits*, London: George Allen & Unwin, 1948, p. 180).

Chapter 31
Critical Rationalism in Sociology

Michael Banton

In his lectures at the London School of Economics, Karl Popper said he hoped that we students would try out his philosophy in the various fields into which we moved. The career opportunities available to me in 1950 took me into the area of overlap between social anthropology and sociology. Several years of research and teaching passed, during which time I saw little opportunity to draw on Popper's advice until, in 1957, I met Joseph Agassi. The subsequent story has, necessarily, to be autobiographical, but it might have a wider import.

At that time I held a post in the Centre for Co-operation and Research in the Social Sciences in the University of Edinburgh, where we were engaged in discussion about modes of co-operation. It was also a time when holism and individualism were presented as alternative methodologies. I contended that the different social sciences studied different aspects of behaviour, asking different questions. 'Scientific explanation is limited to aspects of behaviour and no action can be explained in its entirety. It is in their ability to isolate aspects of behaviour that the objectivity of the social sciences is grounded' (to quote from an internal text in which I thanked Agassi for earlier comments).

In 1960, in a major aritcle, Agassi wrote about the opposition between holism and individualism, defending 'institutionalistic-individualism', which he described as Popper's great contribution to the philosophy of the social sciences. According to this perspective 'society is the *conventional means of co-ordination* between individual actions. This last view is known as contractualism or conventionalism'. He cited my book *White and Coloured* as an example of conventionalism in proposing a specific sociological explanation.[1]

[1] It was indeed an example of conventionalism, but Agassi's statement that I had proposed a theory turning on a postulated gap in British society astonished me. An earlier writer had explained white people's observance of social distance towards non-whites as a consequence of their beliefs about

M. Banton (✉)
University of Bristol, Bristol, UK
e-mail: michaelbanton58@gmail.com

N. Bar-Am, S. Gattei (eds.), *Encouraging Openness*, Boston Studies in the Philosophy and History of Science 325, DOI 10.1007/978-3-319-57669-5_31

31.1 Race Relations

In 1967 I published a textbook for the study of what were called (and sometimes are still called) 'race relations'. Comparable textbooks almost always start from what are seen as contemporary social and political problems (particularly in the United States of America where the biggest market for such books is to be found). There is a place for books that describe the urgency of these problems, but the task of an academic textbook, I believed, was to conceptualize an academic field, tracing a tradition of inquiry in Popperian fashion. Agassi's criticism of 'inductivist history' (or, to use a later name, 'presentism') influenced the use I made of the history of 'racial' thought, then and in subsequent work. The continuing debate about the source of variation within what many regarded as the human species constituted a framework within which dispute continued. In my textbook I maintained that

> What distinguishes the social scientist's chief contribution from the journalistic commentator's is that he seeks new questions which will open up aspects of the problem insufficiently comprehended in common speech. The obstacle to greater understanding is often not that of finding new answers but of discovering new questions, which, by reformulating the old ones, make possible their solution (1967: 7).

The reference to 'common speech' (or ordinary language) was an echo of my reaction when, as a second-year undergraduate at the LSE, I attended two lectures in a course on 'Race Relations and Racial Problems'. A newspaper might have to employ such expressions, but we knew, did we not, that the biological inheritance of persons did not determine the way they behaved towards one another?

How were race relations to be identified for *sociological* purposes? My answer in my textbook was to maintain that when racial differences were used as a way of dividing a population, the outward differences were used as role signs (1967: 57). Thus a conception of race was being used in the construction of a social category. Reviewing my book, Julian Pitt-Rivers (1970: 340) objected that 'Banton's concept of race is simply the popular British concept raised from the daily life of the British Isles to the status of a universal scientific category'. He was right; no-one can see another's race, it is the phenotype (and particularly the skin colour) that is seen; I should have said that it was phenotype that was used as a role sign.

Looking back on this episode in 2005, I claimed that I had made an interesting mistake, exemplifying Popper's (1963: ix) doctrine that '*all* our knowledge grows *only* through the correcting of our mistakes'. Mine was interesting because it stimulated conceptual improvement rather than simply rejection. Not all mistakes are interesting.

Pitt-Rivers had maintained that 'the biologists' and the sociologists' concepts of race are quite inimical to each other and are connected in fact only by the history of a word. The connexion must be severed and they must be separated'. That, though,

how members of their peer group would judge their behaviour. I had added that they would also observe social distance because they viewed non-white immigrants as strangers who could not be expected to behave in the same ways as members of their white peer group.

was no way to solve my problem. What name for the academic field could one use instead, given its centrality in the political vocabulary? Nor did the biologists use any concept of race in the way assumed. The establishment of population genetics had eliminated any surviving traces of typological thinking in biology.

My concern, and that of the Social Science Research Council Research Unit on Ethnic Relations, at Bristol from 1970–78, was with the development of social science theory. At the same time, my colleague Henri Tajfel conducted experiments in social psychology that led commentators to refer to his explanations as a theory of social identity. I said to myself 'that's a model, not a theory; I must have a better theory than that'. Tajfel was opposed to any suggestion that he should formulate his arguments as a set of propositions. In reaction, I wrote out my ideas as a set of four propositions constituting a Rational Choice Theory of Racial and Ethnic Relations:

(i) Men act so as to attain the maximum net advantage
(ii) Actions at one moment in time influence the possibilities that will be present at subsequent moments, i.e., actions build up or modify social structures which in turn shape and constrain action so that individuals have to choose between a limited number of alternative courses
(iii) Men utilise physical and cultural differences in order to create groups and categories by the processes of inclusion and exclusion
(iv) When relations between individuals who can be physically differentiated are limited by monopolistic practices that prevent exchange or give one group an advantage, boundaries are drawn in ways that create racial categories (Banton 1977:3).

Seven years later this argument appeared in expanded form in a book, *Racial and Ethnic Competition* (Banton 1983). Its line of reasoning was taken, of course, from neo-classical economics, but, since many sociologists misunderstood or rejected that kind of reasoning, it would have been better had I presented it as part of the theory of collective action. One of its merits was that it located research questions, the *explananda*, within a larger conceptual scheme.

There was also a connection with the dispute over holism and individualism. As Agassi wrote, when the individualist contends that only individuals are responsible actors, the holist retorts that society is more than merely a collection of individuals. In sociology this was sometimes referred to as the micro-macro problem. For some of us it was resolved by Max Gluckman's argument that the working of macro forces can always be discerned at the micro level if the researcher searches for them. This was an illustration of a bottom-up as opposed to a top-down approach. My rational choice theory was bottom-up.

31.2 The Sociology of the Police

To interpret behaviour as influenced by the sign value of an attribute was to locate an observation within a theoretical frame larger than any particular society. This strategy also characterized the research I had been undertaking on relations between the police and the public (on which I had reported in *The Policeman in the Community* (1964)). Assessing the subsequent history of 'police studies', Robert Reiner (2015: 315) wrote that 'as the first empirical research based book on the topic it virtually created the field'. Unlike most of its successors, its central concern was with peace-keeping. My research had started in Scotland and then moved to the United States. My encounter with the very different circumstances of policing there led me to tell an audience of police chiefs in California that I had come from 'a policeman's paradise', one in which the Scottish policeman, like the minister, was set apart from other members of the community out of respect for his office. Durkheim had defined the socially sacred as that which is set apart and treated as both intrinsically good and as dangerous. In this sense, the British police, by comparison with their American counterpart, was 'somewhat sacred'.

Reiner thought my book's opening sentence expressed my challenge to some contemporary assumptions: 'A cardinal principle for the understanding of police organization and activity is that the police are only one amongst many agencies of social control'. The other two books with which it was compared by Reiner were framed by the authors' experience of the contemporary United States of America and its society. My ambition was universalistic, to rise above particular times and places. It was a foretaste of the way that, in 2016, I would seek to distinguish the study of sociology from the study of social policy.

Thus, when I wrote an entry on 'Police' for the *Encyclopaedia Britannica* (14th ed., 1974), I wrote: 'The police function is universal in society, though it is only in the larger and more complex states of modern times that full-time officials are appointed with special responsibilities regulated by law.' Later on, I stated 'Though it is unreal abstraction – like the idea of a frictionless pendulum – the notion of a self-policing community can serve as a model when studying what is required of the police'.

Of course, many circumstances have since changed, as was illustrated after half a century in a newspaper headline 'Police failing on diversity, figures reveal: Minority ethnic applicants less likely to get job in more than two-thirds of forces' (*The Guardian* 2/1/2016). At the time of my research there was no record of the ethnic origins of police officers or job applicants, no targets, no supervision of the appointments process, and no thought that Her Majesty's Chief Inspector of Constabulary could be someone recruited from outside the police (in this instance from the post of Rail Regulator). An equally new attitude was the acceptance of the need for the police to be representative of their publics expressed by forces that had been unable to meet their targets. The British police were being called to account in ways previously unthinkable. There is a wider public recognition of the point that I was making, but there has been little follow-up in sociology.

31.3 Superseding Race?

In the early 1980s, I tried, following on from Pitt-Rivers, to develop a distinction between the popular idea of race and the scientific concept of race, as in the title of a lecture delivered in 1975 that was published two years later. Rejecting any focus on the biologists' and the sociologists' concepts, I reflected on the process of conceptualisation, claiming that it was better to distinguish analytical from folk concepts of race and ethnicity (1979b). Maybe I succeeded in doing this for race, but certainly not for ethnicity. A little later I came to believe that the distinction was better grasped by the contrast drawn by American anthropologists between *emic* and *etic* constructs. Shortly afterwards Brubaker and Cooper (2000) drew much the same distinction when they contrasted categories of practice and categories of analysis.

'Society' and 'behaviour' are both emic constructs, but the multiplicity of meanings given to 'behaviour' can more easily be controlled by the addition of adjectives like observed behaviour, speech behaviour, religious behaviour, and so on. It is less easy to distinguish the meanings given to 'society', either of society in general or particular societies as opposed to particular states.

The difficulties with respect to 'race' were illustrated by a paradox in the American Sociological Association's 2002 statement on the 'Importance of Collecting Data on Race'. It was prepared in response to a current political dispute, and, while recognizing that use of 'race' no longer had any validity in the academic field in which it had originated, it defended a mode of data collection that implied the opposite. I wrote:

> For several reasons the intellectual challenge was, and remains, more difficult than the political challenge. One of them is that states have obligations under international law that require them to use the words 'race' and 'racial'. The perception of a conflict between scientific knowledge and public practice has arisen because scientists and legislators have different objectives and use different vocabularies in order to attain them. The scientists say, in effect, that 'once some of our predecessors thought that race might be a useful concept in biology; now we know that there is no place for such a word in our vocabulary'. The legislators say, in effect, that 'we know that the word race has misleading associations that we hope to dispel by educational measures, but at the present time its use is necessary to the discharge of our international and domestic obligations' (Banton 2015: 3).

I concluded that the fundamental distinction must be that between practical and theoretical knowledge. The vocabulary of the former necessarily consists of emic constructs, words with multiple and changing meanings, suited to the resolution of practical problems. The vocabulary of the latter needs words that, in a technical context, have single meanings understood by the specialists and are therefore etic constructs. This claim seems consistent with much that others have written about epistemology but has not, so far as I am aware, appeared in published form.

If the claim is accepted, it means that expressions like 'race relations' should be accepted as the constructs of ordinary language so long as they retain their currency. They are the constructs of their time yet, as journalists and other commentators often succeed in 'cleaning up' ordinary language, they may come to seem out-dated (like my use of the male pronoun in my initial description of a rational choice theory, quoted above). The process by which such expressions are superseded may be a slow one. Where research is founded in experimentation concepts with a single, technical, meaning supersede those that have been found inadequate can be more rapid because the criteria are independent of time and place.

To begin with, the practice of regarding race as a social category had to be subsumed in a larger theory of social categories. So I formulated 16 propositions as the beginning of a theory of social categories, both physical and cultural, and the significance attributed to them (Banton 2011).

Since then I have gone on to claim that the distinction between the two kinds of knowledge is relevant to other fields of sociology as well, including those of kinship, religion, and nationalism, plus the difference between the study of sociology and the study of social policy. Practical knowledge is diverse. Theoretical knowledge in sociology has to be knowledge about behaviour.

31.4 Religion

My argument was anticipated, in part, by Max Weber, for in his famous essay on 'Objectivity in Social Science and Social Policy' of 1904, Weber compared sociological with historical writing, noting that 'the language of the historian contains hundreds of words that are ambiguous constructs' (Weber 2004: 398). He contrasted this with the sociologist's need for words with but a single meaning, words that are now sometimes called univocal constructs in contrast to multi-vocal constructs.

The title given to Weber's English-language collected essays, *The Sociology of Religion*, has contributed to the field's focus on religion as a social institution. Study of *The Oxford Handbook of the Sociology of Religion* (Clarke 2009), and perusal of volumes on library shelves, shows that many contemporary works in this field start from an emic construct of religion and explore it without questioning its function as defining a field of study and as generating further problems to be researched. The emic construct of religion is very important in the realm of practical knowledge because some states recognize a state religion and rely, or have relied, on organised religion to serve particular social functions. They then treat such religions in a special manner for the purposes of taxation, so that the definition of religion becomes an important matter of practice.

Practical considerations make it the more difficult to identify existing sociological knowledge in this field, as in the fields of 'race' and 'kinship', because, in each case, the primary definer is an emic construct. The definition of the field as the study of religion, instead of the study of the behaviour that is identified as religious in ordinary language, has, with a notable exception, impeded the discovery of good research problems. That discovery demands imagination, the perception of possible connections that have hitherto been overlooked. It implies a rejection or modification of prevailing ideas. If the pursuit of such a possibility produces valuable new knowledge, that is its own justification.

There is an exception to this generalization. *Acts of Faith* (Stark and Finke 2000), may be presented as 'explaining the human side of religion', but it is, for the most part, a plan for the comprehensive study of religious behaviour drawing upon rational choice theory. It concludes with a list of 99 propositions phrased so as to be testable, and 34 definitions. It constitutes an agenda for any critical rationalist interested in the sociology of religion.

31.5 Sociology

From Popper and Agassi I had concluded that critical rationalism as a *Logik der Forschung* was primarily concerned to promote the growth of objective knowledge by seeking deductive explanations. My encounter with the difficulties arising from the multiple meanings vested in the word 'race' led me to conclude that it was necessary to separate the search for theoretical knowledge from the search for practical knowledge and that this differentiation had a place in critical rationalism.

Popper had encouraged his students to try out his doctrines in new fields to see if they were more successful than prevailing doctrines in the pursuit of new knowledge. Agassi would have seconded this, although the reasons he gave for preferring individualism to holism related to the world of practice, not that of theory[2].

As a potential academic subject, late nineteenth-century sociology drew much of its inspiration from contemporary philosophies of history and from the view expressed by John Stuart Mill (1898: 595):

> The fundamental problem, therefore, of the social science, is to find the laws according to which any state of society produces the state which succeeds it and takes its place.

Something of the same view continues in writing like that which asked whether the election of Barack Obama as president of the United States marked a transition to a 'post-race society'. It could reach no clear conclusion because there was insufficient agreement on the criteria for deciding the issue. The approach from critical rationalism, by contrast, starts from observations about actual social relations, particularly of behaviour that does not quite fit with current sociological explanations and therefore evokes intellectual curiosity. It seeks deductive explanations.

Perhaps the leading characteristic of critical rationalism as a philosophy in social science is its focus on the growth of objective knowledge. This growth occurs within particular fields or subjects. Indeed, it is the search for answers that creates the fields. Some prove to be dead ends, and others develop. Knowledge about the past can be important to the present day research worker in giving him or her a perspective on research design.

---

[2] In his 1960 opposition of Holism and Individualism he is primarily concerned to assess the two approaches in terms of their ability to explain intended institutional reform or the absence of reform (Agassi 1960: 261). Popper's essays on problem-solving (written for a wider readership) were likewise mainly concerned with political problems.

No philosophy of research can cast very much light upon one of the key factors, the research worker's imagination. This may be illustrated by reference to the work of Erving Goffman whom I met on joining the Edinburgh department. He had returned from fieldwork in the Shetland Islands, having taken theoretical ideas from W. Lloyd Warner and Everett C. Hughes while adding something quite new. In his book *The Presentation of Self in Everyday Life*, he formulated concepts for studying the manner in which individuals tried to manage the impression they created on others. We published an offset-photo edition of this book from the Social Sciences Research Centre. For several years Goffman's innovation was ignored, in part because it seemed too close to the satirical writing of Stephen Potter in his very popular book *The Theory and Practice of Gamesmanship: Or the Art of Winning Games Without Actually Cheating*, of 1947, and the author's equally entertaining books on *Lifemanship* and *One-Upmanship*. When, some years after its publication, I reread Goffman's book, I was astonished to notice how much it had influenced my own focus on interpersonal relations without my realising this.

Several of Goffman's books were collections of essays that started from observational studies. They influenced subsequent theoretical writing but cannot be usefully expressed in the form of propositions and did not create any new school of thought. Some of his books, like *Asylums*, had a great influence in the world of practice as well as that of theory. If their approach was bottom-up, it did not go very far 'up'. Goffman did not create a new field within sociology, but brought new perspectives into several existing fields.

Much current sociological writing relies upon social surveys in which the respondents offer explanations of their own behaviour; most surveys are unable to explore the influences of which the respondents may not be conscious or influences that, for lack of suitable words, they cannot identify. Information on the theoretically critical issues is often lacking. The United Kingdom's Economic and Social Research Council has been told to give special consideration to research proposals likely to have 'social impact'. Practical knowledge is prioritized.

At the time I was thinking about the emic/etic contrast, I had some graduate students from Malaysia to advise. Together with one of them, I developed an experimental technique that could be used in an otherwise conventional kind of opinion survey. Subjects were asked to say how they thought a representative figure would respond in certain imagined but realistic situations (Banton & Mansor 1992). Hypotheses were tested, and the research contributed to the growth of theoretical knowledge, but the technique has not been taken up and developed by other researchers. Currently, the most promising research in this field is that of political scientists in the USA, some of whom have demonstrated the potential of the theory of games for the study of inter-group relations (e.g. Habyarimana et al. 2009). Research of this kind may result in the recognition of behavioural sociology as a new specialism alongside behavioural economics.

So mine is a minority position in several respects. It claims that the implementation of a critical rationalist perspective entails recognition of the difference between the worlds of practice and theory reflected in two kinds of vocabulary. The vocabulary of theory is subject to continuing revision with the discovery not only of better concepts, but also of better ways to organize research, so that new fields of inquiry are established and the boundaries of academic disciplines have to change. The growth of theoretical knowledge has implications for the vocabulary of practice, but, as the case of 'race' in law illustrates, revision of that vocabulary may be the responsibility of others.

If a particular conception of methodological individualism was Popper's great contribution to the philosophy of the social sciences, it reinforces the view that this whole line of philosophising arises from the belief that the growth of knowledge in the social sciences could be accelerated were the practitioners to pay more attention to philosophical assumptions.

31.6 Conclusion

Popper wrote memorably about objective knowledge and about problem-solving, but he did not consider differences in the kinds of problem or discuss the relation of objective knowledge to other kinds of knowledge. Nor did he consider whether the growth of objective knowledge is helped by any transition from a vocabulary employing words with multiple meanings to ones that use words with only a single meaning. So I am left wondering whether Joseph Agassi and others interested in the development of Popper's doctrines would accept my work as contributing to the shared endeavour that can be called critical rationalism.

In sociology, I believe I am the only practitioner to have used the expression 'critical rationalism' as a name for a distinctive research philosophy within the field. To me, it appears potentially useful for clarifying points at issue in some disciplinary disputes. What is the position in other subjects? Is there any accepted definition of critical rationalism?

References

Agassi, Joseph. 1960. Methodological individualism. *British Journal of Sociology* 11 (3): 244–270.

———. 1963. Towards an historiography of science. *History and theory: Studies in the philosophy of science*, Beiheft 2. 'S-Gravenhage: Mouton.

Banton, Michael. 1959. *White and coloured: The behaviour of British people towards coloured immigrants*. London: Cape.

———. 1964. *The policeman in the community*. London: Tavistock.

———. 1967. *Race relations*. London: Tavistock.

———. 1977. *Rational choice: A theory of racial and ethnic relations*. SSRC Research Unit on Ethnic Relations, Bristol: Working Paper 8.

———. 1979a. The idea of race and the concept of race. In *Race, education and identity*, ed. Gajendra K. Verma and Christopher Bagley, 15–30. London: Macmillan.

———. 1979b. Analytical and folk concepts of race and ethnicity. *Ethnic and Racial Studies* 2: 127–138.

———. 1983. *Racial and ethnic competition*, Cambridge University Press. Reprinted 1992, Aldershot: Gregg Revivals.

———. 1987. *Racial theories*. 2nd ed. Cambridge: Cambridge University Press. 1998.

———. 2005. Finding, and correcting, my mistakes. *Sociology* 39 (3): 463–479.

———. 2011. A theory of social categories. *Sociology* 45 (2): 189–201.

———. 2015. *What we now know about race and ethnicity*. New York/Oxford: Berghahn Books.

———. 2016a. Reflections on the relation between sociology and social policy. *Sociology* 50 (5): 993–1001.

———. 2016b. Practical and theoretical reasoning: Race, kinship, religion, *under submission*.

Banton, Michael, and Mohd Noor Mansor. 1992. The study of ethnic alignment: A new technique and an application in Malaysia. *Ethnic and Racial Studies* 15 (4): 599–613.

Brubaker, Rogers, and Frederick Cooper. 2000. Beyond identity. *Theory and Society* 29 (1): 1–47.

Burns, Tom. 1992. *Erving Goffman*. London: Routledge.

Clarke, Peter B., ed. 2009. *The Oxford handbook of the sociology of religion*. Oxford: Oxford University Press.

Habyarimana, James, Macartan Humphreys, Daniel N. Posner, and Jeremy M. Weinstein. 2009. *Coethnicity. Diversity and the dilemmas of collective action*. New York: Russell Sage Foundation.

Mill, John Stuart. 1898. System of logic, ratiocinative and inductive, People's Eighth Edition. London: Longmans Green.

Pitt-Rivers, Julian. 1970. Race Relations as a Science: A review of Michael Banton's Race Relations. *Race* XI (3): 335–342.

Popper, Karl R. 1963. *Conjectures and refutations. The growth of scientific knowledge*. London: Routledge.

———. 1972. *Objective knowledge: An evolutionary approach*. Oxford: Clarendon Press.

———. 1999. *All life is problem solving*. London: Routledge.

Reiner, Robert. 2015. Revisiting the classics: Three seminal founders of the study of policing: Michael Banton, Jerome Skolnick, and Egon Bittner. *Policing and Society* 25 (3): 308–327.

Stark, Rodney, and Roger Finkle. 2000. *Acts of faith: Explaining the human side of religion*. Berkeley: University of California Press.

Weber, Max. 1964. *The sociology of religion*. Boston: Beacon Press.

———. 2004. The "objectivity" of knowledge in social science and social policy. In *The essential weber. A reader*, ed. Sam Whimster, 359–404. London: Routledge.

Chapter 32
Worker Bees and Slave Labor

Hilde Hein

Work connotes activity, effecting or bringing something about. The worker does something physically and/or mentally, and whatever is "worked" has been acted upon. Work is a denial of passivity, and thus a manifestation of force or power. Indeed, the association of work with force sometimes invites a further conjunction between work and violence.[1] John Locke, for example, identifies labor with "subduing" the earth and thereby rationalizes private property: "The Law Man was under, was rather for *appropriating*. God Commanded, and his [man's] Wants forced him to *labor*. And hence subduing or cultivating the Earth, and having Dominion, we see, are joined together. The one gave Title to the other. So that God, by commanding to subdue, gave Authority so far to *appropriate*. And the Condition of Human Life, which requires Labor and Materials to work on, necessarily introduces *private possessions*."[2]

Hannah Arendt makes a similar connection between work and violence, observing that destructiveness is a necessary feature of fabrication: "Material is already a product of human hands which have removed it from its natural location, either killing a life process, as in the case of the trees which must be destroyed in order to provide wood, or interrupting one of nature's slower processes, as in the case of iron, stone, or marble torn out of the womb of the earth. This element of violation and violence is present in all fabrication, and *homo faber*, the creator of the human

[1] Georges Sorel sought to discover the conditions that enhance productivity. His book, *Reflections on Violence* is classified by libraries within the literature of work and the labor struggle. In it Sorel maintains that the conditions of productivity are identical with those exhibited in a general strike and thus render violence both necessary and inevitable. Thus, he argues, work is inextricably associated with violence historically, if not logically. (Georges Sorel, *Reflections on Violence* (1908; repr., Mineola, NY: Dover Publications, 2004).)

[2] John Locke, "Of Property," in *The Second Treatise of Civil Government*, Great Books in Philosophy (1690; repr., Amherst, New York: Prometheus Books, 1986), sec. 35.

H. Hein (✉)
Women's Studies Research Center, Brandeis University, Waltham, MA 02454-9110, USA
e-mail: hilde@brandeis.edu

N. Bar-Am, S. Gattei (eds.), *Encouraging Openness*, Boston Studies in the Philosophy and History of Science 325, DOI 10.1007/978-3-319-57669-5_32

artifice, has always been a destroyer of nature."[3] She points out, in addition that: "Since his (man's) productivity was seen in the image of a Creator-God, so that where God creates *ex nihilo*, man creates out of given substance. Human productivity was by definition bound to result in a Promethean revolt because it could erect a man-made world only after destroying part of God-created nature."[4]

The very concept of work is thus linked to rupture and the notion of mastery; for it is only through work that man transforms and domesticates nature, coercing her (*sic*) to serve his ends and bending her to his will. This is a classical representation of power relationships: but women's work does not adhere to this pattern. As a female experience exhibiting power does not entail violence or domination. It is not essentially a coercion of one by another; nor does it imply the subduing of nature. Convention depicts nature as feminine as "Mother Nature"; hence women are taken to be "akin to nature" and nature is archetypically female. This attribution has shaped women's social history and individual experience. Since she is "in attunement" with, rather than transcendent of nature, it is unremarkable that her most characteristic labor should be perceived not as violating, but as conforming with nature. Just as nature – "mother nature" - is addressed as female, yielding the fruit of her womb, so the labor of women in childbirth is taken to conform to her fundamental nature. While men perceive their own relation to nature as requiring their forcible extraction of products; by digging, unlocking, unveiling, detaching, isolating, and other violent exertion, women are pacifically implicated in the creative unfolding of the product itself. It is an amplification of their own substance. From men's external vantage point women appear to be a receptacle or a hospitable and temporary host. But women perceive themselves to be actively gestating and guiding a product of their own creativity to fruition. Moreover, it is not a woman's selfhood that is realized through this experience (albeit that is enhanced), but rather a newborn child which moves on to its own free and independent perfection. The familiar metaphors applied to male work are inapplicable here. There can be no question of appropriation; for the task is creative. A new life comes to be. Neither is this a masterful execution of intellect or will. It is a wholly integrated commitment employing all of the woman's faculties and energy to the accomplishment of an end that is consuming and transfiguring. Far from being self-aggrandizing the process is assimilative and adapts the agent to a historically human as well as natural community. According to its most ardent eulogists, that is the condition to which all work aspires. But since women's agency has never been taken seriously as work, her singular activity – albeit designated "labor" – has not been part of a conceptual model for a philosophical understanding of work.

The question, "What is work?" has no single answer since every response is framed in temporal, social, technological or economic terms. Even were that not the case, exclusion of women's experience of work would be an unpardonable oversight in an investigation that purports to be of universal scope. But, before elaborating a

[3] Hannah Arendt, *The Human Condition*, 2nd ed. (1958; repr., Chicago: University of Chicago Press, 1998), chap. 19.

[4] Ibid.

potential women's conceptualization of work let us reflect upon the distressingly partial treatment of work that tradition has preserved. Probably the most celebrated and possibly most elaborate discussion of work is by Karl Marx, albeit it is obscured both by the systematic ambiguity of the concept that he offers and, consequently, by the ambivalence he displays toward it.[5]

Marx's intention was not primarily to clarify the meaning of the concept of work philosophically; rather, he aspired to activate a dispossessed class of people and to engage them in a struggle for liberation from exploitation. In its unalienated form, work is taken to be an exchange by and among human beings, individually and as a species with the world about them. Work is thus a mode of self-expression.[6] It is a way of being in the world through interaction with it, or rather, becoming by way of acting in the world. Thus, it is also a mode of self-creation. Some authors have chosen to emphasize the impression imposed upon the world, its transformation and coercion to conform to human purposes (cf. Locke, above.) Others have concentrated upon the reflexive effect that work has upon the agent. In the appropriation of nature, mankind (*sic*) generates its own human nature. The "humanizing" of the world is presumed to take place as the raw materials of nature are trans-substantiated into cultural artifacts. The same process raises man from sheer animal existence to true humanity. Work is thus not merely practically rewarding, but also definitive of human identity. So depicted, human identity must be understood as male identity since woman's work is not transcendent of nature, but is instead an expression of it. Women do not humanize the world through procreation; nor are they raised thereby to a personally aggrandized state. However they do fulfill themselves as women through motherhood, since without it they are simply incomplete. Their performance of non-reproductive work is irrelevant.

The notion that work is expressive of humanity persists with reference to the creative arts. Art objects are designated "works" that reflect the free agency or genius of their creator. Paradoxically, we thereby simultaneously dignify and malign both the objects produced and their authors since we also designate people by reference to their (unfree) professional commitment or paid employment. This reference to payment implicitly reflects the motivation of the effort – "for services rendered"- while it underscores the impact that the endeavor, whatever its products, has upon our imagination. Work is most commonly understood in its alienated form. Far from expressing our nature, work is not what we want to do, but what we are obliged to do for the sake of survival. Only the lure of external reward persuades us to pursue it. In this sense work is intrinsically undesirable and displeasing. It is not self-fulfilling except indirectly since it provides us with the means to obtain whatever will gratify us, including survival. Engaging in work is thus fractionating, for only rarely does it represent our unique being or real interest. It is mostly instrumental, and, to the extent that the objectives we strive for are not our own, we are not merely

---

[5] K. Marx, *Capital: The Process of Capitalist Production*, ed. F. Engels, trans. S. Moore, E. B. Aveling, and E. Untermann, vol. 1, *Capital: A Critique of Political Economy* (1897; repr., C. H. Kerr, 1906), pt. III.

[6] Ibid., vol. 1, chap. 7.

cut off from ourselves, but in fact subordinated to the will and interests of others. Hence work is a contradiction that normally dehumanizes the worker even as it "humanizes" the world.

Much of Marx's early writing analyzes the etiology of alienated labor and its attendant social consequences. He blames capitalism and industrialism and urges a proletarian revolution that would end the need to turn ones labor into a saleable commodity. The abandonment of property would, he argues, permit work to become a gratifying mode of self-expression instead of a mere means of survival. Until that happens, productivity must continue to be a Promethean struggle for mastery over nature, a challenge to the gods.[7] As such, work is not to be altogether eliminated, but only the work that has been degraded, shriveling body and soul, distortive of human nature and disruptive of human community. In its "good" sense, work is to be reclaimed. Marx was neither the first nor the last to distinguish positive and negative elements of work and to see these as differently manifested by distinct segments of society. Marx, however, outlined a program to overcome those differences or at least equalize their effect through radical political and economic change.

Hannah Arendt maintained that Greek society also distinguished between a "good" and a "bad" form of work. She designated only the former as "work," while the latter was identified as "labor", albeit this terminological distinction is employed irregularly.[8] And, as Arendt admits, modern work has taken on many of the features of labor so the categories have become barely distinguishable. The initial difference is metaphysical, arising out of the conception of what it means to be truly human. Man – and the term must be understood in both its gendered and generic sense – is properly himself only when he acts as a free member of the polis. He transcends his animal nature by separating himself from the realm of private necessity. This was accomplished in ancient Greece through the institution of slavery and the physical arrangement of the Greek household. Women and slaves were consigned to the private sphere of "labor," where they occupied themselves with the necessities of consumption and survival. Greek men, i.e. those who qualified as Athenian citizens, were occupied with the intellectual and cultural life of the city. It is noteworthy that the public/private division cannot be drawn here in terms of physical and mental (or spiritual) dimensions although work designations often are expressed in those terms. But cultural life is not exclusively mental, and surely was not invariably an indoor activity among the Greeks. Conversely, domestic management and the provision of the necessities for survival required considerable mental skill (as it does today.) More to the point might be the distinction between production of consumables for immediate use, commonly carried out in the domestic sphere and those

[7]According to Harry Braverman, the labor processes and modes of production fostered by industrialism remain the same in socialist as in capitalist systems. Therefore, he argues, it is industrialism as such, rather than the issue of the ownership of the means of production that must be analyzed in order to get at the roots of alienation through work. Merely improving the conditions of work will not suffice. (Harry Braverman, *Labor and Monopoly Capital: The Degradation of Work in the Twentieth Century*, 25th anniversary edition (New York: Monthly Review Press, 1998.)

[8]Arendt, *The Human Condition*, pt. III.

occupations that address society more broadly. Plato's laddered achievement of immortality explained in the *Symposium* is relevant here.[9] He affirms that the generation of biological offspring is an inferior form of self-perpetuation, while the creation of theory, political systems, or art aspires to the immortality of the eternal and immutable Ideas.

Significantly, slaves and women are excluded from the public realm; the former because they have revealed themselves explicitly as less than human through their choice of life itself, even as mere survival in captivity, over the opportunity to meet a glorious and heroic death. Women are predestined by their biology to the procreative and nurturant role. And, while Plato makes some concession to selected women as "guardians" in his ideal republic,[10] he presumes that even the best of them will always be inferior to the best of their male counterparts. Effectively, he is making a statistical observation that a few women will excel over the worst of men, and their functional service should not be wasted. (History has looked too generously upon Plato's alleged egalitarianism. He is no forerunner to feminism.) Be that as it may, the ancient separation of private and public spheres, and its relegation of women to the former, differs significantly from the similarly designated split that emerged with modern industrialism.

Industrialism rewarded distinct types of work differently, albeit the result of the work performed in the two realms might not turn out to be qualitatively different. Since the industrial revolution, private life has been associated with the home and with intimate personal "domestic" relations while public life is the domain of production and institutionalized commercial and political exchange. Public life is governed by contract and law; private life by sympathy and affection. In the public sphere work is contractually engaged and remunerated. Indeed, the fact of payment is what renders work "productive", since there is not invariably a concrete or tangible product. Also, only their volume distinguishes some factory productions destined for commerce from those individually made in and for consumption in the home. It is claimed that things are fabricated at home for their use value, while those made in the public sphere are commodities for exchange (but they are no less "used.") The latter may therefore be regulated and subject to "quality control." It does not follow that they are of superior quality but only that their production is standardized. Must it follow that domestic production is haphazard? Certainly not. Work in the private sphere demands as much, if not more attention and expenditure of energy as work in the public sphere; but it is unregulated and defined chiefly by the values of its performers and designated beneficiaries; hence it no longer "counts" as work. Alienated work is prescribed for and in the interests of generalized others and so demands rewards distinct from its own fulfillment. Self-assigned tasks, reimbursed only through the satisfaction of their achievement do not qualify as work according to this definition. And this designation is irrelevant to the actual degree of misery or alienation that confinement within the private sphere might impose.

[9] Plato, "The Symposium," in *Complete Works*, ed. John M. Cooper and D. S. Hutchinson, trans. A. Nehamas and P. Woodruff (Indianapolis, Ind: Hackett Pub, 1997).

[10] Plato, *The Republic*, trans. H. D. P Lee and M. S Lane (London: Penguin, 2007), 451–57.

Only that which is explicitly prescribed by another, appropriated by another, and paid for by another is work. Hence the housewife who is presumed to be motivated by love, and who cares for and provides for the wellbeing of her family without compensation, albeit she works her fingers to the bone – is not "working". It is irrelevant that the outcome of her exertion exhibits the same continuity of impulse, design and disposition that industrially programmed labor is meant to display. Technical progress in the latter domain actually dictates the multiplication of simplified procedures to be carried out by different agencies, often in disparate locations, so that a radical separation exists between the planning and the execution of a task, and the worker may be totally ignorant of the ultimate purpose of the object she or he is producing. Given these circumstances, workers rarely contribute constructive criticism to its fabrication. The housewife, who oversees her project from inception to end, is able to adapt and modify it en route and to introduce inspired improvements or accommodations as contingencies such as material scarcity or surplus arise. This flexibility endows her work with originality and personality, but interferes with the standardized uniformity and replicability that the logic of industrialism demands.

Thus neither effort expended nor the quality of its product define work, but only the systemic conditions under which it takes place.[11] Work can be done in the home, for example on a consignment basis where it is not integrated into the domestic life of the producer but is deliberately encapsulated. The worker is paid per unit produced or hours spent-and these hours are set apart, uncontaminated by mundane household occupation. When one "works", one only works. The experience of satisfaction or dissatisfaction in ones work is also irrelevant to its identification. The fact that a housewife does not literally choose her occupation (albeit she may have married voluntarily) is not sufficient to establish her activity as work even if she dislikes it. Sheer odiousness does not qualify as definitive. There is a puritanical bias against identifying any agreeable activity as work, but misery is neither a necessary nor a sufficient condition: Neither would pleasure in repetitive factory operations suffice to negate their identification as work. The definition of work is independent of the agent's subjective state. It is the fortunate worker who is gratified both by payment and by the actual performance of the task performed, but the presumption is all too common that to count as "work" an activity must be unpleasant.

The displeasure associated with work is complex. It is not exclusively due to alienation, although that is a signature element. There is also a dimension whose origin is biblical: "In the sweat of thy brow shalt thou earn thy bread," God says to Adam, as if banishment from the Garden of Eden were not a sufficient punishment of his disobedience.[12] Work is understood as a curse placed upon mankind. The parallel travail for women is that they are, henceforth, to bring forth children in pain and suffering. The word "labor" remains in use in the English language to refer to at least a portion of this process. More hopefully, work is sometimes understood as

---

[11] They may, however affect the price of its purchase.

[12] Gen. 3:19.

a road to salvation. "By their works ye shall know them." Sometimes work is an instrument of atonement or repayment of an obligation, but even where thus gratifying, it is not in itself an admissible pleasure. In fact, there is something suspect about anyone's claiming direct gratification from work. Surely it must at least be a form of escape or compensation for a prior offense.

Perhaps this is why the creation of "works of art" is often not taken seriously as "real work". Ancient Greeks had no such reservations about aesthetic enjoyment, nor about hedonism in general. Indeed, their confinement of "workers" to slavery in the non-political realm of necessity enabled free citizens to indulge pleasurably and guiltlessly in studious and aesthetic contemplation. The modern western tendency to contrast work with play and leisure derives from their historically double root. From the Greek tradition comes contempt for work and exaltation of leisure; from the Christian source, the condemnation of frivolous non-utilitarian preoccupation and a glorification of work. While they disagree on the specification of what actually is work and what qualifies as play, they are fully in accord that the two are properly opposed to one another. One effect of this polarization is a further dichotomy of social classes and their respective roles. Certain enjoyments are reserved for only the privileged, but even the inferior class is entitled to its own form of gratification – now perceived largely as recreational relief. Contemporary life reduces all work to "earning one's living", thereby standardizing and sterilizing it. The same reasoning designates diminished compensatory models of recreation and relaxation reminiscent of play only in their presumption to gratify a vacuous spirit with aimless experiences and sensations.

Historic analyses of play, ranging from Kant[13] and Schiller[14] to Huizinga[15] and Ducasse,[16] regard play as stylized or rule-governed behavior which is endo-or autotelic, including its purpose within itself. It is thus distinguishable from purely aimless activity and also from work oriented toward a goal external to itself. Work is instrumental, while the value of play is presumed to be intrinsic. In fact, however, most defenders of play have attributed a utilitarian function to it, whether as a rehearsal for life experience and obligations (learning spatial estimation, preparing for motherhood) a release of excess energy, a safely contained psychic re-experience that enables mastery of threatening situations, or as training of one's mental and physical faculties. So qualified, play becomes almost indistinguishable from work, except for the permitted presence of pleasure, which renders it slightly suspect. This representation actually underestimates the seriousness of engagement in play as activity that involves intense commitment and fully integrated personal investment.

---

[13] Immanuel Kant, *Critique of Judgment*, trans. J.H. Bernard, Dover Philosophical Classics (1790; repr., Dover Publications, 2012), sec. 54.

[14] Friedrich Schiller, *Letters Upon the Aesthetic Education of Man* (1794; repr., Whitefish, MT.: Kessinger, 2004), chap. 14.

[15] Johan Huizinga, *Homo Ludens: A Study of the Play-Element in Culture*, 2014. Beacon Press (1971).

[16] Curt John Ducasse, The *Philosophy* of Art, (NY: Lincoln, MacVeagh, Dial Press, 1929).

Work is likewise degraded by this dichotomized representation. If its primary identification is the external purpose for which it is performed, then anything undertaken for some reward other than the endeavor itself (e.g., grades, money, acceptance, love) becomes work. Effectively all work is bartering one's skills or parts of oneself (a form of prostitution) for some other benefit. Even those activities previously performed for their own satisfaction (playfully?) are transformed by remuneration into work. Informal pursuits such as recreational athletics, art creation, and the provision of counsel or companionship are merging with their professional counterparts. In the confusion, the actual significance of work is lost: It becomes increasingly alienated and diminishing of its performer, while play is exalted as a counter-balancing and therapeutic distraction. Eventually, play becomes a tool exploitatively used to extract work and stimulate consumption which, in turn, generates more work. Work and play thus become parasitic antagonists, locked into a relation of mutual dependency and discord. Depicted deceptively as a foil to work in its non-utilitarian spontaneity and hedonism, play is denied the moral approbation awarded to work, sometimes commensurately with its onerousness.

Work is presumed virtuous, not merely as a means to discipline the spirit, but inherently, much as endurance of pain and suffering is admired as moral fortitude. As such, work is more frequently recommended to other people than sought for oneself. But there is a peculiar form of moral self-righteousness that people exhibit relative to their own work. This is manifested by the student or working parent who requires that all noisy diversions be halted in the presence of the exalted laborer. As work becomes virtuous, virtue, in turn, becomes a kind of work, rigorous and contrary to inclination – elected for the sake of the reward that lies beyond it. The worker enjoys the prestige of good citizenship. The community applauds him and despises the derelict and the unemployed. Having a job, being a steady and reliable worker, providing for one's family – marks a person as a good and substantial member of society. And society rewards such persons through the rhetoric of approval. More accurately, it depicts them in positive contrast with the unfortunates found unemployable, willfully inactive, or who are denied access to work. These poor souls – the sick, retarded, and variously "defective" – are cut off from the dignity and status associated with meaningful work. We have shifted 180 degrees from the ancient Greek exaltation of leisure and its contemptuous relegation of work to those goaded by necessity. In today's world, work is a privilege awarded to those who conform to social norms and withheld from those who deviate from them.

Classical literature sometimes depicted work as uplifting, while later history recorded it as brutalizing and degrading. In the past, social critics protesting the dehumanization of labor often pointed to the animal-like conditions in which workers were forced to live and work. Humanitarian legislation inspired by the dangerous conditions of the early phases of industrialism refers in graphic detail especially to the treatment of women and children who worked as expendable beasts of burden in mines and factories. The bare subsistence level of their maintenance did, in fact, reduce people to the status of animals, but its critics rarely denounced work as such – only its occurrent oppression and exploitation. Much of that labor was subsequently relegated to machines, which accomplished more rapidly, precisely and inexpensively the same movements that had been performed by human arms, legs, backs and fingers. That displacement of humans (and animals) by inanimate devices persists today, but its critics are less disturbed by the debased condition of the workers displaced than by their social redundancy. Work has become rationalized. Conceived as wholly goal oriented, work is abstracted according to the logic of the tasks to be accomplished. Machines no longer mimic the actual behavior of the humans they replace. Their task is end oriented – structured to produce an objective in the most efficient and expedient manner. Humans thus become accessories to and appendages of the machines they "service". The workers, like the machines, are expendable and replaceable elements.

This collapse of agents and instruments is not confined to those centers where concrete material objects are produced. The logic of the assembly line is as readily applicable to abstract occupations and personal services as it is to the production of cars, pre-fabricated houses, and cans of tuna fish. It is a matter of operationalizing a task in terms of the objectives to be achieved and the most logically expeditious methods and devices for achieving them. The model is universal, for logic has no dialects. It can build cars, bombs and houses, educate children, run a farm, make your marriage work, make your money work for you, transform you into a more assertive person, and also rationalize civic systems, organize space excursions, and fight wars at a distance. Human beings become non-responsible components of systems, filling the interstices between hardware and software. The dehumanization that some critics anticipate is deeper than that which formerly addressed a reversion to bestiality. It does not fear neglect of the "higher" faculties, but rather their deliberate mutation to suit systemic needs. Neither is it simply economic.

The industry of shaping people is carried out by schools, therapeutic institutions, religious organizations, laws, and through advertising and the media. It is propagated by private organizations and governmental agencies, by courts and legislatures. Even the private circle of the family is penetrated via fashions in childcare, discipline, nutrition, hygiene and personal grooming. The family is the funnel through which society refurbishes and replaces itself.

Initial doubts about the advantages of automation pertained to the displacement of workers, and indeed, it has produced a surplus of the unemployed. But automated performance of routine repetitive tasks is widely applauded. Many, including Marxists, celebrate science and technology as liberation of the "higher" human faculties. They presumed that, freed of the tedium of the work that machines can perform, men would be emancipated to follow the "higher" pursuits of their choice. Everyone would choose to be creative artists, gardeners, cooks, and the like, and rejoice in fulfilling one's own calling. The actual consequence is that necessity becomes "need", and expands from want into obsession such that satisfaction is never achieved. Indeed, dissatisfaction is stimulated by advertising. The demand for consumption increases and more things, including "experiences" become objects for consumption. The manipulated craving is inexhaustible albeit the world's resources are not. Yet, the insatiable human appetite is continuously stimulated and continuously unappeased. Through our induced expectations of "normal" living we face the endless dissatisfaction of Tantalus.

In its early phase mechanization was confined to replicating the current procedures of human physical labor. The work of a single human arm could be multiplied by a conjoined series of contrived "arms" performing the identical task. There followed an articulation of labor that was not imitative of human agency, but calculated from its destined endpoint through sequenced steps that machines could accomplish according to a different logic. Humans became mere enablers and attachments to the machines they "serviced." Work was commodified as fragmentation following this "rationalized" division of labor entered into every facet of human experience.[17] Alienation is not a disease of the working class alone; it became an epidemic that defies class divisions. It affects rich and poor, and with the irony of the Midas touch, turns the sources of happiness and pleasure into instrumentalities of misery. The failure of "success" to bring about happiness has turned the classic formulas of work, virtue and progress into hollow mockery.

From a "worker's" perspective, the distress of employers does not merit much sympathy. They consume in vast quantity, but "like the lilies of the field, they toil not". However, the concept of productivity, like that of property, has been expanded and dematerialized. To be a productive member of society, one does not necessarily bring forth concrete objects. On the contrary, those who literally do so remain at the bottom of the social hierarchy. And, ironically, the most obvious instance of production, i.e., giving birth and raising children, is not considered to be productive labor at all. To "count" as productive, a function or service must fit within a specified system and contribute to the flourishing of that system. The odd expression "gross national product" illustrates this point. It has nothing whatsoever to do with things produced, but rather with monetary exchange. The greater the circulation of currency and credit, the greater the GNP; but anything one makes or does gratis or barters for materials or goods or services does not contribute to the GNP and therefore has no official existence either as work or as productivity. This, once again, is why the labor of the housewife is not perceived as work. She contributes to the efficiency of the system, not merely by preparing a man to function in it and by "bringing forth" additional workers, but also by assuming responsibility for consumption, a role of no small importance that has been increasingly given over to women and now to children. But these contributions are indirect, and their performance is informal and not directly regulated.

Productivity is not defined in terms of the literal output of labor, which may be non-existent, nor in terms of effort expended. Identical activities are considered work in some contexts and as "unproductive" amusement in others. The operative definition of productivity is derived in terms of the agent's involvement in the system of goods, services and functions that constitutes society. That entails the subjection of one's activity to scrutiny and standards determined within the system and hence a sacrifice of individual autonomy. Paradoxically, the logic of this definition implies that creative individuality is counter-productive and that a degree of passivity is a condition of productivity. What began with a definition of work as activity and self-expression (if not self-actualizing) turns out to be irreconcilable with a

[17]The practice persists today in the subdivision of intellectual and social operations.

socially sanctioned notion of work. If productivity is what keeps the social system going, then those most invested and committed to it must be the most productive. These tend to be the high level executives and corporate managers rather than the workers at the bottom of the scale. Thus, no matter how many "units" a worker produces, he is only marginally "productive". An artist is not productive until she or he sells a "work of art", and the subsistence farmer who feeds only his family has no product to sell – and so is not a "worker." Among the most "productive" lines of work are some that are almost entirely abstract – investment in "futures", image making, management consulting, to mention a few, are all well remunerated professions. They augment the gross national product even as they do not produce bread or shoes or fertilizer, but operate upon the economy in terms of the masses of people and services distributed. Theologians of form, the practitioners of these abstract vocations are removed from the small parishes that comprise the system that encompasses them all.

I do not mean to minimize the genuineness of abstract work. I am among those who perform it and fully appreciate its satisfactions. My doubt is that such labor entails a conquest, or that it necessarily takes place at the expense of something or someone else. To work is evidently not simply to be active or to do something. It involves a deliberate project and is irreducible to sheer exertion. Thus activities taken to be "purely natural" or spontaneous are frequently not classified as work.[18] Sometimes they are designated "play". The production of art except under conditions of negotiation for payment or patronage falls into this dubious category although any artist will surely deny that art is simply a spontaneous emission. And this is why women's production of children and subsequent nurturance of them is also not classified as work albeit women are defined in light of their fulfillment of the sacred profession of motherhood. At the same time, that achievement is hardly self-actualizing; indeed it may be self-annihilating. Where women's reproductive capacity is considered a natural resource to be utilized as social property, it is clear that notwithstanding the skills and commitment the woman brings to that occupation, it will not be seen as her "work" (although it is her mission.)

Alienated workers, slaves and women are thus similarly cut off from both the experience of work in its honorific sense and the respect which such productivity evokes. This is the result of denial and violation. As Marx describes him, the worker's exclusion from ownership of the means of production forces him to sell his labor as a commodity in the marketplace.[19] He is therefore cut off from the fruit of his own effort and must exchange his means of self-expression for bare survival. Since the worker is competing with others who are likewise dispossessed, he is also denied the companionship of those with whom he has most in common. Alienated from both himself and his fellows, he is degraded to the condition of an instrument. Woman's situation is analogous. Denied access not only to the means of production,

[18] The terminology of "worker bees" and "slave" insects is metaphorical, used to characterize what appears (to humans) to be a servile interaction among organisms whose motivation we judge only by analogy to our own.

[19] Marx, *Capital: The Process of Capitalist Production*, vol. 1, chap. 6.

but also to the market where labor is contracted, she has little choice but to sell that function which alone is socially granted to her – reproductivity. Society maintains women through the institution of marriage, relying on individual husbands to share their earnings in exchange for sexual service and the wife's ability to bear, raise and nurture a family. This blunt (now somewhat obsolete) description of what takes place fails to clarify how women's situation differs from that of conventional (i.e. men's) alienated labor, a burden that some women must also bear in addition to that dictated by their gender.

A man, compelled to sell his labor, may protest and rebel against the indignity to his self. Having been "reduced" to a state of servitude, he is morally entitled to self-righteous hatred of his "oppressors". Whether or not he is successful in "throwing off his chains", he is encouraged to believe that human (i.e., male) self-realization lies in the struggle to transcend such limitations and to overcome both nature and social oppression. Women have no such recourse. They are consigned to be nature, not to overcome it. A woman's self-denial is not due to a misfortune of circumstance, but is a metaphysical necessity. Her predestined "work" is not to rebel against or transcend nature, but to fulfill it. Hence her work is not "work" as socially defined; but is obedience to the prescription of nature. As such, it is a fulfillment of her womanhood, but not of her self. To dispute that destiny would be to "denature" herself. The right of revolution simply has no place in this context. A growing number of women have chosen to "violate nature" and to claim the route of male transcendence as their individual right. However, such personal rebellion fails to challenge the conventionally contrived definition of work and of woman's exclusion relative to it. It simply demonstrates frustration and proclaims the capacity to "act like men."

Feminists have long urged women to take charge of their own lives and work. But if that exhortation signifies simply to act as men do, projecting oneself into the world by way of deliberate and intentional action, the consequences would be (and are) disappointing, if not disastrous. Some proponents of women's "liberation" advocate equalizing the conditions of labor for men and women, citing women's historic exclusion from many professions as the ground of women's oppression. They also look to the advance of technology, the availability of child care, household and other domestic conveniences as critical to women's equality. Women are claimed to have been denied access to "productive" work for two major reasons, their physical weakness and the necessity of their primary commitment to their family role. But neither reason is persuasive. Women's relatively smaller stature has never prevented their actual work capability; moreover they are frequently better qualified to perform delicate tasks than men. Rhetoric notwithstanding, women workers the world over have invariably made up in endurance, resourcefulness and industry whatever deficiencies they might actually have in physical strength. As labor has become increasingly mechanized and automated, it is anyway becoming independent of brawn. Relatively few tasks demand physical capacities that exceed those of ordinary women. On the contrary, their more efficient insulation and lower center of gravity renders women structurally more fitted for the performance of certain types of activity than men are. In addition, women's increasing athletic participation, adequate nutrition, and less restrictive dress fashions permit them to become increasingly powerful and agile physical specimens, in no way inferior to their male counterparts.

The one remaining task to which they are biologically fitted to the exclusion of men is childbearing, but this does not justify denial of women's productive role in the work force. Neither does it entail women's exclusive responsibility for childrearing and family nurturance. Apart from the actual biological event of birthing children and lactating, women are no more biologically rooted in family life than men, who surely have a stake in family wellbeing. Families are a social institution, and their prolonged intimate cohesion is of comparatively recent origin. Some form of benevolent custody of dependent infants is obviously necessary, but it is not self-evident that the caretakers must be identical with the biological mothers or that they must be women at all. Maternity may require periodic absences from other forms of productive work, but not more so than common male occupations such as military service. Effective use of birth control renders such periodic absences more or less predictable and can coordinate them with work schedules. There is, therefore, no inherent reason for the almost universal practice of holding women solely and exclusively responsible for prolonged childcare. Practicality and convention no longer dictate a lifetime's commitment to family rearing. With longer life expectancy for adults, smaller family units, and decreased infant mortality, most women can anticipate full and functional work-lives. Relatively minor changes in the customary modes of employment including minimal adjustment on the part of men would be sufficient to accomplish this.

But if change is to be more than nominal, the relationship between production and reproduction must be re-examined as well as that of women's experience relative to both. Women's mere presence in a conventional workforce will not suffice to equalize the status of men and women nor enhance the lives of either. Equal pay for equal work is surely a desideratum, particularly if equal work were a reality, but it would not, alone, rectify the disequilibrium that centuries of separate and unequal work distribution have left behind. Apart from the difference in their compensation, women who are employed in "productive" work have a different experience than that of equivalently employed men; and this shapes their understanding of what it is to work at all. Men are educated to regard work as, if not personally fulfilling, then at least as central to their role as provider. They must accept its monopoly over their time and energy as legitimate. Activities outside of work may be personally more satisfying and rewarding, but they constitute "leisure" occupation – hobbies and avocations. Even though home and family may be the professed object for the sake of which a man works, his return to it at the end of the day is a divestment of his "serious" self. He leaves his "public" identity behind, and retreats to the "private" world of family life. In consequence, he learns to fragment himself and to rank his own activities as hierarchical. He must "play off" the segments of his life against one another. If work is fatiguing and frustrating, there is the expected comfort of home; if family is disappointing, escape into work is a socially approved alternative. A man's total submersion in and dedication to work is commonly applauded, far outweighing the charge of neglect that his family might level against him. Yet, release from the pressure of work in the form of domestic indulgence is also acceptable.[20] Indeed, such comforts are owed to him in acknowledgement both of success, should he be a "winner" and as compensation for defeat should he be a "loser".

A woman who enters the work-force rarely has similar options. Regardless of its deliberate choice, work is not a permissible mode of self-fulfillment, but is a surrogate that sometimes demands apology. Work is not perceived as an expression of her "real" nature; neither is it a legitimate deflection of the energy that should normally be expended upon family responsibility. Unlike men, who are expected to divert their libidinal energy into work, women are not permitted to sublimate the primitive impulses requisite to the propagation and survival of the human species. Were they to find complete satisfaction and fulfillment in work, they would cease to be "women". Working women must either denature themselves to become "career women" – or assume an additional role and juggle their dual lives accordingly. Above all, the prevailing ideology prohibits women from rendering themselves inaccessible or allowing their work to interfere with their accessibility to men. Thus, the choice of work, however confidently made – and even where it is made (as it frequently is) as a matter of necessity, is perceived and often subjectively experienced as a partial abdication. The problem of reconciling roles arises for women in a manner that rarely confronts men. For most women – even those vocationally inspired – the call of the species is so deeply indoctrinated that they are incapable of disregarding it.[21] They fret over family obligations. The effect is that women's lives are painfully and unremittingly subject to superficial interruption and distraction.

Externally, women do not receive the respect that would protect their attention to "serious" work. Interference with household chores is commonplace. It is in itself amorphous, repetitive and interminable, but this shapelessness is amplified by lack of remuneration. Indeed housekeeping is pronounced a "labor of love." Domestic work is continuous; it rarely permits either escape or fulfillment. Consequently, the punctuation that marks men's labor is absent for women. Women's work is continuous and demands constant attention, whether as laborer, wife, or mother. Albeit necessary, payment can neither define nor determine its value. Neither can misery; for, notwithstanding its repetitive drudgery, familial life as the fulcrum of human interaction is the center from which all production radiates and to which it returns. Its burdens are a contingent result of circumstance. Women's work is thus insignificant only by attribution, not in terms of its necessity, difficulty or actual merit. It is miserable to turn out babies and service them like automobiles, but having and rearing children is not a procedure that should be operationalized on the model of the assembly line. Neither do the joys of parenthood justify denying the status of work to it. Bearing and raising children is work that women do. Sometimes it is enjoyable; sometimes not. Yet, it is not a job like tending orchards or mining coal, or waiting on tables, whatever the superficial likenesses one might allege.

---

[20] Recreational indulgence is, likewise, altogether removed from both work and family responsibility.

[21] Many women, upon reaching a "certain age" happily childless, begin to have unexpected yearnings that appear to have biological origins.

The work that women do is not aptly opposed to either play or leisure. It is not unusual that women choose "outside" employment because it affords them the companionship and stimulation that men seek in sports and recreational activities. For many women a "job" is a break from isolation, an opportunity for agreeable human contact, and a necessary source of material compensation. Women tend to have a less articulate concept of leisure than men do because their activities are more integrated and less categorically insulated. A man's weekend pleasure in "puttering" about the house is the woman's ordinary housework. She neither collects payment for doing it, nor follows external orders. She may accomplish it in hours snatched between stints of employed labor. But in her case it is not leisure activity; it is socially required and so is interpolated in her ordinary routine. En route to "work", she mentally cooks the evening meal, and as she bathes the baby, she may resolve a problem carried home from work. Where men have a single primary responsibility to which all else is subordinate, women are taught that whatever needs to be done is their responsibility. This expectation is exported from the private sphere to the workplace and often even to volunteer commitments. Thus women tend to adhere less rigorously than men to exclusively those work obligations to which they are contractually committed. Women are readily persuaded to pick up additional tasks, and do whatever seems necessary to bring a job to completion. This is not because women have a low self-image or are excessively docile, but because the work calls for closure. Work is viewed holistically not because women have achieved a higher level of moral development, but because women's lives normally require the responsible undertaking of amorphous tasks with unknown outcomes and little or no recognition or reward. There is little point in articulating the "duties of my station" for these are transient abstractions irrelevant to the demands of actual life.

The analytic partition of work into divisible tasks that can be completed by different persons or machines in different places is often defended as the rationalization of work that makes possible its efficient and economic achievement. Women's alleged analytic disability prohibits their equal participation in the work force; hence their "retraining" to become "productive" is more than a matter of mastering requisite skills. They must learn to disintegrate activity into separable components and commit undeviatingly to routine procedures. This undoubtedly renders procedures more publicly assessable and replicable. In a mechanical context it is a measure of efficiency. A standardized element of a machine that becomes defective can easily be replaced by another which performs the identical function. That model of labor now prevails in the public consciousness and it extends to human employees as well as to inanimate components. Women emerging from pre-industrial, unsupervised confinement must adapt. However, since efficiency is itself a relative notion that varies with one's goals, it is not self-evident that the rationalized mechanical model is most effective for all purposes (e.g. the provision of medical care and treatment.)

To cite an example, a precept of rationality requires that undivided attention be given to the performance of a single task at one time. Therefore tasks must be ranked in order of priority and successively accomplished under conditions of total concentration. The care of small children, however, entails constant surveillance but does not require undivided attention. Most women habitually attend to various matters simultaneously, shifting their focused concentration as the need arises. One can interrupt a discussion to count stitches of knitting or diaper a baby and return to one's task without losing the thread of discourse. It is neither necessary nor desirable to sit stationary with hands folded while settling important policy matters. In fact, men frequently resort to compulsive, meaningless gestures such as shredding paper, bending paper clips, or paring fingernails in such situations.[22] Yet they ridicule the women who shell peas or peel potatoes or knit while discussing the options of their dependents. Family cares do not disappear during coffee breaks. Men have had the luxury of ignoring everyday complexity because women have assumed that responsibility. This is not a consequence of women's lesser efficiency, but of men's privileged oversimplification. That privilege relies upon the service of subordinates, but the habitual activity patterns exhibited by those subordinates are not esteemed on their own merits, but only as means of service.

An examination of work from a woman's point of view and as a consequence of her historic experience would therefore differ from that derived from our male dominated tradition. Most significantly, work would not be distinguished from other human activity in terms of its positive or negative affective quality. Some work is agreeable; some is not. No work is always pleasant. But neither is any activity invariably enjoyable. Most activities cease to please us if they become obligatory or routinely expected – regardless of payment. Whether one does or does not enjoy one's work has nothing to do with its being work. The presence or absence of remuneration is also not a defining feature of work, although it may be a central factor in determining whether or not to perform it. Most people must work in order to survive. They work for the sake of payment but it is not the payment that renders what they do work. No behavior as such and independently of its consequences would invariably count as work. I have argued that the housewife who cooks dinner for her family is working every bit as hard as the chef who cooks for the clientele of a restaurant. But what of the amateur "gourmet cook" who prepares an elaborate meal to delight and entertain his friends? His expenditure of effort may be enormous, but it is disputable that he is working. A prostitute is surely working for her living, and yet we might disagree that the same is true of the conventional wife whose sexual service is legally awarded to her husband. We rarely consider a sexual encounter as work if it is willingly and enthusiastically embarked upon by persons who are lustful or in love. The presence or absence of pleasure, as of payment, is irrelevant.

Productivity in the sense of creating surplus value is not an identifying feature of work for most women, although many women are now employed in the wage labor force. But all women are initially oriented toward the creation of use value and are likely to regard productivity as simply a means to that end: "The linkage between the capitalist mode of commodity production and the process of the maintenance and reproduction of labor-power (individual consumption) assumes two aspects. To the husband it appears as the wage. To the wife it appears as the world of goods and

[22] Lighting cigarettes and filling pipes were the diversions of choice when smoking was a permitted indulgence.

services, the raw materials for the necessary labor largely performed by her. The sex division of labor in the maintenance and reproduction of labor-power under capitalism thus involves a fundamental mystification. The husband's necessary labor on the job appears to lose all its useful character for him; he sees it only in terms of exchange-value embodied in the wage form that buys the 'means of subsistence'. The wife, conversely, confronts the world of capitalism as if it were composed of pure use-values (including the husband's wage) waiting to be consumed in the useful labor necessary to maintain and reproduce the family."[23]

Whereas in a male context work becomes an abstracted activity whose very rewards are abstract – i.e., money, status, power-, women's work converts even these abstractions back into concrete objects to be utilized for living. Thus a component of women's work is to spend money. As in the past, when women fabricated the domestic necessities of daily life out of natural ingredients, so today women use the money paid in wages to themselves or other family members as the raw material which enables them to provide their families with the same goods and services. Some women enjoy shopping; others hate it – but regardless of their attitude, consumption is a significant component of what society expects of them. The family wage has displaced the natural resources that fed and clothed the work force of the past. However, the responsibility of women to reproduce workers and to nourish and socialize children has not changed fundamentally. Thus, a man's reward for work becomes the instrument of her work to the woman with whom his living is shared. But, since she, too, is frequently a wage earner and is privy to his economic struggle, she cannot ignore the symbolic role that money plays in the world, and so its abstract status must also be familiar to her.

This status cannot be disassociated from the tasks that wealth performs and its effect upon those who possess it. The gulf that men perceive between labor and its means and goals is indistinctly meaningful to women because they remain closer to the transition at every step of the way. Especially those women cut off from the capitalist means of production must continue to produce use value. The less they have to exchange, the more ingenious must their own productivity be. And the more alienated the labor that supports them (either through their own effort or that of another), the greater the demand to expend themselves and their ingenuity to utilize whatever resources are accessible to them. Innovation is a part of their work, while the male conceptualization of work often stops short of the imaginative fulfillment that women's role as nurturer includes. As customarily apprehended, work is standardized objectifying activity that strives for closure in the form of a reward, be it material, moral, abstract or social in nature. Conceived by women in consequence of their historic experience, the process and product of work are not sharply distinguished. As the sexual segregation of work roles and domestic activity diminishes, it may be that men will come to appreciate closer integration of work with productive living that has been the common experience of women. Some people view this trend as a return to a pre-capitalist feudal society, but it does not entail the abandonment of

---

[23] Lise Vogel, "The Earthly Family," ed. Paul Buhle, *Radical America* 7, no. 4–5 (October 1973), http://library.brown.edu/cds/catalog/catalog.php?verb=render&id=1124976284380235&colid=20

modern industrial technology. It will, however, result in considerable modification of labor allocation and social rationality. It will, for example, affect the temporal division of the "work" day and week, and the design of locations where work is performed.[24] Specifically, it might curtail the trend to locate work in designated compounds where people spend prescribed segments of their day (or week), while complexes of an equally institutional nature are maintained for off-hours of leisure and personal interaction. If these central facets of lived experience are to become more closely integrated, their current spatio-temporal segregation is counterproductive.[25] But reversing the separation of work and domesticity will entail many additional adaptations – e.g., of transportation and communication networks, fuel and energy distribution, housing and urban design, and so forth. It is, consequently, a conceptual shift whose realization will have radical practical impact.

But that is precisely the point of considering how a woman's perspective might reflect upon the classic conceptions of philosophy – for philosophy is the basis of our social reality. It should not persist as elevated discourse, dustily littering the shelves of silent libraries. Neither are philosophical ideas simply ad hoc justifications of existing material realities. Ideas shape our experience of reality and must, in turn, be altered by it. Within the prevailing tradition, endorsed by prevailing philosophy, women have been perceived (by themselves as well as by men) as biological perpetuators of the species and thus as a naturally productive resource. As such, they have been mined or cultivated, like bees or silkworms. Their conduct is held to be spontaneous and undeliberate (albeit malleable) and to lack the rational intention that male philosophers consider central to their own work. Yet, it is neither true that biologically based conduct must be mindlessly expressed (as any athlete could testify) nor is it the case that women, free to choose their own lives, would confine themselves to reproduction and nurturant activity. If what women do and think were appreciated seriously as work, some of the confusions and contradictions now associated with work would disappear. The rigid polarization of work and play, work and leisure, work and pleasure, work and waste would end. Productivity would acquire new meaning. And, we might all experience the satisfaction of useful and gratifying work – our own and that of others.

[24] To a great extent this has already happened as workers of all genders work from home, in temporary offices, cafes, at the beach, in their pajamas, and at all hours of the day or night.

[25] Long commuting is not only wasteful of resources, but maintains the separate and unequal family roles, where women are confined in place and fathers rarely see their children.

Chapter 33
The Pitfall of Hypostatization and the Reality of Social Things

Fred Eidlin and Barry Eidlin

33.1 The Problem of Hypostatization

Fred Eidlin died on October 28, 2016, before being able to complete this manuscript. His son, Barry Eidlin, was able to edit the final version for clarity. Hypostatization or reification[1] refers to an abstraction being "treated as if it were a concrete, real event, or physical entity, confusing words and things, treating abstractions or hypothetical constructs as if they were concrete entities existing in the 'real world'. It means confusing concepts and categories used to probe reality with reality itself. Hypostatization has occurred when concepts, frameworks, and theories freeze into supposed representations of reality that reality cannot shake (Engel 1995).

However, the very attempt to avoid hypostatization by denying existence to social things can encourage a tendency to hypostatize the constructs social science imposes on social reality. I will argue that, without a realistic ontology of the social, it is difficult to envision how social reality might "kick back" at the constructs social science seeks to impose upon them and render them problematic.

Fred Eidlin (deceased)

[1]The term "reification" is used much more frequently in the social sciences than the term "hypostatization." The two terms overlap in meaning, however there is an important difference between them. Hypostatization is simply a logical fallacy—treating an abstraction as if it were a real thing. "Reification" means hypostatization, but it also means more. The term has been introduced into the social sciences mainly by scholars inspired by Marx. For Marx and Marxists, "reification" is not only a logical fallacy, it is part of ideology. Workers falsely believe the miserable conditions in which they live to be natural and unchangable. They will therefore not think about trying to change them. When the workers' consciousness is raised, they will realize that their conditions are not natural and unchangable, and will be motivated to change them. In this paper, I am interested only in the logical fallacy of hypostatization, not in reification in the Marxist sense. So I will use the term "hypostatization" unless the Marxist usage of "reification" is intended.

B. Eidlin (✉)
Department of Sociology, McGill University, Montreal, QC, Canada
e-mail: barry.eidlin@mcgill.ca

N. Bar-Am, S. Gattei (eds.), *Encouraging Openness*, Boston Studies in the Philosophy and History of Science 325, DOI 10.1007/978-3-319-57669-5_33

All science attempts to simplify and typify the reality it seeks to represent. In this sense, abstractions and hypothetical constructs are a necessary part of scientific inquiry. However, it is the notion of a reality independent of the constructs science attempts to impose on it that compels scientists to abandon or modify these constructs as inadequate representations. This paper advances a remedial approach to hypostatization. It advocates a practical ontology of the social, and attention to formulating falsifiable hypotheses. I will argue that lack of a clear sense of social reality encourages a seductive psychological disposition to hypostatize. Without a clear sense of what is "out there" in the real world, concepts create worlds of their own without protest from reality (Foley 1937). Improvement of our sense of social reality is the only effective means of combating is hypostatization. The clearer a researcher's sense of social reality, the more falsifiable his/her hypotheses can be, the better reality can kick back and shatter hypotheses the researcher seeks to impose on it, and drive research forward. In the final section of this paper, I will thus explore various sources of orderliness in social reality that can serve as objective constraints on the hypotheses social science seeks to impose on them.

33.2 Stubbornness of the Tendency to Hypostatize

Warnings about hypostatization abound in social science training. Nevertheless, social science literature is awash with hypostatization. Most social scientists are aware of the problem, but few worry much about it or think of it as important to their work. Accusations of 'hypostatization' or "reification" occasionally crop up in social science literature, usually in criticism of the work of others. Someone else is said to be hypostatizing or reifying.

Some hypostatizations are so easy to recognize that even those who have never heard the word "hypostatization" will feel that something is amiss. "Ronald Reagan's best friend is freedom. It did most of the work for him in Geneva. It was on his shoulder when he was walking Mikhail Gorbachev down toward the lake. It was tiptoeing around the room in the Chateau Fleur d'Eau and may even have whispered in Gorbachev's car (*Time magazine*. Dec. 2, 1985).

Hypostatization becomes trickier to deal with where words refer to social things that are widely believed to exist independently of the beliefs of individuals. Such words might include "religion," "democracy," "freedom," "totalitarianism," "political power," or other abstract terms of this kind. Although definitional disputes about such things are common, few question their reality. Social scientists typically hypostatize such terms without the slightest awareness of what they are doing. Some social scientists have no qualms about quantifying terms, the meanings of which are so unclear that one may wonder just what it is that is being measured. They too become victims of hypostatization (Coppedge et al. 2011: 247–267; Munck and Verkuilen 2002: 5–34; Freedom House 2016).

I will attempt to show how deep the roots of hypostatization run, and that hypostatization is far more pervasive in social science than is usually realized. It can be managed to some extent, but is impossible to escape completely.

How we talk about a thing determines how we come to see it, what we come to believe about it, and as a result, how we come to react to it. The reality we live in is largely constructed by the language with which we describe it, and there is a natural psychological tendency to assume that words, at least nouns, refer to real things existing in a real world outside of language. We tend to believe that the concepts used in social science, such as 'political system', 'elites', 'interest groups', 'bureaucracies', and 'capitalism' correspond to real social things existing in a real social world. Such things are commonly believed to be as real as stones, trees, and thunderstorms.

The tendency to hypostatize even extends into our experience of nature. Astronomical constellations like the Big Dipper appear to exist as discrete objects in the real world, but in fact, they are hypostatizations. The stars in constellations may be real, but they have no relationship to each other. Their reality as constellations exists entirely in the human mind. Perception of the Earth as flat, and of the stars and planets revolving around the Earth are also hypostatizations. It took great efforts to convince not only theologians and ordinary people, but also scientists, that the Earth is round, and revolves around the Sun in a double motion. The Round Earth Theory was highly counterintuitive, and there is a strong rational component underlying geocentric and heliocentric perceptions. When Christopher Columbus set sail, the Flat Earth Theory integrated the experience of people living at that time far better than the theory that the Earth was round. The theory of Copernicus and Galileo that the Earth revolved around the Sun also ran counter to the common sense of people living at that time. Everyday experience confirmed the view that the Sun revolved around the Earth. The idea that the Earth made a full revolution on its axis every 24 hours also flew in the face of lived experience.

Before the Scientific Revolution ushered in by Galileo and Newton, all of nature was anthropomorphized, and hypostatized, that is, represented entirely in human terms. Thunderstorms happened because Zeus was angry. The sea was rough because Neptune was angry. Modern science replaced such anthropomorphism and hypostatization of nature with the view that Nature followed her own laws that humans needed to discover. Whatever governed such things as the motions of the Earth, planets, and stars, and the origins of species was independent of human belief. Reality no longer depended on what people thought or said about it. Scientists create hypotheses and theories and test them against a reality assumed to be independent of what people think about it.

33.3 Hypostatization and Social Reality

If the social sciences seek to apply this method of hypothesis formulation and testing, there needs to be some sense of the the reality against which hypotheses and theories can be tested. What kinds of things do social scientists observe and speculate about when they investigate social and political reality? How can social reality be made to refute the hypotheses and theories that social scientists attempt to impose on it?

Natural scientists do not worry about such questions. The objects they investigate are real in the sense that they possess such properties as mass, extension, heat, light, and velocity. Even in their most sophisticated work, natural scientists rarely go beyond common sense about reality. It does not matter to natural scientists if philosophers have difficulties proving that a real world exists. It does not matter to them if philosophers have difficulties in justifying the knowledge claims of science.

Nor is it difficult for us to grasp the reality of subjective states of mind such as those of which poets write – love, fear, anger, etc. We know the reality of such things through direct experience. While it may be difficult to pin down just what they are, we nevertheless have a good sense of what kinds of reality they represent.

But what about the realities social scientists study – languages, bureaucracies, universities, traditions, social systems, elites, and governments? Few doubt the reality of such things any more than they doubt the reality of natural objects or feelings. People go about their daily lives talking about social things as if they existed, rarely pausing to wonder about the natures of the things they are talking about.

Sometimes, common sense is all that is needed for the conduct of social research as well. We can sometimes speak about social objects and processes, without running into difficulty. However, our understanding of social and political reality, that is, our ontologies of social things can be highly problematic, often without our being aware of it. When social scientists study things like governments, markets, or power, what is it that they look at? Whether aware of it or not, they make assumptions about the things they are looking at, and such assumptions shape the decisions they make about the kinds of data appropriate for their research. Sometimes, in the process of trying to operationalize a concept, they discover how confused, how inadequate commonsense ontology can be in the domain of social reality.

Many writers have cautioned against confusing words and concepts with things in the real world. S. I. Hayakawa writes: "The first of the principles governing symbols is this: The symbol is not the thing symbolized; the word is not the thing; the map is not the territory it stands for (1978: 25). Benjamin Lee Whorf draws attention to our tendency to confuse our ways of thinking about the world "out there" with a neutral labeling of natural things. We mistake concepts for the things in themselves (1956:152, 238). Alfred North Whitehead calls this "accidental error of mistaking the abstract for the concrete [...] the Fallacy of Misplaced Concreteness" (1953: 64–74). Quine refers to "object-posting" modes of thought (1959: 155). Nevertheless, despite all cautioning of this kind, and despite widespread awareness, at least in principle of the problem of reification, we continue to mistake concepts about things for the things in themselves (Handler 1984:55).

No amount of such cautioning solves the psychological and practical problem even in the natural sciences. Biases and distortions are inherent in any system of naming, classification and cognitive organization. Any decision to attach a name or classification to a thing implies an important tacit decision (and bias) not to call it or classify it as something else: "terrorist" or "freedom fighter," "occupation" or "liberation," "annexation" or "reunification?" Names, concepts, and classifications are theoretical in character and systematize reality. Once absorbed, they have a kind of autonomy within our systems of thought and considerable influence on it. We gather facts and count as facts only what our pre-existing cognitive apparatus is set to perceive.

Some of the founding fathers of social science were aware of the problem of hypostatization and paid considerable attention to it, yet none was successful in resolving it. Karl Marx was first to identify the problem. According to Marx, a tendency to believe that superstructural elements have a reality and permanence which is actually illusory is, in fact, a component of ideology. As John McMurtry puts it, Marx saw the categories or ideas of ideological thinking as "not only 'emancipated from the world" by virtue of their indeterminateness, but just as importantly are 'transformed from predicates into subjects' from general properties ascribed descriptively to the self-subsistent movers of the world (1976:136). Consciousness raising supposedly exposes such reified categories as the mere ideology that they are. "Where speculation ends – in real life – there real positive science begins: [...] Empty talk about consciousness ceases, and real knowledge has to take its place" (Marx and Engels 1948:15).

Later on, however, other thinkers noticed that Marx's own categories, such as class, the proletariat, and capitalism, turned out to be hypostatizations themselves (See, e.g., Weber 1949:103; Mannheim 2013:67–9, 124,248–9; Hughes 1958:77,81–82, 87–89,93). For example, although it can be shown that the notion of class is incoherent, Marxists continue to speak about classes as if were a real social thing. The same can be said about other key Marxist concepts.

Max Weber had a more sophisticated understanding of the problem of hypostatization than did Marx. His ideal-typical method was a reaction to the tendency in nineteenth-century social thought to treat such entities as nations and classes as irreducible social wholes existing in the real world. The social world, Weber insisted, is infinitely rich in atomic details that are in constant flux. The social entities that characterize so much of nineteenth-century social theory, he insisted, do not exist in the same way trees and stones exist. They result from imposition of the social scientist's perspective upon the infinite richness and constant flux of social reality (in Weber 1949: 111; 1976: 55–80). Ideal types, Weber insisted, are not descriptions of objects existing in the real world. They are, rather, instruments which the social scientist creates for purposes of investigating the social world (Weber 1949:90).

However, Weber's very efforts to avoid hypostatization opened a back door to a new form of hypostatization. Whatever framework or ideal type a researcher adopts, it will act as a searchlight, picking out facts that are relevant it. These facts, selected by the framework, will confirm, put flesh on the framework; in other words, they will hypostatize it. Weber insisted that ideal types are neither hypotheses nor descriptions of objects existing in the real world. But a central principle of modern science is that there is a reality outside of human language and belief systems. We learn about reality by making hypotheses about it and allowing reality to refute them. By refusing to recognize the reality of social things, Weber unwittingly rejected the possibility of falsification, the engine that drives the growth of knowledge. Despite Weber's warnings to the effect that ideal types are only constructs that pick out, indeed exaggerate aspects of social reality, Weber's view supports the tendency to hypostatize them. Surely enough, social scientists influenced by Weber tend to hypostatize ideal types such as bureaucracy, totalitarianism, and types of authority, treat them as if they referred to entities existing in the real world.

Social scientists are told to be sensitive to the fact that ideal types are artificial constructs that pick out, indeed exaggerate certain aspects of social reality. However, the warning that ideal types are merely constructs, that they do not correspond to real things in a real world is apparently not adequate. The disposition to hypostatize is too deeply rooted to simply evaporate upon being exposed.

33.4 Examples from Contemporary Political Science

"Political power" is a good example of a concept that is easily understandable in many different contexts, but often ends up hypostatized. When one attempts to pin down the meaning of political power, it turns out to be elusive. The concept has a wide range of meanings which are not reducible to one single meaning. As long as the concept is allowed to remain vague, the illusion may persist that the different meanings of political power all refer to the same thing. Such vagueness opens the door to hypostatization of the concept.

It is easy to understand what it means to say that the President of the United States or the boss of a political machine is a powerful individual. However, does political power refer to an ability to coerce someone to do what he/she would not otherwise have done, as some have suggested (Dahl 1961)? Does it refer to an ability to shape the scope of decisions to be made as others have suggested (Bachrach and Baratz 1962; Schattschneider 1960)? Or is it an ability to persuade others to do what they should have had enough sense to do without being persuaded, as Richard Neustadt defines it (1990: 10, 28)? Is political power a push-pull relationship like a physical force? If not, how are we to measure it? In the 1950s and 1960s, political scientists unfruitfully invested much effort into finding ways of measuring power. There are many good books and articles about political power, but it remains unclear what one should look at or what to measure in researching it.

Another example of a hypostatization is provided by the highly influential ideal typical concept of "totalitarianism" during the Cold War. This hypostatization long defined the essence of Soviet-type political regimes. It blinded researchers to dynamic processes driving fundamental transformation of such regimes. Such dynamic processes drove the Communist regimes in Czechoslovakia and the Soviet Union towards the Prague Spring and Perestroika. Many advocates of the totalitarian model also argued that totalitarian regimes of the right, such as Mussolini's fascist regime and Hitler's National Socialist regime shared the same essential characteristics. The totalitarian model shaped the research and analysis of Communist regimes for more than a generation and blinded specialists to all kinds of aspects of social reality driving change in Soviet-type regimes.

33.5 Kicking the Furniture of the Social World

Since Kant, it is widely recognized that there is no pre-ordained, non-distorting correspondence between thought and reality. The facts and laws of nature do not contain their own order and meaning which impose themselves upon the human mind. Rather, it is the human mind that tries, with varying degrees of success to impose order and meaning upon nature. "If," as Henry Aiken puts it, "the reason in things is the same as the reason which we acknowledge as the standard of valid thinking about any object, this is only because we ourselves have preordained what conditions any object must meet if it is to be counted by us as 'real'" (1956: 15). We see what we expect to see.

This means that, although preconceptions give rise to distortions in knowledge, it may not be possible to eliminate them. Preconceptions (or prejudices) cannot be separated from knowledge since they are an integral part of it. Nature and society cannot be relied upon to falsify the hypotheses imposed on them—even in the natural sciences. There is never any guarantee of congruence between the structures of nature and society and the structures of meaning and order that we impose on them.

Some scholars have argued that we construct not only the social world but even the world of nature. Even in the natural sciences, they contend, since different theories construct different realities, they are not even about the same reality. To be sure, the histories of the natural sciences abound with examples where the appeal to an "objective reality" has decisively tipped the scales for one theory or hypothesis over another. It was, for example, an appeal to such objective reality that tipped the scales in favor of the heliocentric theory of the universe over the geocentric theory, and in favor of Einstein's theory over Newton's. However, the histories of the natural sciences also abound with examples in which the "facts of nature" were inconclusive, not only on competing scientific theories but even on competing metaphysical and magico-religious explanations.

As we have seen, the social sciences face the additional problem of frequent unclarity about the nature of the furniture of the social world. The "kick test" usually serves well as a definition of reality in the natural sciences. If we can kick it, or it kicks back, it is real. The kick test can often be applied to social reality as well. An entrepreneur who does not believe in the Laws of the Market will nevertheless bump into the reality of The Market. Someone who refuses to believe in the reality of a social taboo will nevertheless be ostracized. The vagueness of much of our ontology of social things can make it difficult to study this kicking process.

In some areas of the social sciences, as Jarvie puts it, "[w]hereas the way the physical world is is not a function of the way we perceive it to be; the way the social world is [...] is heavily influenced by the way we perceive it to be [...] (1981: 223)." The Oedipus Effect provides a good example. If a renowned expert on the stock market predicts how the stock market will move, people acting by these predictions will change the way the stock market moves. That the social world is 'heavily influenced by the way we perceive it to be' is important, and it sometimes adds a dimension of confusion to the objects of study in the social sciences.

How then is it possible to be a sociological realist, one who assumes that social things exist in an objective sense, without falling into the traps of hypostatization, empty, sterile arguments about the meanings of concepts, and impoverished pictures of social reality?

The social sciences must make use of models, metaphors, analogies, constructs and ideal types, in their attempts to organize and make sense of the social world. This dependence on constructs does not differentiate them either from the natural sciences or philosophy. As Popper puts it, "all good science and all good philosophy consist of lucky oversimplification' or if you prefer the term, idealization (1974: 976)." Any model used to interpret reality can easily come to be confused with that reality. Such hypostatization can be found in the natural sciences, as well as the social sciences, and provides part of the explanation for the dogmatism and passionate disagreement running through the history of the natural sciences. After all, whatever theory one adheres to, it will always be possible to find facts that "fit" that theory and, in "fitting" it, "confirm" it. The resulting illusion of "fit" and "confirmation" help explain our natural psychological tendency to forget that the facts were selected in light of the theory in the first place. Every successful fitting of a construct or description to uninterpreted facts creates an amalgam of fact and interpretation.

Psychologically, fact and interpretation fuse. 'Signal' emanating from the world provides ongoing confirmation of a seemingly natural fit of fact and report of fact. We remain unconscious that the facts were selected by preexisting concepts, theories, and interpretive frameworks. We refer casually to "reports of facts" as "facts," as if these reports were faithful replicas of the facts. We see them just as Galileo saw the mathematical formulations of science as faithfully replicating the underlying mathematical structure of nature (Burtt 1954: 74–75), and just as Aristotle's "basic premises" were statements describing the essences of things. Since the fact and its interpretation or meaning have become psychologically fused or reified, we are biased against different constructions of the same facts.

Martin Spencer has even gone so far as to argue that, despite the prevailing nominalism of social science, and despite widespread awareness of the "fallacy of misplaced concreteness," social science cannot avoid hypostatization. Social science, he contends, practices an "operative realism" with respect to its constructs, and "the cognitive conditions of social science are such that a realistic use of constructs is inevitable" (1982: 121). Spencer argues that once a construct has been applied to reality, it organizes that reality both logically and psychologically. Once a particular construct has been used to describe reality, that construct becomes part of reality.

J. O. Wisdom has similarly pointed out that there is a wide variety of schemata underlying social science. It is easy for such schemata to become hypostatized in Spencer's sense. The reality of the social world we see is constructed by the schemata so that psychologically we perceive the schemata as describing different and competing realities. However, Wisdom shows that these different schemata, rather than creating different realities, can easily be seen as complementary, as providing different angles on the same reality. Dogmatic social scientists may try to assimilate differing schemata to the one they believe to be the only correct representation of reality. But, as Wisdom shows, we are psychologically quite capable of seeing the different schemata as complementary angles on the same reality (Wisdom 1980, 1981).

33.6 The Nature of Social Reality

I propose to retain the ideal of formulating hypotheses that presume to describe objectively states of affairs in the social world and are falsifiable; that is testable against reality. Hypotheses should be formulated in such a way as to invite social reality to kick back, to falsify, to shatter them. To get a handle on how this might be done, we need to look more closely at the nature of the social things which will kick back, falsify, shatter hypotheses. What is it that does the kicking? What are the sources of the commonalities, patterns, invariances, recurrent features in the social world? Once this simple question is asked, it is easy to recognize a wide variety of such sources.

Some sources of order in the social world are mechanical, and others fit mechanical imagery without distortion. A typical description and explanation of social phenomena are likely to contain different types of variables, including, among other things, mental phenomena, habits, reasons, and social structures. All can be treated as real. All can be described objectively and can serve as constraints on hypothesizing. The test is, in Bronowski's words is: "Will the concept work? Does it give an unforced unity to the experience of men [sic]? Does the concept make life orderly, not by edict but by fact (1975: 41)?"

Objective facts, selected in the light of preexisting expectations, frameworks, concepts, and theories, are never known to us free of interpretation. Moreover, reality is structured loosely enough so that different and even contradictory accounts of the same facts are common even in the natural sciences.

Attempts to address this problem by "purifying" the language can only lead to its impoverishment. Language is unavoidably biased. Every description is an ordering of reality, and every decision to order reality in one way is simultaneously a tacit decision not to organize it in some other possible way. Without bias, there is no order and without order there is chaos. The attempt to clarify concepts leads ultimately to the confusions, contradictions, vagueness, metaphysical theories, etc., embedded in the culture that produced the concepts. Legislating conceptual clarity and uniformity suffocate inquiry, as illustrated by so much of positivistic social science.

It is not the construction of reality by interpretation or the problem of meaning or the problem of open systems that distinguishes science from ideology. What distinguishes science from ideology is an attitude that statements about reality should be falsifiable. They should invite reality to kick back.

Acknowledgements Fred Eidlin passed away on October 28, 2016. We are fortunate and grateful to be able to publish here his last work. We thank his son, Barry, for helping us bring it to print, and for making sure that the nearly final draft here included was indeed Fred's last.

References

Aiken, Henry D. 1956. *The age of ideology*. New York: New American Library.

Bachrach, Peter, and Morton S. Baratz. 1962. Two Faces of Power. *The American Political Science Review* 56 (4): 947–952.

Berger, P.L., and T. Luckmann. 1966. *The social construction of reality: A treatise in the sociology of knowledge*. Garden City: First Anchor.

Bronowski, Jacob. 1975. *Science and human values*. New York: Harper & Row.

Burtt, E.A. 1954. *The metaphysical foundations of modern science*. Garden City: Doubleday Anchor Books.

Coppedge, Michael, John Gerring, David Altman, Michael Bernhard, Steven Fish, Allen Hicken, Matthew Kroenig, Staffan I. Lindberg, Kelly McMann, Pamela Paxton, Holli A. Semetko, Svend-Erik Skaaning, Jeffrey Staton, and Jan Teorell. 2011. Conceptualizing and measuring democracy: A new approach. *Perspectives on Politics* 9 (June): 247–267.

Dahl, Robert A. 1961. *Who Governs?: Democracy and Power in an American City*. New Haven: Yale University Press.

Engel, S. Morris. 1995. What is the fallacy of hypostatization? *Inquiry: Critical Thinking Across the Disciplines* 14 (4): 42–51.

Foley, John P. 1937. The fallacy of hypostatization. *Journal of General Psychology* 16: 491.

Freedom House. 2016. *Freedom in the word: Anxious dictators, wavering democracies. Global freedom under pressure*. New York: Freedom House. https://freedomhouse.org/sites/default/files/FH_FITW_Report_2016.pdf. Retrieved, 21 August 2016.

Gewirth, Alan. 1973. Can men change the laws of social science? In *Modes of individualism & collectivism*, ed. John O'Neill. London: Heinemann.

Handler, Richard. 1984. On sociocultural discontinuity: Nationalism and cultural objectification in Quebec. *Current Anthropology* 25 (February): 55.

Hayakawa, S.I. 1978. *Language, thought, and action*. 4th ed. New York: Harcourt, Brace & World.

Hughes, Henry Stuart. 1958. *Consciousness and society: The reorientation of European social thought, 1890–1930*. New York: Knopf.

Jarvie, I.C. 1972. *Concepts and society*, 91–127. London: Routledge & Kegan Paul.

———. 1981. Social perception and social change. *Journal for the Theory of Social Behavior* 11 (October): 223–240.

Mannheim, Karl. 2013. *Essays in the sociology of knowledge*. Vol. 5. New York: Routledge.

Marx, Karl, and Friedrich Engels. 1948. *The German ideology*. New York: International Publishers.

McMurtry, John. 1976. *The structure of Marx's world-view*. Princeton: Princeton University Press.

Munck, Gerardo L., and Jay Verkuilen. 2002. Conceptualizing and measuring democracy: Evaluating alternative indices. *Comparative Political Studies* 35 (February): 5–34.

Neustadt, Richard E. 1990. *Presidential power and the modern presidents: The politics of leadership from Roosevelt to Reagan*. New York: The Free Press.

Popper, Karl R. 1974. Replies to my critics. In *The philosophy of Karl Popper*, ed. P.A. Schillp. La Salle: Open Court.

Popper, Karl R., and John C. Eccles. 2012. *The self and its brain: An argument for interactionism*. Berlin: Springer.

Quine, W.V. 1959. Meaning and translation. In *On translation*, ed. Reuben Brower, 148–172. Cambridge: Harvard University Press.

Schattschneider, E.E. 1960. *The Semisovereign People: A Realist's View of Democracy in America*. New York: Holt.

Spencer, Martin. 1982. The ontologies of social science. *Philosophy of the Social Sciences* 12: 121–141.

Whitehead, Alfred North. 1953. *Science and the modern world*, 64–74. Cambridge: Cambridge University Press.

Weber, Max. 1949. *The methodology of the social sciences*, translated and edited by Edward A. Shils and Henry A. Finch, with a foreword by Edward A. Shils. New York: The Free Press of Glencoe.

———. 1976. *Roscher and Knies: The logical problems of historical economics*. New York: The Free Press of Glencoe.

Whorf, Benjamin Lee. 1956. *Language, thought and reality*. New York: Wiley.

Wisdom, John O. 1970. Situational individualism and the emergent group-properties. In *Explanation in the behavioral sciences*, 271–296. Cambridge: Cambridge University Press.

Wisdom, J.O. 1980. Schemata in social science: Part I. *Inquiry* 23: 445–464.

———. 1981. Schmata in social science: Part II. *Inquiry* 24: 1–19.

Chapter 34
On Being a Popperian Economist

Lawrence A. Boland

When I began my economics graduate studies at the University of Illinois in the 1960s, I lived in an apartment in the married graduate student housing. The apartment directly above was occupied by a graduate philosophy student. He particularly did not like my loudly playing music. As a matter of severe revenge for the loud music, he recommended that I take a philosophy seminar from someone named Joseph Agassi. Not realizing that the recommendation was a matter of revenge, I did sign up to the seminar. It was one of the best classes I took in graduate studies.

By the middle of the third year of my graduate studies, I had finished course work and had passed the qualifying exams. Since my qualifying exams were passed, I was supposed to be writing my thesis but nothing was happening. It fortunately happened that I was often visiting Joske in his small office in the philosophy department to discuss what I learned in his seminar. And fortunately, during a Thursday visit, Joske told me I had to sit down and write the thesis, and give him a copy of it on the following Monday. On the following Monday I did drop a 270 page handwritten manuscript on his desk.

This thesis was about so-called capital theory with a little Popper thrown in. After it was typed up by my then wife, I convinced my PhD supervisor to schedule a final exam on the thesis – and so I submitted my thesis to the whole PhD committee. The exam was held but the committee rejected my thesis. Fortunately, one of the members of the committee objected and forced the committee to give me a second chance. The committee said they would let me write another thesis if I focused only on applying Popper's *Logic of Scientific Discovery* (*LSD*) to economic model building.

L.A. Boland (✉)
Simon Fraser University, Burnaby, BC, Canada
e-mail: boland@sfu.ca

N. Bar-Am, S. Gattei (eds.), *Encouraging Openness*, Boston Studies in the Philosophy and History of Science 325, DOI 10.1007/978-3-319-57669-5_34

34.1 Testability in Economic Model Building

In the early 1960s, economic model builders all thought it was necessary to assure their models were testable. At the time I did not know why they all thought this was necessary. I incorrectly presumed it was due to Popper. Actually, it was due to the common critique in the late 1930s of mathematical model building. The critique claimed such mathematical models could only produce tautologies. And for this reason, economic model builders continued to think testability was required merely to prove the critics wrong.

In *LSD* I did find a discussion of the degree of testability which Popper called the dimension of a theory. I decided to apply that to economic model building. Specifically, I decided to make the purpose of my thesis to develop a quantitative measure of testability (I called it the P-dimension) and apply it to numerous economic models. Since model builders all thought all models required testability, just how many observations would it take to satisfy this requirement for any particular model? That is, how many observations would it take to construct a potential refutation of a given model? For example, it would take three observations to prove that one is not looking at a plotted straight line. The simplest model I looked at was a two-equation linear Keynesian model involving two observable variables – aggregate national income, and aggregate national consumption – and two mathematically necessary unobservable parameters. For this model I determined that it would take six observations to logically refute this model. I then considered a slightly more elaborate Keynesian models by adding another unobservable parameter and calculated the P-dimension and determined that for this slight change the number of required observations increased to 10. The P-dimension was also affected by the degree of the model's solution equation. The next model I considered added just one more observable variable and one equation which then raised the P-dimension to 51. The p-dimension goes up fast as we add equations and variables. For example, eventually I looked at a model that my thesis supervisor (Hans Brems) had constructed which had six equations and six observable variables. By my calculations, whenever his model was actually false and we did not know the values of its eight parameters, it still would take 24,283 observations to construct any refuting observation set! I also looked at a more elaborate and popular model of the day that had as one of its equations a Cobb-Douglas production function, and depending on the actual value of one of that equation's key parameters, it could require 475,017 observations! Of course, my main conclusion was that economic model builders are not actually serious when they say they are concerned about the testability of their models. For those interested, my thesis can be found in Chapters 2 and 3 of my 1989 book.

The graduate economics program in which I studied was a special program promoting mathematical economics. So, as a result I graduated as a mathematical economist. I did not take me long to realize that there is more to economics than mathematical model building. When I was hired at Simon Fraser University, given the nature of my PhD thesis, I was actually hired to be an economic methodologist. I am not sure what they thought this was but I was allowed to create and teach a seminar in economic methodology. Given how much I liked the seminar I took with Joske, I designed it to be like his. There were no lectures and students took turns presenting something of interest to them. It was all very Socratic.

34.2 The State of Methodology in Economics in the 1960s and 1970s

As we all know, academics is a 'publish or parish' enterprise. So, once I got to Simon Fraser, I set about trying to publish articles about methodology of economic model building. Little did I know, the chances of publishing economic methodology in the late 1960s were virtually impossible – as a result, I had 55 rejections before my first article was published. Unlike what I faced, before 1963 it was not uncommon to find articles published about economic methodology. The 1959 article by Joske and Kurt Klappholz was an example. And given the availability of published articles on economic methodology, the 1962 meetings of the American Economics Association included a popular session on the 'Problems of Methodology'. The session included papers by the well known economic methodologists up to that time: Fritz Machlup, Andreas Papandreou and Sherman Krupp – and one philosopher, Ernest Nagel. Three other economists who published articles on methodology were the discussants: Chris Archibald, Herbert Simon and Paul Samuelson. The session is most remembered for Samuelson' discussion of Milton Friedman's famous 1953 methodology essay. The session was published in 1963 – and after that, for economic methodology, a grand nothing.

For a brief moment in 1973 a real methodology article was published in the same journal. The article was written by my friend and former student, the late Stanley Wong. The reason for publishing an exception to the apparent prohibition rule seemed to be that Wong's PhD advisor at Cambridge was Joan Robinson and Joan for many years harassed George Borts, the *AER* editor, for not publishing anything other than mainstream articles. I think Borts thought by publishing Wong's methodology article he could at least get Joan of his back for a while.

34.3 The State of Methodology in Economics After 1979

As for publishing methodology articles, except for Wong's singular article, no more were published until my article about the critiques of Friedman's 1953 essay was published in 1979 in the *Journal of Economic Literature* (*JEL*) – *JEL* happens to be a journal that does not accept submissions. Borts and the other editors of major economics journals continued to reject submitted methodology articles – including my submitted methodological article about criticisms of the mainstream economics' single behavioural assumption of maximization. Fortunately, my submitted article was eventually published in the *AER* but only after Borts was replaced by Robert Clower.

The situation for publishing methodology books was slightly better as there were two published in the mid-1970s – both by Cambridge University Press and both had authors who were philosophers. However, for books by economists, things began to change when in 1978 Wong published his PhD thesis and then in 1979 when two more were published by authors who were economists.

In 1980 the book situation began to change dramatically. From its beginning in 1962, Mark Blaug's history of economic thought textbook had included a chapter about methodology. The chapter was then expanded in the 1968 edition of the textbook. Blaug decided to further expand the methodology chapter as a 1980 book just on economic methodology. His methodology book turned out to be very popular and thanks to its success, a major economics publisher (Geo. Allen and Unwin) decided to published methodology books. I submitted a proposal and it turned out, so did Bruce Caldwell who submitted his PhD thesis on economic methodology. They offered contracts to both of us. After this, methodology for a long time became something that the book publishers would at last at least consider.

34.4 Methodology in Conferences Starting in the 1980s

During the early 1980s, the leaders of the History of Economics Society decided to add methodology to their list of history of economic thought topics to consider in their yearly meetings. This continues today. In 1989 Henry Woo, a businessman in Hong Kong created an organization to promote discussion of economic methodology – he called it The International Network for Economic Methodology (INEM). I and 79 other academics were invited to be founding members – the list of founding members is published in every issue. Mr Woo's organization also published a journal, *Methodus*, that was devoted to publishing only articles about economic methodology. Interestingly, there were very few philosophers among the founding members. In mid-1994 it was decided to rename *Methodus* – it became the *Journal of Economic Methodology*. Today it continues publishing only economic methodology articles. However, I suspect that this renaming event was really a hijacking by a group of historians of economic thought prompted by a subgroup sympathetic to analytical philosophy and in particular a group opposed to anything dealing with Karl Popper. The extent to which they would consider Popper was only if it was the cartoon charaterization of Popper that was created by Imre Lakatos – the phony view of Popper being a 'naive falsificationist'. The phony view is the one that claims scientists are only interested in testing and refuting theories. This is despite Popper explicitly rejecting the Lakatosian falsificationist characterization of his philosophy of science. He did this in the introduction to his 1983 publication of his previously unpublished *Postscript* where he says, [pp. xxxiv–xxxv] (emphasis in original): "Am I really the man who had naive falsificationism as the linchpin of his thoughts? Is the Kuhnian paradigm true? May I 'legitimately be treated as' a 'naive falsificationist', even though Kuhn admits, after looking at *The Logic of Scientific Discovery*, that, as early as 1934, I was *not* one? [...] Tests are attempted refutations. All knowledge remains fallible, conjectural. There is no justification, including, of course, no final justification of a refutation. Nevertheless we learn by refutations, i.e., by the elimination of errors, by feedback. In this account there is no room at all for 'naive falsification'".

Even today, Imre's falsificationist view of Popper persists in economic methodology discussions mostly because of Blaug's promotion of that view. Blaug unfortunately never read Popper's works before he learned about Popper from his discussions with Lakatos and hence Blaug's confusion. In other words, Blaug's view of Popper is fundamentally ignorant. In Chapter 20 of my 1997 methodology book, I discussed the two common competing views of Popper. On the one hand, there is Lakatos' so-called Popperian falsificationism view, and on the other hand the view that most students of Popper promote. They promote the view that Popper advocated the same view of learning that Socrates advocated in Plato's early dialogues – namely, learning is error correction. And of course, for Popper, Science as a conjectural process of learning is devoted to criticism.

Today, analytical philosophers who have recently discovered economics as a subject to study usually agree with Blaug's view that Popper is a falsificationist. Many of them in effect follow Blaug by claiming that economists do not go about testing and refuting economic theories so either economics cannot be a science or Popper is wrong. This common error is partly due to these philosophers of economics not having any idea about how economic model builders go about their business. I have argued many times that such philosophers and some heterodox critics of mainstream economics do not understand modern economic research. Today, modern economic research is all about model building – and, testing and refutations both happen regularly in economics. Until about 1980, building a mathematical model was always seen to be a way of making one's theory rigorous. Today, however, economists no longer distinguish between theories and models – they instead often use these words interchangeably. And instead of distinguishing between theories and models, they see two types of model: theoretical and empirical. With both types, model builders choose assumptions with a purpose more related to mathematical convenience. And in the case of empirical models, rejection is not infrequent such that Popperians might easily count such rejections as refutations. In the case of theoretical models, when a model fails to prove what it is intended to prove, the rejection is a refutation. In other words, Blaug was wrong about economists. Testing and refutations is a regular affair today.

34.5 The Recent History of Economic Methodology

For many years and roughly every other year, INEM has continued holding meetings about methodology. They once were held in conjunction with the History of Economics Society meetings. Today, they are usually held independently. Apart from INEM conferences, there have also been a few independent economic methodology conferences. These were held to discuss specific topics, such as Popper's or Friedman's view of economic methodology. First, in 1985, there was a conference held in Amsterdam to honour the retiring methodologist Joop Klant. And, apparently, since he sometimes referred to Popper, the conference was about Popper and economic methodology. The conference was followed by the usual conference volume in 1988 which I reviewed in 1990. Strangely, for some unknown reason I was deliberately not invited despite my well-known Popperian oriented publications and reputation in economic methodology. Moreover, at that time, I was the only such recognized economic methodologist other than Klant.

In December 2003 a second methodology conference was held. This one was held to celebrate the fiftieth anniversary of the publication of Friedman's famous essay on economic methodology. And again, despite my 1979 *JEL* article about that essay being one of the most cited methodology articles, I was not invited. The conference's papers were subsequently edited and published by a prominent analytical philosopher of economics in a 2009 book which I reviewed for *Economics and Philosophy* in 2010.

I note the two conferences because they illustrate what has happened to the community of academics who claim to be studying economics methodology. This sad situation exists I think because my field of economic methodology has really been hijacked by analytical philosophers. And seeing this, I think it might even explain first why the few older economists who did talk about methodology never cited a philosopher to back up their views. And second, I think this probably explains why economic methodology was for so long been banned from major economics journals. The problem is that too often analytical philosophers think they have a monopoly on clear thinking. I saw this first-hand at a seminar series held at my university. In the 1980s, a group of professors set up a weekly seminar where other professors would take turns discussing their current research projects. We had faculty from the science departments as well as the social sciences and humanities. I learned a lot from these. But we had a problem. One of the frequent attendees was from the philosophy department. He continually tried to correct the various presenters telling them they were wrong about this or that according the current opinion of (analytical) philosophers. Apparently, he did think analytical philosophers have a monopoly on clear thinking. Eventually, we had to explicitly ban the philosopher from the seminar. After that, we all got back to learning and enjoying our weekly seminar.

34.6 Small-m vs. Big-M Methodology in Economics

Thirty years ago a prominent economist raised a distinction between what he called 'small-m methodology' and 'big-M methodology'. He was explicitly identifying a difference between what the very few economic methodologists like me talk about and what analytical philosophers of economics talk about. What he called small-m methodology is basically what I have been writing on from the beginning of my publication career. Specifically, what I have been identifying are the various answers to simple questions about the reasons given by economic model builders for assuming what they assume? In contrast, analytical philosophers of economics worry about big-M questions such as: What do economic model builders mean by 'realism' or by 'realistic' assumptions? Can the assumptions used to build a model be justified? What is the cognitive status of economic theory? And so on.

While a small-m methodologist will usually be aware of the views of many philosophers of science, few of these small-m methodologists have studied philosophy. Similarly, most big-M philosophers of economics rarely know much about economics beyond the level of a beginning economics class. And when big-M methodologists look into the writings of economists, it is only with an interest compatible with analytical philosophical studies as we experienced in the more general case with the banned philosopher from our seminar.

The primary result of the hijacking of the community of economic methodologists by analytical philosophers is that it is now again more difficult to get articles on economic methodology published in mainstream economics journals. Fortunately for me, book publishers still find publishing my books about small-m methodology to be a viable enterprise.

The few of us who do understand Popper have repeatedly pointed out that attempting to test a theory is just one form of criticism, but this does not seem to matter much to the followers of Blaug and Lakatos. Nor does it seem to matter to them that testing is not the main issue in Popper's philosophy of science, particularly when it is applied to economics. For a long time, particularly in the history of economic thought or methodology conferences, Popper's views were rejected on the grounds that he did not understand anything about economics. But Popper actually explained his view of economic methodology, as is evident in Chapter 8 of Mark Notturno's 1994 book (that contains a collection of Popper's papers including one presented to the department of economics as Harvard in the 1960s).

The analytical philosophers of economics continue to maintain a primary perspective where methodology is a normative study, a study concerned mostly with appraising the methods of building theories and economic models. Apparently, the primary purpose for appraising models and theories is to choose the best one based on conventional criteria. Moreover, given the usual philosopher of economics' acceptance of the Lakatosian falsificationism view of Popper, they seem to think that Popper is merely advocating falsifiability as a criterion to choose between theories and models as a matter of appraisal.

Today, one can still find conferences about economic methodology. However, the hijacking of methodology means that virtually all of them are dominated by the interest of big-M methodologists. Rarely if ever does one find anyone discussing the burning small-m issues of modern economic model building. While, thanks to Blaug, in the 1970s and 1980s one did find some interest in Popper's philosophy of science, also one did find a growing discontent with any discussion of Popper's philosophy of science. And this growing discontent seemed to culminate in the 1985 conference discussed above. While ordinary economists were never discontented with the discussions of Popper, almost all analytical philosophers of economics reject even mention of Popper, and particularly when the discussion is in terms of the Lakatosian falsificationist version of Popper's views. As a result, little of what is discussed in today's economic methodology is of interest to anyone who does understand Popper's critical rationalist philosophy of science. Too bad.

34.7 On What I Learned

As I finish my eighth book on economic methodology and as I teach our 'writing intensive' class where half of a student's grade depends on writing skills demonstrated, I always keep in mind what I learned from Joske in our frequent discussions in his small office at the University of Illinois. As he told me and I tell my students, understanding your audience is most important because you need to anticipate the question that is likely to be asked after every idea you put forth. And if you do not immediately answer that question in the next sentence or paragraph, you will lose your intended audience. Hopefully, my students will have learned from me as much as I did from Joske.

References

Blaug, M. 1962/68. *Economic theory in retrospect*, 2nd edn. Homewood: Irwin.
———. 1980. *The methodology of economics*. Cambridge: Cambridge University Press.
Boland, L. 1979. A critique of Friedman's critics. *Journal of Economic Literature* 17: 503–522.
———. 1981. On the futility of criticizing the neoclassical maximization hypothesis. *American Economic Review* 71: 1031–1036.
———. 1982. *The foundations of economic method*. London: Geo. Allen & Unwin.
———. 1986. *Methodology for a new microeconomics: The critical foundations*. Boston: Allen & Unwin.
———. 1989. *The methodology of economic model building: Methodology after Samuelson*. London: Routledge.
———. 1990. Understanding the Popperian legacy in economics. *Research in the History of Economic Thought and Methodology* 7: 265–276.
———. 1997. *Critical economic methodology: A personal odyssey*. London: Routledge.
———. 2003. *The foundations of economic methodology: A Popperian perspective*. London: Routledge.
———. 2010. Review of Mäki (2009). *Philosophy and Economics* 26: 376–382.
———. 2014. *Model building in economics: Its purposes and limitations*. New York: Cambridge University Press.
Brems, H. 1959. *Output, employment, capital and growth: A quantitative analysis*. New York: Harper.
de Marchi, N. 1988. *The Popperian legacy in economics. Papers presented at a symposium in Amsterdam*. Cambridge: Cambridge Univ. Press.
Hollis, M., and E. Nell. 1975. *Rational economic man*. Cambridge: Cambridge University Press.
Klant, J.J. 1979. *The rules of the game: The logical structure of economic theory* (trans: Ina Smart). Cambridge: Cambridge University Press.

Klappholz, K., and J. Agassi. 1959. Methodological prescriptions in economics. *Economica* 26 (NS): 60–74.

Mäki, U. 2009. *The methodology of positive economics: Reflections on the Milton Friedman legacy*. Cambridge: Cambridge University Press.

Notturno, M. 1994. *The myth of the framework: In defense of science and rationality*. London: Routledge.

Popper, K. 1934/59. *Logic of scientific discovery*. New York: Science Editions.

Rosenberg, A. 1976. *Microeconomic laws: A philosophical analysis*. Pittsburg: University of Pittsburg Press.

Stewart, I. 1979. *Reasoning and method in economics*. London: McGraw-Hill.

Wong, S. 1973. The 'F-twist' and the methodology of Paul Samuelson. *American Economic Review* 63: 312–325.

———. 1978. *The foundations of Paul Samuelson' revealed preference theory*. London: Rouledge and Kegan Paul.

Chapter 35
Capitalism at a Crossroad

Raphael Sassower

35.1 Classical and Neoclassical Capitalism

Ever since the Age of Reason, the advocacy of free enterprise—capitalism—has been commonly associated with democracy, as if there is a mutual support between their frameworks, not to mention their adherence to Enlightenment principles (Agassi 1988). When equality and freedom are conjoined many authors have argued that the success and protection of the one is necessarily at the expense of the other (de Tocqueville 1945). Some economists of the nineteenth century worried about the ways in which equality and freedom were to interrelate, at least insofar as economic success would be viewed adversely as an act of selfishness with the unintended consequences of harm to others (Sassower 2013). Yet advocates of capitalism repeatedly claim that this conjunction of principles is as likely (or perhaps more clearly) to be found in free markets as in the halls of parliament, protected as much by market mechanisms as by democratic legislation (Friedman 1982; Soros 2000). More recently efforts have been made to present *The Theory of Moral Sentiments* (Smith 1976) and *An Inquiry into the Nature and Causes of the Wealth of Nations* (Smith 1937) of Adam Smith as one system: He presented his *The Wealth of Nations* as an effort to supplement his *Moral Sentiments*. Already there a balancing act is in evidence, if not a nuanced juxtaposition of one system—the market—on top of or within the contours of the other—social sentiments that predate market institutions and the state. Smith took it for granted that minimal government is the best way to handle human affairs, yet he insisted that it is no utopia: he mentioned that the Hanseatic League approached his ideal (Bk. III, Ch. 4). At the same time, Jean-Jacque Rousseau defended democracy as the best political order, and likewise insisted that it is no utopia: the city of Geneva had a regime that approached his

R. Sassower (✉)
Department of Philosophy, University of Colorado Colorado Springs,
Colorado Springs, CO, USA
e-mail: rsassowe@uccs.edu

N. Bar-Am, S. Gattei (eds.), *Encouraging Openness*, Boston Studies in the Philosophy and History of Science 325, DOI 10.1007/978-3-319-57669-5_35

ideal (1978, Bk. I, Ch. 6). The two ideals clash when the collapse of the free market requires government intervention far beyond a minimal level. Defenders of free enterprise, such as the Austrian economist Friedrich Hayek, claim that this is never the case (1944): in principle, he said, the quickest way to get market recovery is to let it alone, no matter the price to individual companies or private banks (in our recent US experience). His arch-critic John Maynard Keynes asked, how long should we wait to see the market recover before asking the government for help? (1964) He greatly appreciated the insights of the French physiocrats in regards to the circulation of money in the economy and its ongoing growth potential (Grospelier and Sophister 2005), and in line with this insight he focused on when and how the government should intervene to bring about growth. With government stimulus demand will increase, which will allow the entire market to rebound, paying off the initial government subsidy through increased tax revenue from wages of the growing number of taxed workers (Stigler 1965).

The disagreement concerns not only the way to emerge from a recession, but also the cause of recessions and their prevention. The cause of the Great Depression, the advocates of free markets say, is the panicky response of governments to the market collapse by raising tariff walls, thus causing international trade to collapse and a run on the banks. Their critics blame governments for not having interfered in the ailing market sooner with vast capital to shore up the financial markets and increase employment (Parker 2007). The disagreement is not limited to markets and the business cycles. The broadest testing ground is a clash of ideologies, one of which advocates freedom first and the other equality first. The preference of freedom over equality—liberalism (and neoliberalism)—is the advocacy of individual freedom, best expressed as minimal government economic activity or intervention in the market, signaling an idealized free enterprise system (where Smith's "Invisible Hand" has its focus on the invisibility of the hand rather than the need for any hand to be there at all). This preference rests on the supposition that whereas equality with no freedom is quite possible but inherently flawed (usually as uniformity), freedom breeds equality in the sense of equal opportunity. This latter supposition may be empirically irrefutable until a totally free market is instituted, which may be impossible. Alternatively, it is empirically refuted (Agassi 1977). Similarly, the preference for equality over freedom—egalitarianism (and socialism)—rests on the fact that freedom is useless without equality: "The law, in its majestic equality, forbids the rich as well as the poor to sleep under bridges, to beg in the streets, and to steal bread" (France 1894, Ch. 7). The liberals rely on market mechanisms; the egalitarians rely on the mechanisms of democratic control (including constitutional checks and balances). Liberals deem the reliance on governments naïve (Hayek 2011, Introduction *et passim*); egalitarians deem the reliance on the market mechanism misguided and in bad faith. The popular view today is part liberal and part egalitarian. We expect that governments should maintain relatively free markets and institute mechanisms by which to establish equality. Keeping markets relatively free can be seen in terms of civil society and avoiding hyper-concentration of capital; keeping the promise of equality can be seen in terms of promoting equal opportunity for all and eliminating the worst inequalities (e.g., progressive taxation). Supporters of such ideas refer to them jointly under the label of the welfare state (derogatorily known as the paternalistic state or as the Nanny State). Government agencies in charge of welfare are supposed to see to it that the level of welfare for all citizens is above a certain minimum (Stanley 2013), perhaps also to combat poverty traps, especially by providing education for all (Agassi 2009). In Canada (e.g., Ontario), the law requires of the state provide mental health care for all. Some liberals see in egalitarian moves such as these a real threat to freedom. They do not deny that the free market leaves many problems with no solution. They insist, nonetheless, that it is not for the government to offer solutions to them, since this will necessarily create excessive bureaucracy that will unavoidably encroach on freedoms beyond a tolerable limit. Hayek referred to the system of patronage, others referred to diverse charitable institutions, especially churches; this discussion is secondary.

In modern democracies both parties agree: the primary role of governments is to protect the Enlightenment principles of freedom and equality. The competition for government attention in regards to the support of either or both has become more exaggerated in the twentieth century, when two World Wars threatened to eliminate democracy altogether—not simply as a philosophical ideal but much more as cherished reality in the European and American continents. Bertrand Russell has offered three venues where governments have democratic roles to play: Security, justice, and preservation (Russell 1949, 54ff). Under security he included economic prosperity and the means by which it can be achieved; he viewed it as government-induced activity sumpplementing personal initiative (e.g., minimizing barriers to entry into the market, loan guarantees, and short-term subsidies). Not quite a Keynesian, but definitely a promoter of an active government, however minimal, Russell championed government functions when security matters are viewed as not exclusively militaristic.

Russell's suggestions come regularly under fire: it was perceived that in his plan governments do too much or too little, never quite finding the right balance between preserving equality among its citizens in social and personal matters alike (cooperation as well as healthy competition), while pretending that unfettered freedom is ultimately the answer to all the ills that plague democracies. The urgency of the problem may be eased, since at higher levels of capitalist developments the balance between entrepreneurs (freedom) and welfare (equality) can more easily be achieved because of national surpluses. Already in the nineteenth-century Karl Marx insisted that the ideals of freedom and equality are within reach and only capitalists and the political system prevent their achievment. The Austrian economists agreed with him on the possibility half a century later, although they advocated capitalism with a vengeance. The two extremes agreed at least on this: the more surplus is produced through growth, the more wealth can be spread around to the needy. Likewise, they appreciated combining personal initiatives with collaborative efforts, from the knowledge industry to banking, and they wanted it settled quickly. As long as Smith's idealized village was postulated, where moral sentiments were shared and promoted by the Impartial Spectator (either God or one's conscience), the ideal market could indeed thrive for the benefit of all. Even Marx agreed, but, he added, capitalism is not viable, since capital keeps concentrating in few hands, and market collapses keep worsening. Liberals, egalitarians, and compromisers agree that the two principles of freedom and equality should complement each other rather than remain at odds, yet the two extremes put all their weight on one of them and the compromisers feel the wrath of both extremes.

While neoliberals keep putting all the weight on freedom, others continue to support the Marxist critique of classical capitalism. They do so at least on three fronts. First, capitalism does not embody the Enlightenment ideals of freedom and equality; instead, it uses such rhetoric to hide its intentions to commodify the world and all its contents so as to preserve a particular biased view of the world to benefit one group—the bourgeoisie—at the expense of all others. Instead of freedom and equality we have seen nothing but exploitation and injustice wherever capitalism is dominant. Some exeggarated hostility to capitalism is not exclusively Marxist as it infuses moral condemnation to its economic analysis. The right response to it is not to deny it but to seek ways of improving matters. One suggestion has been a democratic welfare state (that is entrepreneural as well insofar as it sets infrastructure, like the Internet, for market participants) but it's not fully accepted by some moralizing critics (Lazzarato 2012).

Second, capitalism causes more harm than good to the community of workers and owners alike and the environment. At its wake, large footprints of poverty and suffering, hunger and ecological devastation are apparent. According to Marx, unfettered competition will not allow worker-unions to counterbalance capitalist wage-exploitation, an argument already offered already by Smith (1937, 66). When the authorities use force to support strikebreakers, their conduct offers tremendous support for Marx's analysis. When government violence stopped, workers' real wages rose significantly. Marx's view—that capitalism must run its course until the revolution will terminate it—was plainly refuted in the USSR. Marx's view of capitalism was narrow, not appreciating its elasticity, as we currently see in China. Consequently, Marxist-like critiques have made primarily moral claims about economic realities, emphasizing the internal contradictions of capitalism rather than focusing on test-cases (Cassidy 2010).

Third, capitalism has co-opted democracies to do its boding, bailing its privately-owned banks no matter their own risky behavior and mistakes and no matter the cost to the rest of the public (in taxes and long-term national debt). This recent experience would reaffirm Mraxist critiques about the power relation between capitalists and the state rulers. Instead of capitalism being the handmaiden of democracy and the expression of the ideals of the modern liberal state—as Smith and his disciples had envisaged—modern capitalism is a sham to its own ideals and an excuse to extract economic power through political and legal institutions of the state without the political warrant for its legitimacy. Smith's double-vision of a market open to all (personal freedom + equal opportunity) has become a bastion of power for the so-called 1% of the very rich (Occupy Wall Street). Even apologists of the current system would have to admit that the moral and political consequences are undermining democracy globally (Stiglitz 2002).

35.2 Contemporary Debacles

It is very difficult to put general ideas to test. Consider the first famous large-scale government intervention meant to help restore the US market after its massive collapse in 1929. That was the New Deal that created, among many other institutions, the Tennessee Valley Authority and the Works Projects Administration (among many other, still operational agencies and social programs). Was it a success? Was the intervention justified? These questions are still under dispute. The ideal test—scientifically speaking—would be to have two similar collapsing economic systems, one undergoing a thorough government intervention and one left alone (as the control group). Of course, such an experiment is out of question, since it requires a sufficient consensus in each of the two countries to follow the experiment through or to divide one nation into two spheres with simultaneous alternative policies. This is impracticable, since modern liberal democracy depends for its functioning on representatives who voice the needs of their constituents and attempt to meet them in complete disregard for (and at times contrary to) economic theory. It was quite astonishing that during the New Deal and its aftermath American politicians used liberally their interpretations of Keynes who suggested that increased market demand would increase supply (with government infusion of capital to expand the country's infrastructure, as mentioned above). Economic theories are fodder to politicians' arsenal of rhetorical devices so as to justify whatever they deem will bolster their popularity, help them raise funds, and get them reelected. When government infusion of capital causes a bubble that bursts, it denies responsibility and washes its hands off the mess to which it contributed. When the government waits too long to intervene, it can always claim not to divine what direction markets will take in the long run. Besdies, how much is enough intervention?

Since World War II the situation changed. First was the appearance of the idea that the debate is too ideal: there is no utterly free market system (Popper 1966). Free markets do exist, such as those precarious setups that engage adventurous traders who operate temporarily, mainly in no-man's lands. They are hardly marketplaces, much less free-market-systems. They are too frail and marginal to count as systems. The free-enterprise advocates advise governments to defend the market by imposing laws that guard fair trade, such as preventing fraud, breech of contracts, collusion (anti-trust), dumping, etc. This lends itself to conflicting results: regulating the market in order to deregulate the activities of its members. Incidentally, already in 1762 Rousseau outlined this expectation in his formulation of the principle of the social contract to which all citizens are supposed to surrender some of their individual freedom in order to recapture much more than they gave up (through the legitimacy of the state of which they are now members in good standing, a legitimacy that permits the state to impose rules and regulations on all). What exactly should citizens give up in return for security? This remains an unsolved problem. Perhaps it is unsolvable.

We may agree, for example, on government (legal) intervention to impose truth and trust in contracts. Advocates of minimal government oppose current laws that impose truth in advertising. They claim that free markets do this better on their own especially in the era of digitally-run social-media when everyone plays a self-appointed police officer (ideally in practice in the scientific community.) The advocates of free markets claim that this feature of the markets turns them vigilant enough to render superfluous all government regulations other than the fulfillment of contracts (Bastiat 1964; Friedman 1980). Obviously, the principle behind this view is that whatever the market mechanism can do well enough should be left to the market mechanism; only what it cannot do should be imposed externally. The principle is reasonable enough as it follows from the liberal principle that advocates legislation in the citizen's interest to keep. It would indeed be nice were the market mechanism able to stop all fraudulent conduct with no police intervention. Yet we must concede that the market mechanism is not that honest or efficient. When it fails, what functions does it peform well enough? This issue is at the heart of the ideological clash discussed above so that the very involvement of government agencies continue to be scrutinized. It is important to note, therefore, that ideological as the dispute remains, arguments for and against each side are empirical as well and so in need of empirical tests (Tullock 1994, 84). These are too slow to come when possible at all, especially since this is tacitly denied (Albert 2009, 90; Boland 1997, 42, 54, 120).

Economists and governments of diverse sorts agree that market incentives are at times insufficient to encourage the desired amounts of investment, in general or in certain branches. To improve this, governments offer incentives, general and specific. The general ones include raising import tariffs, lowering taxes and interest-rates, and reducing administrative constraints. Governments also invest huge sums in products that they "dump" on the free market (domestically or internationally, agricultural or manufactured), thus causing gross distortions in commodity-markets, since dumping is as much a form of protection as tariffs. Dumping is a largely specific encouragement in a particular segment of the market. Advocates of free markets theoretically oppose encouragement but add that the least harmful practical intervention is financial. There are many ways to offer financial incentives; they may be tax breaks and loan guarantees tied to specific ends, such as government participation in the preservation of the environment. These incentives are supposed to work more directly and precisely than auctions. Governments tend to ignore their unintended consequences. Advocates of free markets, especially in the Hayek mold, keep asserting that the unavoidable presence of unintended consequences renders government intervention always of questionable value. Fredrick von Mises offered an argument that speaks specifically against almost any government intervention: intervention creates a bureaucracy and all bureaucracy is self-perpetuating and wasteful (2007, 36). The advantage of tax-breaks over direct interventions in the economy then is that there is already a bureaucracy in place, even in the case of the ongoing agricultural subsidies.

Are all incentives wrong interventions in the market? If yes, are they all equally wrong? Otherwise, it makes sense to grade them for *the degrees of their undesirability* (Kamensky and Morales 2006). And then it makes sense to ask what is the optimum situation of a mixed economy of free market and of government intervention, as is practiced in some parts of the world. Studies of this optimization are the new territory in economic research (Ikeda 1997; Weber 2001).

A free-market oriented US administration offered recently incentives for mortgages (for first-time buyers) that created a housing-market bubble that burst and caused the recent recession (2008–2012). It began in 2003, when President George W. Bush signed into law a plan designed to reduce taxes and stimulate economic growth. Few economists supported it; most of those who disapproved of it did so on ideological grounds, not in fear of market collapse. When it did, it took everybody by surprise. Most significantly, it was the unintended consequence of a policy of a non-interventionist administration that spilled over into the Obama administration. This eventually led to a most radical intervention with mixed results. The intervention came in the form of bailing out ailing private banks and mortgage companies and raised the general recognition among economists of the need for reevaluation of capitalist ideals in the contemporary context: should markets be left to their own devices even during their weakest moments? The failure of Lehman Brothers Holdings, the fourth largest investment bank before its bankruptcy in 2008, had an immense ripple effect on the rest of the financial markets. Had the administration intervened early enough, providing Lehman Brothers liquidity in the form of cash or credits or guarantees to avert bankruptcy, it might have averted the global meltdown that followed. To repeat, counterfactuals of this kind ("what if") are of interest only in hindsight. The Lehman scare promoted both Republican and Democratic administrations to initiate the Troubled Asset Relief Program with an estimated cost of close to one trillion dollars (depending on how funding is counted). Was this the best medicine to an ailing economy? Did it work as well as predicted? Generations of scholars will continue debating the answers to these questions (just as they do about the New Deal). The debate began when Noble Laureate economist Paul Krugman (2012) claimed in an interventionist manner that the government should have pumped even more of its resources to prop the economy much earlier, and economists that oppose government intervention, such as David Stockman (2013), who decried the "corruption of capitalism in America", not to mention George Soros (2008) who claimed that the administration should have demanded much earlier the rewriting of mortgage contracts to avoid the market's collapse.

The complexity of modern capitalism bears little resemblance to classical and neoclassical models. As Robert Reich argues in *Supercapitalism* (2007), the standard alignment of the (democratic) state with capitalism has changed: mega-corporations operate independently of the state, and in some cases they even work against state's interests, ignoring standards of pollution, employment, child-employment, workplace safety, and minority and gender pay equality. These corporations have a strong lever: they threaten to leave any state that puts obstacles in their ways, and with this threat redefine the boundaries of the market. It's hard to see why states care so much about this threat, since these corporations already systematically evade paying taxes, using off-shore tax-havens to shield their exposure to local or national taxes (as seen recently in the case of Apple in Ireland). More than that: the current situation has allowed multinationals to redefine concepts, such as self-interest and collaboration, in order to align them to whatever their interests dictate. All this is well beyond both capitalist and socialist theories and the ideals they embody. Smith had hoped that the self-love of individual entrepreneurs would be benign and aligned with the wellbeing of the community as a whole. Marx envisaged a socialist economy with decentralized capital and coordinated policies that harmonizes interests and operations of individuals and the state apparatus. He expected the harmonization to work so well that the market would be self-sustaining and the state would partially wither away (as superfluous). Contemporary ("post-capitalist") markets endorse parts of both idealized models as a simple reality of the economic conditions under which they operate, yet they are still struggling: in some cases individual initiative is essential, in others the implementation of any innovation requires the coordination of multiple groups of researchers and workers (Sassower 2009). Less confined by ideology, mixed-economies or hybrid-economies allow complex markets to weave seamlessly between the one and the other ideals, pragmatically adopting the best policy, whether dealing with the entertainment industry (Lessig 2008) or China's state-capitalism. However dictatorial China's political system remains, its economy is more efficiently run that under a staunchly socialist economy of Chairman Mao.

The tension between the two ideal models of capitalism and socialism falls apart when pragmatic policy-makers willingly use whatever means work to mitigate a catastrophe—perhaps this is why the rhetoric is so loud: to camouflage the devious disregard for ideal models and for the ideologies they express. Faith in the ideal models is lost since neither was ever tested in its ideal form, since any time a test is proposed it is always a hybrid intervention, in partial loyalty to the one ideal or the other because of political concerns and the compromises that they demand.

The label of "postcapitalism" is thus a shorthand term for the pragmatic adaptation of classical or neoclassical economic theory to current environments, where the quality of life is not left to the Invisible Hand to control. This includes human rights concerns as well as the increased expectations of individuals in democratic societies to improve the quality of their working lives, not to be exploited or alienated from their work or from fellow-workers or their environments (Buber Agassi 1978–79). We witness the spread of the demand for satisfaction from work: of the Volvo type—where workers rotate around the assembly-line so as to have a full view of the car as a whole rather than just this or that part (Buber Agassi 1985), or of the Microsoft kind (and many other later digital industry giants, like Google)—where workers have no set hours or cubicles, but roam freely so as to energize their creative juices. Admittedly, this is a simplification or a rose-colored view of things; expecting the system to approach a mixed or hybrid market ideology and practice is increasingly becoming more common and as such a staple of postcapitalism. Collaboration happens then because it is efficient and fun, because profit is not the only measure of success. Otherwise, such generous movement as Open Source and Copyleft (that leave important inventions in the public domain) would be unthinkable or sheer utopias. The very idea of intellectual property as private, highly and legally protected, as an incentive, has been thus widely challenged and has given rise to an ideological debate quite different from the one the Enlightenment leaders engaged in (Lessig 2008).

35.3 Postcapitalism

Some unplanned pragmatic developments were initiated because of two large-scale conflicting tendencies, the increased power of international corporations and the rise of movements for improvements of the quality of life. The ideology behind the corporations is a century old with roots in capitalism; the other has roots in various humanist movements, including Marxism (Russell 2001 Ch. 19; Althusser 1964).

One example that illustrates how ideals and ideologies must change in a post-capitalist world, a world where idealized capitalism and socialism find common ground rather than being at odds, is the one associated with the "knockoff economy" or the imitation economy. This segment of the economy includes the fashion industries, chef-inspired recipes, football strategies, stand-up comedy routines, and open-source programming. It is the refusal to protect intellectual property. Kal Raustiala and Christopher Springman (2012) suggest that rather than undermine the pace of progress of competitive free-floating ideas, the knockoff economy spurs it at an increasing rate: the copying of fashion styles reduces the time-span of their popularity, demanding faster turnover of new styles, revving up the ingenuity process (contrary to the standard rationale for securing intellectual-property rights). The inducement to invent rests on the recognition that ideas are bound to be superseded faster than recognized by the copyright protection laws, proving superfluous (if not downright reactionary) the monopoly power that intellectual-property rights guard. The free-flow of ideas and recipes of chefs, likewise, ensures a healthy inventive environment that honors creativity.

The knockoff economy is guided by unbridled copying frenzy, and by a social etiquette that demands proper attribution of originality, viewing as poor taste copying without acknowledgement. There is a certain code of behavior reminiscent of Smith's description of social sentiments overseen by an Impartial Spectator. This is also reminiscent of Hayek's assumption and hope that capitalist ideologues should portray self-interested behavior accompanied with adequate contribution to some community fund in line with Smith's image of the market as a small village where mutual care is sacrosant. This setting requires more than honor among thieves, as all participants in it are aware of the conditions under which market behavior must remain self-regulated and so self-policing. It also illustrates the viability of holding onto two seemingly opposing ideals: individual competitiveness and collaborative recognition of shared knowledge. Just because we share in common ideas and inventions does not mean we lack initiative to excel and be creative. At times our rewards might be financial and at other times it is the distinction of being the first to use a technique or ingredient. With this in mind, the market might become less alienating and more inviting to novices and first-comers who may begin by copying (with attribution) but eventually contribute their own creations as artists have traditionally done. But how significant can this knockoff economy be?

However marginal cooperation may be, it behooves us to notice that all the classical models of the economy postulate consumers, realizing their given preferences as best they can, entrepreneurs or firms, maximizing their profits, and governments, defending and participating in these markets. Donors are either consumers who realize their preferences or firms who contribute under the rubric of public relations expenses. This is an idealization that we may tolerate or not, but at some stage we have to do better than focus on the gift economy at the margins.

But cooperation is not marginal. Edward O. Wilson, in his *The Social Conquest of Earth* (2012), broadened the general methodology of evolution as envisioned by Charles Darwin (as the natural selection of new mutations) to multi-level evolution, where the unit of possible evolution is not only the individual within a species but also the species itself. This he finds in antspecies that have survived millions of years because of their ability to collaborate in warding off outside predatory forces. Undermining the standard view of individual self-interest and selfishness, Wilson offers an additional dimension of human behavior that is cooperative, even self-sacrificing, perhaps in Rousseau's sense of the "noble savage." This view, incidentally, is also found in Jewish tradition that presents individuals as endowed with good and evil propensities. While Smith took selfish conduct as no obstacle to the cause of the community, and Social-Darwinists ignore the community and focusd on individual success, Wilson's view helps bridge the gap. Following him, market-forces can more easily transport from the selfish to the cooperative depending on specific conditions and contexts. This is in line not only with the knockoff economy that economists may dismiss as insignificant but also with the ideas of Paul Rosenstein-Rodan (1961) of the "big push," global intervention to build infrastructures in developing nations. This idea is becoming increasingly significant as the gap between rich economies and poor ones endangers world peace.

Significantly, the knockoff economy and the big push rely on increasing information flows that can promote collaboration on the way to revitalizing market activity, thus bridging between free-market and Keynesian models. Stiglitz disparages the disregard of economists for global asymmetries of information (2002). Most models ignore systematically the obvious fact that the strongest foundations for human actions comprise a mix of egoism and altruism, perhaps because traditional philosophy takes altruism to be a kind of egoism. Newer models of hybrid economies may represent market-behavior more usefully (because they replace the ideal "rational man" with modes of irrational behavior). The neoclassical model assumes a permanent state of dynamic equilibrium. Clinging to it is objectionable because of its removal of externalities or exogenous factors; these factors directly influence participants in the economy. It is likewise clear that all idealizations suffer the same fate. Thus, the assumption of free entry to the market leaves this idea of entry unstudied—what are the barriers? And how daunting are they? The assumption of total knowledge excludes the study of the growth of knowledge and its assymetric proliferation in markets. Clinging to models of equilibrium or even partial equilibrium does not reflect the persistence of ongoing disequilibrium states (Boland 1986), and this approach overlooks some of the reasons for business cycles. Similarly, the difference between market satisfying and satisficing is left out of the picture (Lipsey and Lancaster 2000). Of course, studies of these neglected factors abound, and this will improve the field, yet they leave the two ideal models intact as the only ideological options. We need a new, alternative ideology that takes cooperation seriously.

The Keynesian model is more realistic than the neoclassical one in that it allows for government intervention in order to overcome the hard times of economic lows (when unemployment rises to unacceptable levels). What it recommends may have been useful during the Great Depression but less so during the Great Recession; lessons of the past may be insufficient for setting policies under different conditions. The testability of the Keynesian model has been strained by political calculations that accounted for economic conditions in social, moral, and most importantly, electoral terms rather than in economic terms alone. Similarly, in the aftermath of World War II, the Marshall Plan or the European Recovery Plan provided American infusion of funds to rebuild Europe. This rebuilding was a (foreign) government intervention in the name of democracy and capitalism. It is reminiscent of Rosenstein-Rodan's theory of the big-push (1961) that requires the formation of an infrastructure by forces that do not abide by the laws of the market alone. It is difficult if not outright impossible to test any economic model for its functionality and reliability. As long as political leaders and bureaucrats are insufficiently scientifically-minded, this will remain so, and it will only change if electorates become more critical of their representatives. Until then, we are left with a situation where relevant data are ignored and where frameworks of these data are exploited to fit prefigured political purposes. If anything matters to these representatives beyond immediate practical results, it is their unshakable ideological commitments. This is obvious, for example, in the cases of the utterly undemocratic installation of technocrats in 2012 both in Greece (Lucas Papademos) and in Italy (Mario Monti) in response to the Euro-zone's financial crisis, setting aside theories or ideological convictions in the name of austerity-measures to repay outsized debts primarily to Germany (Lazzarato 2012, 158–160, 172).

The postcapitalist model may help here: its recognition of these underlying situations and of the immediate problems that they raise, not simply as theoretical obstacles—is this too greedy-capitalistic-sounding, or is it too socialist-looking?—but as obstacles inherited from historical processes, may turn out to be crucial for improving matters significantly. Comingling salient contributions of the different parties may become part of a new ideology, offering new workable policy recommendations. This is paramount in the pragmatism of postcapitalism, pragmatism in deference to the human condition that is sometimes self-interested and at other times collaborative (in Wilson's sense). This is a Popperian and postmodern-pragmatist approach that recognizes the difference between free theorizing and the democratic practice that should be cautious as errors may cause human suffering (Popper 1966 Ch. 24; Sassower 2009). This is as well an application of Popper's methodology to the social sciences (Sassower 1988).

To go slightly to more detail, Keynes minimized the difference between his new views and the standard, neoclassical model. He added to the exogenous list of tastes what he called liquidity preference; at a certain moment trust in the market is so low that investors decline investing no matter how low the prices of shares go. The practical application of this principle is that market cash-flow dries up and the fastest way to revive it is infusion of cash. Keynes stressed that any way of doing this is fine, but the most efficient way is through government's investment in infrastructure (deficit spending). This is called fiscal policy. Friedman said the more efficient way to do this is through monetary policy, the treasury or central bank should increase the supply of money (by lowering interest rates or loan guarantees) to spur investment and increase employment. Keynes declared building the infrastructure more important than keeping the government in check, while Friedman declared the opposite. But, as Don Patinkin has put it, "the picture which emerges from all this is that by the 1940s the Chicago school itself had, quite understandably, been influenced by Keynesian monetary theory" (Patinkin 1969). This theory suffered from clinging to the old ideology: consumer preferences are exogenous (influenced by external factors) because they are supposedly free of market considerations, while investors have only money preferences. As Patinkin also noted, even the preference of one sort of money over another was left almost entirely unstudied as if it were of no import. When the analysis of market collapse pinpointed the trouble as cash-flow shortage, this should have raised a discussion as to the possible options that governments have. The conduct of the Bush Administration shows the poverty of discussion, and this poverty is due to clinging too much to old ideologies.

Thinking in postcapitalist terms doesn't only reflect more accurate observations of human conduct, but it is also more useful for political purposes (Sassower 2010). Endorsing a more open-minded discussion of the economy requires the rethinking of the ideologies used to legitimate it. Ideologies count, if only because they are significant heuristics that highlight how deviations from them may result in new realist proposals; so we need a new ideology. The belligerent theft of ingenious proprietary industrial processes may have to be protected for a time (say a year or two), while the copying of ideas and themes should expand under the "fair use" clause of copyright legislation. Similarly, governmental interventions in markets should be encouraged under certain conditions. This requires admission of economists that the neoliberal ideology is too restrictive, even blatantly hypocritical and unworkable. A reformulation of ideological convictions should improve openness and encourage younger participants, the millennial generation that was brought up in the neoliberal Digital Age. The exposure of the new generation to the Internet and the numerous gadget digital technologies to which they have access have formed certain expectations in their minds about what counts as appropriate market behavior and the economic conditions under which they interact in and outside the marketplace. They are less concerned with taxation as previous generations were due to the rise in their expectation from governments to play important roles in ensuring fair and prosperous participation of newcomers the way they remix music and ideas without the fear of legal restrictions. In short, the new generation finds it easier to understand that the current economy is neither Smith's village nor Marx's utopian commune, but rather a global environment in which cohousing is practical, privacy is precarious, and for-profit surveillance ubiquitous (Sassower 2013).

There is no need to assume perfect knowledge or too much about human nature and conduct to appreciate how economic and political variables comingle. An ideology based on the belief in incremental changes or piecemeal social engineering might be viewed as *pragmatic postcapitalism*; it will incorporate the proinciples of critical rationality and postmodern contextualized judgment; and it will be open to tests that in turn should lead to their reform so as to accommodate new circumstances. To speak of privacy or prosperity in a political vacuum is to blind future generations to potential social injustice. Insight regarding the circulation within a political economy of goods and services, ideas and money, is as relevant today as it was for the physiocrats of yesteryear and more recently for the Keynesians: all facets or variables of the economy and the political institutions that frame and protect it are somehow relevant. To tweak one element without any attention to its possible (intended and unintended) consequences elsewhere in the economy is to ignore both theory and experience, and this is lamentable if not dangerous in the global economy of the twenty-first century.

Acknowledgements Special thanks to Joseph Agassi whose contributions and suggestions for this essay have been inavluable; his continued support for my studies in this area date back to my graduate years at Boston University.

References

Agassi, Joseph. 1977. Discrimination, statistical. In *International encyclopedia of the social sciences*. New York: Macmillan.

———. 1988. Neoclassical economics as 18th century theory of man. *Fundamenta Scientiae* 9: 189–202.

———. 2009. Popper's insights into the state of positive economics and of welfare economics. In *Rethinking Popper, Boston studies in the philosophy of science*, ed. Zuzana Parusnikova and Robert S. Cohen, vol. 272. Dordrecht: Springer.

Albert, Hans. 2009. *Ökonomische Theorie als politische Ideologie: Das Ökonomische Argument in der ordnungspolitischen Debatte [1954]*. Tübingen: Mohr Siebeck.

Althusser, Louis. 1964. *Marxism and humanism*, Cahiers de l'I.S.E.A., June 1964.

Bastiat, Fredric. 1964. *Economic sophisms* [1845]. Translated by Arthur Goddard. New York: The Foundation for Economic Education, Inc.

Boland, Lawrence A. 1986. *Methodology for a new microeconomics: The critical foundations*. Boston: Allen & Unwin.

———. 1997. *Critical economic methodology: A personal oddisey*. London: Routledge.

Buber Agassi, Judith. 1978–9. Alienation from work: A conceptual analysis, *Philosophical Forum*, 10: 265–94.

———. 1985. *The evaluation of approaches in recent Swedish work reforms*. Stockholm: Department of Psychology, The University of Stockholm.

Cassidy, John. 2010. *How markets fail: The logic of economic calamities*. New York: Farrar, Straus and Giroux.

de Tocqueville, Alexis. 1945. *Democracy in America [1835 & 1840]*. New York: Vintage Books.

France, Anatole. 1910. *The Red Lily* [1894]. Translated by Winifred Stephens. New York: John Lane Company.

Friedman, Milton. 1982. *Capitalism and freedom [1962]*. Chicago/London: The University of Chicago Press.

Friedman, Milton, and Rose Friedman. 1980. *Free to choose: A personal statement*. New York/London: Harcourt Brace Jovanovich.

Grospelier, Joelle, and Senior Sophister. 2005. What could have prompted Keynes to call Montesquieu 'the Real Equivalent of Adam Smith, the Greatest of French Economists'? *Student Economic Review* 19: 3–15.

Hayek, Friedrich A. 1944. *The road to Serfdom*. Chicago: University of Chicago Press.
———. 2011. *The constitution of liberty [1960]*. Chicago: University of Chicago Press.
Ikeda, Sanford. 1997. *Dynamics of the mixed economy: Toward a theory of interventionism*. London: Routledge.
Kamensky, John M., and Albert Morales, eds. 2006. *Competition, choice, and incentives in government programs*. Lanham: Rowman & Littlefield.
Keynes, John Maynard. 1964. *The general theory of employment, interest, and money [1936]*. New York/London: Harcourt Brace Jovanovich.
Krugman, Paul. 2012. *End this depression now*. New York: W. W. Norton.
Lazzarato, Maurizio. 2012. *The making of indebted man: An essay on the Neoliberal condition* (translated Joshua David Jordan). Los Angeles: Semiotext(e).
Lessig, Lawrence. 2008. *Remix: Making art and commerce thrive in the hybrid economy*. New York: Penguin.
Lipsey Richard G. and Kelvin Lancaster. 2000. The general theory of second best [1956–7]. *The Review of Economic Studies*, 24: 11–32. Robert E. Kuenne, ed. 2000. *Readings in social welfare: Theory and policy*, 48–72.
Parker, Randall. 2007. *The economics of the great depression: A twenty-first century look back at the economics of the Interwar Era*. Cheltenham: Edward Elgar.
Patinkin, Don. 1969. The Chicago tradition, The quantity theory, and Friedman. *Journal of Money, Credit and Banking* 1: 46–70.
Popper, Karl R. 1966. *The open society and its enemies [1945]*. Princeton: Princeton University Press.
Raustiala, Kal, and Christopher Sprigman. 2012. *The knockoff economy: How imitation sparks innovation*. Oxford/New York: Oxford University Press.
Reich, Robert B. 2007. *Supercapitalism: The transformation of business, democracy, and everyday life*. New York: Alfred A. Knopf.
Rosenstein-Rodan, Paul. 1961. Notes on the theory of the Big Push. In *Economic development for Latin America*, ed. H.S. Ellis and H.C. Wallich. New York: St. Martin's.
Rousseau, Jean-Jacques. 1978. *On the social contract* [1762]. Edited by Roger D. Masters, translated by Judith R. Masters. New York: St. Martin's.
Russell, Bertrand. 2001. *Freedom and organisation, 1814–1914 [1934]*. London: Routledge.
———. 1949. *Authority and the individual*. Boston: Beacon.
Sassower, Raphael. 1988. Economics according to Popper. *Philosophy of the Social Sciences* 18: 383.
———. 2009. *Postcapitalism: Moving beyond ideology in America's economic crises*. Boulder & London: Paradigm.
———. 2010. Is homo economics extinct? *Philosophy of the Social Sciences* 40: 603–615.
———. 2013. *Digital exposure: Postmodern postcapitalism*. Hampshire: Palgrave Macmillan.
Smith, Adam. 1937. *An inquiry into the nature and causes of the wealth of nations* [1776]. Edited by E. Cannan. New York: The Modern Library.
———. 1976. *The theory of moral sentiments* [1759]. Indianapolis: Liberty Classics.
Soros, George. 2000. *Open society: Reforming global capitalism*. New York: Public Affairs.
———. 2008. *The new paradigm for financial markets: The credit crisis of 2008 and what it means*. New York: Public Affairs.
Stanley, Richard T. 2013. *Freedom, Common Sense, and the "Nanny State"*. Bloomington: Universe.
Stigler, George J. 1965. *Essays in the history of economics*. Chicago/London: The University of Chicago Press.
Stiglitz, Joseph E. 2002. *Globalization and its discontents*. New York/London: W. W. Norton.
Stockman, David A. 2013. *The great deformation: The corruption of capitalism in America*. New York: Perseus Books Group.
Tullock, Gordon. 1994. On the trail of homo economicus, ed. Gordon L. Brady and Robert D. Tollison. Lanham: George Mason University Press.

von Mises, Ludwig, and Bettina Bien Greaves. 2007. *Bureaucracy [1944]*. Indianapolis: Liberty Fund.

Weber, John. 2001. *Socially mixed economies: How social gains develop in opposed systems*. Lanham: Lexington Books.

Wilson, Edward O. 2012. *The social conquest of Earth*. New York/London: W. W. Norton & Company.

Chapter 36
Notes on Revisiting Klappholz and Agassi's "Methodological Prescriptions in Economics"

Lucien Foldes

I reconsider the paper 'Methodological Prescriptions in Economics', *Economica*, February 1959, by Kurt Klappholz and Joseph Agassi. I criticise the thesis that "there is only one generally applicable methodological rule, and that is the exhortation to be critical [...]." I compare the methodology of physics with that of economics, and discuss whether typical economic 'laws' are testable hypotheses. I consider whether an approach to economic policy based on welfare economics is biassed. In conclusion, I sketch an approach to economic methodology and suggest some rules of method that may be useful to the working economist.

1. I welcome the opportunity to contribute an essay to the *Festschrift* in honour of Joske's 90th birthday, although my choice of topic is limited both by my unfamiliarity with recent work on Critical Rationalism and by health problems which at present limit my access to literature. So I have decided on an ancient, but still quoted, contribution of Joske's to a topic in economics, and at the same time make a gesture *In Memoriam* of my late friend Kurt.

The article under discussion was a product of the renewed interest in the methodology of economics which found expression in the well-known, not to say notorious, seminar on Methodology and Testing which ran at the LSE in the 1950s. Both K and A were members, as I was, although I was considered by the dominant faction (which did not include K&A) as heretical and was accused of such un-Popperian sins as Essentialism, Conventionalism and even Inductivism. (If I give the impression that the atmosphere of the discussion was at times coercive, rather in the style of Agitprop, that is how I remember it.) The main programme of the seminar was of course to get rid of the *a priorist* methodology propounded by Lionel Robbins (Robbins 1945), and to replace it by a Popperian (Popper 1959, 1963) approach

I am indebted for comments on an earlier draft to Adam Ostaszewski, Sven Rady, Jörn Rothe and Carol Foldes (Carol Hewlett).

L. Foldes (✉)
London School of Economics and Political Science, London, UK
e-mail: l.foldes@lse.ac.uk

N. Bar-Am, S. Gattei (eds.), *Encouraging Openness*, Boston Studies in the Philosophy and History of Science 325, DOI 10.1007/978-3-319-57669-5_36

based on conjectures and refutations. These ideas are summarised in the paper under review.

2. But to begin at the beginning. K&A state their view that 'there is only one generally applicable methodological rule, and that is the exhortation to be critical and always to subject one's hypotheses to critical scrutiny. Any attempt to reinforce this maxim by a set of additional rules is likely to be futile and possibly harmful.' Taken at face value, this statement seems to deny that there is such a subject as methodology, so why read on? However, a clue that there is more to the thesis is afforded by the fact that somehow Popper's methodology seems to be exempt from the general stricture. Pursuing this line of thought, I have found the following brilliantly concise statement (albeit written at a later date) of Agassi's methodological position in his Apologia (Agassi 1993, 2008),[1] which suggests that the plea for a critical attitude may be taken as a coded invitation to adopt Popper's approach, perhaps with minor modifications.

> Please indulge me (*sic*) two paragraphs on the content of *Logik der Forschung*, since it is such a philosophical landmark. It raises two questions, discusses two traditional answers to each and presents alternatives to them. The questions: *One*: David Hume's problem of induction: how does theoretical learning from experience take place? *Two*: Immanuel Kant's problem of the demarcation of science: what theory is scientific? The inductivist-empiricist answers: *One*: theoretical learning from experience occurs when theories are backed by experience. *Two*: theories are scientific when they are backed by experience [...]. The new criticalist answers: *One*: theoretical learning from experience occurs when theories are refuted by new experiences. *Two*: theories are scientific when they are given to empirical tests that may lead to their refutation in case they are false. By the inductivist-empiricist view science is the set of all presently empirically backed theories plus all known empirical information; [...] by the criticalist view science is all known refutable hypotheses and all known possible empirical refutations to them, potential and actual, true (whose truth refutes some hypotheses) and false (whose falsity corroborates some hypotheses).

This passage goes beyond my remit, which is limited to economics. However, I shall venture some amateurish remarks. Agassi's discussion is limited to empirical science, but I prefer to adhere to the dictionary definition of science as '(pursuit or principles of) systematic and formulated knowledge', and to add a qualifier when a particular science is considered. This incidentally avoids the frequently heard question whether economics really is a science. Next, as to Agassi's question 1, consider first the problem of testing a deterministic theory, i.e. one which predicts a specific outcome from an experiment. Presumably the attraction of a 'rejectionist' approach to testing is that refutation by given evidence can be regarded as purely a matter of logic, whereas a (universal) theory can never be proved to follow logically from a body of evidence, however extensive. However, it seems perverse to limit the concept of learning to episodes of rejection. Clearly learning from experience occurs all the time even when there is no stated hypothesis to be confirmed or refuted. More important, the attempt to reduce learning to a purely logical procedure ignores its subjective aspect. The observation that a given situation or experiment usually has a

---

[1] Agassi (1993) pp. 170–171. The passage quoted is repeated with unimportant changes at p. 212 in Agassi (2008).

particular outcome creates an expectation that the same pattern will be repeated, at least if circumstances are not expected to change significantly, and in practice this expectation is usually justified. Everyday life would be very difficult if it were not so. In short, inductive learning, although subject to logical rules, is concerned with the dynamics of belief. (cf. De Finetti 1937). Further, it is well known that in the case of a statistical hypothesis, a finite set of data can at best yield an estimate of the probability that the hypothesis is false, or alternatively that it is true, which negates the attempt to reduce testing to an exercise in (non-probabilistic) logic; both rejection and acceptance become matters for decision, requiring consideration of practical criteria. Which brings me to my last comment under this heading, namely that the philosophical discussion of learning as summarised above seems to have little to say about the problems which face the scientist in practice, such as the selection of data and the design and interpretation of experiments.

Turning to Agassi's question 2, one wonders why a demarcation criterion should be sought at all, beyond the obvious requirement that an (empirical) science should try to explain some facts of experience. Here we seem to be concerned with the eternal human quest (so often disappointed) for reassurance, for firm ground, for *un doux et mol chevet.* (Montaigne, *Essais* III, xiii.) No demarcation for science in general is likely to be successful, because different sciences have different problems, standards and requirements. I shall return below to the question of a demarcation criterion for economics. To sum up this excursion into methodology *in general*, I venture to suggest that it offers little beyond what is evident to common sense.

3. Returning now to the article by K&A, we find the following further passage: '[...] scientific statements accepted without debate [...] are of little scientific interest. Scientific interest is aroused by controversial statements, and by the controversy surrounding them'. Again, 'the law of demand, the quantity theory of money, and similar propositions are of the utmost importance for economic policy. They are comparable, however, not to the laws of physics which are now the subject of controversy, like quantum theory, but rather to Gray's law of conductivity [...] which is obviously significant technologically. This law is as undisputed today as the law of demand, and equally manifest in everyday experience. Yet this is precisely why Gray's law is not of intellectual interest in current research in physics. Similarly, the present wide acceptability of the law of demand renders it an inadequate example of the significance of economic science. The significance of economics, as of any other science, lies in its search for new truths.' This comes pretty close to declaring that economics as a whole is not of intellectual or scientific interest, though it will hardly be disputed that economics has discovered new truths, or at least truths which were new when discovered.

Incidentally, going beyond my competence again, my impression is that quantum mechanics has aged a bit since the K&A paper was written and is now generally accepted (apart perhaps from its ontological implications) because it has survived a great variety of experimental tests. This surely does not deprive quantum mechanics of its interest for research. Indeed, because of its wide acceptance, it can be used as the basis for explaining further phenomena, e.g. the difference between superconductivity and 'classical' conductivity. Going further, I am uneasy about K&A's

tendency to dismiss technological questions as lacking scientific or intellectual interest. Many, perhaps most, theoretical enquiries are suggested by problems or observations encountered in applications. Indeed theory and application cannot be rigidly separated. It is also clear that what is interesting is largely a subjective matter; a result which seems obvious to a specialist may come as a revelation to an outsider and point to new connections between different subjects. In any case we need a better criterion of what is interesting than simply what is new and not yet well understood.

4. I have the impression that the passage quoted above is not seriously about economics, but rather about physics. This suggests the need for a brief review of some of the obvious differences between these subjects which are relevant to methodology.

As a general point, it is necessary to bear in mind the distinction between theories concerning processes which are stationary in the sense that historical time plays no part—e.g. theories which are modelled by autonomous differential equations—and non-stationary processes, which include all of economic history. Einstein could cheerfully declare 'for us physicists, the distinction between past, present and future has no meaning other than that of an illusion, though a tenacious one.' (Quoted in Honderich (1995) page 806.) Indeed, in the stationary case, if the laws of motion are deterministic, then once they are fully understood, repeated observation yields no new information, provided that initial conditions are suitably adjusted. (The situation is more complicated in the case of stochastic stationary processes, where under suitable conditions a single sequence of repeated observations yields information about the asymptotic time average of the process). Suffice it to say that we economists are not in the happy situation which Einstein evokes, since every day brings numerous changes influencing economic variables, and this casts doubt on any attempt to simply transcribe methodological rules suggested by physics. In economics, the date to which an observation relates is always significant.

Some other differences must be noted, even in a brief discussion. Physics deals with an environment which is more or less constant, or at least does not respond 'consciously' to what we do; to adapt Fred Hoyle's splendid phrase, there is in fact *no* bastard in the cloud. Thus it is possible to distinguish relatively sharply between observations which agree with our conjectures and those which do not, (although there remain well-known ambiguities concerning observational errors, probabilistic conjectures and interference effects). Economics, by contrast, deals mainly with decisions by people, in particular decisions about the allocation of resources, and the interaction among the results of these decisions through markets, governments and other social mechanisms. The design and operation of these mechanisms is itself a subject for economic analysis. A further essential feature of the economic process is that, at any time, decisions about present actions are influenced by expectations of future developments, and the expectations of different actors are liable to be interdependent. What occurs is affected by what is expected to occur. (Admittedly the analysis of these effects is often 'fudged' by an assumption of 'rational' expectations or something similar, but that is essentially an acknowledgement of our ignorance about the way in which expectations are actually formed, which no doubt is a major cause of present dissatisfaction with the state of the subject). A further point is that the outcome of decisions is dependent on the particular rules governing markets and other institutions, and these rules are the results of a historical evolution which must be studied if a correct understanding of the economic process is to be achieved. Finally, the scope for repeated observations under similar conditions is limited because history is ever-changing and irreversible.[2] Experiments in economics for purely scientific purposes are rare (outside the artificial setting of the psychological laboratory), and conditions for repetition of an experiment are liable to change if it is known that repetition is intended. So, when Max Planck said that he decided to study physics rather than economics because economics is too difficult, he clearly had a point.

5. Writers on economics often prefer to say little about the definition of their subject but rather plunge *in medias res*. Lionel Robbins clearly has some sympathy with this approach when he says, a trifle contemptuously, that 'economics is not one of those social sciences which are always discussing method before proceeding to deliver the goods' (Robbins 1945, pp. 115–116). Nevertheless he devotes a whole book to an attempt to demarcate economics from related subjects and to establish a special methodology for it. I shall not review his arguments here since they are well known among economists. My view is that his effort is not entirely successful, mainly because of the great areas of overlap between economics and related problems in psychology, biology and various mathematical subjects such as statistics and decision theory etc. Besides, one wonders whether demarcations are of much use other than for departmental organisation and curriculum design. Nevertheless, Robbins has identified a core of problems which are essential to economics, namely the problems of resource allocation, which in a market economy means the theory of supply and demand and its various generalisations and applications. It seems reasonable to attempt a methodology for this area, although the differences from related fields limit any extension. I shall return to this topic below.

The article by K&A avoids any need to define its domain of relevance because it is organised as a series of disputes with earlier authors, who may be assumed to be talking about economics. Mostly these critiques are excellent, but they are rather arcane and I do not wish to review them after the lapse of more than half a century.[3]

---

[2] In this connection it is amusing to recall Poincaré's discussion (Poincaré 1905, Ch. IX) of Carlyle's statement (*Past and Present*, Ch. 1, book II) that 'Nothing but facts are of importance. John Lackland passed by here. Here is something that is admirable. Here is a reality for which I would give all the theories in the world.' Poincaré comments that 'this is the language of the historian. The physicist (Bacon) would most likely have said 'John Lackland passed by here. It is all the same to me, for he will not pass this way again." Where does the economist stand on this? Maybe King John's journey was the one on which he lost his treasure in The Wash, in which case it would be relevant to the economic history of England, and its effect on the money supply might call for discussion. On the other hand this unique event is hardly of significance for the validity of any law of economics.

[3] However, one apparent error should be noted. K&A assert (p. 61) that 'a law, to be true, must correctly refer to all the relevant factors (affecting its validity), which must of course be finite in number (since otherwise we could not mention them all).' This runs counter to the common practice of lumping together in an 'error' term a set (possibly infinite) of 'small' factors which are

On the other hand, the lack of a constructive statement about economic methodology is a deficiency. The authors try to combine a literal interpretation of their slogan 'all you need is criticism', which makes methodology an empty box, with an insistence on an (updated) Popperian methodology which is more or less a trans-scription from physics. This is inconsistent and unhelpful.

However K&A do offer some remarks about laws of economics, and it is of interest to pursue this topic briefly. Of course the concept of law is not without its difficulties—see the articles "Laws, natural or scientific" and "Science, problems of the philosophy of" in the Oxford Companion to Philosophy. Of the alternatives on offer, I like Ramsey's suggestion that laws are 'a consequence of those propositions which we should take as axioms if we knew everything and organised it as simply as possible in a deductive system'. This fits in rather well with Robbins' characterisation of certain economic laws as 'simple and indisputable facts of experience [...] they are so much the stuff of our everyday experience that they have only to be stated to be recognised as obvious'.[4] So why not take them as axioms right away? Matters are not quite so simple, and I leave the point there for the moment.

6. I now turn to a discussion of some alleged laws mentioned in the K&A article. Recall that their discussion centres on whether these laws and related propositions are tautologies, and more generally whether they are testable.

First, the 'Law of Demand'. This is regarded by K&A as generally accepted (and therefore boring), but they do not remind us of its content. Presumably what is meant is either (a) the 'law' of diminishing marginal utility, or (b) the observation that demand curves generally slope downwards. Definition (a) is common among older authors.[5] There are various problems of interpretation with this proposition and some exceptions (e.g. addiction), but if suitably qualified it can reasonably claim the status of a law. Note that it does not follow from the modern theory of demand based on indifference curves unless utility is interpreted as 'cardinal' and the utility surface is strictly concave. As to (b), this does not in general follow from indifference curve analysis because it neglects the income effect.[6] Also, once one

assumed collectively to follow a particular statistical distribution. More generally, I am rather persuaded by a remark of Imre Lakatos (made in connection with mathematical theorems) that there is always an infinite number of 'hidden lemmas'. Be that as it may, I leave the reader to whet his critical teeth on this one.

[4] Robbins (1945), p. 78–79. To quote Poincaré again, (Poincaré 1902, Ch.IX): 'L'expérience est la source unique de le verité: elle seule peut nous apprendre quelque chose de nouveau; elle seule peut nous donner la certitude. Voilà deux points que nul ne peut contester.' Leaving aside whether this assertion is really incontestable, the reason for quoting it in French is that 'expérience' can mean either 'experiment' or 'experience'. Given the physical context, the English translation naturally chooses the former rendering. However the quotation from Robbins suggests that, in economics at least, the wider concept is more appropriate for deciding the acceptability of a law.

[5] 'The law of demand is to the effect that the utility afforded by any increment of any kind of desired object diminishes with increase of the amount possessed', W.E.Johnson in Palgrave's *Dictionary of Political Economy*, quoted in J. N. Keynes (1917), p. 245. Johnson regards this 'law' as a psychological datum rather than as a theorem of economics.

[6] The possibility that the income effects may outweigh the substitution effect, so that demand increases with price, is often presented as paradoxical, but it may be quite natural for some

leaves the abstract world of indifference curves and allows for dynamic, speculative and behavioural considerations there are lots of familiar exceptions to (b), so that this 'law' is little more than an observation of what usually happens and is at best testable in particular situations.

The next alleged law to be considered is the law of diminishing returns. The usual distinction must be made between returns to scale and returns to a variable input, but in either case it is often difficult to say precisely what the law asserts about a particular industry or to show that the relevant convexity properties hold at some points on the production surface (however that is defined). These difficulties aside, there are plenty of examples to show that many industries operate at levels where there are increasing returns to scale and even increasing, or at least constant, returns to variable factors. The question whether this behaviour is consistent with equilibrium (apart from situations of monopoly or oligopoly) is at the root of the theory of monopolistic competition. Attempts have been made to subject this theory to testing, but it is doubtful whether it was ever meant to be testable, or whether 'rejection' would ever lead to its abandonment or serious modification. It is really an enquiry into the logic and limits of competitive models, and any attempt to make it statistically testable would probably require such distortion of the theory as to make the interpretation of tests quite uncertain. On the whole, I am inclined to agree with Neville Keynes that the law of diminishing returns should be regarded as a 'bare physical fact' about the productivity of labour applied to land, and not as a true economic law (J. N. Keynes 1917, p. 85). Note that in the competitive case we can consider a Law of Supply to the effect that supply curves generally slope upward, with various reservations as in the case of demand.

A third example of a possible 'law' is the Quantity of Money equation $MV = PT$. It is generally acknowledged that, with appropriate definitions of the variables, this is a mathematical identity. To make the equation into a predictive theory requires hypotheses about the behaviour of the variables. Until quite recently the usual interpretation was to assume that the velocity of circulation V is relatively stable and that an increase in the quantity of money M would lead to an increase in transactions T until capacity constraints were reached, with a proportional increase in prices P thereafter (with no doubt an intermediate phase to allow for aggregation effects). With this interpretation we seem at last to have an example of a testable theory. And recent experience suggests a decisive rejection. Broadly speaking, central banks have created enormous quantities of cash, but rather than stimulate trade or raise prices the additional sums have largely been added to idle balances. This is not the place to discuss the causes of this behaviour, which are in any case familiar from the financial press. The point of interest here is that we have a testable hypothesis and a decisive rejection, at least in the short run; (over a longer period the vast accumulation of liquidity may yet lead to hyper-inflation, with the traditional interpretation of the Quantity Equation coming into its own). Of course the experiment with

---

'commodities', e.g. saving or leisure. Thus saving (=demand for future goods) may well fall as the rate of interest rises. Again, demand for leisure (=hours not worked) may well increase if the wage rate (opportunity cost of idleness) falls, i.e. if the supply curve of labour is downward sloping.

ultra-loose monetary conditions has not been undertaken in the interests of science, and it is amusing but sad to note that the failure of the policy has not led to its abandonment but (at least in some cases) to a feverish quest for ever more extreme measures in the same direction (negative interest etc.). From the methodological point of view, the natural inference is a rejection of the hypothesis of constancy of V for the period in question, but not a rejection of the Quantity Equation, even though formally the equation is being tested along with the hypotheses about V and the other variables. This turns out to be typical of the interpretation of tests where an economic law is involved.[7]

The economic laws considered so far have in common that they are basic building blocks for economic models yet, when correctly stated and hedged about with reservations, offer little scope for useful tests of general interest. As Robbins suggests, they represent more or less obvious facts of experience. The discovery of situations in which the laws seem not to hold does not lead to 'rejection' of the laws in general, but rather to the characterisation of these situations as exceptional, possibly because some parameters appearing in the laws take unusual values.

There are other examples of regularities which are so obvious once recognised that testing seems irrelevant (apart perhaps from measurement of parameters, which can be regarded as testing although such a presentation yields no special insight). An example is the demonstration that banks create money (and do not merely circulate a fixed amount of currency). Given a description of the way that banks operate, a very little algebra puts the matter beyond doubt. Of course it is necessary to state some assumptions, e.g. to specify reserve ratios and that banks can find customers to borrow their available funds. Similar remarks can be made about the national income multiplier.

7. One begins to wonder whether there are any non-trivial hypotheses in economics which follow the Popper/Agassi paradigm. Well, there are, up to a point. Consider for example Karl Marx's prediction that capitalism would lead to the progressive impoverishment of the proletariat (I cannot bring myself to say 'immiserisation'). It appears to be a consequence of a serious theory, namely Marx's Labour Theory of Value, and it is a bold, testable hypothesis. There are of course ambiguities and reservations, e.g. concerning the disappearance of much of the proletariat as Marx knew it, and the question whether the hypothesis refers to absolute or relative impoverishment. On the latter interpretation, the experience of recent years offers some confirmation of the hypothesis. (This might be regarded as a pity from Agassi's point of view, since rejection allegedly yields the best learning). However, the hypothesis relates to a unique historical sequence of events which cannot be repeated, and moreover a different result might be obtained if a different period were considered; clearly it would not be satisfactory to alternate between 'acceptance' and 'rejection'. It seems that, in the case of a historical process, the language

[7]This procedure may be regarded as comparable to the usual interpretation of the results of an experiment which seems to violate the law of conservation of energy, cf. Poincaré (1905), esp. pp. 185 and 159–160, also the articles "Poincaré" and "conventialism" in the Oxford Companion to Philosophy.

of conjecture and refutation, while not incorrect, is not very helpful. Better insight can be obtained by comparing the evolution of variables of interest—say, the share of labour in the national product—in various sub-periods, relating this evolution to explanatory variables in each period, and perhaps using this analysis to form conjectures about future trends. The emphasis in such work is on the collection and interpretation of data to obtain a correct record of events and the inter-dependence of changes in the relevant variables in different periods. In short, the emphasis is on 'understanding' what has actually occurred rather than on conjectures about what might have occurred under different conditions. (Naturally I avoid defining 'understanding', just as K&A avoid defining 'criticism').

8. This discussion, while rather unsystematic and incomplete, does prepare the ground for a sketch of a possible 'landscape' of economic methodology. Briefly, I see economics as comprising two main related parts, called (economic) theory and (economic) history.

Theory comprises (mathematical and other) models, which may be deterministic or stochastic (or perhaps of other types). Models are abstract, i.e. devoid of empirical content, and deductive. They are identifiable as 'economic' models by the names of the variables appearing in them (prices, money etc.) and by the types of relations postulated among the variables. Specification of a model typically includes a statement of some structural conditions, e.g. assumptions about preferences, production possibilities, market structure etc., which usually incorporate one or more 'laws'. Usually there are also assumptions about 'input' variables, in particular a statement about initial conditions. From these input variables and structural conditions, relations among 'output' variables are derived by logical inference. These relations may be indexed by time, also by 'state of nature' in the case of stochastic models. The inferences will usually assert relations among the output variables, and possibly also (theoretical) predictions of future values.

History comprises all relevant empirical data, such as records of events and statistical series, and even forecasts of future events. It is appropriate to stress the importance in economics of the routine work of collecting and preserving data, without which there cannot be adequate explanation of events. We do not usually have the possibility of filling in gaps by repeating experiments, nor can we rely on light reaching us across the vastness of space to tell us what happened long ago. This is not to belittle the role of imaginative reconstruction and even folk memory, but reliably recorded facts are better.

Given the sets of models and history, we consider *explanations* (of historical data by means of models). Much—maybe too much—has been written about the nature of economic explanation, notably in the nineteenth century *Methodenstreit* which opposed the German historical, or inductive, method to the English abstract, or deductive, approach.[8] The following sketch tries to avoid this dichotomy.

Suppose that we are interested in the development of a certain collection of (possibly random) economic variables during a given period. We are given the historical record up to the present. We suppose that our stock of models contains at least one

[8] For a brief survey, see J. N. Keynes (1917), Ch. I.

which formulates hypothetical relations among the variables in question (strictly, among the theoretical variables which have the same names as the empirical variables for which we have records). Given past and current observations of the variables, we use our model to formulate theoretical relations among the variables and to generate theoretical predictions for a suitable future period. If in due course the observed values of the variables are found to agree adequately with the theoretical relations and predictions, we are happy and go on to the next job. If not, we modify the model, usually by altering some parameters or introducing new explanatory variables. Complete rejection of the model is unusual since, as we have seen, the building blocks of economic models usually include 'simple and indisputable facts of experience'. Nevertheless models are sometimes abandoned and replaced by new ones when they are considered inadequate, perhaps because of new insights or due to changes in the economic environment.[9] Whatever the extent of agreement between prediction and observation, the learning process will involve recording the results, reviewing the adequacy of the data and the model, and no doubt repetition of the procedure for a further period. A trivial and obvious account, but one with a very different emphasis from the Popper/Agassi version of scientific learning.

9. Much ink has flowed over the question of the 'value freedom' (*Wertfreiheit*) of economics. Note that the preceding discussion relates entirely to 'positive' economics, where the answer appears clearly in the affirmative. But freedom from value judgements does not imply freedom from bias. As all students know, economic theory is largely organised around a number of simplified benchmark models—perfect competition, monopoly, Keynesian macro etc.—and these serve as the basis for the analysis of real-world situations. This is essential because of the complexity of the economy and the great variety of practical situations. However the habit of treating real situations as departures from ideal situations does introduce a bias into the analysis, which can become exaggerated if the structure of the economy changes while the benchmarks remain the same. (Why do we have to spend so much time learning about perfect competition when there is so little of it about?) Now this bias can lead to the implicit introduction of value judgements when the discussion turns to political economy, i.e. to economic policy. It arises particularly in applications of welfare economics, with its optimality theorems which establish a correspondence between competitive equilibria and Pareto optima. (Note that these are existence theorems, not practical recipes for finding the income or wealth distribution making a given optimum into an equilibrium; but often the distinction is blurred). The presentation of the theorems sometimes creates the suggestion that welfare for all can be improved by implementing a competitive regime and then moving to an appropriate distribution; which in practice is not possible, not least because the necessary information is lacking, and certainly not if the necessary redistribution is not actually undertaken. One is often left with the impression that 'competition is best' because it will lead to an optimal allocation of resources, never mind the possible distributive consequences. The so-called compensation criteria involve similar sources of bias if compensation is not actually paid. Besides, adoption of a particular

[9] Compare the discussion by Poincaré (1905), p. 160.

measure on the grounds that the gainers can over-compensate the losers may pre-empt alternative measures which are even more desirable. Thus in practice the application of results from welfare economics tends to become a vehicle for the promotion of 'free market' policies which can be represented as having advantages for the allocation of resources while neglecting their distributive consequences. What would help to validate this approach would be a sort of law of large numbers for policies satisfying 'allocative' criteria, i.e. a result showing that repeated use of such policies would eventually lead, at least with high probability, to allocations making almost everyone better off. But I am not aware of any such result and an attempt to establish one might just add to the list of failed Ph.D.s.

10. I now return to the original issue of the existence of useful methodological prescriptions in economics. I have so far left aside the qualification implicit in K&A's assertion that 'there is only one *generally applicable* methodological rule [...]' This qualification unfortunately makes their thesis quite ambiguous. May we not seek further rules of more limited applicability but perhaps of greater practical use to the working economist if not of great interest to the visiting philosopher? Such rules, I suggest, are embedded in the prolonged training which we give to students before admitting them to the ranks of qualified economists. After all, any teacher knows that adolescents are usually endowed with plenty of critical attitude, but that alone does not equip them to be researchers or policy advisers. Since acquisition of knowledge of appropriate methods is inseparable from the process of study as a whole, it is not possible in a short note (and perhaps not at all) to make a complete list. But in any case, the box of rules is far from empty. The more specific the problem under discussion, the more detailed are the available rules, though of course they are subject to criticism and revision. But even at a rather high level of generality, there is useful advice available. To give a few examples, consider the following:

(a) Consider separately the determinants of demand and of supply when analysing price formation, distinguishing the periods over which the various factors operate.
(b) Accompany any point forecast with an estimate of its uncertainty, or give a probabilistic forecast.
(c) When devising a policy, remember that people respond to incentives, often in undesired ways; for example, price controls are liable to create shortages and black markets. More generally, watch out for feedback and other systemic effects of a policy designed to bring about a specific consequence.
(d) When framing recommendations for policy, try to separate factual analysis of the consequences of action from value judgements.
(e) Be wary of formulations of policy problems which tend to predetermine the outcome of analysis, for instance comparisons of an actual situation with a standard defined by an idealised situation such as perfect competition, particularly where no actions are available to mitigate undesired side-effects such as redistribution of income.

(f) Economic experiments in the real world (as distinct from the psychological laboratory) are not carried out primarily in pursuit of theoretical knowledge, but rather for profit or to influence the behaviour of an economic system, and outcomes may be ambiguous and take a long time to appear. Besides, decision-makers commonly have various, often not clearly specified, hypotheses in mind. This usually makes the classification of a particular outcome as rejection or confirmation ambiguous.

Some of these proposed rules are so well established as to appear trivial, others are more difficult to apply or controversial, but all address situations which commonly face the working economist and which have sometimes been neglected with unfortunate results. Many other examples could be given. It is a boring list, and for my pains in drawing it up I have been compared to Polonius. But no matter, if I have shown that useful methodological advice can be given. Criticism is no doubt necessary, but not sufficient.

References

Agassi, Joseph. 1993. *A philosopher's apprentice: In Karl Popper's workshop*, 1st ed. Amsterdam/ Atlanta: Editions Rodopi B.V.

———. 2008. *A philosopher's apprentice: In Karl Popper's workshop*, 2nd edition. Amsterdam/ New York: Editions Rodopi B.V.

De Finetti, Bruno. 1937. La prévision: ses lois logiques, ses sources subjectifs. *Annales de l'Institut Henri Poincaré* 7 (1): 1–68.

Honderich, Ted, ed. 1995. *Oxford companion to philosophy*. Oxford: OUP.

Keynes, J.N. 1917. *The Scope and Method of Political Economy*. 4th ed. London: Macmillan.

Klappholz, Kurt, and Joseph Agassi. 1959. Methodological prescriptions in economics. *Economica* XXVI (101): 60–74.

Poincaré, Henri. 1902. *La Science et l'Hypothèse*. Editions Ebooks libres et gratuits.

———. 1905. *Science and hypothesis*. London/Newcastle-on-Tyne: Walter Scott Publishing Co, Ltd. Project Gutenberg EBook.

Popper, K.R. 1959. *The logic of scientific discovery*. London: Hutchinson.

———. 1963. *Conjectures and refutations*. London: Routledge & Kegan Paul.

Robbins, Lionel. 1945. *An essay on the nature and significance of economic science*. 2nd ed. London: Macmillan.

Chapter 37
Joseph Agassi, the M²T Seminar, and His Influence on My Work

Richard G. Lipsey

I am a professional economist with only a rudimentary knowledge of some basic methodological principles. Most of these I learned from Joseph Agassi, either directly or through his friend and my colleague, Kurt Klappholz. I never met Karl Popper but, through Joseph, we were also introduced to many of Popper's ideas. All of this happened through the vehicle of the LSE Staff Seminar on Metholology, Measurement and Testing in Economics that became known as the M²T Seminar. In it we half dozen founding members, as well as several later additions, learned much from each other as we sought to apply what we were learning about methodology to our own work in economics. Since I have nothing new to offer by way of methodological insights, I have been asked to discuss how Joseph influenced my subsequent work. I hope to persuade readers that his influence was not insignificant.

37.1 M²T and Agassi

I have told in several places the story of how I became one of the founding fathers of the M²T Seminar. (See in particular Lipsey 1997.) So I can be brief here.

As an undergraduate in honours economics at the University of British Columbia in the late 1940s one of the prescribed books was Lionel Robbins' *Essay on the Nature and Significance of Economic Science.* It impressed me greatly until I came to his discussion of the place of empirical data in economics. There I read: "The propositions of economic theory, like all scientific theory, are obviously deductions from a series of postulates. And the chief of these postulates are all assumptions involving in some way simple and *indisputable* facts of experience [...]. (1935, 78 Italics added). If the premises relate to reality the deductions from them must have

R.G. Lipsey (✉)
Simon Fraser University, Vancouver, BC, Canada
e-mail: rlipsey@sfu.ca

N. Bar-Am, S. Gattei (eds.), *Encouraging Openness*, Boston Studies in the Philosophy and History of Science 325, DOI 10.1007/978-3-319-57669-5_37

a similar point of reference" (1935, 104). I was surprised at reading this and thought that it could not be right. Surely, I thought, scientific truths (as I then thought of them) could not be arrived at by mere introspection based on personal experience; empirical observation should play a more important role than just illustrating propositions that intuition tells us are true.

In 1953 I entered the LSE as a PhD student. I attended Lionel Robbins' great Wednesday afternoon seminar first as a student and then from 1955 as a staff member. My contemporaries and I admired Lionel for his many great qualities, but we found ourselves increasingly dissatisfied with his methodology. Assessing theories by whether or not their assumptions appealed to our intuition seemed increasingly unsatisfactory. We cast around for an alternative method of assessing the increasingly complex theories to which we were being exposed. A small group of us met informally for a year to discuss matters of methodology. One of our original members, Kurt Klappholz, introduced us to Joseph Agassi, or "Joske" as we came to know him. Joseph in turn introduced us to Popper's works. Although Popper was on the LSE staff, being allergic to cigarette smoke, which the young staff produced in quantity, he never came to the Senior Common Room and was unknown to those of us who did not already have a knowledge of formal methodology. After a year, we decided to create a seminar devoted to the new (to us) methodology of testing and refutation that we had discovered under Agassi's tutelage, and so M²T was born.

The primary methodological influences on we economists were Popper, as seen through Agassi's eyes, and Agassi's own many insights, as well, a bit later on, Imre Lakatos. Although many of Lakatos' ideas have been criticized, I always found the concept of an untestable core and a testable protective belt helpful and used it in my essay "IS-LM, Keynesianism and the New Classicism". To Joseph, more than any other single person, we owe our conversion from the Austrian methodology as expounded by Robbins to a view that the mark of a good theory was its ability to pass tests where there was a real change of finding conflicting evidence.

Although I am not a methodologist and have not followed the modern controversies among that profession, what we learned in M²T times has helped me enormously in subsequent work and although there are no doubt many areas in methodology where I am uninformed, I believe that if modern economists knew as much methodology as Joseph taught us, their work would be much more productive than it often now is. In the remaining part of this essay, I will illustrate this by noting a number of methodological ideas that we mainly learned from Joseph and that I have put to what I hope is good use.

37.2 An Introduction to Positive Economics

In 1960, fired with the enthusiasm that I had developed through the M²T seminar, I decided to write a text book that would expound our methodological ideas in contrast to those that certainly prevailed in the UK, and probably in many other places as well. At the same time, I volunteered to teach the first year economic course in

LSE's newly revised BSc Econ degree; so the teaching and writing complemented each other. It should have taken me about two years from start to publication but, as discussed below, I ran into the problem of the Keynesian identities, which delayed me for about a year. The first edition was published in October 1963 and ran to only 559 pages. I was still working on the index when the limousine called to take us to Southampton where my family and I were sailing for a sabbatical year as visiting professor at UC Berkeley California. I called the book *An Introduction to Positive Economics* partly in an allusion to positive versus normative and partly to emphasise testability. It was known as *IPE* for years and only latterly as 'Lipsey'. Some economists misconstrued the title as indicating that I was a logical positivist, which I certainly was not, and I did write at the time an essay (Lipsey 1964) explaining my main objectives in writing the book.

The IPE's introduction began with a long chapter on methodology, stressing the importance of testing and evaluating theories according to their empirical relevance rather than their intuitive appeal. It also dealt with many of the questions that had plagued my students earlier, such as: How can we have a science concerning something so erratic as free-will based human behaviour? The microeconomics half was divided into four parts. Part 1 stated such basic material on how competitive markets worked that I did not think a chapter on criticisms and tests was called for. So I ended instead with a chapter on applications which showed how useful the theory was and how well it stood up to being exposed to data in its everyday use.

The other three micro parts, the theory of demand, production and distribution, covered material that was open to debate in places so I ended each part with a chapter on criticisms and tests. The object was to show students that what they learned was not necessarily absolute truth. Instead it might be replaced by better knowledge in the future—this in contrast to the revealed-truth approach of many of the then-current textbooks. In the chapter on the theory of demand I took on Lionel Robbins' contention that economic measurements were of little use to economists other than to record transient historical events. I have discussed this issue in greater detail in Lipsey (2009) an essay written to commemorate the 50th anniversary of the publication of *The Nature and Significance.* The macro half of the book dealt with Keynesian theory, but devoid of the then ubiquitous definitional identities, about which more below. It ended with two chapters on a recurring interest of mine, the relation between economic analysis and economic policy. The book was an instantaneous success, being reprinted 5 times in its first three years, being translated into over 15 foreign languages, and becoming an established text in several places beyond the UK, such as Spain, Italy and India.

The book was so different from any existing text that I am sure it would never have succeeded in today's world where the course lecturer, or a committee of his peers, selects a text and students are told to buy it. In the 1950s and 1960s the typical UK instructor would give his own set of lectures and tell students at day one: "There are several good textbooks around and you should complement my lectures by reading the text that best suits your needs; here is a list of some of the good ones." So when I wrote an unorthodox book, students could, and did, vote with their wallets to make it a market leader within months, long before some instructors had opened its covers. Indeed it was published some weeks after the beginning of the 1963–1964 academic year, which would spell certain death for a modern textbook but which saw its first printing selling out before the end of that academic year.

The UK edition evolved along the lines I had established for the first five editions but came under increasing market pressure from adopters as the selection system for textbooks superseded the previous textbook anarchy. Under relentless market pressure, about which I am not proud for succumbing to, the chapters on measurements and tests were dropped. The methodology chapter was slowly downgraded on the grounds that much of it was now old hat, rather than revolutionary. However, when I see some of the methodological confusions among today's economists, I am not sure that this was good advice. At the 8th edition, the publisher was changed from Weidenfeld and Nicolson, to the Oxford University Press and a co-author, Alec Chrystal, was added. Starting with the 9th edition the book was re-titled *Economics*. The 13th edition was published in 2015. The American edition (Lipsey and Steiner 1966) was always titled *Economics* and went through twelve editions with several co-authors. Although this did not have the same sweeping success as did IPE, it did do quite well. I recall getting a gold bound edition commemorating the one millionth copy sold and it certainly made my first American co-author rich.

I think the first few editions of IPE did have a major influence in persuading at least a generation of non-American students of the value of adopting a fact-controlled view of the subject; that nothing in economics is revealed truth, including the virtues of a market-oriented system, but is subject to alterations in the face of new conflicts with the facts; and that nothing could be learned about the world by manipulating definitional identities. It is not for me to evaluate these claims of influence. But anyone who does try should not be misled by the fact that the later versions of *Economics* are not radically different in approach from its many modern competitors. It is the first four or five editions that need to be studied to see the nature of the non-orthodox approach and to assess its impact.

Be that as it may, I could never have written *IPE* without the inspiration provided by all of the members of the M^{2}T seminar, and the lessons in basic methodology that I got from Joske, either directly from him, or indirectly from the others.

37.3 Methodological Applications

Not surprisingly, since Robins had made intuitively obvious assumptions a major component of economic theory, we spent much of the seminar's first year considering the whole issue of the place of assumptions in economy theory.

37.3.1 The Primacy of Popperian Predictions Rather Than Robbinsian Assumptions

Early on we came to what was to us a great revelation: a theory that was consistent with all states of the universe was empirically empty and so could not explain anything. One aspect of this understanding was that a theory that included an unspecified set of *ceteris paribus* conditions was empty, since any apparent refutation could be explained by one of the items on the (unspecified) list having not remained constant. The economic journals are replete with models that are consistent with all possible observations and a great improvement in relevance would be achieved if editors asked, before publishing any theory, that its author explain what set of observation would conflict with the theory.

Here is a case in point. Not long after our group began our methodological discussions, Kurt Klappholz asked me if I remembered the two of us criticising a student's theory because it could not explain (was not consistent with) some observations that we conjectured. So we were criticising him because his theory was not consistent with all conceivable observations and hence not empirically empty. It was an enormous step for all of us from thinking that a good theory had to be consistent with all conceivable observations to accepting that the more conceivable observations with which it was inconsistent, the more its empirical content. (I still remember how resistant we were when Joske first made such a statement.)

Closely related is the issue of sunrise texts: developing and testing a theory whose only predictions are already well known observations. This is one of economists favourite occupations. The concept of a stylised fact is relevant here: an alleged 'fact' that captures in generalised from some commonly made set of observations. Having stated one such 'fact', theorists often set out to build quite elaborate models that predict that fact and nothing more. We learned from Popper that building a theory that only passed sunrise tests did not add to our understanding of the world. Of course, if the theory also produces some other predictions, either explicitly, or in the nature of some of its empirical assumptions, knowledge can be advanced by testing these. But all too often such added bonuses are non-existent. As the editor of the *Economic Journal*, wrote in his retiring essay (Hey 1997: 4): "It often appears that the model has been constructed for no other purpose than to produce a result which is a stylised fact observed by the author. Simply producing a model that comes up with a desired result is a test of the cleverness of the author not a test of the relevance of the theory."

In all subsequent work my co-authors and I have been careful to ensure that any theory that we propounded made predictions that were capable of being confronted with conflicting observations.

37.3.2 Testing of Assumptions

Early on we had to consider Milton Friedman's (1953) essay on "The Methodology of Positive Economics", which is the only material on methodology that many American economists read. Critical in this context was Friedman's contention that only a theory's predictions, and not its assumptions, should be subject to test. After long consideration, we rejected that contention and much of the rest of Freidman's methodology. (For details of our collective views see Klappoholz and Agassi (1959).) Mark Blaug, who joined the M²T seminar not long after its inception, has criticised "[…] the license that Friedman's 'methodology of positive economics' gave economists to make any and all unrealistic assumptions, provided only that their theories yielded verifiable implications. (Blaug 1998: 20)[1] Since one of the key purposes of behavioural assumptions is to link a theory's predictions to observable behaviour, the view of the extreme irrelevance of assumptions reduces theory to mere operationalism. It then does not make sense to ask which of the assumptions is causing the trouble if a theory that seemed to accord with the facts no longer does so.

I have argued in several places that "[...] if empirically correct predictions were deduced from a set of empirically false assumptions, this called for further serious study, not complacency." (Lipsey 2008, 526).

But can we test behavioural assumptions? Here is an example that I have used in several places. (See for example Lipsey 2013.)

> It is commonly assumed in micro economics that firms have positively sloped marginal cost curves. Statistically estimated cost curves show, however, that many such curves are often horizontal. It makes sense to ask what caused this difference between the assumption and the empirical evidence. It is then found that the assumption is in fact derived from two prior ones: that the law of diminishing returns (or variable proportions) holds and that the fixed factor is subject to an equality constraint so that the firm must use all of it as output is varied over the short run. If so, the ratio of the fixed to the variable factor must vary as output varies, causing costs to vary according to the law. But evidence shows that in many (possibly most) firms the fixed factor is subject to an inequality constraint so that the firm can alter the use of its fixed factor up to its maximum capacity by leaving some it unemployed. In this case, the use of the fixed and variable factors can be altered as output varies, allowing the ratio of the two to be held constant at its optimal value and so avoiding the conditions under which the law holds.
>
> Consider what we have done here. We started by investigating a commonly made assumption about the shape of a firm's cost curve; one that is often used in both micro and macro theorizing. We found that the evidence was in conflict with the assumption. When we sought the source of the problem, we found that this assumption was in fact a prediction derived from two prior assumptions and that one of these, that the fixed factor was subject to an equality constraint, was often not correct. Now we have learned to be cautious of any theory for which the shape of the firm's short-run cost curve […plays a critical part].

[1] For a full discussion of this issue from an economist's point of view see Blaug (1992: 91–105).

37.3.3 Robustness of Assumptions

In his discussion of assumptions, Friedman did not recognise the several different senses in which assumptions are used in economics and the different implicit ways in which they should be assessed. Some assumptions are merely designed to simplify the problem without altering it in substantial ways. For example, the neoclassical theory of production takes place on the head of a pin, in which spatial location does not matter, and in many actual cases it does not. But if we are considering competition among retail establishments, their locations in relation to each other matters a great deal and needs to be introduced into the theory if it is to cover variables that matter. (For a full study of these issues see Lipsey and Eaton 1977.) These two cases illustrate the important requirement that assumptions that are patently counter factual be robust in the sense that they can be relaxed without seriously altering the theory's predictions.

A modern example of this issue can be found in the literature on what are called General Purpose Technologies, GPTs. My co-authors and I have defined this concept as follows: *A GPT is a single technology, or closely related group of technologies, that is widely used across most of the economy, is technologically dynamic in the sense that it evolves in efficiency and range of use in its own right, and is an input in many downstream sectors where those uses enable a cascade of further inventions and innovations.* (Bekar et al. 2016). Early on, the introduction of one of these GPTs, electricity, was shown by Paul David (1990) to be associated with a slowdown and then an acceleration of productivity growth. It soon came to be believed that this macro sequence of slowdowns followed by booms might be a general result of the introduction of all new GPTs. Several theories were developed to produce that result, usually using patently counter-factual assumptions to do the job. (See for example Aghion and Howitt (1998) and Helpman and Trajtenberg (1998a, b) These assumptions were not robust because removing them removed the phenomena being considered. Later theories (especially Carlaw and Lipsey 2011) were constructed with assumptions that captured as many as possible of the characteristics of these technologies and then asked if the patterns of slowdowns and booms emerged as predictions. In these theories, the pattern occurred only sometimes, thus raising the interesting question of what were the circumstances that led to this pattern of macro behaviour, a question that could not be studied in theories that were expressly constructed always to produce that behaviour.

37.3.4 Naïve Versus Conditional Refutations

Early on many of us became naïve falsificationsts. (I like to think that I was not one but cannot point to printed evidence and so must accept that I was probably one these.) Of course theories in the natural sciences do for all intents and purposes get refuted. For example, the evidence against that biblical account of the age of the earth and the simultaneous creation of all species is refuted by an overwhelming body of evidence. But such finality is rare in the social sciences. Whatever we now believe, there is always the possibility that new evidence, or a new interpretation of existing evidence will upset one of our currently cherished beliefs. Where we have conflicting theories the best that can usually be done is to assess the balance of evidence for one theory over the other, and where there is only one theory the test can always be against the hypothesis that the behaviour in question is determined by a random process. One of our founding members, Chris Archibald, wrote a valuable article reflecting much of our thinking on this issue (Archibald 1967).

37.3.5 Two Alleged Dichotomies

I had been raised on two dichotomies that Joske taught me to look at in a more nuanced way.

Normative and Positive Statements

The division between normative and positive statements was one of the great early simplifying distinctions in my life. But under Joske's persistent criticisms I came to see that this distinction was not as clear as I had thought at first. Eventually I came to the conclusion that I stated in the 3rd edition of *Positive Economics* (1971, 5, n.1).

> Having grasped this distinction, the student must beware that he does not turn it into an enquiry-stopping, dogmatic rule. From the fact that positive economics does not include normative questions (because its tools are inappropriate to them) does *not* follow that the student of positive economics must stop his enquiry as soon as someone says the word ought. Consider the statement: 'It is my value judgment that we *ought to have* rent control because controls are *good*.' Now it is quite in order for you as a practitioner of positive economics to ask 'Why?' It may then be argued that controls have certain consequences and it is these consequences which are judged to be good. But the statement about the consequences of rent control will be positive testable statements. Thus the pursuit of what appears to be a normative statement will often turn up positive hypotheses on which our *ought* conclusion depends [...].
>
> Philosopher friends have persuaded me that, when pushed to limits, the distinction between positive and normative becomes blurred, or else breaks down completely. The reason for this is that when examined carefully most apparently normative statements reveal some positive underpinning (e.g. 'Unemployment is worse than inflation *because* the (measurable) effects of unemployment on human beings are judged by the majority of adult citizens to be more serious than the (measurable) effects of inflation'). I remain convinced, however, that, at this stage of the development of economics, the distinction is a necessary working rule the present abandonment of which would contribute more to confusion than to clarity. The justification for this view is that although we are not sure what to make of an apparently normative statement (because it may has a positive underpinning) we do know a purely positive statement when we see one.

Knowledge-Based Advice and Value-Judgment-Motivated Decisions

Another of the dichotomies that were important in my early development as an economist was the one that I think I first got from Robbins in *The Nature and Significance,* that economists advised how the world worked while decision makers combined that knowledge with their value judgments to make decisions about which economists qua professionals were neutral. Joske and other philosopher friends criticised this sharp division but I was unimpressed by their arguments until I got inside Whitehall as an economic analyst in the newly formed National Economic Development Office (the so-called NEDY) in 1961–3. When I returned to the LSE, I presented a paper to the M²T seminar to the effect that this strict division between neutral analysists and committed policy makers did not stand up. It was a long time later that I put these reservations into print as Lipsey (1981). Sadly, by including it in a collection of essays directed at Canadian economists, it did not get the attention that I think it deserved since its message is quite general. The argument stretches over 20 pages so I can only mention here its general conclusion.

Robbins could make this clear dichotomy because he believed that economic knowledge was certain knowledge. As soon as we admit that all knowledge is subject to being upset by subsequent observations; that what we think we know now, we know with less than 100 percent certainty; and that economists and policy makers must communicate with each other through the imperfect media of language not mathematics, the relationship between the economic analysist and the policy maker becomes much more opaque. "The economic advisor and the policy maker are involved in complex human relationship, entangled in various uncertainties and communicating with each other through an inevitable haze of emotional reactions. Economists may strive towards an ideal of communicating their knowledge as objectively as possible, but objectivity remains an ideal that guides their actions, not a reality that fully describes them" (Lipsey 1981, 35.)

37.3.6 Testing Existential Statements

Early on in our M²T seminar deliberations we encountered Popper's distinction between a prediction that was refutable by finding conflicting evidence but not provable because one could never rule out the possibility of finding conflicting evidence in the future and an existential statement that was provable by finding the postulated item but not refutable because one could never rule out finding it through some later more embracing study.

This issue never arose for me in a practical situation until much later when I became involved in the development of a new idea in growth theory: the importance of the General Purpose Technologies (GPTs) discussed in an earlier section. Two economists (Moser and Nicholas 2004) studied the patents associated with electricity and found them not to be anything like as far reaching as GPTs were supposed to spread their influence. From these observations they concluded that electricity was not a GPT. However, if there ever was a GPT, it is electricity because it enables the great bulk of all modern technologies that would not be possible without it. (If you doubt this, observe what stops functioning when electricity is cut off by some power failure.) The problem these researchers had was that patents do not cover many of the places in which electricity enabled the development of other new technologies. More generally, the proposition that electricity is a GPT is an existential statement and the correct conclusion of their study was that *their patent data was not sufficient to establish that electricity was a GPT* but *not that the data was sufficient to establish that electricity was not a GPT*.

This led me to consider when an existential statement could be disproved and to conclude that if it was circumscribed enough so that all possible places where it was postulated to exist could be located, then it could be disproved. So to re-use Popper's famous example, the statement that 'angels exist' can be proven by finding one but cannot not be disproved, while the statement that 'angels exist in observable form and are in this room' can be disproved by not finding one after an exhaustive search of the room. The importance of this distinction lies in the criticism of GPT theory by some economic historians. They criticise the concept because they argue that it does not rule out such trivial but widely used technologies as screws as being GPTs. However, one of the characteristics of a GPT is that it does not have close substitutes and it is clear that screws do have these. All of their uses can be enumerated and it can then be shown that in each case there are close substitutes. So the statement that electricity is a GPT cannot be refuted by evidence while the statement that screws are GPTs can be.

We may agree on this but many journal editors apparently do not–after all the refutation of electricity being a GPT was published in one of the most influential economic journals, the *American Economic Review*.

37.3.7 Identities and Behavioural Equations

One of the most difficult conceptual issues that I ever had to grapple with, and one that still confuses many in the profession, concerns the formulation–starting from Keynes himself and running up to the 1970s–of Keynesian macro-economic theory using identities rather than behavioural eqs. I well remember sitting in the LSE cafeteria and hearing a famous American economist, who was much older, and we presumed much wiser than us, saying that one of Keynes' great contributions was to 'discover' the important set of identities that underlay his theory.

Indeed during the period from the end of WWII through the 1960s, the Keynesian definitional identities permeated the text books at all levels with empirical propositions being 'derived' from their manipulation! The first year textbook version took the form of the famous diagonal cross diagram. Expenditure was divided between an endogenous component that varied with national income, Y, and an exogenous component that was independent of income. In the simplest version the endogenous

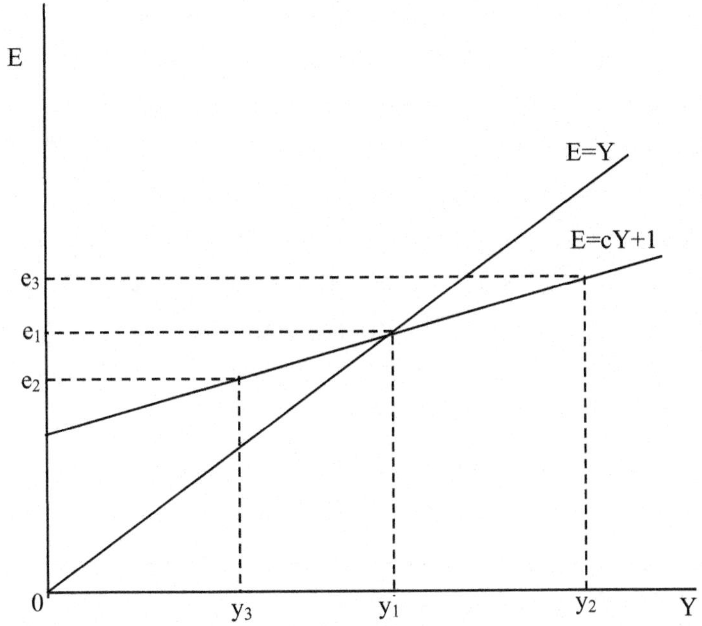

Fig. 37.1 The diagonal cross

component was consumption expenditure, C, while the exogenous component was investment expenditure, I.

$$E = C(Y) + I \tag{37.1}$$

and in the simplest linear form.

$$C = cY \tag{37.2}$$

where c is the fraction of income spent on consumption. Then by substitution

$$E = cY + I \tag{37.3}$$

which was the expenditure function.

The 'model' was then completed by the definitional identity

$$E \equiv Y. \tag{37.4}$$

(Indeed these two variables are defined as identical in national income accounting.)

Plotting (37.3) and (37.4) produced the diagonal cross diagram as shown in Fig. 37.1. It should have been obvious that this was not a valid prediction-generating model by observing that if E and Y were just two different symbols for the same thing, (37.4) could be substituted into (37.3) to yield

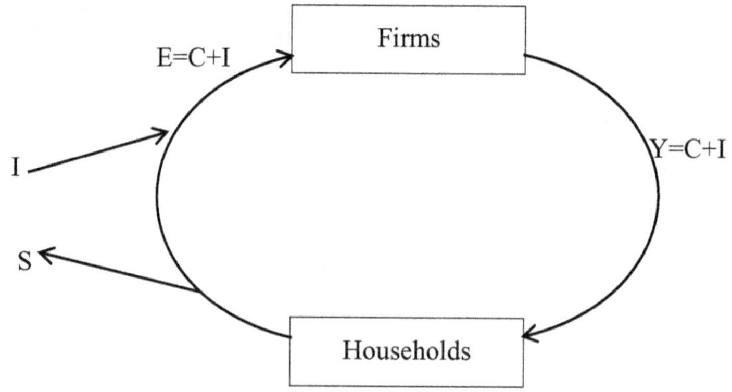

Fig. 37.2 The circular flow of income and expenditure

$$(1-c)Y \equiv I \tag{37.5}$$

which states that, given I, there is *only one possible value for* Y, $(I/(1\text{-}c))$, which is $y_1 \equiv e_1$ in the figure.

One other common Keynesian proposition of the time is implicit in the above. The fraction $(1 - c)$ represents the fraction of income not spent on consumption, which in Keynesian theory is the fraction saved, i.e., $(1 - c)Y \equiv S$. So (37.5) can be rewritten as

$$S \equiv Y \tag{37.6}$$

telling us one of the early Keynesian dogmas: that savings must always be equal to investment, not just in equilibrium but always. Of course there is nothing wrong in defining them that way but it is an error to then think this tells us anything about real world expenditure on saving and investment when they are not defined to be the same thing.

I had been exposed as a graduate student to Bill Phillips' water machine model of the circular flow of income and expenditure between firms and households. Firms create income for households by producing the national income, Y, which is divided been consumption and investment production. Households allocate their expenditure between consumption, which creates income for firms, and savings. The income and expenditure related to consumption circulates in a closed circuit while investment is modelled as in injection into this circular flow and savings as a withdrawal from it. In the machine, these two flows are only equal when income is unchanging. When it is rising, the injection of investment expenditure exceeds the withdrawal of savings and when it is falling the opposite relation holds between savings and investment.

When I came in 1961 to begin the macro half of *Positive Economics* I laid out a numerical example of the circular flow model as in Fig. 37.2 and tried to reconcile the flows with the Keynesian identities. Try as I might, I could not do this and I spent much time in an intellectual fog. A recourse to the major texts did not help because they all used the identities with neither hesitations nor qualifications. Finally, I realised that the identities could not tell us anything that was happening in the actual circular flow model nor in the actual world. All of this may sound quaint to today's readers but to me it was a great insight and it took much time to convince my colleagues that the text books were all wrong on this issue.

I wrote this up in an article in which my RA and I surveyed the literature to locate the many unacceptable uses of the identities. I started the paper with the following list of incorrect statements all drawn from the contemporary literature Lipsey (1972, 303–4).

> *I. The Static Model in Equilibrium*
>
> 1. The equilibrium of a Keynesian model is given by the intersection of the aggregate demand function and the 45° line which expresses the accounting identity $E \equiv Y$.
>
> *II. The Static Model in Disequilibrium*
>
> 2. Although people may try to save different amounts from what people try to invest, this is impossible; thus actual realised savings must always equal actual realised investment.
> 3. Out of equilibrium, planned savings do not equal planned investment, and it follows from (2) that someone's plans must be being frustrated; thus there must be some *unintended* savings or dissavings, and investment or disinvestment.
> 4. The simultaneous fulfilment of the plans of savers and investors can occur only when income is at its equilibrium level in just the same way as the plans of buyers and sellers can only be simultaneously fulfilled at the equilibrium price.
>
> *III. The Dynamic Behaviour of the Model*
>
> 5. Whenever savers plan to save a different amount than investors plan to invest, a mechanism works to ensure that actual savings remain equal to actual investment, in spite of people's attempts to make it otherwise. Indeed, this mechanism is the source of dynamic change in the circular flow of income.
> 6. Since the real world, unlike the simple textbook model, contains a very complex set of inter-relations, it is not an easy matter to see how it is that savings stay equal to investment even in situations of the worst disequilibrium and the most rapid change.
> 7. The dynamic behaviour of a Keynesian circular flow model in which disequilibrium implies unintended investment or disinvestment can be analysed by moving upwards or downwards along a diagonal cross diagram in the 'step-fashion' illustrated in Fig. 37.1 [Figure not shown here].
>
> Each of these statements can be found in the majority of standard treatments of national income theory and there are few books that do not make some of them. The statements would appear to be regarded as part of established knowledge in the subject. Yet every one of them is wrong.

The paper went on to show how each was mistaken in predicting real world behaviour derived from the identities. Here I only mention one of the many absurdities. In Fig. 37.1 where income and expenditure are defined to be the same thing and are shown by the 45° line, there is only one level of income consistent with the way the curves are drawn in the figure, y_1. Since the identity means that we have the same variable on each axis, there cannot be actual points off the 45° line. Usually implicitly, but sometimes explicitly, the expenditure function was interpreted as a desired function whose points could lie off the 45° line while all actual points had to lie on that line. But then the text books invariably went on to consider what was happening when income took on other values, such as y_2 or y_3 in the figure and expenditure was not equal to income, all of which were ruled out by the definition that Y and E were the same thing. In the paper we go into the many confusions that stemmed from trying to do the impossible of dealing with disequilibrium situations when no actual points can lie off the 45° line, as well as dealing in detail with the other errors listed above.

The resolution of this problem was to regard E as a desired expenditure and Y as GDP actually produced. Equilibrium income would then occur when people desired to purchase the whole GDP, no more and no less. Thus $E = Y$ became an equilibrium condition stating that agents desired to purchase just what was produced, no more and no less.

My research assistant, John Stilwell, and I co-authored an appendix to my paper titled "A Short History of the Middle." We surveyed the pre WWII savings-investment debate in which many of the correct points were made and contrasted this with the almost total lack of the correct analysis in the literature of the several decades following WWII. Even Paul Samuelson got it wrong when he wrote: "measured S and measured I stay identical even during the worst disequilibrium – by virtue of some people experiencing unintended losses or gains and/or experiencing unintended investment or disinvestment. (Samuelson 6th edn. 261). It appears that according to Samuelsson that during a disequilibrium real behaviour is being affected in order to keep the identity fulfilled! We concluded with the following observations (Stilwell and Lipsey, 339–340):

> We have, therefore, the two facts that the majority of modern elementary and intermediate textbooks, together with parts of much more advanced literature, treat the Keynesian identity as an inexorably se1f-establishing equilibrium condition, and that the pre-war saving and investment debate, although giving rise to much confusion, contained a number of articles which correctly exposed the errors and changes – notably those of Robertson, Haberler and Lutz. How can these be reconciled? The error did not enter the most influential textbooks, Benham and Samuelson, until relatively late. The fourth edition of Benham (1948) was free of the errors that first appeared in the fifth' edition (1955). Benham's book was, however, innocent of Keynesian economics until 1955 so that the errors are introduced the first time Benham attempts to expound Keynesian theory. The case of Samuelson's extremely influential textbook is more interesting. Samuelson's book was of course Keynesian in spirit from the outset but it is error free until the fourth [1958] edition.
>
> Perhaps the error in its modern form originated somewhat independently of Keynes. It might have stemmed from the combination in most textbooks of a chapter dealing with elementary national income accounting, and a chapter dealing with income flows where the simplest model, that of one injection, one withdrawal, system is considered, the quantities being labelled respectively investment and saving. The definitions used in the accounting chapter may have crept over into the latter parts of the other.
>
> Alternatively it is possible, and perhaps more likely, that by the middle fifties the salutary effects of the debate which followed the publication of the General Theory had been forgotten and the reputation of Keynes led to the inclusion in textbooks of the crudest expressions of the errors. This is only speculation; the fact which we have is that the customary exposition of simple macro-static models is worse that it was twenty years ago.

My Article was rejected by *Economica* and, preoccupied with other things, I did not try to get it published elsewhere, (for which I am now sorry). Then, when a festschrift was announced for Lionel Robbins' 80th birthday, I dug out the article and it was included in his volume–and there met a fate that was at the time usual for pieces in conference volumes: almost totally ignored.

But I did get it right in the first edition of *Positive Economics* where I introduced macro theory in terms of circular flow diagrams in which *S* and *I* were equal only in equilibrium; and where E = Y was an equilibrium condition expressing an equality between what is actually produced, *Y*, and what agents *desire* to purchase, *E*. I also added an appendix "National Income Accounting" where I explained in detail the distinction between the definitional identity of *E* and *Y* in national income accounting and the use of these as desired variables that were only equal in equilibrium in economic theory. I concluded (IPE 1st edn, 360): "The student needs to operate in a theory of income with concepts of savings and investment which do not make these two magnitudes the same thing. The student need not worry further about the use of saving and investment that makes those two magnitudes identical."

Given IPE's instant success as mentioned in Section II, I like to think that I had some part in the transformation over the next couple of decades from using E ≡ Y as an equilibrium condition to interpreting that identity as the way ex poste measurements were made and using E = Y as an equilibrium condition.

37.3.8 Internally and Externally Driven Research Programs

Another important methodological issue was not discussed directly in the M²T seminar, although all of its substance was implicit in our deliberations. I came upon this issue directly when I was asked to give a plenary lecture to the meeting of "The International Network of Economic Method" at The University of British Columbia in June 2000, subsequently published as Lipsey (2001).

I chose as my subject the mathematizing of economic theory in the second half of the twentieth century. Most economists were aware of how much the power of mathematical analysis allows them to do that could not be done with verbal and geometrical analysis, the main analytical tools when I was an undergraduate in the long-ago 1940s. So I concentrated on what had been lost. One important loss has been much of the knowledge of the world about which today's students are theorising. Such knowledge that was once part of the education of economists has gone to clear time for learning more and more complex mathematical analysis. While students of Industrial Organisation can develop complex theorems in Game Theory, many of them have only the slightest knowledge of the world to which these models are meant to apply. In a different graduate-school box, students of advanced micro economic theory see the world's production capacity as typified by a single representative price-taking firm (i.e., one without market power) who encounters constant returns to scale is it expands its activities. Microeconomics text books once had lot to say about the highly important phenomenon of increasing returns, but since these create non-convexities that are not easy to incorporate in the analysis based on the assumption of universal convexity, they have been largely eliminated from the text books. For every 100 students in either of these groups who can handle the complex theory to which they have been exposed, maybe two or three have some idea of how firms actually behave: how they set their prices, taking account of what they can and cannot change; how their behaviour is constrained by governments, how much by the need to pacify shareholders, and how much by the need to shade behaviour in the light of strong public concerns about environmental issues. In short, those in micro economic theory spend so much time studying models of the workings of the hidden hand they have no time to study the workings of the visible hand–the title of Alfred Chandler's great book (1977) on how businesses actually do behave.

Much though I believe that Joske would have joined eagerly in such a discussion if it had been relevant then rather than now, the point I want to raise here is the distinction between two types of research program. I define an internally driven research program, an IDRP as a research program that is driven by its own internal logic; investigators seek to understand problems created by the models that they are using, rather than deriving their problems from observations. In contrast, an "externally driven research program" (EDRP), is one that is driven by, and constrained by, observed facts.

An IDRP is not always conducted without any appeal to facts. It often begins with some factual question and later, if any tests are used at all, they typically take the form of trying to make the model track the data. Since investigators are usually free to alter the detailed specifications of such key components as variables, and lags, tracking the facts is a very weak test. With a bit of ingenuity almost any model can be made to track a set of "relevant" facts.[2] Many of the fads and fashions that sweep economics are aspects of one internally driven research program or another.[3]

Early post WWII growth theory was typical of an IDRP in that at least one the originators, Evsey Domar, had empirical concerns—about the possibility of maintaining full employment in a capitalist economy – concerns that were raised by the experiences of the Great Depression and the Second World War. His simple model (Domar 1946) assumed a fixed ratio of capital to labour and produced a knife-edge equilibrium, departure from which caused the economy either to fly into unctrollbable inflation or uncontrollable depression. Robert Solow (1956) in a path-breaking article showed that if substitution between the two factors of capital and labour was allowed for, the model could settle into a balanced growth path. This was an

[2]A particularly worrying habit of economists, when they are making models track the data, is to try many alternative specifications and then publish only the one that has the highest t statistic (or the best value of whatever test of goodness of fit is relevant to the techniques being used). Then, with a straight face, report these statistics as if they were genuine tests of the adequacy of the model.

[3]This is not to say that internally driven programs never produce interesting results; they sometimes do. The probability of their producing empirically relevant results—even when there is a demand that their models be made to track the data—is, however, quite low.

important contribution but it lead to an IDRP that lasted nearly two decades. What if there are more than two factors of production? What if there is more than one output? What is the most efficient way to move to a balanced growth path if the economy is off it originally, and so on and so on for more than 15 years.

At the end of it all this effort the Nobel Prize winning economist, Amartya Sen, had this to say about the enterprise that had absorbed the attention some of the subjects greatest minds for so long.: "The policy issues related to economic growth are numerous and intricate. [...] While the logical aspects involved in these exercises are much better understood now than they used to be, perhaps the weakest link in the chain is the set of empirical theories of growth that underlie the logical exercises. Possible improvement of policies towards growth that could be achieved through a better understanding of the actual process of growth remains substantially unexplored. [!] It is partly a measure of the complexity of economic growth that the phenomenon of growth should remain, after three decades of intensive intellectual study, such an enigma. It is, however, also a reflection of our sense of values, particularly of the preoccupation with brain-twisters. Part of the difficulty arises undoubtedly from the fact that the selection of topics for work in growth economics is guided much more by logical curiosity than by a taste for relevance. The character of the subject owes much to that." (Sen 1970: 33)

Economics is replete with this type of IDRP. Someone comes up with an interesting theory that predicts a surprising result. One or two researchers seek to test the prediction, while most seek to determine if the model is robust in the sense that the prediction will survive making the model from which it is derived more complex in any number of ways, including making it 'more realistic'. Soon this activity becomes dominant and rounds of theorising ensue. Eventually, the original issue is forgotten while efforts go into making more and more elaborate models that began with the original predictor. Eventually, a consensus evolves that this effort is going nowhere and it peters out – often with a conference summing up and asking why so little came from so much effort. Whereas Joske and the other members of the M²T seminar could have told them, once the IDRP rather that the EDRP nature of the program had been established, that this would be the end result.

37.3.9 Economic Policy with and Without Welfare Rules

Finally I come to my biggest issue. It began with my undergraduate training, was fully discussed at M²T, and for me is still bearing unexpected fruit. Classic textbook economic policy is based on a highly idealised model of the working of the price system. It is a static equilibrium model with given technology and in which all markets contain so many buyers (households) and so many sellers (firms) that no one of them can exert any significant influence on the market – what economists call perfectly competitive markets. Everyone must accept the market prices as they are set by the impersonal forces of demand and supply (i.e., set by the collective behaviour of all buyers and sellers each motivated by self-interest). A. C. Pigou, the founder of formal welfare economics, argued in his path breaking book *The Economics of Welfare* (1920) that this model produced an optimal allocation resources that maximised the sum of wellbeing.

During our time in the M^{2}T seminar, the profession was just going over from the verbal geometrical analysis of this model, to a formal mathematical proof of the existence of the model's equilibrium as pioneered by Arrow and Debreu's formulation of it as a general equilibrium mathematical model. Pigou's argument for the optimality of this model was replaced by this more formal proof, which required many conditions beyond what was in Pigou's treatment, such as that there be a market for every single good and service; that there are no externalities where behaviour in one market has effects beyond that market (think pollution!), that there be no public goods such as parks where one person's use does not preclude another person's use. The formal treatment derived what came to be called the two fundamental theorems of welfare economics that graduate students devote much effort to proving. According to Mark Blaug who, although not a founding member, joined our seminar shortly after to become a frequent contributor to our ideas: "The first fundamental theorem states that, subject to certain exceptions—such as externalities, public goods, economies of scale, and imperfect information—every competitive equilibrium is Pareto optimal." (Blaug 2007: 185) A Pareto optimal allocation of resources is one in which it is impossible to make any change that would result in making someone better off without making someone else worse off.[4] Note Blaug's irony here as he knows full well that the 'exceptions' to which he refers are found almost universally in real economies. "The second fundamental theorem states that every Pareto-optimal allocation of resources is an equilibrium for a perfectly competitive economy, provided a redistribution of initial endowments and property rights is permitted; alternatively expressed, every Pareto-optimal allocation of resources can be realized as the outcome of competitive equilibrium after a lump-sum transfer of claims on income." (Blaug 2007: 185)

Standard, text-book economic policy consisted at that time (and much of it still does) in identifying the existence of things that would prevent the establishment of a Pareto efficient allocation of resources in the model; calling them 'distortions' or 'market failures' and then advocating their piecemeal removal wherever they were found in the real world (i.e., virtually everywhere). (This analysis used to be part of what was then called 'welfare economics' but is now called 'public economics'.) This policy had already been given a severe blow on its own terms by the publication of my first article, "The General Second Best" (Lipsey and Lancaster 1956), We showed that even in this abstract model, the piecemeal establishment of any number of the conditions needed for a full optimum would not necessarily make people better off (actually or potentially) as long as not all of the conditions were fulfilled. In other words the conditions for maximising any objective function do not provide rules for piecemeal increases in the value of that function when its maximum is not achieved.

[4] In this context a person's welfare is indicated by his utility function that orders for him/her all possible consumption states.

A deeper criticism, and one we discussed in the seminar, is that it is odd to argue that the conditions needed for an optimum allocation of resources in a model so far from reality should be a guide to what should be done in the real economy. Indeed, the members of the seminar were in general agreement with the messages of the several critical books on welfare economics written before the two fundamental theorems had taken fully over from the older form of analysis (e.g., Little 1950 and de Villiers Graaff 1957). These books argued in depth that the conditions required for the abstract model to produce an optimum allocation were so very far from the reality of any real market economy as to make it a poor guide for real-world economic policy. One of our founding members, Kurt Klappholz, gave a series of lectures on welfare economics and the irrelevance of the standard approach to policy derived from it. (I urged him to publish this excellent set but he had a severe writer's block that possibly came from being the only one of his family to survive a Nazi concentration camp.) It is an interesting comment on the point raised in an earlier section that when welfare economics was taught verbally and graphically, students were routinely exposed to criticisms of its applicability. But when it became mathematised in the form of the two fundamental theorems, students get almost no exposure in the text books to criticisms similar to those in the earlier books that argued that the conditions necessary for an optimum allocation of resources in the neoclassical model are so far from reality that they provide an imperfect or even are irrelevant guide for policy.

Most real policies have effects on both efficiency and distribution, there being winners and losers. The second fundamental theorem is often taught to students as a way of separating issues of efficiency from those of allocation. You suggest a change that raises efficiency and then notionally make lump sum transfers from the winners to the losers so that no one losses. Of course the idea of doing this in practice is, as Baumol (2002, 143) says a "fairy tale" that should be discarded. It is, however, a tale that all too many students repeat when they say that the distributional effects of this efficiency policy are to be offset by lump sum transfers. Even when real-world payments are made to compensate losers, these do not (indeed could not) take the form of lump sum transfers. So in reality, if not in the classroom, efficiency and distribution are inevitably intertwined when any actual policy measure is considered.

There is a vast amount of literature relating to second best and economic policy, some of which is summarised and critiqued in Lipsey (2007) long after it was published. Fortunately, we can bypass this material and begin in 1990 when I had a major epiphany on these issues that took me well beyond any M²T discussions. In that year I was asked by the Canadian Institute for Advanced Research to lead their international program on "Economic Growth and Policy," which I did in 1990, resigning as leader after it was fully set up and continuing as a member well into the 2000s. My great insight was that technology and technological change had to be understood in a much more nuanced form than in the way it occurs in all macro growth models – whether as exogenous (as in the Solow growth models) or endogenous (as in the Romer and Lucas models) – where it appears as a scalar parameter or variable, It came as a great insight to me that we need to treat technology as endogenous and something that needs to be understood in detail, as did Nathan Rosenberg (see e.g., 1982, 1994) rather than as a single broad aggregate, as do virtually all models based on an aggregate production function. It took a long time to see how much of what is taught in economics depends on technology being exogenous and how many cherished propositions change as soon as its endogeneity and complexity are accepted. (There is no space to go into all of these changes here, changes I have discussed in some detail in Lipsey 2013.)

Stated boldly: because we live in a dynamic rather than a static world, the concept of a static, optimal allocation of resources has no policy relevance. Even if we were given full information about everything that can be known, an optimum cannot be defined, let alone attained, because of endogenously generated technological change undertaken under conditions of pervasive uncertainty. In a world of uncertainty, agents must make what are at best informed guesses about the values of different lines of activity, including R&D. Different agents will make different guesses, even under identical conditions. In such a world, static maximisation is impossible and, as I have many times observed, firms are more correctly seen as groping into an uncertain world in a manner that is profit seeking but not profit maximising. Here the concept of a static optimum allocation of resources is undefined since no one can know in advance the actual or expected outcomes of each possible resource allocation.

Many of the 'distortions' that are seen as undesirable sources of market failures in the neoclassical theory are the very things that drive economic growth in a dynamic, evolving and uncertain world. They are thus to be encouraged not supressed. For just one example, firms compete with innovations designed to produce the large temporary profits that are needed to persuade them to undertake the risky ventures – the vast majority of which fail – of developing and introducing the new products, new processes and new forms of organisation that are the essence of economic growth.

The surprising fact is that so many economists have accepted the view that the conditions that would make a wholly imaginary static world optional – including zero profits in all lines of production – are likely to be the conditions that would improve matters in the dynamic real world. The critical contrast between the policy advice of neoclassical and evolutionary theories is well stated by Mariana Mazzucato, the R. M. Phillips Professor of Economics at the Science Policy Research Unit, Sussex England (2015: 5–6): "According to neoclassical economic theory […] the goal of government policy is simply to correct market failures […] once the sources of failures have been addressed […] market forces will efficiently allocate resources [...Instead] nearly all the technological revolutions in the past–from the Internet to today's green tech revolution–required a massive push by the State. Silicon Valley's techno-libertarians might be surprised to find out that Uncle Sam funded many of the innovations behind the information technology revolution."

Many economists have argued that if we abandon the policy rules derived from welfare economics and the neoclassical theory of the optimum (static) allocation of resources, as e.g., Mazzucato had advocated, there can be no rationally based economic policy. (See for example Ng 2017.) In disagreeing (see Lipsey 2017a, b), I join Mark Blaug in arguing that although the appearance of scientific certainly is removed when the welfare rules are no longer seen as the underpinnings of economic policy, economists still have much to add to policy discussions – just as they have been doing for decades. Just because we cannot be absolutely sure that anything we do will unambiguously raise the entire community's welfare, policy need not be without rational guidance, being based on a mixture of evidence, theory and judgement. As Mark Blaug put it (2007, p. 202–3: italics added): "Virtually all economists are in favor of road pricing because they believe that the potential Pareto improvements created by reducing road congestion and saving travel time greatly outweigh the costs of installing the necessary hardware, including the enforcement costs of policing whatever system is installed. There will be many gainers but there will also be many losers,… Here once again we have the classic difficulty of separating efficiency from equity. What do economists in fact bring to such an argument? First of all, a considerable familiarity with the facts regarding the use of public and private vehicles. Secondly, a considerable familiarity with the facts regarding family income and transport expenditure patterns, including the possession of private cars, allowing for accurate estimates of the price elasticity of demand for more or less fuel-efficient vehicles, not to mention the price elasticity of demand for gasoline. Thirdly, considerable experience with survey evidence in large-scale social experiments comparing families and individuals with unequal access to transport to gauge the effect of, say, a miles-travelled tax rather than a fuel-based tax, possibly varying across people with different risks of causing accidents. *None of this will provide neat answers to the revolutionary introduction of road pricing. All it will do is to add one more highly informed voice to the squabble and that, I say, is what modern welfare economics is about and ought to be about, rather than teaching and learning a set of mathematically expressed fundamental theorems*."

I completely agree with Mark. This is just what I do when involved in the many policy issues that occupy my time–and something that I do without any recourse to the welfare conditions for achieving an optimum allocation of resources in static and imaginary world.

37.4 Concluding Remarks

Philosophers debate many subtle and esoteric issues on methodology, issues that are beyond my ability to understand fully – at least given the time I must devote to my own speciality. But surely it is not without interest to methodologists to see, as illustrated in this essay, that many relatively straight forward methodological principles that might be found in a methodology 101 course can, if understood and applied properly, have big effects on economics and, I suspect, would also do so on many other social sciences.

References

Aghion, P., and P. Howitt. 1998. On the macroeconomic effects of major technological change. In *General purpose technologies and economic growth*, ed. E. Helpman. Cambridge, Massachusetts Institute of Technology Press.

Archibald, G.C. 1967. Refutation or comparison. *The British Journal for the Philosophy of Science* 17 (4): 279–296.

Baumol, W.J. 2002. *The free market innovation machine*. Princeton: Princeton University Press.

Bekar, Clifford, Kenneth Carlaw, and Richard Lipsey. 2016. *General purpose technologies in theory, applications and controversy: A review*. Simon Fraser University Working Paper, 16–15.

Blaug, Mark. 1992. *The Methodology of Economics*. 2nd ed. Cambridge: Cambridge University Press.

Blaug, Mark. 1998. Disturbing currents in modern economics. *Challenge* 41 (3): 11–34.

———. 2007. The fundamental theorems of modern welfare economics, historically contemplated. *History of Political Economy* 39 (2): 18–207.

Carlaw, Kenneth, and Richard G. Lipsey. 2011. Sustained endogenous growth driven by structured and evolving general purpose technologies. *Journal of Evolutionary Economics*. 21 (4): 563–593.

Chandler Alfred, D. 1977. *The visible hand: The managerial revolution in American business*. Cambridge, MA: Harvard University Press.

David, Paul. 1990. The dynamo and the computer: An historical perspective on the modern productivity paradox. *The American Economic Review* 80 (2): 355–361.

de Villiers Graaff, J. 1957. *Theoretical welfare economics*. Cambridge: Cambridge University Press.

Domar, Evsey D. 1946. Capital expansion, rate of growth, and employment. *Econometrica* 14 (2): 137–147.

Eaton, B.C., and Richard Lipsey. 1997. *On the foundations of monopolistic competition and economic geography: The selected essays of B. Curtis Eaton and Richard G. Lipsey*. Cheltenham: Edward Elgar Publishing.

Friedman, Milton. 1953. The methodology of positive economics. In *Essays in positive economics*. Chicago: University of Chicago Press.

Helpman, E., and M. Trajtenberg. 1998a. A time to sow and a time to reap: Growth based on general purpose technologies. In *General purpose technologies and economic growth*, ed. E. Helpman. Cambridge: Massachusetts Institute of Technology Press.

———. 1998b. Diffusion of general purpose technologies. In *General purpose technologies and economic growth*, ed. Elhanan Helpman. Cambridge: Massachusetts Institute of Technology Press.

Hey, J. 1997. The economic journal: Report of the managing editor. *Royal Economics Society News Letter*, 3–5.

Klappholz, K., and J. Agassi. 1959. Methodological prescriptions in economics. *Economica* 26 (101): 60–74.

Lipsey, Richard G. 1963. *An introduction to positive economics*. London: Weidenfeld & Nicolson. 2nd edition 1966. 3rd edition 1971.

———. 1964. Positive economics in relation to some current trends. *The Journal of the Economics Association* 5: 365–371.

———. 1972. The foundations of the theory of national income: An analysis of some fundamental errors. In *Essays in honour of Lionel Robbins*, ed. B.A. Corry and M.H. Preston. London: Weidenfeld and Nicolson.

———. 1981. Economists, policy makers and economic policy. In *Economic policy making in Canada*, ed. D.C. Smith. C.D. Howe Institute: Montreal.

———. 1997. An intellectual autobiography. In *The selected essays of Richard Lipsey: Volume I: Microeconomics, growth and political economy. Volume II: Macroeconomic theory and policy*. Cheltenham: Edward Elgar Publishing.

———. 2000. IS-LM, Keynesianism and the new classicism. In *Macroeconomics and the real world, volume 2: Keynesian economics, unemployment, and policy*, ed. Roger E. Blackhouse and Andrea Salanti. Oxford: Oxford University Press.
———. 2001. Successes and failures in the transformation of economics. *The Journal of Economic Methodology* 8 (2): 169–201.
———. 2007. Reflections on the general theory of second best at its golden jubilee. *International Tax Public Finance* 14 (4): 349–364.
———. 2008. Positive economics. In *The New Palgrave: Dictionary of economics*, vol. 6. Basingstoke: Palgrave Macmillan UK.
———. 2009. Some legacies of Robbins' An essay on the nature and significance of economic science, *Economica* 76(1), "Special issue Robbins's Essay at 75″: 845–856.
———. 2013. Some contentious issues in theory and policy in memory of Mark Blaug. In *Mark Blaug: Rebel with many causes*, ed. Marcel Boumans and Matthias Klaes. Cheltenham: Edward Elgar Publishing.
———. 2017a. Generality versus context specificity: First, second and third best in theory and practice. In *A symposium of second and third best theory: Criticisms and application*, ed. Richard G. Lipsey and Yew-Kwang Ng. Forthcoming in *The Pacific Economic Review* 22(2).
———. 2017b. Economic policy with and without maximizing rules. In *A symposium of second and third best theory: Criticisms and application*, ed. Richard G. Lipsey and Yew-Kwang Ng. Forthcoming in *The Pacific Economic Review* 22(2).
Lipsey, Richard G., and Alec Chrystal. 2015. *Economics*. 13th ed. Oxford: Oxford University Press.
Lipsey, Richard G., and B. Curtis Eaton. 1977. The introduction of space into the neo-classical model of value theory. In *Studies in modern economics*, ed. M.J. Artis and A.R. Nobay, 59–96. Oxford: Basil Blackwell.
Lipsey, Richard G., and Kelvin Lancaster. 1956. The general theory of second best. *The Review of Economic Studies* 24: 11–32.
Lipsey, Richard G., and P. O. Steiner. 1966. *Economics*, 1st ed. New York: Harper & Row. 13th Edition 2007 with C. Ragan. New York: Addison Wesley.
Little, I.M.D. 1950. *A critique of welfare economics*. Oxford: Clarendon Press.
Mazzucato, Mariana. 2015. *The entrepreneurial state: Debunking public vs private sector myths*. New York: Public Affairs.
Moser, P., and T. Nicholas. 2004. Was electricity a general purpose technology? Evidence from historical patent citations. *American Economic Review* 94 (2): 388–394.
Ng, Yew-Kwang. 2017. Theory of third best: How to interpret and apply? *The Pacific Economic Review* 22 (2).
Pigou, Arthur C. 1920. *The economics of welfare*. London: MacMillan and Compnay Limited.
Robbins, Lionel. 1935. *An essay on the nature and significance of economic science*, 2nd ed, rev. and extended. London: Macmillan.
Rosenberg, Nathan. 1982. *Inside the black box: Technology and economics*. Cambridge: Cambridge University Press.
———. 1994. *Exploring the black box: Technology, economics and history*. Cambridge: Cambridge University Press.
Sen, Amartya, ed. 1970. *Growth economics*. Harmondsworth: Penguin.
Solow, Robert M. 1956. A contribution to the theory of economic growth. *Quarterly Journal of Economics* 70 (1): 65–64.

Part VI
Philosophical Encounters

Chapter 38
To Transcend the Popper-Agassi Impasse

Nathaniel Laor

> *A: I dreamt I was a butterfly, said Chuang Tzu.*
> *B: I dreamt I was human, said I.*
> *A: I dreamt the world was my oyster, said Falstaff.*
> *B: We do the best we can, anyway, whatever it is.*
> *A: I wish I were more human, say we all.*
> *B: I wish I were a butterfly, said Borges. (Agassi 2008, p. 312)*

38.1 A Domestic Scene

At fully 92 years of age, Sir Karl Popper veritably leaped from his chair. Striding across his living room, Sir Karl rushed to pull from a loaded bookshelf, Joseph Agassi's *Science in Flux* (1975). Slowly he walked back again, with the book already wide open in his hands, pointing with his accusing finger at the specific lines he was so anxious to show me, as he sank back into his chair to face me again. Here, he announced, was demonstration in clear refutation of my assertion that he, Karl Popper, is deeply respected and acknowledged by his former apprentice.

This scene goes back about a quarter of a century ago, and its focus was a paper published half a century ago. The issue at hand was Popper's declaration to me on the phone that due to his former student's ingratitude, he was unwilling to contribute to Agassi's *Festschrift* then planned by I. C. Jarvie and me.

Prior to the encounter described above that took place at Popper's home, I had made numerous phone calls to Sir Karl. Mrs. Mew, his assistant-and-secretary, repeatedly intercepted these. Upon hearing about the target of my request, she conveyed to me his final negative response and the conclusion that there is no point in my personally conversing with Sir Karl. Each time I would pass through London, I tried my luck and only hit the same brick wall yet again. As the deadline for the

N. Laor (✉)
Departments of Medical Education and Psychiatry, Sackler Faculty of Medicine, Tel Aviv University, Tel Aviv, Israel

Department of Philosophy, Lester and Sally Entin Faculty of Humanities; and Child Study Center, Yale University, New Haven, CT, USA
e-mail: nlaor@me.com

N. Bar-Am, S. Gattei (eds.), *Encouraging Openness*, Boston Studies in the Philosophy and History of Science 325, DOI 10.1007/978-3-319-57669-5_38

Festschrift was approaching and Mrs. Mew kept imploring me to stop calling Sir Karl as he would not change his mind on the matter, I promised her to abide by her request provided that I hear the refusal from Sir Karl himself. Next, I heard was Sir Karl pounding at me with his deep voice "I shall never contribute etc." I asked matter-of-factly whether he was ready to explore with me the refutability of his statement. "Come at once" was his answer – the answer I had intimately known and responded to with great enthusiasm ever since I had met Professor Agassi and wished to become his student.

Two hours later, I found myself seated in front of the warmest, most charming, and intellectually vibrantly engaging person. I had heard so much about him. His ideas had become foundations of my growing self and culture since my young adulthood. Authoritatively, he read out loud Agassi's clear words as they appear in his "Sensationalism" (1966, in Agassi 1975, p. 110): "[…] even he [Popper] agrees that accepting a report is, for the time that it is accepted, considering it to be true. My own alternative is that observation reports ought to be accepted as a task, as something we should try to explain, and this does not exclude the possibility that we should explain that piece of information as based on an error." Agassi, claimed Popper, must have known that he (Popper) had stated that much and that, therefore, there was no room for Agassi's dissenting sentence. In one step, so it then seemed to me, Popper demolished both my description of Agassi's attitude toward him as well as my attempt to mollify his resolve in denying Agassi's *Festschrift* his contribution. I promised to take this lesson back to Agassi and get his response. Agassi listened carefully to my report of the meeting with Popper and promised that if Popper's contribution to the *Festschrift* be that above mentioned comment on of him, he would offer his response in writing and if found insufficient, to add his apology.

My attempt to facilitate a critical dialogue between my two heroes of critical rationalism was quickly aborted. The historical evidence endures in print: the correspondence between Popper and Agassi (1992–1993) at the end of Agassi's second edition of his *A Philosopher's Apprentice* (2008) and the lacuna left by Popper in Agassi's *Festschrift* edited by Jarvie et al. (1994). Ever since then, as had been the case for decades before this reported event, Agassi's puzzlement and inability to formulate Sir Karl's neglect and accusation of plagiarism and ingratitude was gnawing at him and he ever so often would bring the problem up in discussions with me, hoping for an explanation for Popper's attitude and behavior toward him. I, who had witnessed them both in person as well as interacting through me, venture to offer two kinds of explanations, psychological and philosophical, both to the teacher's behavior and the student's inability neither to comprehend nor to walk away peacefully.

38.2 Setting the Philosophical Stage

The classical psychological explanation would be quite straightforward and refer to the psychoanalytic theory of the Oedipal relationships that prevail between Father and Son figures. These days, the logic of the Oedipal relationships does not need explication. We all know how much love-hate ambivalent feelings this logic could breed; and aggression may grow quite intense, unchecked. Nowadays, it is commonly received that the ambivalent feelings are never fully put to rest, as Freud had surmised, once the mature Super-Ego is established on the ruins of the Oedipus complex. As Hans Loewald (1980), the eminent psychoanalyst, amply explicated and clinically demonstrated, they continue waxing and waning, pretty much throughout one's life. Hence, it could be expected that even between two mature individuals, one could observe the logic of the Oedipal conflict resurfacing occasionally. The question remains as to what brings great critical minds to remain so mired in their relationship with such meager success at an open discussion or interpretation. To refer to a neurotic fixation (Fuller 2011) as the sole explanation for the complex phenomenon as I encountered it, seemed to me insufficient and superficial, on both the psychological and the philosophic level.

Popper was able to deny his contribution to Agassi's *Festschrift* despite his utter enthusiasm to do so upon hearing from me about Agassi's response to his critique; he retreated from the commitment he gave me to consider writing a paper for the festive volume by repeatedly bringing up his loyalty to the promise he had made to his late wife "on her death bed" "not to ever reconnect with Agassi". These were his last words justifying his position while inviting me to revisit with him whenever in London. Agassi maintained hope of understanding better the impasse between him and his revered teacher, perhaps even of reconciliation, as he continuously was writing about Popperian philosophy and his debt to his teacher. But Sir Karl repeatedly accused him of ingratitude and plagiarism (Letter from 1.8.92, in Agassi 2008, p. 350): Popper viewed him as one that had built a career out of a distortion of his work, unfair criticism and plagiarism. Agassi describes numerous attempts to respond to Sir Karl, but when recently asked: "Have you tried to interpret the situation or reformulated it for both of you?" he retorted: "He was my teacher and the one to set the rules". I have recognized the answer as one I had heard in the Yeshiva expected from students by authoritarian Rabbis. However, Popper was the champion of the *Open Society*. Could he have avoided recommending and, indeed, fostering any different logic of Master-Student relationship? In addition, Agassi, who emphasizes personal autonomy, for students of science and philosophy as well as for the layperson, and endorsed Popper's recommendation for bold criticism, what might have been the obstacles on the way for a free critical dialogue?

Popper described (1945, ch.10, 1998) the Pre-Socratic conduct the critical tradition that started out with Thales and his student Anaximander who disagreed with him openly. Tolerance for criticism changed the world of ideas. The grouping of the Pre-Socratics into diverse schools of thought awaited Aristotle but his classification was not enough to explain the critical institutional change that occurred and to internalize its import. Aristotle remained authoritarian. Would the intellectual recognition of the import of critical reasoning impart critical non-authoritarian reasoning? Would the endorsement of the principles and theory of critical rationalism necessarily lead to some critical non-authoritarian philosophical practice and/or life style? This question constitutes the heart of the problem of eudemonism in the

Socratic sense. Has Popperian philosophy grappled with the problem? Before we proceed to philosophical discussions, let me still go back to psychology.

Upcoming student-teacher rules of conduct were informally developing in Popper's circle in the 1950s and early 1960s. The intensity of both critical exchange of ideas and interpersonal relationships find their expression in conference, essay collections, biographies, and personal reports, oral and written. Allow me a brief detour for the exploration of a different Master-Student relationship, in hopes of shedding some light upon both the psychological and the philosophical concern before us. I shall focus briefly on another circle of students, established by Freud in Vienna a few decades before Popper's entry into the scene. It is known that Freud's relationships with his prominent disciples were aggressive once basic differences of opinion emerged between them. Adler and Jung had to leave the Master's Circle and form their own dissenting schools. Indeed, it seems rather natural that the striving for autonomy would lead offspring-students to separate and individuate from parents-teachers; the distancing move of a student from a teacher may prove psychologically and intellectually fruitful from a developmental point of view. What is less known, however, is that at least in one case Freud expressed fear of a creative and admiring disciple of his, Viktor Tautsk, even without having detected any dissension from his theories. According to Helene Deutsch, who was one of Freud's most devoted disciples, the Master refused to analyze Tautsk and entrusted him into her analytic hands, while he still maintained in his own hands the supervision over the case. Freud confided in her and told her about his fear of Tautsk, a philosopher and a psychoanalyst. During his analytic conversations with Tautsk, Freud reported, he was feeling that while strolling in the park together with him Tausk was reading his (Freud's) mind and drawing ideas out of it, thus stealing the Master's innovations. Here Freud divulged not mere neurotic counter-Oedipal feelings but also sheer paranoia. Tautsk reported in his analysis about his sense of frustration with Freud, as Freud was still supervising Tautsk's psychoanalysis, getting the reports about Tautsk's feelings and guiding Helene Deutsch's analysis of him. This did not continue for a very long time. Tautsk's life thereafter abruptly ended by suicide. Paul Roazen wrote in his *Brother Animal: The Story of Freud and TYausk* (1969) and his *Helene Deutsch: A Psychoanalyst's Life* (1985) (based on her *Confrontations with Myself*, 1973). It was a vivid report and analysis of the complex relationship between the three, and intense feelings harbored so deeply by these psychoanalysts who had made self-reflection and awareness their ideal, and yet so abysmally failed to apply it in this case, and with the most grave and tragic result. When despair met envy and paranoia, reflection came to naught, and analysis crumbled into a death trap.

Freud realized the inability to apply psychoanalytic knowledge to one's own self. His solution to this pitfall was to add to the training leading to the certification of students to become psychoanalysts the requirement that they undergo analysis by one of their teachers. In the vein of the critical tradition, Thomas S. Szasz, the psychoanalyst and classical individualist, raised serious objections to Freud's epistemological and methodological views upon which he had founded his psychoanalytic technique, the idea of the transference and its application in practice, particularly as applied to the psychoanalysis of trainees (Szasz 1958, 1963). He referred to Freud's criticism of his own theory of transference upon which he had based his psychoanalytic technique in general and training analysis in particular. Szasz declared this a mere authoritarian safeguard, an obstacle to the growth and development of analysand and analyst alike. He clearly stated that this logic places the view of reality as perceived by the psychoanalyst above that perceived by the analysand. Experientially and epistemologically, the situation thus constructed of necessity gives rise to a rigid logic—of imposition. If the analysand is a sensitive individual, the powerful party—the analyst—imposes on the analysand his view of reality thus becoming an aggressor. Paranoid feelings may then prevail in the analyst-analysand relationship and get out of hand. Karl Popper (1963, p. 38) and Ernst Gellner (1985) rejected that logic as irrational and immoral.

The philosopher Michael Polanyi (1958) would not care much about authoritarian safeguards in the domain of epistemology and methodology of science. For, he hails tacit knowledge as a cornerstone of scientific knowledge and with this gets the Master's position fully proofed. Polanyi would not worry about the capacity of the Master to impose tradition of thought, because Polanyi saw no other way than the authoritarian and initially irrational. What would be Einstein's fate in Polanyi's lab? Could he develop under Polanyi to become a revolutionary thinker? Yes, once he has absorbed the whole of tradition, thus himself turning Master. Polanyi solved the traditional conflict between individualism and collectivism by holding to individualism, buttressed by irrationalism and authoritarianism leaving, even if attenuated, the conflict alive, and for that matter, so did Freud in *Totem and Taboo*. No need to worry about conflicts, as they are part of human existence, said Georg Simmel. The root of lingering conflict lies in inherent basic ambivalence – the single most important and enduring of Freud's discoveries that he considered essential of human nature (see end of "Analysis Terminable and Interminable". See also Wisdom 1984). Popper's individualism, in the hands of Agassi's formulation, ie institutional individualism, may be considered as a formulation that aims to intellectually transcend said ambivalence, although, individuals confronting the various philosophical options presented herein, may remain ambivalent about whatever choice of theory. Not to get ahead of ourselves, however. Here, suffice it to say that Popper's philosophy as formulated by Agassi presents a diametrically opposed alternative to Polanyi, since it aims to open the dialogical space for autonomous individuals to exchange critically ideas within democratic social institutions. Hence, The Popperian logic shows the way how within the scientific or philosophic lab both dissent and development of individuals, students and teachers, as well as of intellectual traditions and frameworks could coexist, complement and contribute each to the other.

Popperian philosophy could be presented as both in response and as an alternative to both Polanyi and the critique of Freudian psychoanalytic impasses, epistemological, moral and personal. Let us then word the questions before us this way: Will intellectual philosophical knowledge à la Popper bring about a change in the student's practice, lifestyle, or even person? Would such a change be effected if we aim only at the student's intellectual practice and identity? Would such a change do without affecting intimate and social practice, lifestyle and identity? For, if not, are we not raising intellectual standards unnecessarily? Do we not thereby transgress our boundaries and set ourselves up for failure? If we ignore the problem, however, are we not thereby facing the risk of trading ideas without any capacity to change personal, let alone social traditions of thought and behavior? Is that not particularly so if eudemonia cannot be trusted to deliver the goods? It is against this backdrop of these philosophical problems that the particular case of the Popper-Agassi relationship remains troubling.

38.3 In Search of Rationality

At times, I can readily detect paranoid feelings within the logic of the Popper-Agassi relationship. I did observe the fear in Popper's eyes when expressing his final unwillingness to engage in the debate with Agassi. I did identify his suspicion that by the addition of a few words Agassi could improve his presentation of his new suggestions. He may have well himself known by then what are the missing few words in the passage he pointed to me—words that may have been on his own mind too, but never stated. After all, the philosopher and his apprentice knew each other all-too-well. It took Agassi a pause of a few minutes to explain to me what was at stake and how he would retort, if criticized by Popper concerning the passage in question. I took Popper's refusal to collaborate with the suggested process for a critical dialogue as method to resolve the intellectual and personal impasse as a sign of both concession and avoidance. To repeat, he may have understood very well what was implied in the passage, i.e., that Popper does not allow for false *explicanda*; that Agassi takes not the report but its having been reported as the *explicandum*: and its having been reported is explained by its being a good approximation. If this is so, then it is my conjecture that in this case Popper's accusation of Agassi as a plagiarist is rooted in nothing but Popper's concrete reading of Agassi's passage, instead of taking the possible intended meaning. It is then his refusal to read Agassi's passage in a more sophisticated way that stopped the dialogue between them. Otherwise, he was claiming for himself the explicitly by then as yet unstated philosophical insight, i.e., the sophisticated reading of the same passage. This way he could comfortably accuse Agassi of plagiarism and more—of having raised in him turmoil such that his late wife found it necessary to make him promise to avoid. My claim is that the logic here is that of envy and paranoia bearing tremendous hostility, with little hope for reconciliation for the frustrated parties involved.

I have raised this hypothesis concerning the paranoid Popper-Agassi relationship at a Jerusalem gathering in honor of Agassi's 85th birthday, went on to congratulate him, and commend him on his supportive educational contributions to students, indeed, his commitment and affection to both his students and to the intellectual world. One of Agassi's Israeli students, a Philosophy Professor at my University, criticized my contribution as un-Popperian as it did not honor Agassi by criticism. Agassi concurred. The esteemed professor went on to report upon his own fear of a face-to-face critical dialogue with Agassi, albeit, while writing philosophy, he went on to report, it was only Agassi's image that he was continuously beholding and arguing with. I, too, at the beginning of my studies under Agassi, felt as intrusive his pressing pace, his capacity for one-upmanship and his precise and punctual instruction; he was going much beyond my capacity to own my own ideas comfortably. Here, I found my colleague's self-report in agreement with my hypothesis about the phenomenon of paranoid feelings so often prevailing in the teacher-student relationship. Could joining the perception of an admired teacher as authority with such paranoid side effects, still allow critical rationalism to exert any transformational effect on interlocutors, particularly young novices? If we accept that such are the possible psychological events within the teacher-student relationships, could critical rationalism help transcend them, and, if so, how? Over the years, all such feelings within my relationship with my teacher dissipated and dialogues with him have become playfully relaxed and solely enriching. So was my dialogue with Popper. Could I ascribe this development to the critical rationalist theory that we have been sharing, or is that theory a mere red herring in this case as well as in any other? Critical rationalism *qua* theory could be a mixed bag that may harbor facilitators as well as safeguards against open criticism, and students as well as teachers may enjoy this or that level of autonomy and varying levels of reflective adherence to their role. Fear, envy and paranoia may instigate regression to safeguards against openness and in one's level of adherence to one's role and commitments, against one's better philosophical interests and judgement, thus paralyzing the process of learning. Does critical rationalism entail or foster a theory of education and a theory of pedagogical practice of critical rationalism? Or is it still ruled to this day and age, by and large, by mere principles guiding critical argumentation (dialectics) and paradigmatic legends concerning the master's pedagogical practices? Such as Bartley reports (1982) in his "A Popperian Harvest", for example, very much in line with Plato's report in his *Symposium* about Socrates restraining his instincts given the attempt of Alcibiades at a homoerotic seduction. We are back with the yet unresolved problem of eudemonia.

In April 1992, I showed Popper my intended contribution to Agassi's *Festschrift* (Laor 1994). It had to do with the demarcation between two versions of rationality, hard and soft, as they already appear in Popper's writings. To my mind, the soft version made Bartley's critique of Popper (Bartley 1962) superfluous. The hard version functions as intellectual justificationist safeguard, and the soft one is no more than an invitation for critical dialogue. In Popper words: "*I may be wrong and you may be right, and by an effort, we may get nearer to the truth*" [1945, vol.2, ch. 24, p. 225, italics his; quoted in my *Op. Cit.*, p. 109]. I then applied the analysis and identified in Popper's, Bartley's and Agassi's some remnants of the hard version of rationality. I also applied the idea to religious experience and Popper, who stated he had not written enough on religion, said he found himself in agreement with my formulations. I reported about the exchange to my teacher.

Just as my own paper appeared in Agassi's *Festschrift* (Jarvie and Laor 1994), I learned that, following his "Author's Note, 1993", in his "Introduction" to *The Myth of the Framework* (Popper 1994, pp. xii–xiii), Popper quoted from his own *Open Society* the same very italicized statement (as it appears in the previous paragraph) graphically presented in a two line stanza. He explained he had originally italicized these words of his "in order to indicate that I [Popper] regarded them as important. For these two lines were an attempt to summarize a very central part of my moral article of faith. The view that they summed up I called critical rationalism". Popper went on to claim that "as far as I know, none of my critics showed any interest in them [these italicized lines], or quoted them". "I never found anyone who had taken notice of the two lines that I had intended as my moral *credo* – […] to exclude any possibility of a dogmatic interpretation of 'critical rationalism'. And he added that his papers collected in this volume "are written to defend rationality and rational criticism": It is […] an attitude that I have tried to formulate (perhaps first in 1932) in […these] two lines".

I consider these statements as revisionist safeguard betraying bad faith: Popper had already been exposed a year earlier to my interest in these very two lines quoted within the context of my critique of his dual theory of critical rationalism. I based on the idea conveyed in these two lines the formulation of Popper's soft version of critical rationalism. To my surprise, despite his previous agreement with my formulation and critique, this time around, Popper associated these lines with the need to "defend rationality and critical rationalism" and with "a confession of faith" that had been originally presented "in a nutshell". This position constitutes the core of the hard version of critical rationalism.

My reaction was with puzzlement and concern. Why did Popper call to the aid of critical rationalism the domain of faith? What does a moral injunction have to add to his critical stance reflected in his methodological individualism as foundation for the open society? Last but not least, why did he suppress the facts about our meeting, known at the time to him and to my teacher, and claimed that "the critics of my *Open Society* and of critical rationalism were, it seems, blind to these two lines [quoted above from that book]"? It seems that, against both my intention and, as I view it, Popper's best interest, I, too – for Popper, a second-generation student – was meant to be drawn into and got entangled in the logic of the very impasse I am here attempting to transcend.

Agassi, in his "Search for Rationality" (1982), said the following: "The new religion I ascribe to Popper is not quite born of out of despair; rather, it is the counsel of despair itself. That is to say, it takes despair as self-understood and unshakable" (Agassi 1982, p. 242). In the end of his paper, Agassi recommends Russell's "Free Man's Worship" as a metaphysical framework for Popper's philosophy, thus complementing it with a "manifesto of hope born out of despair, clinging to both despair and hope most heroically". Agassi viewed Russell's ideas as expression of current religion, current scientific ethos, while he claims that "Popper's philosophy of science gave it substance" (Ibid p. 248). Are we allowed to assume that the lines from Popper's 1994 publication quoted above are a late critical response to Agassi's statement about his "new religion"? For Popper clearly stated that his hope is a moral obligation expressing his "faith in peace, in humanity, in tolerance, in modesty, in trying to learn from one's own mistakes; and in the possibility of critical discussion" (Popper 1994).

In a most recent publication (Jarvie 2016, p. 303-4), Ian Jarvie relates to Popper's statement of his creed as it also appears in Mariano Artigas (1999); see also Gattei (2002). Jarvie underscores what he refers to as the "almost never discussed" "moral dimension" of methodological individualism. He claims that Popper's view amounts to the prescription of hope as a moral commitment that comes to counteract "certain fatalism", the result of the view that collectives act in a way that is irreducible to individual actions. The moral injunction of hope comes to support one's responsible actions in society. However, if one recommends methodology to render the expected fruitful explanatory outcome in the social sciences and its technological application to achieve social change, why would Popper take his refuge in a moral safeguard? Popper explains: "[…] we need not accept that there is […] a historical tendency for things to become worse. *The future depends upon ourselves*. It is we who bear all the responsibility […]. *It is our duty to remain optimists*" (Popper 1994, p. xiii). I find the moral injunction based on an unnecessarily defensively inflated presupposition. We could bear all the responsibility only if the future depends solely on us. Why would the philosopher who taught about unintended consequences of social change take such a move and ask us to follow his faith? Perhaps he did so due to his concern lest the soft formulation implied in the two lines quoted and elaborated on by their author for the first time "after half a century" (*Ibid*), is too weak to stand against both the attack of irrationalism against argument and the growing rate of the breakout of violence (*Ibid*).

Over the past decades, I have heard times and again from Agassi the statement concerning the moral injunction of hope against the despair rooted in skepticism without the mitigation of philosophical justification. I have myself taught so to my own students at the Department of Philosophy, Tel Aviv University. I thought that this is also what Agassi claimed at the end of his "In Search for Rationality". Revisiting the paper, I have recently realized I was utterly mistaken. Agassi's writing differs widely from his oral statement. The latter may have followed Popper's, yet, *pace* his own oral statement, the written one is more typical of him and in line with his own philosophy that widely differs from Popper's. Let me explain.

Agassi does not claim that hope is a moral obligation in the face of the necessary fatalism and despair that may arise for the philosopher-suitor of truth upon the discovery that Truth is unattainable, or that one is confronted by forces much greater than one is, or that collectives are irreducible to individuals, yet collectives do not act, only individuals do. Popper, to repeat, introduces optimism and hope as moral-attitudinal foundation for individual responsibility in society. For Agassi, however, hope is not a moral demand. According to Agassi, although we are fallible and susceptible to error, even to undetectable error of which we could hardly be sure of, hope without proof is consistent with the moral injunction to offer the benefit of doubt to oneself or to the other, and to do so wherever possible. Hence, the best we could hope for is a critical dialogue with our fellow humans, and this way we may hope to improve. Although Popperian, this is not Popper's explicit view. My own view of soft rationality is similar, yet, rather than a moral injunction, I suggest, suffice it to be offered as a methodological recommendation. This turns my view both Popperian and Agassian. In contradistinction to this type of position, Popper meant his later discussion of hope as a confession of faith. This leaves Popper as founder, according to his own self-selected definition, of a school of thought. This is, to repeat, a safeguarded version of critical rationalism (Laor 1994).

38.4 An Independent Philosopher

Popper thought that he had re-founded the pre-Socratic intellectual tradition. In order for behavior to turn into customs and traditions ascribed to groups of thinkers, and then for these customs and traditions to be worded as regulative ideas, first some distance must be taken between behavior and its perception as habitual and repetitive; only then could reflection come in to the aid of wording. Aristotle, in his *Metaphysics*, made the first-step contribution when presented the pre-Socratic thinkers into schools of thought. He created for them a specific space in what Popper would later refer to as World 3 (Popper 1972, ch. 4). Enter Popper: he characterized the pre-Socratics as ushering in his own self-styled philosophy, as those who allowed their students to dissent critically from their teachers, thus constituting the cradle of Western philosophy. This more reflective characterization places the pre-Socratic school on a higher level of abstraction within Popper's world 3. Yet, although (according to Popper) critical thinking as such appears to have its products delivered straight into world 3, as institutions, theories and methods it may take time to explore the implications and change a whole culture. A wider and thoroughgoing change may need to await the identification of new problems in their various contexts as well as developments of further degrees of cognitive abstraction. Popper in Chapter 10 of his *The Open Society and Its Enemies* (1945) observes that much when he describes how, at the time of the Greek Great Generation, political institutions and traditions became objects of critical contemplation. So were other diverse institutions – institution of education not excluded. The philosophical discussions concerning the role of the teacher *vis-à-vis* that of the student and the teacher-student relationship, introduced these subject matters into world 3 as well (Popper 1972). Hence, of necessity the philosophy of the social domain and the philosophy of education became closely inter-related. Popper observed that much. He, thus, recommended an openly critical education – sharp and aggressive criticism – founded upon the autonomy of students and the autonomy of the realm of ideas. Although traditions have an important role in the Popperian philosophy, the advancement of learning requires bold criticism too. Popper recognizes all too well the intricate tensions that may emerge within the context of teacher-student relationships. Therefore, as a necessary safeguard he recommend friendliness in the relationships, on the understanding that unlike kinship, friendship "each party may terminate autonomously" (Popper 1945, ch. 25, p. 276).

Agassi tells the story of the termination of his assistantship with Sir Karl and of his departure from London to Hong Kong. One would expect that the declared distancing would bring about a change in the rules of the game between the philosopher and his apprentice. To begin with, what had been the rules of the game between them during the period of Agassi's studentship and of his assistantship? It transpires from Agassi's report that Popper had considered all of his student-assistants' critique of his views as owed to him. Does this not make good sense? After all, it was the master who had introduced the student into his own world and laboratory of ideas, exposes them to him only in order to educate the student in the critical method of dialogue and the fruit of that very dialogue. As the student accepted the role of offering genuinely his or her criticism to the master, then these fruit, of necessity, belongs to the recipient of the criticism, i.e., to the master, particularly when the criticism is favorably accepted. Agassi told me that he had accepted Popper's expectation while studying and working under Popper. But Popper himself, according to Agassi (2008, p 106) denied he expected that much. The crux of the matter is the right of the apprentice to claim credit for ideas and for their publication as their author. Once accepted, breaking the rule would be considered a breach of trust and immoral and its outcome could be formulated as plagiarism. To his last communication with Agassi (letter from 1.8.92 in Agassi 2008), Popper had not given up on accusing his former apprentice turned internationally acclaimed philosopher, who has done more than any thinker, including his other former students, to underscore his debt to Sir Karl and explain Popper's crucial important position in our history of philosophical ideas highly relevant to this day and age, Popper had not given up on accusing him of plagiarism, past and future. Granted that a student who is taking his training in philosophical criticism may be put under such rules for the game, would the same hold for the relationship between master and assistant? The answer is in the positive if the assistant is considered an apprentice only. Indeed, this does not hold in scientific laboratories throughout the western world, where assistants get credit for their contribution in joint publications. Does this expectation end upon termination of the teacher-assistant relationships? Agassi tells (*Ibid.*) about his attempted discussion with Popper concerning his critical review (1974) of Popper's *Objective Knowledge* (1972), before submitting the paper to *Philosophia* (Agassi 1974): Popper wished to hear his critique provided Agassi promise that once aired before him, it not be published. Agassi's refusal to make such a promise unless persuaded he was mistaken and if not that the critique be published, brought about an argumentative lengthy match, to which the wives of the two philosophers were silent witnesses. Why could not Agassi walk away? He did publish his review of the book and thus drew his teacher's wrath. Why did he remain puzzled and perturbed? Did he wish to re-establish Popper as a well-meaning teacher-father-figure accepting his prodigal son? Is it because he needed to repair the master's image as genuinely living what he had been preaching?

Most recently, Agassi raised, yet again, "this issue" during one of our customary Friday afternoon meetings. He could not find any of my explanations exhausting his puzzlement. The next conversation we had took place only a few hours following that earlier discussion. Agassi who contracted some blood sepsis was in the hospital with high fever and medicated. He was as philosophically alert and acutely curious as ever. He took up the conversation where we had dropped off. Why was Agassi unable to interrupt physically and distance mentally from the encounter just reported? Why could he not interpret to Sir Karl the problem situation, philosophically if not psychologically? The answer that Agassi reiterated: Popper was the teacher, setting his own rules for the game. This could not suffice. For, at best it left both Agassi and me in a continuous state of wondering.

It may well be that, one could understand Agassi's wondering and repeated surprise when faced with what he perceived in his teacher's behavior inconsistency with his (Popper's) critical rationalist philosophy. One may understand it as rooted in Agassi's adherence to eudemonism, i.e., the Socratic idea promoted by the Stoics and widely endorsed by various philosophical schools of thought, that there is no gap between whatever one rationally considers as good and one's actual behavior. The classical alternatives to this view are the various theories of "the divided human self", its parts being in continuous conflict—soul and body, good and evil, super ego and id. Popper hypothesized "the strain of civilization" and thus his theory of human action ought to have included recognition of this very conflict. *The Open Society and its Enemies* may have meant to offer this recognition of the conflict as well as Popper's prescription for overcoming that conflict by his siding with progress *via* cautious experimentation. However, Popper may have wished but never explicitly offered social cure for the conflict that may lie also in the bosom of individuals simply by virtue of how they play a role in social institutions.

Agassi's views that underscore the attraction (not the burden) of civilization, of searching for novelty and progress and considering this attraction as unproblematic, presents a philosophy, an alternative to Popper's that conflicts with it. Whereas Popper viewed a burden any operation against a basic attraction and the need to safeguard progress via reason by the injunction of hope and social responsibility, Agassi views various opportunities for progress and regress, all may seem differently to different people at different times, attractive or repulsive, and individuals may wish to offer the benefit of the doubt to their fellow humans and maintain critical dialogues with them in the hope to improve. He, however, identifies a task here that comes with social responsibility, the task to lower as much as possible any standards that come in the way of innovative problem solving (Agassi 1981). Agassi's pedagogical philosophy sharply differs from Popper's and is in line with his general philosophy that I view as more than merely critical notes to Popper's. His uncompromising stress on openness in the teacher-student relations as well as in the family is Russellian, his view of ontological declarative statements as *a priori* tentative as well as his rejection of the dichotomy between truth by Nature and truth by Convention is Einsteinian, and his hope, solely anchored in human compassionate dialogue, is Buberian. The integration, within a full-fledged modified conventionalist methodology all the way as well as some other innovations, e.g. the field of the philosophy of technology (Agassi 1975, ch. 12, 13, 1985) is solely Agassian.

This view of Agassi as an independent philosopher, post-Popperian indeed – like Steve Fuller's view (2011) of Agassi's philosophy is as "Popper 2.0" – competes with two prevalent views of his philosophy: (a) as mere notes and elaboration on Popper's (to which such add-ons Popper at times declared plagiarism), and (b) as adhering to Popperian basic assumptions and principles in the most committed way, even more than Popper himself (in this case, any statement on Agassi's part that would start by "Even Popper did not…" would constitute an accusation); in world

3, the only locus of encounter for the ideas involved in Agassi's dissent, no irritation or insult could be possible at all. Since flesh and blood, philosophers are involved persons; their identity plays a part as well. The latter two views on the Agassi's philosophy as it is relating to Popper's are as prevalent as they are, to my mind, faulty, yet they may give rise to the feeling that we are facing a drama of identity snatching, unravelling over a few decades between two heroic philosophical figures, thereby developing to surprising heights the logic of paranoia. For consequence of the forced meeting, in literature and in person, between two people claiming to hold the "Popper" philosophical identity, could be likened to the violent suffering and feud aroused by the encounter of two "Jesus's" on a psychiatric ward. The encounter was actually devised as an experiment that indeed aroused exactly such intense displeasure (Rokeach 1964).

38.5 Philosophical Transcendence of the Interpersonal Impasse

Have we improved our understanding of the Popper-Agassi impasse? How could we philosophically understand Agassi's continuously perturbed wondering about their failed last encounters? I suggest, Agassi's implied eudemonism prescribes the expectation from Popper to act according to his own injunctions for social responsibility and hope. Finding refuge in the strain of civilization and the attraction of a closed world are unacceptable *explicanda* for the student for his teacher's behavior. However, both participants in the dialogue may have been reacting to what, within the Popperian teacher-student framework, they experienced as the collapse of the vision of their common enlightened society: Popper, due to the way he was experiencing Agassi's both demand for openness and inability to maintain what Popper may have considered their commonly received traditional critical pedagogical civility; Agassi, due Popper's both clinging to their old pedagogical tradition and unwillingness to uphold, and thus destroy, their progressive critical dialogue as well as Popper's recommended open and critical teacher-student relationships. Popper cleaved to the social contract of tradition, whereas Agassi expected adherence to principles and ideals. Popper felt betrayed. Agassi found himself disillusioned. From a eudemonic point-of-view Popper could not be perceived but as betraying his philosophy or else behaviorally refuting eudemonia. Agassi's alternative was to go on denying that conflict and ambivalence are necessarily inherent to the process, thus holding on to eudemonia. For our heroes, the experience took colossal proportions.

Popper described the threat of a similar colossal breakdown experience in his *The Open Society and Its Enemies* (Ch. 10 p. 177): "The breakdown of the closed society [...] must have had the same effect upon the citizens as a serious family quarrel and the breaking up of the family home is liable to have on children". (In a note appended to this particular line, Popper likens the "strain" in his diagnosis to the "discontent" in Freud's). The alternative, for Agassi, would have been to withdraw and give up on the paradigmatic citizen of his own open society.

To note, Popper held to a view of human advancement despite the inherent conflict with pulls of regression and movement toward progression that lead to stains of development. He certainly was not a eudaimonist as he was all too aware and therefore suspicious of the regressive capacity of humanity (e.g. the Nazi phenomenon) and supplemented his critical rationalism with an ethics to support a theory of advancement of society via institutional change. His innovations, however, did not touch upon the nuclear institution of society, i.e. of the family, and its possible advancement, let alone, advancement of the members of that institution within its own framework. The expectation that such a formulation be offered by Popperian philosophy is not outlandish since, even within our open society, the family often shows up as the most natural closed institution. The logic of the master-apprentice relationship could be viewed as derivative of its participants' taking part in the family. Indeed, according to Agassi (2008, p. 103), in the first phase of their relationship, Popper and he were "extremely close" and Popper then "announced that we were tied by a sort of family relations".

Without a philosophy that addresses openly the transition from the characterization of their relation as simply a teacher-student relation to a gradation of a transition from teacher-student relation to collegial relation the participants in the pedagogical experience would not be able to promote a critical dialogue concerning standards and maps (Jarvie 1972) for their developmental road. I would argue that in the face of the lack of such standards and maps, a dialogue like the above mentioned one, between Popper and Agassi, would run very quickly into an impasse building an emotional cathartic process. The catharsis might be successful or may fail, as the particular case would allow. One may safely expect that similar problems occurred also with other of Popper's close students.

Agassi has demarcated between science and technology (Agassi 1985) and characterized gradations of rationality, in his *Towards A Rational Philosophical Anthropology* (1977, pp. 97, 365–6) as well as in his recent lecture to the Department of Philosophy, Tel Aviv University ("The Apology of Socrates in the Light of Recent Developments" (Hebrew), 2016), where he holds fast to a utilitarian version of eudaimonism, however, with less regard to its application. He considers knowing as thoroughgoing and therefore binding without any need to justify action by reference to moral conflict or freedom from conflict with evil, for example. (He is against all form of moralism and cites Hume to have said: "Reason is, and ought only to be, the slave of the passions" (*Ibid*, p.97). Agassi, *pace* Popper, maintains an optimistic view of humans "living in exchange and transformation; man is by nature, a progressive animal" (concluding sentence of his 1977) – a view diametrically opposed to Popper's. For Agassi this implies amoralism, i.e., the moral indifference of desire.

Agassi has related to theory and principles of intervention to promote rationality, he has done so mainly via his recommendations for lowering of standards of performance and then the gradual elevating of them through minimal socio-behavioral interventions and extension of their field of application. These could not have been of much help in Agassi's continued dialogue with Popper on his philosophy unless he had been willing to formulate and endorse lower standards concerning the rationality of both participants in the dialogue and/or applying some different type of insight as to how to promote them within the Popper-Agassi teacher-post-student informal institution by a collaborative effort.

But the Popper-Agassi impasse might have been insurmountable precisely because of their polarization on matters ethical. The logic of their above-mentioned polarized views, once interpersonally abstractly fixated to, calls for its formulation as paranoia *à la* Fried and Agassi (1976). That is to say, the mutual suspicion of the other of the intention to spiritually harm: Popper, from his own philosophical perspective, would endorse the view of Agassi as preaching against his views and imposing distortions on his philosophy due to Agassi's hidden irresponsible siding with unchecked desire; Agassi, from his own philosophical perspective, would view Popper as distorting his views and attempting to impose on his person by suspecting his natural inclination for progress, thereby threatening to impose a choking puritanical ethic within their relationship (Agassi 2008, pp. 103, 273). Their polarized positions could perhaps be presented as myth *à la* Lévi-Strauss (Agassi 1977) – the formulation whereof lying beyond the scope of this essay. Furthermore, let us note that an interpersonal diagnosis, like the one proposed above, may point beyond any form of individualism, classical or otherwise – an elaboration on this issue too goes beyond our scope.

It had been Agassi's last request for an egalitarian critical discussion with his teacher: "A discussion on ethics" (Agassi 2008, pp. 123, 157). The following event ensued (it occurred about sixty years ago): Sir Karl invited his student to present his viewpoint, so goes Agassi's report, and he responded that it was (and "it is still is", says Agassi in his autobiography, *Op. Cit.*, p. 124) "modified hedonism". Popper abruptly explained away this position as sheer hedonism without ever showing curiosity as to what Agassi meant by his own term. Agassi never responded and the rupture of their teacher-student relationship became almost irreversible. They parted with this complex issue, i.e. their diametrically opposed viewpoints and the suppression of a curious critical discussion, hanging unresolved between them. Incidentally Agassi does tell in his autobiography what he may have responded had he been asked:" [modified hedonism] is the recognition of the moral duty to make the most of the little we have; in appreciation of our fortune we ought to enjoy every bit of it" (*Ibid*). Agassi's formulation here is in line with Robert Stevenson's words in his anti-puritan essay, *An Apology for Idlers* (1877): "There is no duty we so much underrate as the duty of being happy" This, precisely, is Agassi's brand of eudaimony, one that views reason cued by happiness rather than by conflicts with evil.

Soon after that, following a lecture Popper delivered in Brandeis University, another aborted dialogue occurred while the teacher and his student were given a lift in a car *Op. Cit.*, ch. 4, pp. 125–126). The subject matter was their above-mentioned interrupted dialogue on ethics. Agassi, in retrospect, acknowledged that he had possibly made an error due to his own inability to reformulate and modify the application of his standards for critical dialogue and their implementation within their relationship. Though belatedly, he could come up with alternative responses he would offer to repair the ruptured dialogue. (On a different occasion he sincerely asserts: "I should have been more indulgent but I was not capable of it", *Op. Cit.*, p. 304. Neither was Sir Karl capable of reciprocating. Nonetheless, concerning this event, Agassi reflected (turning a moment of "being stuck" into a position of acceptance and accessing an alternative strategy to transcend the impasse): "[...] should I ever meet him again, it would be my turn to tender an apology. And I promised myself that if I ever did, I would." In the second edition of his autobiography, originally published in 1994, Agassi adds one more line to the concluding paragraph of chapter 4: "By now the gates are closed".

Agassi's thoroughgoing and binding view of knowledge cannot be straightforwardly assumed to effect practice without the support of a complementary philosophical theory of application. Hence, even though modified I still would consider it *naïve eudaimony*. Struggle for progress – in philosophy or otherwise – as Agassi cites Herman Blake, for example, to have recommended (1977, p. 370) is not enough. What is called for at this juncture is a more meticulously detailed theory of critical rationalist education and dialogue that includes a theory of knowledge application, such that would constitute a more *sophisticated eudaimony*. Such a theory still awaits exposition.

Concluding this paper, some further questions may come up: Is / how is role-modeling beneficial? Do eudaimonism and student's autonomy yield the same answer(s) to this question?

Could such developments and complementary methods of teaching, in themselves, induce critical philosophical change and development? If so, could Popperian philosophy grapple with these types of processes, modes of learning and their consequences, account for them and take responsibility to promote them in an integrated fashion, and if so, how? More so, could this experience be critically reflected upon and serve as promoter for one's new (or modified) identity? What are the commitments of teachers and of students for their mutual explication of the problems they are facing within such newly defined teacher-student problem-situation? To repeat: Possibly, since Popper has not developed his views on education much beyond a set of methodological recommendations and values rooted in his critical and fallibalist theories, he could not have used a map to navigate in world 3 and negotiate with his philosopher-student on a joint quest for new pedagogical ideas, let alone ideas concerning egalitarian teacher-student relationship in general, and their own in particular. Agassi may have felt trapped in the same lacuna and his own eudaimonism, and, within the context of Popperian philosophy, both philosophers refrained from theoretical philosophical reference to psychology, in shared rejection of psychologism, the view stating the possibility of the translation (reduction) of all philosophy and social science into psychology (so that there is no need, for example, to refer to social institutions or to world 3). This, however, is mere safeguard. Exploring and developing the contribution of critical philosophy to the psychology of learning (touched upon by Berkson and Wettersten (1984)) as well as of the mind, of the developing identity and self, may prove invaluable and reveal unexplored territories to critical rationalists and to their students. The Popperian revolutionary philosophy may need to present a wider scope of culture and the participants therein.

Agassi intended to sketch a program for such endeavor in his *Towards a Rational Philosophical Anthropology* (1977).

Agassi's philosophy that, among others, bridges the gap between the domains of ideas and of their implementation (Agassi 1985) may serve as example when we think about ways out of *naïve eudemonism* and of various kinds of blunder in teacher-student interaction herein described. Popper's critical rationalism as theory of educating the mind, even if revolutionary at the time, and innovative in its critical-workshop technology, may still be limited, unless it invites further reflection and abstraction within the pre-Socratic tradition, Such philosophical development, particularly à la Jarvie (2005), may yield a more critical mind and a resilient identity that could contain and operate both friendly openness and critical boldness *vis-à-vis* a reality in flux.

References

Agassi, Joseph. 1966. Sensationalism. *Mind* 75: 1–24.

———. 1974. Postscript: The futility of fighting the philistines. *Philosophia* 4: 163–201.

———. 1975. *Science in flux*. Dordrecht: D. Reidel Pub. Co..

———. 1977. *Towards a rational philosophical anthropology*. The Hague: Martinus Nijhoff.

———. 1981 [1973]. Standards to live by. In *Science and Society*. 491–501. Dordrecht: D. Reidel Publishing Company.

———. 1982. In search of rationality—A personal report. In: *In Pursuit of Truth*, ed. Paul Levinson, 237–248. Atlantic Heights: Humanities.

———. 1985. *Technology: Philosophical and social aspects*. Dordrecht: D. Reidel Pub. Co..

———. 2008 [1994]. A philosopher's apprentice: In Karl Popper's workshop (2nd). Amsterdam: Rodopi

Artigas, Mariano. 1999. *The ethical nature of Karl Popper's theory of knowledge*. Bern: P. Lang.

Bartley, William W. III. 1962. *The retreat to commitment*. New York: Knopf.

———. 1982. A Popperian harvest. In *In Pursuit of truth*, ed. Paul Levinson, 249–289. Atlantic Heights: Humanities.

Berkson, William, and John Wettersten. 1984. *Learning from error*. La Salle: Open Court Publishing Co..

Deutsch, Helene. 1973. *Confrontations with myself: An epilogue*. Oxford: W. W. Norton.

Fried, Yehuda, and Joseph Agassi. 1976. *Paranoia: A study in diagnosis*. Dordrecht: D. Reidel Publishing Co..

Fuller, Steve. 2011. Philosophy as failed patricide. *The European Legacy* 16: 977–980.

Gattei, Stefano. 2002. The ethical nature of Karl Popper's solution to the problem of rationality. *Philosophy of the Social Sciences* 32 (2): 240–266.

Gellner, Ernest. 1985. *The psychoanalytic movement: The cunning of unreason*. London: Paladin.

Jarvie, Ian C. 1972. *Concepts and society*. London: Routledge & Kegan Paul.

———. 2005. Workshop rationality, dogmatism, and models of the mind. In: *The Mind as a Scientific Object*, ed. David M. Johnson and Christina E. Erneling, 471–486. Oxford: Oxford UP.

———. 2016. Popper's philosophy and the methodology of social science. In: *The Cambridge Companion to Popper*, ed. Geoffrey Stokes and Jeremy Shearmur, 284–317. New York: Cambridge University Press.

Jarvie, I.C., and N. Laor, eds. 1994. *Critical rationalism, the social sciences and the humanities*. Dordrecht: Kluwer Academic Publishers.

Laor, Nathaniel. 1994. Religion and rational philosophy: Coming of age. In: *Critical Rationalism, Metaphysics and Science*: Essays for Joseph Agassi, vol. I, ed. Ian. C. Jarvie, and Nathaniel Laor, 91–129. Dordrecht: Kluwer Academic Publishers.

Loewald, Hans W. 1980. The waning of the Oedipus complex. In *Papers on Psychoanalysis*, 384–404. New Haven/London: Yale University Press.

Polanyi, Michael. 1958. *Personal knowledge*. Chicago: Chicago University Press.

Popper Karl, R. 1945. *The open society and its enemies*. London: Routledge & Sons.

———. 1972. *Objective knowledge: An evolutionary approach*. Oxford: Clarendon Press.

———. 1963. *Conjectures and refutations: The growth of scientific knowledge*. London: Routledge and Keagan Paul.

———. 1994. *The myth of the framework: In Defence of science and rationality*. London: Routledge.

———. 1998 [1958–1959]. Back to the presocratics. In *The World of Parmenides: Essays on the Presocratic Enlightenment*, 7–32. London: Routledge.

Roazen, Paul. 1969. *Brother animal: The story of Freud and Tausk*. New York: A.A. Knopf.

———. 1985. *Helene Deutsch: A psychoanalyst's life*. New York: Anchor Press/Doubleday.

Rokeach, Milton. 1964. *The three christs of Ypsilanti*. New York: Knopf.

Szasz, Thomas S. 1958. Psycho-analytic training. A socio-psychological analysis of its history and present status. *The International Journal of Psycho-analysis* 39: 598–613.

———. 1963. The concept of transference. *The International Journal of Psycho-analysis* 44: 432–443.

Wisdom, John O. 1984. What is left of psychoanalytic theory? *International Review of Psycho-analysis* 11: 313–326.

Chapter 39
Joe Agassi, Intellectual Irritant

Rafe Champion

This tribute begins with a recollection of my friendship with Joe and Judith, which led to rewarding contacts with Charlie Sawyer, Cherie Hoyt, Jancis Long and Mark and Ieva Notturno. Intellectually, Joe encouraged me to cast caution to the winds and unpack the implications of Popper's ideas on objective knowledge and non-justificationism without being distracted by the arguments about whether objective knowledge exists. He contributed directly to economics in partnership with Klappholz and indirectly through his pupil Larry Boland, who in turn mentored Stanley Wong. The story moves on to some little appreciated aspects and implications of Popper's thoughts; the synergy between Popper and the Austrian school of economics; the role of the moral framework of society; his early engagement with the problem of paradigms (his criticism of conventionalism); his warning about the danger of Big Science; his influence on Gordon Tullock. There is a reminder of the treasure trove of material on the website of Joe's friend and collaborator Ian Jarvie. In view of their close partnership, this is a tribute to Joe as well as to Ian. There is some criticism which Joe will see as the most enjoyable part of the paper. And last but not least, no tribute to Joe would be complete without a mention of his life partner and collaborator Judith.

39.1 The Personal Side

My first memory of Joe dates from the Popper Centenary Conference in 2002. The mobile enrolment centre appeared from nowhere on a trailer at the last minute before the designated time to enrol. Then Joe and Judith held court on the lawn in front of the university for their students and others. It was a magic occasion, with

R. Champion (✉)
Independent Scholar, Sydney, Australia
e-mail: rafe.champion@gmail.com

N. Bar-Am, S. Gattei (eds.), *Encouraging Openness*, Boston Studies in the Philosophy and History of Science 325, DOI 10.1007/978-3-319-57669-5_39

nothing to do and some time to spend in the mild late July sunshine. I seem to recall that there were chairs for Joe and Judith and a few others while most of us lounged on the grass to enjoy the relaxed conversation of the old friends and colleagues. Larry Boland was there, Ian Jarvie, John Wettersten, Bill Berkson and others who I don't recall.

Joe and Judith became special friends in 2010 when they invited me to stay with them in Tel Aviv after I completed a "mixed faith" tour of Israel. That was the first of several visits to the house in Levi Eshkol Street, not far from the beach. This was a very enjoyable place to stay with stimulation and relaxation in turn. In addition to Joe's conversation there was a continual flow of visitors. And the books. So many books that it seemed as though the bookshelves would hold up the ceiling if the walls collapsed. Joe said most of his books were out of sight in the basement and he had not reached the point of an elderly associate who stored books in his refrigerator. For relaxation there was a stroll to the beach or the supermarket, and B grade movies on TV in the evening. For their parts, Joe and Judith derived some amusement from my Australian accent.

Perhaps the most significant visit occurred in 2011, en route to Paris to catch up with one of my sons, on the way around the world to Boston to meet another son. The visit was timed to coincide with the departure of some other guests who were keeping the bed warm for me. Arriving in the evening I met a friendly stranger in the lounge. This was Charlie Sawyer. Cherie was upstairs finishing the packing while Judith was making up the bed for the new arrival. Charlie and I soon discovered that we both grew up on dairy farms and this bond of shared experience infused a congenial conversation which continued until Charlie and Cherie set out to catch a 5 am plane to Boston. In the course of the evening I mentioned that I would be passing through Boston and as Charlie went out the door he turned back and said over his shoulder "If you come to Boston, be in touch". And so it came to pass that I stayed at Boxford among the woods, the ponds and the squirrels for an enchanting week or so, the first of many.

But there is more! At the first visit, on the brink of departure from Boston for George Mason University, near Washington DC, Cherie suggested that I might like to get in touch with Jancis Long, a younger associate of Joe from his time in London. This was a long shot and I wondered what might come from an email from an unknown Australian who happened to be passing by. In the event, Jancis drove for an hour out of Washington and took me to lunch to talk about our mutual interest in Karl Popper and his associates. Years before, prompted by Joe (another important intervention by our intellectual irritant), she pursued a study of Imre Lakatos when he was a Stalinist aparatchick in Hungary (Long 1998).

That is only half the story. Over lunch Jancis suggested that I might like to get in touch with Mark Notturno, another Popper scholar in Washington. She gave me a phone number and an email address. To my mind Notturno was a somewhat enigmatic figure. I knew his books and I knew that my friend Colin Simkin met him in passing at Popper's home. I phoned the number with some trepidation, then after we talked for an hour or so one or the other of us suggested that we had better get together for coffee. The next day we met for coffee and conversation, punctuated by serious tennis-watching, pool and occasional recourse to medicinal Bourbon until it was too late to take the train back to George Mason and I stayed overnight. Next day Thanksgiving dinner turned into another late night and another overnight stay and the house in Bloomingdale DC became another home away from home.

39.2 The Intellectual Irritant: Unpacking Ideas

"Somebody may be original in being systematic, even if he is not successful in his effort to be systematic" (Agassi 1968). The paper is about novelty in general, with Popper's ideas as a major example along with avant-garde art and some examples of scientific discovery. The takeaway idea is the notion of originality emerging from the simple discipline of being systematic about an idea or insight and taking it as far as it will go.

Prompted by that thought I embarked upon the task of drawing out the consequences of Popper's theories of objective knowledge and "non-justificationism" across a range of problems and issues. The work on objective knowledge did not progress to publication and it lived on my website until it moved into the collection *Reason and Imagination* (Champion 2015). The following is a slightly edited form of the introduction and summary.

39.3 Objective Knowledge

This article is written to encourage literary intellectuals who may feel threatened by Lord Snow's scientists who "have the future in their bones" and who know all about the second law of thermodynamics. People need to be reminded that we do not live by bread and technology alone; we live by the values, traditions and myths which are embedded in our literature and are studied in the humanities. Our intellectual heritage is a mix of good and bad ideas and if they are not subjected to ongoing criticism there is a risk that the bad may drive out the good. Popper's task in *The Open Society and its Enemies* was to examine the work of some revered figures, notably Plato, to identify bad ideas which can be eliminated without necessarily damaging the status and reputation of the authors.

Popper's theory of objective knowledge breathes fresh life into the study of values, myths and traditions. This theory goes to the root of the problems of the social sciences and the humanities. His ideas about world 3 of objective knowledge have aroused little enthusiasm up to date, reflecting perhaps the time that new ideas need to germinate and bear fruit. I will show how this theory illuminates and unifies problems in the scope and methods of philosophy, in some aspects of moral and political philosophy, in the theory of literature and criticism, in the social sciences and in psychology.

Section I contains some background on Popper's ideas, explaining why they have not penetrated to the educated public. Section II sketches the theory of objective knowledge and some of its history. Section III treats Russell's method of logical analysis and argues that the valuable part of this method consists of teasing out the objective content of scientific theories, not the process of clarifying concepts as is usually believed. Section IV argues that Wittgenstein's "forms of life" may be regarded as the objective contents of traditions. These traditions exert plastic control over our activities and they can be subjected to rational (critical) scrutiny as soon as we become conscious of them. Section V argues that morals have a similar kind of existence and this enables them to exert a plastic control over our actions. They cannot usefully be described as true or false, but the acceptance or rejection of specific values can be controlled by critical discussion and can be a matter of critical preference between alternatives. Section VI examines the nature of creative literature and shows how T.S. Eliot's ideas about the social function of poetry can be illuminated by Popper's theory. Section VII suggests that this theory can contribute to a model of explanation in the social sciences; this is explained with reference to Durkheim's problem of social order and Weber's problem of social change. Section VIII pursues the idea that psychology needs to be revolutionised by looking at the brain as an organ that enables us to interact with objective knowledge in the form of theories, traditions and values.

39.4 Non-justificationism

The concept of non-justificationism is peculiar to Popper and Popperians and it has made next to no impact in the academic community at large. Exceptions are Weimer (1979), Smith (1982), Butos (1987), Lester (2000) and Barry Smith's exposition of fallibillistic apriorism (Smith 1996). Bartley in particular took on the idea and made it his own (1964, 1984).

Again I followed Joe's hint to pursue the implications of the idea in several directions. One is the theory of literature with a rejoinder to the deconstructionists. They threatened to outflank their critics by their robust rhetorical techniques which ensure that they would be a force that rival schools of criticism would have to reckon with for some time to come. I suggested that the rival schools do not need to be intimidated by the pretence of philosophical sophistication on the part of Derrida et al. but instead they should embrace the approach of critical rationalism and become united in self-criticism.

Another paper took up Hayek's turn to non-justificationism in his last book *The Fatal Conceit* (Champion 2013a). A paper presented to the Australian Skeptics recruited non-justificationism for the cause of draining the swamp of prejudice and superstition by "cracking the dogmatic framework of western thought" (Champion 2015 Chapter 11). Yet another suggested a resolution to the "foundation of knowledge" problem which perplexed Carl Menger, the founder of Austrian economics (Champion 2013b Appendix VI). Menger intended to follow his foundational work with more volumes but instead he wrote a polemic on methodological issues and did not continue the main line of his theoretical work. He wanted to account for the foundations of scientific knowledge in economics and he did not succeed, possibly because he lacked a theory of fallible or conjectural knowledge. Without the benefit of such a theory he resorted to devices like "the rule of cognition" to make his case in the face of difficulties with empirical evidence as the foundation. With the benefit of a theory of conjectural knowledge he could have pressed on with his research program, appealing to the explanatory power of his theories rather than justification by any special method.

39.5 Economics

The philosophy and methods of economics became a growth area in the 1980s but some years later a survey concluded that nothing of much value had emerged, although when Popper is read as a critical rationalist, his ideas are robust and relevant (Hands 2001). The lost years which were dedicated to positivism, paradigm theory and the methodology of scientific research programs might have been saved if people had learned from Klappholz and Agassi (1959). They explained that Popper's approach was all about robust criticism, with the implicit message was that there was no need for people to specialize in the philosophy and methodology of economics. People doing economics should be problem-centred, critical and imaginative, and that did not call for books, chairs and conferences on philosophy and methodology. From that perspective the new specialty of philosophy and methodology of economics could even be seen as a diversion rather than a contribution to the field.

Larry Boland is one of the more helpful contributors to that literature. He took an economics degree and found his way into a course of lectures from Joe which changed the direction of his philosophical thinking and his approach to his substantial work in economics (the irritant at work again). He mentored Stanley Wong who made a very significant contribution with his doctoral dissertation, a critique of Paul Samuelson's demand theory (Wong 1978, revised 2006). The book has a helpful Foreword by Mirowski which is available on line (Mirowski 2006).

Wong's chapter on "understanding and criticism" is one of the clearest accounts of Popperian situational analysis in the literature. He explained the situational constraints of a theoretical problem situation and he then looked at the theory of demand (supply and demand) as it evolved in recent times, especially in Paul Samuelson's project from the 1930s to 1950. Samuelson wanted to revolutionize the methods of economics by putting the theory of consumer preferences on a proper scientific basis, eliminating all non-empirical references in the theory. He received the Nobel Prize in economics for his efforts but Wong argued cogently that he did not succeed. The economics profession in general and Samuelson in particular went on as though nothing had happened but if Wong's thesis is robust his critique of Samuelson is a capital achievement, based on Popper's ideas, transmitted through the influence of Agassi and Boland.

39.6 The Synergy of Popper and the Austrian Economists

This is another example of following an idea to see where it leads. The idea in this case is that the most under-rated school of philosophy can empower the most under-rated school of economics and challenge the mainstream of economics to take the Austrians seriously. Popper's ideas suggest that the Austrians are not vulnerable to the standard objection of the mainstream (the Austrians are not scientific) and they demonstrate that the supposedly scientific mainstream has got hold of the wrong end of the stick on the philosophy and methods of science.

An introduction to the Austrian school is necessary to explain why the synergy is significant, if true. The founding fathers in the nineteenth century were native-born Austrians but nowadays most of the "Austrians" are Americans. They represent about 2% of the economics profession and most economists know next to nothing about the Austrian program. The leading features of the school include methodological individualism, the origin of social institutions such as money as the unintended consequences of human action, the salience of dynamic competition and entrepreneurial innovation in the marketplace, the subjective theory of value, recognition of the time factor in social and economic processes, and the uncertainty of human knowledge. The most distinctive feature and the one that has created the most problems in gaining wider acceptance is the epistemological concept of strong apriorism which purports to establish the axioms of the discipline, independent of empirical studies. This aroused the ire of Samuelson who wrote, referring to several authors including von Mises, Frank Knight, Lionel Robbins. "Well, in connection with the exaggerated claims that used to be made in economics for the power of deduction and a priori reasoning…I tremble for the reputation of my subject. Fortunately, we have left that behind us." (Samuelson 1964)

The founding father of the school was Carl Menger (1840–1921), followed by Friedrich von Wieser (1851–1926) and Eugen von Bohm-Bawerk (1851–1914). Major contributors include Knut Wicksell (1851–1926), Ludwig von Mises (1880–1973, Friedrich A. Hayek (1898–1994) and Murray Rothbard (1926–1995). Recent leaders include Ludwig Lachmann (1906–1990), Israel Kirzner (1930–) and a new generation including names such as Ebeling, Salerno, Rizzo, Caldwell, Boettke, Klein, Lewin, Koppl, Herbener, with doctoral programs at GMU, Texas Tech, Baylor, West Virginia, Rey Juan Carlos Madrid, and Francisco Marroquin.

Early in the twentieth century the Austrian ideas appeared to be firmly planted in the mainstream of the economics profession but the rise of Keynes and the logical positivists in the 1930s transformed the situation. After the war the Austrians became practically invisible until the movement staged a revival in the 1970s (Vaughn 1994). As to their scientific status, both positivism and falsificationism seemed to rule out the strong form of apriorism advocated by von Mises and after him Rothbard and the contemporary Hans Herman Hoppe.

Popper's ideas support the Austrian school of economics in three ways. First, his account of the role of methodological conventions and the theory of metaphysical research programs shows that Austrian a priorism cannot be dismissed as "unscientific" as many critics suppose. Some of the so-called *a priori* principles of Austrian economics can be regarded as working assumptions, either methodological or metaphysical postulates, of the kind that occur in all sciences. These need to stand up to criticism but they do not have to be directly testable or falsifiable. They are tested at one stage removed by their capacity to sustain testable theories, progressive research programs and effective advice on policy issues.

Second, Popper's method of situational analysis or the "logic of the situation" is remarkably similar to the approach advocated by von Mises in *Human Action* (1949) and by Talcott Parsons *The Structure of Social Action* (1937). All three were at work in the 1930s developing a framework for the study of economics and the other social sciences which could have:

- maintained sociology and economics as an integrated discipline;
- sponsored partnerships between economists and all students of social institutions – law, politics, literature, religion and cultural studies at large;
- ensured that "high theory" and empirical studies informed, enriched and corrected each other;
- contributed to good public policy, especially by monitoring the results of increased regulation and the erosion of "civic/bourgeois virtues".

There was a window of opportunity for these three leading figures in their respective fields to form a united front across the disciplines of sociology, economics and philosophy to promote the ideas that they shared and to debate the issues where they disagreed. This did not happen, there was no united front and the defective ideas which all three identified in the 1930s became embedded in the rapidly growing community of academics and researchers after the war.

Thirdly, Popper propagated some metaphysical theories which provide a congenial framework for the Austrian approach. In a nutshell, Popper and the Austrians are metaphysical fellow travellers. Barry Smith found that Carl Menger used a set of Austrian/Aristotelian ideas as the framework for his ideas and the ten propositions which he used to define the framework have some overlap with Popper's metaphysical research program (Smith 1990, 1996).

39.7 Popper Versus Paradigms in the 1930s

Popperian exegesis mostly dwells on his arguments with logicians so it took Jarvie's *Republic of Science* (Jarvie 2000) to spell out what he called Popper's "social turn" although it was signalled in *The Logic of Scientific Discovery* (section 5), in Chap 23 of *The Open Society* and section 32 of *The Poverty of Historicism*. The final section of chapter 1 Conjectural Knowledge in *Objective Knowledge* is especially helpful because it describes how Popper discovered a new range of problems after he realised that the quest for justification by way of verification was unsustainable. All theories are hypothetical (conjectural) because any or all may be overthrown although that process can take a long time if the adherents of the ruling program are unwilling to give it up. Popper used the term conventionalism to describe the attempt to retain an established theory against interloping novelties (Newton vs Einstein). This indicates that he was addressing the problems of paradigms long before Kuhn entered the fray.

He became alert to the way that theories can be "immunized" against criticism by means of ad hoc hypotheses, by shifting definitions, ignoring inconvenient observations and even by challenging the competence of rival investigators. Kuhn added the string of incommensurability to the conventionalist's bow and on politically sensitive issues there is recourse to the charge of ideological/political bias. In Popper's view these conventionalist strategies raised the issue of the social nature of science and the norms, traditions and conventions of the scientific community "Thus I was led to the idea of methodological rules and…of an approach which avoided the policy of immunizing our theories against refutation." (Popper 1972, 30).

The next step in the evolution of his ideas came as he applied the critical approach to the test statements of the empirical basis and he recognised the conjectural and theory-laden nature of observation statements. That in turn led to the recognition that all languages are theory-impregnated and that called for a fundamental change in our perception of empiricism which hitherto had located the solid foundations of knowledge in sensory inputs. "It also made me look upon the critical attitude as characteristic of the rational attitude; and it led me to see the significance of the argumentative (or critical) function of language [...]. And it further led me to realize that only a formulated theory (rather than a believed theory) can be objective and to the idea that it is this formulation or objectivity that makes criticism possible; and so to my theory of a 'third world'". (*ibid* 31)

Those very important paragraphs provide the pattern of Popper's progress from demarcation and induction to the rules of the game, to theories of language and the ideas of objective knowledge and the evolutionary link between language and critical. As Jarvie demonstrated in *The Republic of Science*, all those themes were present in Popper's first published work and it took a lifetime to draw out some of their implications.

39.8 What Is to Be Done? Popper's Leninist Turn

Popper's rules for the "republic of science" can be couched in the language of political demands or proposals which he suggested to replace the language of essentialism and historicism in political philosophy. The essentialist explicates the concept of democracy or the state, and the historicist looks at the history of democracy or the state, while Popper and Hayek pose questions about what is to be done – what sort of government do we want, and how do we want to change leaders, what do we consider to be the role of the state and the limits of state activity? Given this approach the task is to discover, formulate, and critically probe the implications and modify those principles which function as the 'rules of the game' in social life (Champion 2013c Chapter 5).

The 'rules of the game' range from the possibly innate rules of grammar, through the tacit knowledge of local traditions and folkways to the rules of games and other codified forms of procedure. They include the laws of the land embodied in common law, statutes and constitutions. This would be essentially an ecological study with the emphasis on unintended 'downstream' effects of changes in the prevailing order. This approach would supplement the methods of conceptual analysis and crude 'positivist' empirical description of social and political systems. It would have the theoretical advantage of linking disciplines and the practical merit of being continually in touch with problems and their possible solutions.

39.9 Ian C Jarvie, Peter Munz and Colin Simkin

Ian Jarvie was such a close friend and colleague of Joe that it is fitting to say a little more about his contribution as a tribute to Joe. We can be grateful for *The Republic of Science* and there is much more to mention including his service as editor of *Philosophy of the Social Sciences* and the unpublished material on his website which spans a remarkable range from studies of film to sociological research and a wide range of philosophical topics. There is a paper on Popper's rationality and situational logic with a long list of problems which Popper addressed in *The Open Society* (Jarvie 1999) and a tribute to Bill Bartley (Jarvie 1989–1990). In the Bartley paper he referred to Agassi's criticism that *The Logic of Scientific Discovery* did not take enough account of the role of metaphysics in the history of science, something which Popper addressed with the theory of metaphysical research programmes, perhaps in response to Joe in his capacity as the intellectual irritant.

Peter Munz and Colin Simkin are included here for a different reason. They became friends of Popper during his exile in New Zealand and they are probably forgotten, so there may never be a better opportunity to remind people about these two stalwart champions of Popper. Peter Munz (1921–2006) was a truly remarkable quiet achiever. Educated in Germany, Switzerland and Florence he arrived in New Zealand as a refugee in 1940. He had the unique distinction of being a student of Popper in Christchurch and of Wittgenstein in Cambridge. Most of his life career was spent in New Zealand, lecturing in history at Victoria University, Wellington. An obituary charts his remarkable achievements in two lines of research during his teaching career and two other lines after he retired (Morris 2016). He published important books on religion and mythology, notably *Relationship and Solitude: An Enquiry into the Relationship Between Myth, Metaphysics and Ethics* (1965) and his other field was medieval history. In retirement he turned to philosophy and evolutionary psychology. *On the Growth of Knowledge: Popper or Wittgenstein* (1985) was a spirited defence of Popper against Wittgenstein and others including Kuhn and Rorty who were being recruited to the cause of relativism and postmodernism. Following Popper's lead, biological and evolutionary themes dominated in *Philosophical Darwinism: On the Origin of Knowledge by Means of Natural Selection* (1993) and *Critique of Impure Reason: An Essay on Neurons, Somatic Markers and Consciousness* in 1999.

Colin Simkin (1915–1998) was an economist. He started a lifelong friendship with Popper while *The Open Society and Its Enemies* was in in preparation and he wrote a charming memoire of that achievement (Simkin 1993, Appendix One). Later, at Oxford, he formed a similarly close friendship with the leading economist John Hicks and his wife which became the subject of another memoire (Simkin 2001). Being a man of strong principles and a lover of classical music he formed a profound antipathy for manly sports and games, especially cricket and rugby which were the dominant religions in New Zealand, although he could be persuaded to watch tennis and swimming on TV with his wife. He kept in close contact with Popper by correspondence (lacking the internet) and he frequently visited Popper in Britain. His friendship meant a lot to Popper and as a mark of respect he was allowed to smoke while they walked in the garden. We became friends and we met almost weekly for the last decade of his life. This revived his interest in writing and he produced a fine introduction to Popper's ideas (Simkin 1993).

39.10 The Challenge of Big Science

The world of science changed out of recognition in Popper's lifetime and it is clear from his unpublished lectures at the London School of Economics that he was very concerned about the emergence of Big Science, driven by government money. President Eisenhower articulated similar concerns in his outgoing Presidential address (Eisenhower 1961). "The prospect of domination of the nation's scholars by Federal employment, project allocation, and the power of money is ever present."

Popper speculated about some influences which could kill science, including the incentives offered to scientists by the newly available money for research. He foresaw a problem of too much money chasing too few ideas, the publication explosion (good buried under bad), angling for money at the expense of good science. In Section 32 of The *Poverty of Historicism* he advanced an institutional theory to account for scientific and industrial progress and he speculated about factors which might impede progress, such as government control of the laboratories and journals. He did not pursue his speculations but he inspired the outstanding political economist Gordon Tullock to write *The Organization of Inquiry* (1965). I am not aware of any reference to Tullock in Popper's published work but some Austrian economists with an interest in Tullock found some 70 pages of correspondence between Popper and Tullock in the 1950s and '60s (David Levy personal communication). Tullock's book sketched a scenario for the evolution of a field of research in a downward spiral to a point where it approached the state of pseudoscience resembling Lysenkoism in Russian genetics and plant breeding. When he wrote in the 1960s he thought that the social sciences were on that path but he considered that the natural sciences were sound.

39.11 Popper on the Moral Framework and Religious Faiths

Popper was sceptical about moral philosophy, which he suggested could be mostly summed up in the Golden Rule. He acknowledged the importance of moral institutions and traditions, especially what he called the "moral framework" of society. In a paper delivered to the Mont Pelerin Society in 1954 he wrote: "Among the traditions that we must count as the most important is what we may call the 'moral framework' (corresponding to the institutional 'legal framework') of a society. This incorporates the society's traditional sense of justice or fairness, or the degree of moral sensitivity that it has reached [...]. Nothing is more dangerous than the destruction of this traditional framework. (Its destruction was consciously aimed at by Nazism)" (Popper 1963 Chapter 17).

Another much-neglected part of the moral framework is the prevailing attitude towards work, business and the cluster of ideas which Deirdre McCloskey (2010) has labelled "the bourgeois virtues" such as honesty, the work ethic, business acumen, prudence and temperance. "European culture in classical and Christian times spurned work and the bourgeoisie. Yet from 1600 to 1800, startlingly, it developed a lively appreciation of the 'bourgeois virtues', from which came the stirrings of enterprise that made the modern world [...]. But after 1848 the artists and intellectuals turned sharply against capitalism. From this, alas, came the events of 1914 and 1917 and all our woe". (McCloskey 2006)

Religions have been the main sources and foundations of moral frameworks and Popper held interesting and nuanced views about God and religion although he was loth to talk about them in public. Late in life he consented to two interviews with rabbi Edward Zerin on condition that there should be no publication while he was alive. A small part of the transcript appeared in the *Skeptic* (US) in March 1998 and it is reprinted in a collection of papers (Popper 2008, Chapter 5).

Another thought-provoking public contribution can be found in a lecture that he delivered in 1940 in New Zealand in a series of university extension lectures on 'Religion: Some Modern Problems and Developments'. One of his lectures is reprinted in *After the Open Society* (Popper 2008). He argued that the dispute between religion and science in the nineteenth century was a thing of the past because it was based on each side trespassing on the territory of the other. Science evolved out of the religious mythology that men first invented to explain the world and because most religions are "true belief" systems there is a strong and unhelpful residue of "true belief" in science itself.

39.12 A Critical Note on Global Warming

Joe's latest book, *Beg to Differ*, written with A. Meidan (Agassi and Meidan 2016) touched on the debate between those who are alarmed about global warming and those who are not.

"Most students of global warming [read '*most of the people with government grants to study global warming*'] agree about most of the relevant facts. The world is getting significantly warmer since the beginning of the twentieth century to date [*by perhaps one degree C*]. This is a trend that will probably not desist overnight. [*it reversed between 1940 and 1970 and it practically stopped in 1998*] One of the main reasons for this trend is the burning of fossil fuel [...]. If this trend goes on the damage for human society will be unbearable". (Agassi and Meidan p. 57). On the other side, "the globe undergoes warming and cooling repeatedly and for thus-far unknown reasons and with thus-far unpredictability [...] all the models have been refuted [...]. Therefore there is no rational basis for the prediction that global warming will continue and cause a global catastrophe". (*ibid* pp. 57–58)

They ask: "Which of these two theories makes the world stranger?"

39.13 The Strangeness of Alarmism

Possibly the strangest thing is that there is any alarm at all. Since 1850 the temperature has increased by about one degree C and the increase in CO_2 in more recent decades has significantly greened the planet. For the best part of two decades since the El Nino in 1998 there has been very little warming (the pause) until the recent El Nino, followed by a rapid decline towards the post-1998 plateau level. As to the role of carbon dioxide and "carbon pollution", the El Nino has nothing to do with CO_2, the temperature declined between 1940 and 1970 while CO_2 increased, and almost all of the models built to project the temperature based on increasing CO_2 have been falsified (Spencer 2014; Michaels and Knappenberger 2016).

Turning to the most famous paper in the literature, the 97.1% consensus (Cook et al. 2013), the ace which President Obama played to push for global action on emissions, it is surprising to read the paper and find no reference to the amount of warming or the need to worry about it. The points of consensus reported in the paper are (1) there has been warming (but who disagrees?) and (2) humans have contributed. Taking account of the heat-island effect, who disagrees with that? As to the amount of human impact, a significant but unreported finding in the project was that only 2% of the sample thought that humans contributed more than 50% of warming. That result was on the university website but not in the publication (Shollenberger 2016 location 66–104). So the investigation appears to demonstrate a consensus that humans caused less than half of the (beneficial) warming of the last 150 years. What follows from that in the world of political initiatives?

Later in the book Agassi and Meidan turned to the cost/benefit aspect of various scenarios. They were alert to the costs of mitigation strategies in terms of hospitals, schools and other desirable things that will have to be foregone to subsidise renewable energy. Here we find some genuinely alarming developments because the costs of mitigation and warming-alarmism are being experienced here and now, in contrast with hypothetical harms in the future based on failed models.

On the cost side, some tens of thousands of people die from starvation each year caused by the increased price of food due to the diversion of corn and other crops from the human food chain into auto fuel. Jobs in power-intensive industries are under threat worldwide and surging power costs threaten the lives of the poor and the old who are liable to freeze to death in winter. It is important to realise that the mega-trillions spent to develop alternatives to fossil fuels are wasted until the renewables are adequate to the task and a new generation of storage batteries is available to cover days when the wind does not blow (or blows too hard) and the sun does not shine (not to mention the nights). Until that time fossil-fuel fired electricity is required to back up the renewables – that means supporting dual power sources and probably doubling the cost of power. The cost of subsidies to support crony capitalists in this area is very large as well (Butos and McQuade 2015).

So the really strange thing is the alarm and the urgency to spend massive sums of money to create a host of real problems. This has many of the characteristics of "solutioneering", a concept named by Roger James in his excellent introduction to Popper's ideas (James 1980). "First. Jump to a solution before clearly formulating what the problem is (or indeed if there is one at all) or how success or failure are to be judged. Achievement of the solution then becomes the goal…I call this mistake solutioneering." (*ibid.*, p. 3). Second, play up the desperate urgency of the situation – there is no time to waste, it is almost too late already – ignore the problems of implementing the solution, under-estimate the cost of the solution and over-estimate the cost of the alleged problem (see Lord Stern's estimate of the cost of warming). If we are dealing with solutioneering in this case, what supports the paradigm of climate science which generates alarm? Part of the answer is: a number of things that Karl Popper addressed and criticized the best part of a century ago (see 39.7). Two sets of influences can be listed, first historical/systematic/background factors and second, immediate or foreground strategies to defend the ruling paradigm, especially by the allocation of government funds.

39.14 The Historical Background Factors

One background factor is the ongoing concern in the positivist/empiricist philosophy of science with "confirmation theory" and the calculation of inductive probabilities of theories. A century after the rise of modern positivism there is still no way to confirm general explanatory theories or to allocate probabilities to them but countless generations of students have been inducted into the arcane lore of confirmation and induction, driven by the unstated assumption that we want to know how sure we are about what we think we know rather than criticising and testing our ideas to find growing points for advances in the field. Students are likely to emerge from these courses without any real insight into the world of science "outside the window" or the role of criticism in it. Add to confirmation bias in the philosophy of science some things which I have treated in a series of sketches under the heading "Joining the dots" (Champion 2016a). These dots are:

- The post World War 2 decline of education at all levels and the need for attention to what Jacques Barzun Called "the house of intellect".
- The radicalization of the environmental movement which grew out of the "Ban the Bomb" movement of the 1950s and the international campaign against nuclear energy which followed.
- The postwar development of big science funded by governments.
- The rise of relatively uncritical "normal science" flagged by T. S. Kuhn and the impetus which he unwittingly gave to it.
- The long-term implications of the radicalization of university students in the 1960s and '70s.
- The role of environmental entrepreneurs in the United Nations, the proliferation of climate relate organizations and the governance of the IPCC.
- Gordon Tullock's account of the way certain motivational factors on the part of scientists and the institutional structure of science can combine to run down the quality of scientific publications to the terminal stage where the point of view becomes a criterion for acceptance for publication (Champion 2016b).

39.15 Defending the Ruling Paradigm

Popper identified a number of "conventionalist strategies" to protect a ruling theory from competitors (see 39.7) and all of those practices have been deployed to marginalize critics of the alarmist paradigm. In view of the great cost of modern research which is almost entirely directed by governments, the most powerful strategies for paradigm defence are the "positive heuristic" (funding favoured programs) and the "negative heuristic" (not funding rival programs). To indicate the money involved, Butos and McQuade (2015) found that climate-related research and development in the US Federal Budget increased dramatically from 1995 to approach $17 Billion in the 2013 budget. A partial accounting showed the trend from $3.6 Billion in 2001 to $9.1 Billion in 2013. Against this, climate alarmists were agitated to find that "big oil" villains supported the anti-alarmist Heartland Institute. This including almost $800,000 from ExxonMobil between 1998 and 2005 (Wikepedia 2016).

Since then Big Oil has found that it is more profitable to get on the green bandwagon and more Big Oil money is flowing to alarmism than skepticism (Morano 2013). If alarmists were concerned about scientists being bought with a few million dollars, how many scientists can be bought with $17 Billion, driven by a government agenda? Recall Popper's warning that scientific progress could be stopped by government control of the laboratories! For an inside view of the US government agenda under the Obama administration, see Carlin (2015). Carlin is a lifelong environmentalist, hiker, whitewater rafter and international traveller who joined the Sierra Club in the 1960s. He is highly qualified in science (UCLA) and economics (RMIT) and he spent 27 years doing environmental studies and cost-benefit analysis in the US Environmental Protection Agency (EPA). Under the Bush presidency he observed practically no political interference with the work of the agency but his experience in the Obama years corroborated Popper's concerns about government control of the laboratories.

Turning to the positive and negative heuristics, a large part of climate science is devoted to model building. This involves thousands of scientists working on supercomputers in collecting data, development and data processing. The cost ensures that there are only government-funded models. They can be called "hot" models because up to date they have predicted temperatures far in excess of reality (Spencer 2014). They provide input to the environmental impact field of climate science, demonstrating the GIGO principle (Garbage In, Garbage Out). Worldwide teams of biologists and others use the unrealistic projections to explore the implications for endangered species, the food and water supply and the proliferation of human diseases and their vectors. Of course the scenarios are alarming and very newsworthy for slow news days in the media. However in view of the assumptions which underpin these exercises they can only regarded as science fiction.

As to the negative heuristic, there is scant funding for the considerable number of non-human influences on the climate, among them:

- Alternative models, less driven by CO_2.
- The relative impact of different greenhouse gases, especially water vapour and the possible role of CO_2 as an inhibitor of warming rather than a positive agent.
- Ocean circulation effects, both the relatively well known El Nino effect and the less explored deep ocean effects.
- Long-term natural oscillations like the Little Ice Age.
- Solar activity which has both direct and indirect effects.
- Negative feedback loops including convection and cloud cover.

It is interesting to note that despite the limited funding in the areas listed above, it was still possible to find almost 100 peer reviewed papers published in each of the last three years (2014, 2015, 2016) affirming the influential role of the sun in climate change (Richard 2016).

39.16 Revisiting the Precautionary Principle

Up to date the precautionary principle has been urged by radical environmentalists to support the "solutioneering" approach to spare no expense and waste no time in mitigating CO_2 emissions. However in view of the shaky basis of alarming projections and the real and immediate cost of mitigation programs, it appears that the precautionary approach calls for a halt to dramatic interventions until we have a better fix on the trajectory of warming and the causal factors in climate change.

39.17 Judith Buber Agassi and Charlotte Bühler

Joe Agassi has been a tireless intellectual irritant for the best part of a century. I have no doubt that a lot of the credit for his productivity and his good humour can be assigned to his equally indomitable companion for most of that long time. No doubt there will be other tributes to Judith in this volume, coming from people who have known Joe and Judith for many decades. As a feminist, a wife and a scholar she is in the mould of Charlotte Bühler, the wife of Popper's most important teacher. She was a truly remarkable woman, as can be said of Judith and I hope they will both be remembered, along with their partners, Karl Buühler and Joe Agassi, the indefatigable intellectual irritant.

References

Agassi, J. 1968. The novelty of Popper's philosophy of science. *International Philosophical Quarterly* 8: 1968.

Agassi, J., and A. Meidan. 2016. Beg to differ. *The logic of disputes and argumentation*. Springer.

Bartley, W.W. 1964. Rationality versus the theory of rationality. In *The critical approach to science and to philosophy*, ed. M. Bunge, 3–31. New York: The Free Press.

———. 1984. *The retreat to commitment*. 2nd ed. La Salle: Open Court.

Butos, William N. 1987. Rhetoric and rationality: A review essay of McCloskey's The rhetoric of economics. *Eastern Economic Journal* 13 (3): 295–304.

Butos, William N., and Thomas J. McQuade. 2015. Causes and consequences of the climate science boom. The Independent Review 20 (2):165-196. http://www.independent.org/publications/tir/article.asp?a=1080.

Carlin, A. 2015. *Environmentalism gone mad: Environmentalism gone mad: How a Sierra Club activist and senior EPA analyst discovered a radical green energy fantasy*. Mount Vernon: Stairway Press.

Champion, R. 2013a. Hayek, Bartley and Popper: Justificationism and the abuse of reason. In *Hayek: A collaborative biography. Part I Influences from Mises to Bartley*, ed. R. Leeson, 213–225. New York: Palgrave Macmillan.

———. 2013b. Guide to The Open Society and its enemies. *Amazon*. https://www.amazon.com/Guide-Society-Enemies-Popular-Popper-ebook/dp/B00C6RHIGO/ref=sr_1_2?ie=UTF8&qid=1483093545&sr=8-2&keywords=rafe+champion.

———. 2013c. Commentary on Hayek. *Amazon*. https://www.amazon.com/Commentary-Hayek-Critical-Rationalist-Papers-ebook/dp/B00E0L8AR0/ref=sr_1_6?ie=UTF8&qid=1482889945&sr=8-6&keywords=rafe+champion.

———. 2015. Reason and imagination. *Amazon*. https://www.amazon.com/Reason-Imagination-thoughts-William-Bartley/dp/1507512112/ref=sr_1_4?ie=UTF8&qid=1483093591&sr=8-4&keywords=rafe+champion.

———. 2016a. *What happened to climate science? Joining the dots*. http://catallaxyfiles.com/2016/10/11/what-happened-to-climate-science-the-dots/.

———. 2016b. *Gordon Tullock on induced curiosity and the decline of scientific integrity*. http://catallaxyfiles.com/2016/10/18/climate-science-7-gordon-tullock-on-induced-curiosity-and-the-decline-of-scientific-integrity/.

Cook, J., D. Nuccitelli, A. Green Sarah, Richardson Mark, Winkler Barbel, Painting Rob, Way Robert, Jacobs Peter, and Skuce Andrew. 2013. Quantifying the consensus on anthropogenic warming in the scientific literature. *Environmental Research Letters* 8: 024024. http://iopscience.iop.org/article/10.1088/1748-9326/8/2/024024/meta.

Eisenhower, D.D. 1961. *Address to the nation*. http://www.eisenhower.archives.gov/research/online_documents/farewell_address/Reading_Copy.pdf.

Hands, D.W. 2001. *Reflection without rules: Economic methodology and contemporary science theory*. Cambridge: Cambridge University Press.

James, Roger. 1980. *Return to reason: Popper's thought in social life*. Wells Somerset: Open Books.

Jarvie, Ian C. Website. http://www.yorku.ca/jarvie/.

———. 1989–1990. *Philosophy and its problems*. http://www.yorku.ca/jarvie/online_publications/WWBIII.pdf.

———. 1999. *Rationality and situational logic in Popper's scientific work*. http://www.yorku.ca/jarvie/online_publications/VIENNAWD.pdf.

———. 2000. *The republic of science: The emergence of Popper's social view of science 1935–1945*. Amsterdam: Rodopi.

Klappholz, K., and J. Agassi. 1959. Methodological prescriptions in economics. *Economica* 26: 60–74.

Lester, Jan C. 2000. *Escape from Leviathan*. London: Macmillan.

Long, J. 1998. Lakatos in Hungary. *Philosophy of the Social Sciences* 28 (2): 244–311.

McCloskey, D.N. 2006. The discreet virtues of the bourgeoisie: How Europe after 1600 half escaped the ancient condemnation of economic life. History Today 56(9). http://www.historytoday.com/deirdre-mccloskey/discreet-virtues-bourgeoisie.

———. 2010. *Bourgeois dignity: Why economics can't explain the modern world*. Chicago: The University of Chicago Press.

Michaels, Patrick J., and Paul C. Knappenberger. 2016. *Lukewarming: The new climate science that changes everything*. Washington, DC: Cato Institute.

Mirowski, P. 2006. Introduction to new edition Stanley Wong. *The foundations of Paul Samuelson's revealed preference theory*. http://citeseerx.ist.psu.edu/viewdoc/download?doi=10.1.1.401.3046&rep=rep1&type=pdf.

Morano, Marc. 2013. *The BIG LIE: Sceptics funded by Big Oil -no, the Alarmists are*. http://www.climatedepot.com/2013/07/24/the-big-lie-sceptics-funded-by-big-oil-no-the-alarmists-are/ 30 Dec 2016.

Morris, P. 2016. *Obituary—Peter Munz*. http://nzbooks.org.nz/2006/obituaries/obituary-peter-munz/.

———. 1963. *Conjectures and refutations: The growth of scientific knowledge*. London: Routledge & Kegan Paul.

———. 1972. *Objective knowledge: An evolutionary approach*. Oxford: The Clarendon Press.

———. 2008. After The Open Society, ed. J. Shearmur and P.N. Turner. London: Routledge.

Richards, Kenneth. 2016. *The sun-climate connection*. http://notrickszone.com/2016/11/21/the-sun-climate-connection-over-100-scientific-papers-from-2016-link-solar-forcing-to-climate-change/#sthash.RsP42PRx.dpbs.

Samuelson, P.A. 1964. Theory and realism: A reply. *The American Economic Review* 54 (5): 736–739.

Shollenberger, Brandon. 2016. The climate wars: How the consensus is enforced. *Amazon*. https://www.amazon.com.au/Climate-Wars-How-Consensus-Enforced-ebook/dp/B01CEZLAM0.

Simkin, C. 1993. *Popper's views on natural and social science*. New York: Brill.

———. 2001. John and Ursula Hicks – A personal recollection. In *John Hicks: His contributions to economic theory and practice*, ed. K. Puttaswamaiah, 5–14. London: Transaction Publishers. http://www.the-rathouse.com/Colin_Simkin_on_John_Hicks.pdf.

Smith, Barry. 1990. Aristotle, Menger, Mises: An essay in the metaphysics of economics. In *Carl Menger and his legacy in economics. Annual supplement to volume 22 History of political economy*, ed. Caldwell Bruce, 263–288. London: Duke University Press.

———. 1996. In defense of extreme (fallibilistic) apriorism. *Journal of Libertarian Studies* 12: 179–192.

Smith, Huston. 1982. *Beyond the post-modern mind*. New York: Crossroad.

Spencer, R. 2014. http://www.drroyspencer.com/wp-content/uploads/CMIP5-90-models-global-Tsfc-vs-obs-thru-2013.png.

Vaughn, Karen. 1994. *Austrian economics in America: The migration of a tradition*. Cambridge: Cambridge University Press.

Weimer, W.B. 1979. *Notes on the methodology of scientific research*. Hillsdale: Lawrence Erlbaum Associates.

Wikipedia. 2016. *The Heartland Institute*. https://en.wikipedia.org/wiki/The_Heartland_Institute. 30 December 2016.

Wong, S. 1978. 2nd ed 2006 *Foundations of Paul Samuelson's revealed preference theory: A study by the method of rational reconstruction*. London: Routledge.

Chapter 40
Critical Rationalism as Therapy

Michael Segre

"You must really be desperate if you turn to me," is how Joseph Agassi (Joske) is said to have greeted some new students. When I first met Agassi I was fifteen years old and indeed desperate, being confronted with the early symptoms of ADD (Attention Deficit Disorder). I did not know then the exact nature of my sickness, nor was I seeking Agassi's help. I met him by chance or, perhaps, by fate.

ADD is a neurodevelopmental disorder striking children and adolescents and reducing their self-control, concentration, coordination, and memory. Its causes are debated, but once diagnosed, it can be cured with medication. A more common form is ADHD, or Attention Deficit Hyperactivity Disorder, which is characterized by excessive activity and impulsivity. Although this was not the case with me, I was putting excessive efforts into carrying out ordinary tasks such as concentrating, reading, writing, studying, memorizing, and relating to other people. I was always tired, and my short sight and astigmatism added to the burden. Since my disorder would not be diagnosed until recently, my only way to overcome its gloomy outcomes was to develop survival strategies. It is here that Joske came to be helpful—and critical rationalism instrumental.

I remember, shortly before having met Joske, being asked to draw a triangle as part of school homework in geometry; I was puzzled to note that such a simple task took me no less than fifteen minutes. The difficulties grew with time, and I needed more and more effort and time to carry out elementary activities, resulting in stress, overwhelming frustration, envy of my successful peers, and depression. My schoolmates started bullying me. My parents and I were aware that something was going wrong with me, but we had no idea what exactly it was, or how to cope with it.

My family and our social surroundings were not the ideal background for an adolescent struck by ADD. At that time—the Sixties of the last century—we were living in Israel, in a pioneering, conformist society, hailing rough manners and

M. Segre (✉)
University of Chieti, Chieti, Italy
e-mail: segre@unich.it

N. Bar-Am, S. Gattei (eds.), *Encouraging Openness*, Boston Studies in the Philosophy and History of Science 325, DOI 10.1007/978-3-319-57669-5_40

intolerant to "deviations." My parents were well educated, refined, traditional, though somewhat disoriented, Europeans. Much as the two environments were the antithesis of each other, both considered me weird. To try to conform I began to pretend, cheat, or lie—all practices that increased my instability; dishonesty is a dangerous shortcut, as I would soon learn.

One of the few persons who did encourage me was my physics teacher, whose name, unfortunately, I do not remember: he was a red-haired religious Jew with profound human insight. After I had miserably failed several tests, he asked me to come and see him. Smiling, he asked directly, in a friendly and reassuring way, "Why do you get stressed, Michael?" Getting stressed was considered a sin in that environment, but his liberating empathy granted me a rare opportunity to be myself. "It happens to me frequently," I frankly admitted. He offered a deal: "Attempt to answer one question, and if you do it right you will get the highest mark, as if you had answered all the questions correctly." Needless to say, I soon started getting top marks. Before the final examination at the end of the year I approached the teacher and asked whether our agreement remained valid. "Do what you can," he answered. So I attempted all the questions, answered them correctly, and got the highest mark. My sensible teacher confirmed one of Agassi's tenets before I had had the opportunity to meet him: easing pressure in education is beneficial. It is thanks to my school physics teacher that I fell in love with the field and later majored in it at the university.

However, as Agassi argues, education is by and large based on harmful pressure (Agassi 2014),[1] and I was about to experience its full burden and outcomes in the study of literature. One nineteenth/twentieth-century Hebrew writer and poet hailed in Israeli schools (excessively, I believe, but I am not in a position to judge) was Hayim Nahman Bialik. For a non-native speaker like me, his high-level Hebrew had always been beyond reach. At the time when I met Agassi, we had to read Bialik's Hebrew translation of Don Quixote, an excruciating, backbreaking experience, and I admit still feeling resentment when writing about it. I lost the battle; from then on I was no longer able to read prose, to say nothing of poetry, and only recently did I manage to read and enjoy a full book of Gabriele D'Annunzio's superb, humorous prose. It was in the midst of my struggle with Bialik's Don Quixote that I met Agassi.

Agassi was at that time a young academic star in the USA; he had just published his *Towards an Historiography of Science* (1963), and the University of Tel Aviv had offered him a visiting professorship. My father, who had served with him in the army, went to pay a visit to his old comrade and friend, and took me along; I surmise he was hoping that Joske would somehow help. We met at Herzliah, where Joske and Judith had been given an apartment by the university and where we were spending the Passover holiday. What struck me was Agassi's openness and unusual way of tackling issues; he did not condemn what was commonly condemned outright, but evaluated issues from a detached, balanced, and occasionally enthusiastic point

[1] Faithful to the view that improving shortcomings is more effective than hopelessly attempting to demonstrate benefits, Agassi argues against applying pressure rather than in favor of reducing it.

of view. We talked, among other topics, about my difficulties. Surprisingly, he did not condemn them either, but frankly pointed out flaws and suggested original ways to overcome them. His suggestions included the idea that I should drop everything and make a trip to the USA (which for me, then, was like going to the moon). His proposals seemed impracticable, but at least honest. An intuitive, concealed, but ever-healthy inner voice murmured that this man was okay. Being confronted with my flaws without flattery, even if harshly, strengthened my trust.

Our contacts remained sporadic for the time being. Joske returned to the USA. International telephone calls were expensive, correspondence was still on paper, and I had difficulties in writing. I tried to change my environment by distancing myself from my family and returning to Europe, but I was obviously carrying my ADD along with me. The result was disastrous. I was failing in whatever I was doing—school exams, driving test, sailing test, and many other relatively simple challenges such as learning to play the piano or playing tennis. I would occasionally sit for many minutes immobile, gazing into a void, fully aware that something was blocking me, yet unable to react. All this increased my insecurity and indecisions. I remember celebrating my eighteenth birthday alone in front of a mirror, drinking a glass of wine, wondering what "my problem" might be, why I was different from others, and if there was any way out. I would never have dreamt that one day I would be writing about all this, liberated from the devil that held me hostage for so many years.

To be sure, I could reach a fair level of proficiency in some activities such as driving a car, with considerable training and patience. I began to learn to live with my handicap. I enrolled at the university, majoring in physics and regarding studies as therapy. The good news was that mathematical fields do not require much memory. The bad news was that they required the ability to think quickly; hence, my undergraduate studies were far from an easy stroll. I was losing ground and, moreover, exams were written and limited in time. I had difficulties in writing and organizing my ideas under pressure, and seldom did I manage to answer all the questions. Yet I did somehow graduate, albeit with enormous effort and poor marks. And I even won a scholarship to undertake research at a nuclear research institute.

It was my stay at that institute that saw the seed of a major turning point. Independently of my disorder, I felt uneasy there, and only many years later did I realize why: for the first time, I was confronted directly with a "closed" society of scientists. According to Karl Popper, the "open society," i.e. the society that encourages criticism, is the most appropriate for the development of science and democracy (Popper 1966). Yet as Agassi later pointed to me out on more than one occasion, modern scientific institutions—and universities in particular—are still closed; they follow tribal, ritual, dogmatic procedures within some paradigm not always rationally fabricated, as so well described by Thomas Kuhn (Kuhn 1966, 1999; I have tried to trace the historical roots of all this in my last book on *Higher Education and the Growth of Knowledge*, 2015). That concealed, intuitive, healthy part of me was sending another message: this is not your habitat—an encouragement to seek the open society. I decided to abandon physics and turn to philosophy, and this gradually brought me back to Joske.

Joske is always available to help, even in the middle of the night, and has an extraordinary ability to tranquilize and encourage. To begin with, he taught me a basic rule, indispensable for critical rationalism, for the open society, for a civilized world, and for one's sanity: intellectual honesty (as part of plain honesty, of course). To enhance criticism and growth, one needs openness and transparency; this is not easy to apply in a still authoritative society that regards criticism as belittlement rather than as help to grow, or—as Popper and Agassi stress—as an expression of esteem. Becoming honest was fundamental for healing. I gradually learnt not to lie, not to cheat, and not to pretend. I also learnt that what society requests, i.e. common sense, is not always the judicious and right thing: *vox populi* is not *vox dei*. I slowly learned to pay little attention to what people were thinking of me, realizing that I could not control their thoughts. As Joske puts it, there is always somebody who does not like you, and in any case that is his problem. I had embarked on a long process of liberation.

A second challenge was to find a way to make a living. I had tried to work in several places, either in the industry or in the civil service, but was not able to fit into any frame; either I had to give up, or I was simply fired. Everybody agreed that I had much "potential," but this potential did not materialize. There was, nevertheless, one task that I was able to carry out reasonably well, and that was teaching—for the moment, giving private lessons. I tended to explain things as if explaining them to myself with all my shortcomings, and was very good at simplifying. Joske encouraged me to turn to an academic career, as a relatively well salaried profession with the advantages of self-employment, granting enough space for an individual with my disorders to implement his abilities. He warned me, nevertheless, that there are two ways to be in academia. The easy way, aspired to by most academics, is to be an "in" guy, with the danger of selling one's soul. The more difficult, insecure way is to be an "out" guy, with the advantage of being independent. I would later become conscious of the importance of this warning.

I decided to embark on postgraduate studies that combined my previous studies with philosophy, and registered for the postgraduate course in the history and philosophy of science at the Hebrew University of Jerusalem. At that time the curriculum was still part of the Faculty of Natural Sciences, and my academic adviser, a well-known mathematician, scornfully wished me success with "all these nice things" as he scanned the list of seminars I had chosen. As C. P. Snow argued in his famous 1959 lecture on the "Two Cultures," (Snow 2001) the sciences and the humanities are two distinct cultures allegedly not getting along (and the second is nowadays less esteemed). Critical rationalism, on the contrary, considers every field that can be empirically tested, whether humanistic or "scientific," to be scientific.

I did much better in my humanistic master's program than in my "scientific" bachelor's. I even became teaching assistant to Professor Eric Mendoza at his history of science laboratory, and for once I did not get fired. When I came to choose the subject of my M.Sc. thesis, I asked Joske's advice, and he suggested that I deal with the debate around Galileo's alleged experimentalism. Alexandre Koyré, one of the greatest historians of science ever, had suggested that some experiments described by Galileo had never been carried out by him (Koyré 1992); his thesis was heavily criticized by leading Galilean scholars (among them Settle 1961; Drake 1973; MacLachlan 1973; for the debate, see Segre 1980). Joske had had the opportunity to meet Koyré personally and both agreed that science is not—as generally believed—a collection of empirical data inducing theories, but an intellectual endeavor subject to empirical testing. Joske's suggestion successfully combined a major intellectual question with my personal Italian background. Unlike many academics, Joske has the gift of smoothing the way for his students.

I followed Joske's advice. Quite miraculously, I soon got a scholarship from the Italian Foreign Ministry to work directly on the sources in Florence and Pisa. I admit that my disorder has always been balanced by a considerable amount of luck, and I am persuaded that in addition to Joske, there is an angel somewhere that is very busy keeping an eye on me. Since this angel is going to reappear below, to avoid confusion let me call him Martin (I was born on the 11th of November, St. Martin's Day). Back in Italy, I rented a small apartment in the former stables of an old farmhouse on the outskirts of Pisa. The grant was generous enough to cover rent, travel expenses, and meals at the student's canteen of the University of Pisa. Most important, I was left alone, a hermit in the middle of nowhere, and did not have to account to anybody: excellent conditions for working on myself undisturbed.

A feature of my disorder is that I could not, and still cannot, think of or do more than one thing at once (I still have difficulties in using a cellphone). A basic challenge was, then, to develop an appropriate strategy. I applied the following simple yet effective procedure: whenever confronted with a task that I found complex, I broke it down and concentrated on each part separately, disregarding the other parts. For example, whenever I had to tidy objects around me, I got confused, tended to mess things up, and could even panic. So to order my desk, I imagined a straight line running from one side of the desk to the other, and removed the objects touched by the line one by one. To order a larger space such as a room, I used a spiral covering its whole area. All this developed my analytical ability. The price to pay was my time. Such a piecemeal approach, I would later learn, is encouraged by critical rationalism, particularly as far as social long-term planning is concerned, so as to prevent fatal errors (Popper 1966, 1:157–159). Long-term planning is yet another arduous challenge for an ADD victim.

A second challenge was to overcome my inertia. I did so by granting myself points for each hour I managed to work, and a reward if I reached the desired number of points. It required self-discipline, but self-discipline is one of the few qualities I have. The main difficulty, however, was reading and writing. At that time I was not able to formulate so much as a letter. To write my thesis, I thus attempted a shortcut: I collected some related articles and copied relevant paragraphs. At first, I experienced feelings of guilt: I was exploiting the work of others and not producing anything original of my own. Yet after I had copied and assembled several pieces, I noticed that the whole scholarly debate was beside the point, since Koyré's critics were talking on a different level than his own. Joske happened to be visiting Italy at that time, and enthusiastically read what I had written. He reformulated my conclusion in philosophical terms: while Koyré was speaking on a methodological level, i.e. considering the role of Galileo's Experiments, his opponents were merely stating that Galileo was performing experiments, without examining their role. Joske started adding written remarks between my poorly written sentences or paragraphs. I later extended the discussion and added footnotes, and the thesis turned to be a contribution to Galilean studies. After having defended it successfully at the university and obtained an excellent mark, Joske suggested that I extend the thesis into a survey and publish it. I was still not ready for such a project, but I sent the text to Clifford Truesdell, who published it in his prestigious *Archive for History of Exact Sciences* (Segre 1980).[2] The article received an unexpected response (for me, its success was somewhat frightening); it was reprinted and translated, and is still being quoted over 35 years later. From all this I learnt that copying is not necessarily a bad thing if done properly.[3] In my case, it was a breakthrough in learning to write.

Today, with a huge database to hand on the web and the option of copying and pasting, most students copy their papers. I do not discourage my students to do so; I only tell them that if they copy, they should mention the source which they have copied and possibly why they have done so, and, most important, state their opinion and try to discuss what they have copied. This is a first step toward scholarship, and a point on which Agassi agrees with me. Writing, as Joske emphasizes, is a heuristic as well as a therapeutic tool. Firstly, it is a substitute for memory, and is therefore particularly helpful to an ADD victim. Secondly, it is the mirror of one's thoughts. Thirdly, and most importantly for an ADD victim, it allows thoughts to be organized and developed. Once organized on paper, thoughts become better organized in one's mind.

The next step was to write a Ph.D. thesis, and I decided to remain within Galilean studies and deal with Galileo's followers. Martin must have once more given a little push, for I won a three-year postgraduate fellowship at Italy's most prestigious and exclusive university college, the *Scuola Normale* in Pisa. The setting was ideal, but writing remained the most difficult challenge. I still had trouble in formulating texts, both in content and form; I had to write and rewrite sentences to reach a fairly satisfactory formulation. Today this can be overcome by using a computer, but at that time PCs had not yet been invented and the only tools I had at my disposal were a pen or pencil with eraser, paper, an electric typewriter, and a thesaurus. Since there is a limit to the corrections one can make on paper, even with a pencil, I had to rewrite the same page many times. It came to be a wearing enterprise, and I got increasingly impatient. I did not wait for the end of the fellowship, but submitted my work, and got my Ph.D.

[2] Clifford Ambrose Truesdell III (1919–2000), professor of rational mechanics at Johns Hopkins University, was the founder and chief editor of the *Archive for History of Exact Science*, as well as of the *Archive for Rational Mechanics and Analysis*.

[3] On that occasion, however, I had not done the job properly. In my early draft, I had copied most of a paragraph from MacLachlan's nice 1973 article, but in my final article I inadvertently omitted to credit the author. I noticed with distress my oversight only after my article had been published, and all I could do was to write to Professor MacLachlan, apologize, and acknowledge my mistake.

I was by then more or less ready to exit the bell jar and embark on an academic career. I soon realized how right Joske's warning had been. Academia is a closed society; to survive in it and eventually make progress is a matter of politics, and politics is not always clean. To become an "in" guy would have been hopeless for me; there was a danger of falling back into intellectual dishonesty and throwing away the few results achieved with much sweat.

Martin was to help once again. Some personal circumstances brought me to Munich, with its celebrated Deutsches Museum—one of the largest science museums in the world, and a prominent center of research and teaching in the history of sciences. The Institute for the History of Sciences at the University of Munich had its seat in the building of the Museum and enjoyed its facilities, above all its rich and well-organized library. It was directed by Professor Menso Folkerts, a leading historian of science and a fine, reserved man, who had built a successful career despite not being the "in" type. The Institute was advertising a position as *Assistent*; I applied and got it, with the privilege of working there for over a decade and making intellectual and personal progress. I had, of course, difficulties in learning German, but was helped by my future wife, Ursula, whom I had just met. Unlike me, she is quick and an excellent organizer; she has always had patience with me and, like Joske, never judged and always helped. Munich's cold and dry weather (detested by many) also helped, since my brain functions much better in a cold environment (I leave the explanation of this empirical data to physiologists). By then, I thus had the ideal conditions to progress as an academic.

I have technology to thank initially for my breakthrough: the PC had just appeared on the market. I had at my disposal a Xerox word processor with the "Wordstar" program. A word processor allows a text to be improved with ease, in the very spirit of critical rationalism that recommends improving what is to hand instead of endeavoring a (utopian) perfection. I could at last reformulate my texts with startling ease; this, in consequence, improved my thinking as Joske had highlighted. I could return to Galilean studies. The library of the Deutsches Museum and the outstanding Bayerische Staatsbibliothek were, and remain, goldmines for researchers.

While my writing was improving, I still had difficulties in reading; my reading was slow, and after a few lines I became restless and had to stop. Joske strongly recommends stopping reading as soon as one feels tension, and discourages reading word by word unless this is enjoyable. I had to teach myself to read or, at least, to develop strategies to get to know the content of a book without having to read it. Joske suggested at first the following quick reading technique, using a timer set at one minute or less: as soon as it rings, jump to the next page. I was, nevertheless, too restless to follow the advice. Yet the relief did come from Popper's (and Agassi's) general approach to research: always begin by asking a question. I formulated questions pertinent to my research for each text I found. The index and the table of contents of a book were helpful, though browsing through a text without reading it often sufficed. If I found an answer, I quoted it and discussed it. Otherwise, I asked myself what question (if any) was being posed by the author. What was his proposition in writing the book/article? What is the thesis of the book/article (the thesis is normally stated in the introduction or even in the subtitle of a book, or in the first sentences of the article)? Is the text sufficiently clear, and worth reading and citing?

I later refined the technique, and I now teach my students—to their amazement—how to know the content of a text, and even more related information, without having to bother to read it. Ask, for example, who is the author (easy to find out, thanks to the internet). What questions does he pose? To what current of thought does he belong? Are the expressions he uses typical of a particular scientific or intellectual trend? Whom does he quote? Is the literature quoted pertinent, or is it quoted ritually, to conform to the paradigm and be "in"? Is the work intended to contribute to science or to further the author's career?—and many other questions. Once a farmer happened see the thousands of books in my private library and asked, startled, "Have you read them all?" "No," I answered, "but I can tell what each one has to say."

Things were gradually beginning to fall into place, both intellectually and personally. To be sure, I was still messing things up. Once I even managed to delete everything I had written on my computer with a single keystroke. Fortunately, I had printed out the material, and while retyping it I spotted many flaws and improved the text. What was needed was method, patience, perseverance, a little humor (taking Woody Allen as an example), and a little courage. My courage has been acknowledged even by war veterans, but I wonder if it is actual courage, or mere thoughtlessness.

Word processing technology improved steadily, and contributed to my work and development. A special program for scholars, "Nota Bene," enabled me to apply some writing procedures learned from Joske more effectively: the end of every paragraph, section, or chapter should relate to its beginning, and the beginning of every paragraph should be related to the previous paragraph. I was able to move sections of text around with ease and make sure that discussion ran logically, smoothly, and clearly. The confusion in my head was diminishing (indeed, in French the term for computer is "*ordinateur*"). I was, nonetheless, still unable to correct directly on the machine. I used to print, correct on paper, insert the corrections into the digital document, print again, and so forth, consuming prodigious amounts of both paper and time.

Writing, as Joske emphasized to me, was, and remains, fundamental to improving my ability to concentrate and coordinate my work. I had slips of papers and pencils scattered everywhere, so that I could promptly record vanishing thoughts (a classic ADD effect). Interestingly enough, many ideas came to me while shaving, and a physician specialized in acupuncture explained to me that this happens when the meridians in the cheeks are stimulated. I also compiled a series of lists to overcome my forgetfulness; I had one list in my wallet to remind me what to take when going out, and different lists for different types of journeys, whether domestic or abroad. I compiled a to-do list which I rearranged every morning, putting the urgent and important things at the beginning. I left large margins of time to enable me to correct my frequent mistakes in planning and in performing. I had to force myself to follow all these lists and disregard my natural tendency to trust my (poor) memory.

And even in checking the lists, I could carelessly overlook some points. These are all effects of ADD, and rigor spared me from blunders. I had to force myself to place important objects, such as keys, wallet, gloves, or umbrella, in the same specific location every time. My survival strategies turned me into a perfectionist; perfectionism is an undesirable feature, but this was the price I had to pay. Munich's excellent public services facilitated the organization of my life. I became so punctual that I could drive even my German friends nuts.

My research led me to discover some unpublished documents with enlightening information concerning Galileo's life and work. I wrote a few articles and submitted them to scholarly journals. I learned not to send a paper as soon I had finished writing it. I let it lie for at least a few days; since the process of thinking is slow, some ideas may come with a delay. This strategy was successful and helped to get my articles accepted for publication. Occasionally I bumped into the referees' network, which protects the closed academic society from "intruders" by rejecting what does not conform its ideas or trends—as often experienced by Joske himself (Agassi 1990 draws the lines of how a reviewer should correctly fulfill his task). Following Joske's advice, I turned timidly to the highest authority in the field, Professor I. Bernard Cohen, who broke down the resistance of the network and even got me a contract for my first book on Galileo's followers (Segre 1991). He was gracious enough to write the introduction to the book and my gratitude to him is more than words can say.

The next step was to write a *Habilitation*. The *Habilitation* is a second doctoral dissertation requested in Germany and some other European countries as a condition for becoming a professor. There is a debate over whether to abolish it, since some consider it a superfluous replica of a Ph.D. Having gone through both procedures, I feel it is still a welcome step in the academic career. While Ph.D. candidates are still students, even if at an advanced stage, *Habilitation* candidates are regarded as the top experts in the subject they are researching, and esteemed accordingly. In my case, at least, the *Habilitation* was a unique, stimulating experience.

Folkerts, my mentor, rightly recommended that I work on a different subject than the ones I had been engaged in till then. Since our Institute was part of the Faculty of Mathematics, I ventured into the history of that field. Joske helped once more to combine endeavor with my personal background, and suggested I study the work of Giuseppe Peano (1858–1932), an Italian mathematician known for the axioms in arithmetic that carry his name, who made a series of contribution to mathematics and linguistics. Peano was born in Cuneo, a neat Piedmontese city at the foot of the Maritime Alps. I had lived in Cuneo as a child, and my family's estate is in Cuneo Province. The inhabitants of the province, the Cuneesi, are considered the "dummies" of Italy with all the imaginable jokes that status entails. It has happened to me more than once, when driving outside the province, that other drivers look at my number plate with the Cuneo initials and laugh in pity. Yet in their legendary "foolishness," the Cuneesi have managed to turn their province into one of the richest regions in Europe, with flourishing industrial, agricultural and tourism sectors and a high quality of life. As an example, what was a confectioner's shop owned by the Ferrero family in the town of Alba in the last century, only fifteen kilometers from my home, has now become a multinational industry (inventing products including Nutella); Michele Ferrero, the late head of the family, was the richest man in Italy (and 20th in the world). Being a "dummy" has its advantages, as Joske repeatedly emphasizes. I felt a personal solidarity with Peano, who, too, was often considered weird.

My *Habilitation* was the most exciting research project in my career. I laid its foundations with Joske's help, during a short visit I made to Toronto (Joske was then teaching at York University). In Munich, Folkerts was always available to advise and help, and the city's libraries were an invaluable source of primary and secondary material. During writing, many tricky questions emerged, and I bombarded Joske with queries (e-mail had just been introduced). He answered patiently, though often briefly. I copied his answers, tried to clarify them or asked for further clarification, and the work expanded and became more and more interesting.

While writing my *Habilitationsschrift*, I realized more and more both the importance of history as a dimension of science, and the importance of critical rationalism as a heuristic and therapeutic tool. Above all, I realized the importance of Popper's warning not to take anything in science for granted. Take, for example, Peano's definition of integral. As undergraduate mathematics students, we were given several definitions of the concept, and I could never understand why: I thought one definition, the simplest, should suffice. Nobody told us that none of the definitions was entirely satisfactory; definitions and theorems were presented to us as ascertained facts. My *Habilitation* research made me realize how dogmatic science and its teaching can be. Mathematics is far from being an exact science (in contradiction, by the way, to Peano's own belief): its foundations have been debated since the pre-Socratic days of Zeno and Pythagoras. This is why Peano's program seeking to give a final, rigorous formulation to mathematics failed. The foundations of mathematics are such a delicate topic that they have virtually been turned into a kind of taboo, and mathematicians avoid dealing with them.

Having completed my research, I was asked to suggest names of experts to evaluate it. Agassi was one of my suggestions. One other member of the evaluating commission was the late Friedrich Ludwig Bauer, the pioneer of German computer science. In an occasional meeting in the corridor of the Deutsches Museum, he granted me one of the highest compliments I had ever received during my career, saying that he had never seen a mathematical work so devoid of mistakes as mine. It must have been, for once, a positive result of my perfectionism. *Pace* Professor Bauer, whose memory has my deep respect and gratitude, I did later find a mistake (perfectionism again), and, of course, I do not exclude the possibility of there being more.

Discussing the thesis in front of the encouraging staff and students of the Faculty of Mathematics turned out to be an easy and pleasurable stroll. I held the requested *Probevorlesung* (test lecture), and the only complaint was that I was speaking too slowly… The thesis was soon published as a book-length article in Truesdell's *Archive* (Segre 1994), and Truesdell sent me a warm letter of congratulation.

After my *Habilitation*, I won several fellowships and visiting professorships. Joske suggested, among other, that I contact Dario Antiseri, a leading Italian follower of Popper, at the LUISS University in Rome (Libera Università Internazionale degli Studi Sociali "Guido Carli"). Dario soon offered me a visiting professorship, and in Rome I had the pleasure of meeting and collaborating with a lively group of Italian Popperian followers celebrating the demise of communism within and outside the Italian university establishment. Yet I still did not have a tenured position. In Germany academics are rightly requested to move on, and do not get a tenured position at the university where they began their academic career. The University of Munich generously prolonged my contracts as long as it could, but could not grant me a life position. I applied all over the place, but in vain; being an "out" guy did not help (at least, so I thought), but Joske repeatedly reassured me that one does not necessarily need to apply to get an academic position. He pointed out that criticism had been viewed as a mark of respect as early as Plato, and that some academics succeed without playing academic politics; this is hard but possible and even has advantages. I admit to not having taken him seriously, but he was right. No closed society is perfectly closed: a new Faculty of Social Sciences was founded at the Gabriele D'Annunzio University in Chieti, and I received a Chair on a silver platter. It is thanks to the strategies Joske taught me that I was able to make a successful academic career.

I have been teaching in Chieti for fifteen years, trying to help students, applying Joske's teaching, and imparting some of the strategies that helped me neutralize my disorder (Segre 2009). I teach them to study without suffering; I begin my courses by saying that in my classes it is strictly forbidden to suffer, and if students feel uneasy they should not hesitate to come and tell me. Following Joske's example, and overcoming the reticence of the secretaries in our department, I give students my home telephone number and tell them that they should feel free to call me at any time; if one calls me say, at 2 AM with an interesting question, I may even consider granting him or her a higher mark. Regrettably, or perhaps fortunately, nobody has ever taken advantage of my offer. Students nonetheless stream to my lessons, and from their thankfulness and the marks they deliver to me, I suppose I help them.

As to my disorder, I had managed to neutralize the beast, but still had not killed it. Occasionally it sent reminders telling me that it was still there, albeit harmless. Recently I talked about it to Joske, who suggested that I talk with Natti Laor, his former student and a psychiatry professor. The latter directed me to Lilach Rachamim, a clinical psychologist working in the Tel Aviv area, and a fine lady. We skyped a couple of times and she soon spotted, "You suffer from ADD." I had never heard the term before. I investigated in Wikipedia and other web sources—and saw myself in the mirror. All the symptoms were there; I finally understood what had been blocking me since childhood. The beast was no longer a ghost: it stood in front of me in flesh and blood, tamed and harmless.

Lilach suggested that I come to Israel to be tested in a specialized center, so I did. When confronted with a computer testing me, I did not have the opportunity to create a strategy (thought I am certain this can be done). The verdict was unmistakable: my ability to "respond correctly and remain focused," as well as my ability to "respond quickly and accurately," were outside the norm in an "extremely severe" way. The examiners asked me how I had managed to survive till then. My answer was simple: thanks to strategies, and thanks to critical rational thinking. When I was then given 20 mg of the drug Ritalin (the scientific name is methylphenidate and it is a stimulant, the same, I am told, that athletes illegally take to enhance their performance), my performance rose to normality.

Doctors tell me that taking Ritalin for one year would boost my response and memory to normal levels. But I do not actually need it. Thanks to the strategies I have developed, I have been able to live a normal life and obtain everything I wanted and more, and people around me consider me a balanced and stable person. Moreover, just as a blind person may sometimes develop the ability to see more clearly than sighted people, my strategies give me the advantage that I can put them at the service of others. Above all, being able to identify the beast and understand all what has been going on in the past fifty years has been the greatest final therapeutic leap. I was reborn. Ursula says I have totally changed, and she is certain that I will never be depressed again; in fact, at times she even complains she cannot keep up with me.

What would have happened had my disorder been diagnosed fifty years ago (doctors tell me that it was already possible)? I can only recall "The Verger"—a beautiful, profound tale by Somerset Maugham (I did not read it, but I have seen the movie on TV). It is about a verger who had been working faithfully for many years in the service of St. Peter's Church at Neville Square, who is fired after the new vicar finds out that he cannot read. Distressed, he longs to smoke a cigarette, but cannot find a tobacco store. He decides to open a tobacco store at that very place, and then goes on to open tobacco stores wherever he sees a need. He becomes a millionaire, and when asked by his astonished banker where he would have been if he could read, he answers promptly, "A verger at St. Peter's Church."

Our Sages teach us that we should give thanks for ill in the same way we give thanks for good. I therefore thank God for all the good, and for all the ill. And I thank Joske, for all the good he has done.

Acknowledgements I am indebted to Alex Arbel, Nimrod Bar-Am, Natti Laor, and my wife Ursula, for reading early drafts of this article and suggesting improvements, and to Alison Moffat for improving my English.

References

Agassi, Joseph. 1963. *Towards an historiography of science*. The Hague: Mouton. Reprinted 1967, Middletown, Conn.: Wesleyan University Press.

———. 1990. Peer review: A personal report. *Methodology and Science* 2: 171–180.

———. 2014. *The hazard called education: Essays, reviews, and dialogues on education from forty-five years*, ed. R. Swartz and S. Richmond. Rotterdam: Sense.

Drake, Stillman. 1973. Galileo's experimental confirmation of horizontal inertia: Unpublished manuscripts (Galileo Gleanings XXII). *Isis* 64: 290–305.

Koyré, Alexandre. 1992 reprint. *Metaphysics and measurement*. Yverdon: Gordon and Breach.

Kuhn, Thomas S. 1966. *The structure of scientific revolutions*. 3rd ed. Chicago: University of Chicago Press.
———. 1999. Logic of discovery of psychology of research? In *Criticism and the growth of knowledge*, ed. I. Lakatos and A. Musgrave, 1–23. Cambridge: Cambridge University Press.
MacLachlan, James. 1973. A test of an 'Imaginary' experiment of Galileo's. *Isis* 64: 374–379.
Popper, Karl R. 1966. *The open society and its enemies*. 2 vols, 5th ed. London: Routledge & Kegan Paul.
Segre, Michael. 1980. The role of experiment in Galileo's physics. *Archive for History of Exact Sciences* 23: 227–252.
———. 1991. *In the wake of Galileo*. New Brunswick: Rutgers University Press.
———. 1994. Peano's axioms in their historical context. *Archive for History of Exact Sciences* 48: 201–342.
———. 2009. Applying Popperian didactics. *In Rethinking Popper*, ed. Z. Parusniková and R.S. Cohen, 389–395. Springer.
———. 2015. *Higher education and the growth of knowledge: A historical outline of aims and tensions*. New York: Routledge.
Settle, Thomas. 1961. An experiment in the history of science. *Science* 133: 19–23.
Snow, Charles Percy. 2001. *The two cultures*. London: Cambridge University Press.

Chapter 41
Dungeons and Dragons: A Lesson in Etiquette from Professor Agassi

Joan Bailey

On the map it's marked: "Here Be Dragons", aka The Academy and more particularly, Graduate School. To my knowledge, even today, there exists no handbook nor traveler's guide for the unwary but eager who enters this region. If one is very lucky, or perhaps miraculously blessed, one may be accompanied by a guide, a pioneer or griot whose experience, wisdom and sagacious understanding, born of successful survival, appears to generously share and educate the novice in how to survive. It might even be hoped that in this education the novice will learn how not only to survive himself but will educate others and in the process one day to transform the region to a more open place. But dragons, I fear, are only too human, and intelligence, justice and wisdom will continue to be necessary gifts for the survivors.

The tradition of graduate student thesis "oral defense" originated as a test of authorial authenticity but it has morphed into something adversarial, subject to the whims of academic politics. The path to my own oral defense was long and convoluted, including changes in thesis advisor and topic. When the day came, I felt satisfied with the work and ready to defend it to the committee. The context, alas, was permeated by departmental power politics, Darwinian struggles for tenure and behaviors appropriate to the full range of mental health. In any case, the complexity of ideologies, ideologues, personalities and power politics was, I have discovered over the years, sadly not unique. What was unique was what followed.

The defense began with a question so hostile as to be inadmissible: "What makes you think anything in your thesis comprises an original contribution to human knowledge?" This should have been met by an objection from my thesis advisor, but, alas, there was silence. It went downhill from there and ended with instructions for me to satisfy the "suggestions" and "recommendations" of two members of the committee. Both were early career seekers of tenure. One was a bully sufficiently arrogant to have put into writing demeaning and dismissive comments, including

J. Bailey (✉)
The College of New Rochelle, New Rochelle, NY, USA
e-mail: jebailey007@gmail.com

N. Bar-Am, S. Gattei (eds.), *Encouraging Openness*, Boston Studies in the Philosophy and History of Science 325, DOI 10.1007/978-3-319-57669-5_41

name-calling. (It was he who asked the first question.) The other was a reasonable and gifted scholar who had taken my work seriously enough to provide comments for solid improvement. These two most assuredly agreed on absolutely nothing. In most such cases this process would drag on indefinitely, resulting in the doctoral candidate "finding a career outside of the Academy". At that time only 11% of entering graduate students completed doctoral degrees.

I had known Professor Agassi as a dear friend and mentor to my then husband and he had patiently and successfully pushed me through many writing blocks as I had written my thesis. He was not at my university but generously shared his wisdom and insights with several of us who were not his "official" students but were the fortunate beneficiaries of his teaching. And his wisdom had kept me from feeling too precious about my research topic-I could write the book after finishing my degree. Yet I was now nevertheless beaten. I went to him defeated and distressed.

We had long conversations exploring the details of the process I had experienced. In the course of our conversation he asked me if I had agreed to sell my soul or give up my citizenship when I entered the ivied halls of my graduate school. I was taken aback by this question. I certainly did not remember doing so. Did I consider the process fair and just? Not really I said. And the dismissive comments were to be taken seriously? I was appalled but believed I must swallow my good sense and attempt to respond. What could I, a mere student, do? His inimitable response was "Ahh, there is the problem. You sound as though you have sold your soul and your citizenship".

Like most students, and venturers in the dragon's woods, I was fearful. The dragon in my own head was indeed the "baddest" of the lot. I, a mere student, should not, could not take on these bigger more important and powerful dragons-not even for my career life. Looking back, I wonder that all I had learned in the sixties about standing up for one's rights did not even enter my thoughts as relevant until he challenged me. That's of course what allows bullies to prevail and oppression to persist: the oppressed internalize their place as without rights. Hey, I had read Orwell, watched Rosa Parks et al. Didn't I learn anything?? Ever the educator, Professor Agassi challenged me to think of myself as a person once more and then to begin to think about my situation and a proper solution. After several days of serious reflection and extensive strategic thinking the light dawned. What would constitute a solution? Was there anywhere a court of appeal? With his prodding and Socratic questioning I (or was it he?) landed on the answer:

> Aha! A bigger dragon! But not to complain he cautioned. Simply inquire what process might exist to register a complaint. Ever the brilliant strategist! And so I sent a letter to the Chair of the Department asking if there were a process for a graduate student to appeal the process of the thesis defense. No answer from the Chair for several weeks. I then moved up the hierarchy to the Dean of the Graduate School and asked the same question of him. Again no direct answer but sometime later a letter from the Department Chair instructing me to respond to the reasonable suggestions and comments of the scholarly reader identified in the letter and to resubmit my edited dissertation. I did so. Sometime after submitting my edited text, I received a letter from the Director of Graduate Studies that my dissertation had been accepted and no further defense would be necessary and a letter from the Dean with congratulations on receipt of the degree. I can now imagine, based on my many years

in academe, the politics behind that conclusion. There could be no doubt that during those weeks of silence a power struggle had been very costly to some of the faculty involved.

I have been very grateful for Joske's generous patience, wisdom and guidance. The lessons about etiquette, the academy, bullies and oppression have been ever present in my own way of being an academic teacher, advisor and later administrator. I like to think I have at least kept some of the dragons in check and have done what I could to help students and some faculty understand personal dignity and justice and to play by the rules, repealing those rules that support oppression. This seems to me to be Philosophy both pedagogical and ethical in Action!

Chapter 42
Drive Around the Block

Charles Sawyer

Photo by Charles Sawyer

C. Sawyer (✉)
Harvard Extension School, Cambridge, MA, USA
e-mail: sawyer@fas.harvard.edu

N. Bar-Am, S. Gattei (eds.), *Encouraging Openness*, Boston Studies in the Philosophy and History of Science 325, DOI 10.1007/978-3-319-57669-5_42

Spring, 1974, *The Pewter Pot*, a coffee house near Coolidge Corner in Brookline, Massachusetts. I am drinking coffee with Joske Agassi, my friend and mentor. I bring up the subject of the 1973 Yom Kippur war and the Egyptian attack across the Suez Canal. I repeat what I have read in the *London Times* about the attack:

> The IDF (Israeli Defense Force) had a plan to thwart an Egyptian crossing, a system that would flood the waters of the canal with a flammable fluid, like gasoline, ignite it and turn the Canal into a flaming moat. But when the Egyptians began their incursion, the Israeli defense malfunctioned and the crossing succeeded. Official Egyptian press reports said that Egyptian commandos had sabotaged the fluid delivery system the night before the attack.

Joske replies in a heartbeat:

> The commandos were Russians, not Egyptian. Egypt does not have a commando force capable of such an operation. The sabotage required espionage in advance. Somebody gave up the secret.

Here is Joske in full: quick, insightful, penetrating, provocative, schematic, challenging and supremely self-confident. This conversation turned out to be the ignition point of a long ride on a rocket sled. All told it would be eight years before the rocket sled – my obsession with the story – stopped for good.

How I came to be sitting there sipping coffee with Joske is a tale worth telling for this occasion.

In September, 1964, I began working on a PhD at Yale University in the Department of History of Science and Medicine. I felt as though I had arrived in the big leagues, but because my mother had died suddenly two weeks before classes began, I spent the entire year skating on the edge of serious depression. Somehow I survived both semesters and fulfilled the requirement for at least one High Honors grade on the way to taking qualifying exams.

My eventual dissertation would be in history of physics in the period from Isaac Newton forward. That meant my thesis advisor would be Derek J. de Solla Price, Avalon Professor of the History of Science and co-chairman of the department. Price made his reputation by applying statistical methods to the growth of scientific publications and discovering patterns of citation cross-references in the scholarly press. His popular work on the subject was *Little Science Big Science*. He held a PhD in Physics from University of London and a PhD in History of Science from Cambridge University.

So in September of 1965, to plan my path through the second year of my PhD studies I met with Professor Price in his rather ostentatious office in a converted mansion at 56 Hillhouse Avenue, New Haven, CT.

Price chain-smoked pipes and I waited while he made a great production of cleaning his pipe-bowl with a small knife, banging the pipe on an ashtray, making loud sucking sounds and producing great clouds of smoke. Finished with that, he turned to me and began the session with a surprising prologue in his thick Oxbridge accent:

> Well, now, Mr. Sawyer, you've completed your first year of graduate studies, which has given you a foundation. Now you are about to start your period of apprenticeship. My

> advice to you is that you look among the members of the department, find yourself a great man, then sweep a spot near his feet and beg, "Please, sir, may I kneel here?"

Of course the only member of the department who might fit that description was Derek J. de Solla Price himself. His message made clear what would be the terms of our future relations: *Master/Slave* it was to be. This relationship was common among PhD candidates and their thesis advisors, but his blunt way of putting it frightened me, notwithstanding the sheepish grin I gave in response.

In the year that followed, my differences with my professors became evident. I considered that history of science and philosophy of science were inextricably bound together. History is narrative and narrative is story and story without meaning makes no sense. Philosophy is meaning. Hence, the two disciplines needed each other. But my professors at Yale would have none of it. They thought philosophy corrupted history, especially when it came to the history of science. Science was dictated by nature and we learned it by avoiding philosophy at all costs. Ironically, the position that philosophy is harmful to history of science is, itself, based on a philosophy of science called inductivism.

As it happened there was a professor in the philosophy department whose work and intellectual reputation I admired from a distance: Professor Norwood Russell Hanson. Hanson had published a book, *Patterns of Discovery*, which pushed hard against the inductivism of my history of science professors. Price and the others hated him and considered him a charlatan, particularly because of the aura of fame and glamor that followed him everywhere. The more I heard about him the more I was determined to meet him.

Hanson was glamorous, all right. He announced his arrival at Yale by flying over the Yale Bowl during a football game and tossing out leaflets introducing himself as the "Flying Professor," flying, that is, upside-down in his Grumman Bearcat, the last generation of piston engine fighter planes of World War II.

As a pilot Hanson was the real McCoy, a Major (retired) in the U.S. Marine Corps who had flown combat missions in the Pacific. When a kamikaze bomber struck his aircraft carrier the USS Franklin in March of 1945, killing 724 and wounding 265, Hanson was in the air on a mission. He ditched in the Pacific and was rescued by another U.S. Navy vessel.

At the start of the spring term I waited outside the room in the Hall of Graduate Studies where Hanson was scheduled to start a weekly graduate seminar in philosophy of science. Down the hall he came, taking big strides in black Wellington boots. He gave me a big smile when I told him I was from the history of science department and wanted to sit in on his seminar.

"We philosophers of science and you historians of science need each other. I'd be delighted if you would join our seminar. I'm sure we have a lot to learn from you."

The class was everything I had dreamed of; when it was over Hanson invited me to join the seminar and offer a paper. That evening when I met my regular circle of friends for dinner I told them, "Today I met a great man."

If I'd had any sense at all, I would never have advertised my attendance at Hanson's seminar in philosophy of science to my history of science professors, but, of course, I had no sense at all when it came to academic politics.

Another thinker who had a big hand in shaping my views was Thomas Kuhn, Professor of History of Science at Princeton, whose *Structure of Scientific Revolutions*, published in 1962, influenced generations of the educated public. Kuhn, even more than Hanson, presented a huge challenge to inductivism. He adopted an obscure term from rhetoric, "*paradigm*", to refer to silent guides that govern what he called "normal science," i.e. the research programs scientists pursue, consistent with the prevailing theories. Theories, of course, are error-prone and some errors may threaten the entire paradigm. That, wrote Kuhn, is how scientific revolutions happen — when "paradigms shift," the phrase that found its way into the popular lexicon.

Back then I thought Kuhn's idea about paradigms was, itself, subject to testing, testing to be conducted by historians of science, who else? But this idea for a research program in history of science was not likely to win the favor of my professors. That didn't stop me from singing Kuhn's praises in our seminars, to my great cost as it turned out.

Price thought to bring me around to my senses. He put a slim volume in my hand, saying, "This is the antidote to Kuhn." It was *Towards an Historiography of Science* by Joseph Agassi. This, in hindsight, was the one good thing Derek Price did for me.

As I neared the end of my second year's study I prepared a proposal for my PhD qualifying exams, following the practice of the department. Because the interests of both graduate students and faculty were too diverse to fit a single battery of qualifying exams, each student proposed exam categories, which, upon approval, the faculty would use as a template for compiling the examination questions.

In two years at Yale I had read an extraordinary variety of science, ancient and modern. I had read Aristotle's *Physics* and his *Metaphysics*. I had read Galen's account of dissecting an elephant carcass. I had read William Harvey on the circulation of the blood, and Darwin's *The Origin of Species*, yet I didn't think I'd learned much about how scientific thinking evolved until I read Kuhn, Hanson and Agassi. I was determined to dedicate part of my PhD qualifiers to this troika. I would do this, I proposed, under the rubric of "Historiography of Science."

Struggling to find a formulation for this rubric that would be acceptable to my professors I went to the office of Professor Asger Aaboe, an historian of ancient mathematics – cuneiform tablets were his thing. I found him reclining on his couch with one arm slung over his eyes. He must have heard me coming because he asked me what my business was without uncovering his eyes.

"Professor Aaboe," I said, "I would like to know what is your definition of the term 'historiography of science?'"

Without the slightest movement, without uncovering his eyes, he answered "I'm sure I haven't the foggiest."

If I'd had any sense I would have understood that proposing the historiography of science as a major area on a PhD qualifying exam in the history of science, that trying to smuggle Joske, whom I had yet to meet in person, onto my qualifying exams, was suicidal. Only later did I learn that I was offering myself as cannon fodder in an intellectual war.

The answer to my proposal came in a letter from Professor Price, announcing the end of my studies at Yale. It didn't even mention my exam plan. It merely stated that my study in the department was being terminated, my fellowship for the next year was cancelled, and I would not be allowed to take courses in the fall.

I see now the demented genius of my clumsy maneuvering with my professors: I had found a way to flunk my qualifying exams without even taking them.

I hadn't even read the letter a second time before I had reached Russ Hanson on the phone. "Meet me at the college around 1:00," he told me.

Soon I was riding with him in his beat-up van, reading the letter to him. "Bastards," he cursed. "They have to understand that graduate students are not vassals, not serfs in their feudal kingdom."

Hanson laid out some options in quick order. "Do you want to study with me? I'll see that you're admitted to the Philosophy Department."

I thanked him for his generous offer but explained that lacking an undergraduate education in philosophy I would need two years' study to fill that gap, and then would have to take qualifiers in his department, followed by thesis proposals. The thought of starting over again was more than I could bear. [Lucky I didn't accept his offer. He died 10 months later when he crashed his Grumman Bearcat into a hillside in bad weather enroute to Ithaca, NY.]

"How about going to Princeton to study with Tom Kuhn?" was his next proposal.

"I think I've had it with the Ivy League," I said, thinking out loud. "Do you know a man named Joseph Agassi?"

Hanson's face lit up. "Know him? I got him his job at BU. I'll call him tonight and tell him about you, I'm sure he'll meet you."

A meeting was set: noon on a blistering June day in Park Square, Boston. How he'd recognize me I had no idea but at the given time I stood on the curb scanning every passing car. A beat-up American sedan roared up to the curb. The driver reached across, threw open the passenger door and shouted, "Agassi! Get in."

Before I could make any pleasantries he told me to hold the talk until he did an errand. A few blocks from Park Square, in the heart of the shopping district, he pulled up in front of Jordan Marsh, two wheels on the sidewalk, two in the street. Leaving the key in the ignition he reached into the back seat, grabbed a shopping bag, and hopped out, saying "If a cop comes, drive around the block." I watched him disappear into the store. No cop came. Agassi back, vroom, off the sidewalk. "Now," he said in a relaxed voice. Thus began a conversation that continues to this day.

Decades later Joske gave me a copy of his just-published book. I asked him to sign it for me. "I don't do that," he replied. "Please," I pleaded. He took the book, scribbled on the title page and handed it back to me. He had written, "Drive around the block. Love, Joske."

In quick succession that summer (1966) I started a full-time teaching job at New England College in New Hampshire, which got me a draft deferment; rented a farmhouse; began by degrees cohabiting with my girlfriend, Joan Bailey, whom I'd met in Hanson's seminar; and became a part-time student at Boston University, trying again for a PhD, this time in philosophy of science. My first course was Agassi's seminar in that subject.

Joske's manner of conducting a seminar bore little resemblance to what I had known at Yale. There was just one condition: every student should come ready to read a page of his or her writing that addressed a problem of his or her choosing. Few students took this seriously. Some responded by asking "What's a problem?" Joske explained that a problem is a question. "Any question?" Well, preferably a question having to do with philosophy. "How big a question? Big, like the question of free will?" Not necessarily, and it's only a page.

The struggles over this format went on for weeks. Joske was unyielding. He cut down topic-based screeds with the question, "What's the problem you're addressing?" I haunted Joske's office, plying him with questions. He told me that mathematicians and physicists wrote on the blackboards in their offices, sketching out ideas, erasing lines of inquiry that proved useless, and that philosophers should do the same, on their blackboards and in their notebooks. He said that writing about a philosophic problem did not mean recording a discovery or articulating a well-formed argument and that I could write a page, then rip it out and throw it away. Overall there was just one standard that always applied: clarity. He pointed me to *Conjectures and Refutations* by Sir Karl Popper. There I found ideas expressed clearly however tentatively; sometimes quickly shown to be wrong, but always clear.

All this was in contradiction to everything I had learned before. At Yale I was taught that every sentence in an historical piece had to be carefully scanned for facts or fact-based claims and that every such item needed incontrovertible evidence to substantiate it. Thus, writing became an agony in which reaching the period at the end of a sentence was an ordeal. In short, I had been schooled to make writer's block a main feature in my working psychology.

Joske opened up a whole new world of writing for me. It was the difference between writing with a chisel on stone tablets versus scribbling on the bag your Subway sandwich came in. You start with sketchy ideas, and move up from there. Eventually, if you read what you've written critically, and, even more important, others read your writing and criticize it, you will reach the point where it might be good enough to submit for publication or present at a conference.

Empowered with this new way of working, I felt liberated. It was just plain thrilling! With Joske's seminar and Abner Shimony's course on Alfred North Whitehead behind me I was ready to begin studying for the PhD qualifiers, which were offered every spring. First I took and passed the exam in Logic and Philosophy of Science, then began preparing to take the last three exams on History of Philosophy, Ethics, and Aesthetics the next spring semester. The department supplied a master reading list for each area and copies of past years' exams.

At Joske's urging I began studying the best surveys on each subject and each major philosopher, which I found in *The Dictionary of Philosophy*, *The Encyclopedia of Philosophy* and the *Encyclopedia Britannica* Eleventh Edition. When needed, I would go to original works. I posed questions based on the old exams, wrote sample answers on my typewriter, and discussed them with Joske. I was nervous about the exams. But, emboldened by my new writing skills, I saw them as a chance to redeem myself from my past failures.

I liked the look of my triple-spaced typed pages from the practice exams I wrote. They looked smart. I asked the BU department if I could type my exam answers. They said no one had ever made such a request, but they offered me the use of a faculty office where I plugged in my portable electric and banged out my exam answers, away from all the blue book scribblers. I submitted my typed answers hopeful that I had shown myself ready to move on to working on a PhD thesis with Joske as my advisor.

To my surprise the examiners unanimously failed me on all three exams. Joske was furious, furious with me and with the examiners. He went to bat for me, declaring at a department meeting that my qualifier exams were the best he had thus far seen. The department responded by appointing a second jury of examiners to review my exam papers and revisit the verdict of the first panel. The new jury affirmed the original verdict. The department chairman then did something unusual: he loaned me the complete examination papers of two students who had taken the same exam, one graded B, one graded A, in order that I understand what were the department standards and thus help me prepare for a re-do exam the following fall.

By a series of small miracles I still have the original carbon copies of my exam answers, plus a transcript of parts of each comparison exam set. These pages survived 40+ years' storage in a New Hampshire barn and came back to me when the barn changed hands. A look at the answers of one question will show not just the problems I faced as a PhD candidate but also the problems Joske faced as a faculty member at universities, and in turn, the obstacles Popper faced getting his work recognized.

Exam question: "Compare the basic views of Parmenides and Heraclitus".

The opening paragraph of the A-graded exam answer:

> Hegel characterized Heraclitus as an advocate of 'becoming' in contrast to those who argued for the reality of either 'one' or 'many'. But, although Heraclitus did make the statement that one is never able to step into the same river twice, it is incorrect to assume, as Hegel did, that the Heraclitean doctrine proposed a synthesis of the one and the many. Plato was closer to the truth, when he distinguished Heraclitus'doctrine from that of Empedocles by remarking that, to Empedocles, reality was one and many in turn (i.e. in cyclic order), while to Heracltus, reality was both one and many [*sic*] simultaneous.

The opening paragraph of my exam answer:

> Parmenides and Heraclitus offered radically different solutions to the fundamental problem of the Pre-Socratic philosophers: the problem of change and permanence. The problem of change and permanence can be stated as asking, Of change and permanence, can one be reduced to the other, and if so which to which? Parmenides and Heraclitus both agreed that one could be reduced to the other but they differed over which one was fundamental.

The A-student never explained the problem that preoccupied the Pre-Socratics. He launched his answer by citing Hegel, as if the key to understanding the Pre-Socratics was to know what Hegel thought of them. (Need I remind readers that Hegel is one of the chief villains in Popper's *The Open Society and Its Enemies*?

[Chapter 12, Volume 2]). By contrast my exam answers were in the tradition of *Conjectures and Refutations*, centered around problems. Joske's tutelage is evident in every paragraph. The A-student paper was a tour de force of erudition combined with regurgitated nonsense like the "TENSION of OPPOSITES constituting the ONE." By contrast, in my paper I condemned scholarship for scholarship's sake as "precursoritus."[1]

Today, reading the two essays I see that the A answer is well written and represents the standards of the department, following the accepted approach to discussing history of philosophy, which is to exercise strenuous scholarship accompanied by displays of erudition. My answer is Popperian in style – it is built on articulating philosophic problems. Looking back, it's clear now that this whole academic travail was less a clash between my examiners and me than it was between Joske and his colleagues and, in a larger sense, between the tradition of the academy and the Popperian approach to writing philosophy.

At Yale I had served myself up as cannon fodder; at BU I played a proxy in a skirmish (Joske vs. his colleagues) in an intellectual war (Popper vs. the establishment).

To prepare for the re-do exam for three of us rejects from the spring exam sessions, I spent the summer practicing writing in the style of the A-student and learning to turn scholarly pirouettes, but it was all for naught. The redo exam raised the bar: more and deeper esoterica were called for, optional questions were made mandatory. Desperate and panicky, I blew my stack in my blue books (I had jettisoned the typewriter), rewrote the exam questions and let fly, answering my own questions! Facing the likelihood of yet another failure I chose to fail my examiners. When I walked out of the exam room I knew that, save for a real miracle, I had turned my back on an academic career.

Nowadays, looking back on the years of study that seemed to lead nowhere, and recalling the heavy burden of those serial failures, I see one thing that explains it all: I was temperamentally and intellectually ill-suited for the life I was trying to build. Hell, I was expelled from kindergarten for trying to trip my teacher! The die had been cast at an early age.

The blessing, the saving grace of those years was, of course, to have studied with Joske and to have become his friend.

Although there would be no academic career, I did not leave teaching forever. Many years later in my late 40s I began teaching software engineering at Harvard Extension School where I also taught a course in the history of blues music at which a guest artist was B. B. King himself, along with his rhythm section. Here again Joske's deep influence showed itself in my approach to lecturing, which relies on intellectual narrative as the best way to secure an understanding of complex ideas. And I learned from Joske to be honest when asked questions to which I had no answer, not to hide my ignorance by flattering the questioner with "Excellent question!" but rather to say, simply, "I don't know."

---

[1]For the complete answers to this question visit *Replaying the Past*, [https://docs.google.com/document/d/1auzudklbB9KsHVi3GZN0SIcW3LWAk1WlCMv_c0j-Ne4/]

After 25 years of teaching at the Extension School, Harvard gave me a chair, a formidable wooden one, not a professorial one, with my name and years of service inscribed across the back.

Luckily, my friendship with Joske was not compromised by my failures at Boston University. The conversation that had begun after "drive around the block" was still going strong, focusing now on my new interest in photography. This new interest came easily and naturally. It offered creative self-expression and easy access to begin making art.

Photography was not one of Joske's interests but art, in general, certainly was. He mentored me as I strove to understand how an image that took 1/60th of a second to materialize on a strip of celluloid could become a work of art. He saw in my first efforts the beginning of an eye for light and form. He encouraged me to persist, so every second Saturday I brought my latest work to his home in Sudbury, MA, along with books of photos by Henri Cartier-Bresson, André Kertesz and other great photographers. We pored over them together, analyzing the compositions, the portrayals of human nature, and the play of light and texture. He led me to the work of Ernst Gombrich, especially *The Story of Art*, which I carried with me when I toured the great museums of Europe. It was a tip from Joske that led me to the Czech master photographer, Josef Sudek.[2]

In the summer of 1971, while visiting Joske and Judith in Israel I had an explosion of photographic activity. On this first-ever visit to that exotic land I saw photographic possibilities at every corner, in every cafe, from every hilltop. Later, back in the US we pored over the first prints I pulled from the swirling wash tubs of my darkroom. Joske saw great possibilities. "You've got a book here, *almost*," he told me. There was a commercial possibility, too: the 25th anniversary of Israeli independence day was approaching in May of 1972. Could I return to Israel and fill that "almost" to make a real book, with only a few months' time to put it all together? I was on the next plane back to Israel, carrying about 50 rolls of Tri-X 35 mm film, and a list of subjects to seek out for the book we would make from my photos and Joske's commentary. Thus started one of the most tumultuous periods of my life, the details of which must wait for another occasion. For this narrative I will report two results:

1. A mock-up of the book we proposed to finish, should we ever have been so lucky as to find a publisher who would rush into production a photo volume by an unknown photographer and an Israeli philosopher. We called it *Israel in Repose* because it captured the country more or less at rest. [Such a publisher we never found.]

[2] http://tinyurl.com/sudekroudice Exhibition catalogue, Roudnice Gallery of Art, Roudnice, Czech Republic, June, 2012.

2. *Israel in Repose*, an exhibition of photos mounted in *Modern Age – Discovery*, the gallery of a big photo-finishing business in Manhattan.

Joske encouraged me to publish my photographs in a collection of postcards, called *Monograph 18*, for which he wrote a short introduction. He supported my efforts during most of the 1970s to find a publisher for my proposed biography of the blues singer and guitarist, B. B. King, until, after 35 rejection letters from publishers, I secured a contract with MacMillan to complete the book. When my editor at MacMillan tried to rewrite my book, Joske supported my decision to withdraw the manuscript and resume searching for a publisher. After several more rejections

I was signed by Doubleday, which published *The Arrival of B. B. King* in 1980. Every page in the final version bears Joske's editorial influence.[3]

Let's return to the moment in 1974 at the Pewter Pot in Brookline when I told Joske about the defense the Israelis had reputedly installed (in secret) along the banks of the Suez Canal, to incinerate the Egyptians as they crossed the canal on Yom Kippur, 1973. That conversation in Brookline continued over several days as we fleshed out the logical components of a complex military and political operation. We considered how such a system could be constructed in secret, how difficult it might be to ignite a film of combustible fluid floating on water, how intelligence agencies from one country related to those of their allies and their enemies. Over several days we built up a version of events that had verisimilitude and tight-binding logic, the sort of thing Joske did brilliantly, a big hypothetical system built around a presumed factual core.

I began building a fictional plot and characters on top of the supposed facts about the defense. I sketched out the ideas in a paper called "The Moat – An Abstract for a Novel," which circulated among friends and publishers.

One publisher who saw the abstract was Nat Wartels, founder and president of Crown Publishers, a leading trade publisher. He offered me an option on the novel: $500. This was a heady start for an unpublished author! Crown was a powerhouse and Nat was a grand old man in publishing, a crusty character who presided over his empire from behind a desk piled with paper to the point of avalanche. If Nat saw promise in me, I thought, I must be headed for great things in publishing. He wasn't very specific as to what he wanted. I had free rein.

I camped out in Joske's house full-time. My writing tools were felt-tipped pens and oversized yellow legal pads, with stiff cardboard backing. I wrote on every-other line, leaving ample room for inserts and corrections. Joske read my jottings, usually daily. My blend of cursive and printed characters was festooned in places with Joske's legible printed corrections. He was that rare thing, a real-time editor. I rarely disagreed with his criticisms and additions. The story took on a life of its own. This being a time before computers, I passed the handwritten pages to a typist; the pile of typewritten pages grew thicker by the week.

[3] Paperback Dacapo Press, 1982, German translation in 1995, Japanese translation in 2015.

Photo by Charles Sawyer

The Moat: Book I by Charles Sawyer with Joseph Agassi worked out to around 34,000 words. It followed two characters, American journalist Charlie Finn and Israeli ex-pat Dave Avivi as they tried to crack open the story of the failure of the fiery defense "from the outside." I dispatched a xerox copy to Nat at Crown, and began work on Book II.

The answer from Nat Wartels at Crown came even before we finished Book II: *Thanks, but no thanks, not the kind of thing we were hoping for. Good luck.* I was not about to let that stop me.

Book II told the story from the point of view of Luba Avivi, an Israeli lieutenant in command of a bunker on the canal, while an analyst at the CIA watched the run-up to the war from a desk at CIA headquarters in McLean, Virginia. We finished writing it months later and sent it on to Crown. To no avail.

Joske and I then rehashed the design of the novel, sparing no self-criticism. It was clear that the story needed a single narrative and that my first effort suffered from the typical first novel shortcoming, i.e. inadequate character development. Truth be told, our characters, Charlie Finn and Dave Avivi, were thinly disguised versions of Charlie Sawyer and Joseph Agassi.

The new edition was a complete rewrite of the novel – roughly a year's dedicated work, all in close collaboration with Joske. The writing excited me and I was supercharged with fierce ambition – ambition to find my place in the world and to make a living doing it. I was otherwise unemployed throughout the effort, living on dwindling resources.

The new, integrated, single-narrative version of *The Moat*, now at 110,000 words, made the rounds with trade publishers, usually without an introduction or an agent. One senior doyen of publishing, Evan Welling Thomas II, wrote the kindest, most encouraging rejection letter I ever received, complete with an invitation to lunch. Over lunch in a fusty Manhattan gentlemen's club he told me that in prior times he would have optioned *The Moat* and begun a close editing of the text. But in the business climate of the day all he could do was tell me he thought I was on the way to being a writer of real substance and that he hoped that I would make *The Moat* realize its full potential. I vowed to rework the book and resubmit it one day.

The manuscript of *The Moat* (version 2.0) sits in a box in my basement, waiting for a day when I might again be charged with the romance of writing a novel about a battle whose outcome might have had a very different outcome, perilous to the state of Israel.

Eventually the story of *The Moat* found a new life after I found a new life with Bistra Lankova, a Bulgarian filmmaker whom I met and married in 1980. Joske played an important part in our decision to marry.

In July of 1980, Bistra was making final preparations to fly home to Sofia to resume her career in television production after 18 months in the U.S. studying at Columbia and Harvard. When we met (for the 2nd time) sparks flew and within a few days we began dreading her return to Bulgaria. Fate forestalled her departure: a toothache became an infection that invaded her jawbone. The treatment stretched into weeks and once the infection was cured, we faced the daunting decision either to part with little chance of a long-term relationship, or for Bistra to stay and we to marry.

Conveniently, Joske and Judith lived a 10-minute walk from my apartment in Somerville, Massachusetts. Bistra and I went to them for help. Joske advised us together and separately. We recognized that the Cold War situation allowed no middle ground: either she stayed and we got married, or else the possibilities of a life together would remain only in our imaginations. Later I learned that Bistra had consulted Joske and Judith in private on the question of my character. She asked them, "Is this man, Charlie, whom I've known for little more than a month the whole man or is there another, hidden, nefarious side of him?" The question is telling: she trusted Joske and Judith to give an honest answer. They assured her that *the Charlie you see is the Charlie you will get.* I am eternally grateful to them for vouching for me and urging us to seize the day. Joske and Judith were among the 10-12 people attending our wedding on October 31, 1980, in White River Junction, Vermont.[4]

Bistra had executive-produced around 15 feature films for Bulgarian National Television. Her education and experience made her the ideal partner to turn *The Moat, A Novel* into a script for *The Moat, A Motion Picture.* Through 6 complete drafts it occupied all of our attention and creative powers for a year.

[4] See *Nieman Reports* http://niemanreports.org/articles/bistra-lankova-nf-80-1944-1988/

We made a pilgrimage to Hollywood, plying every connection to the film business we could conjure. Ivan Passer, a well established director, told me *The Moat* was a compelling story, but Hollywood never makes movies about Israel. Never.

Together, Bistra and I wrote two more complete screenplays. One was a courtroom drama based on a murder in Vermont, and the other was a teleplay commissioned by WGBH, *How Blue Can You Get – A Teleplay*, a bio-pic based on the life of B. B. King. These two scripts won us a fellowship awarded by the Writers' Guild of America/East. But we were unable to make a living at our craft.

Recognizing the reality of our situation I began looking for a way to earn a living. In the 1980s I went back to school and earned a Master of Science in computer science and started a new career in software that continued for 33 years. I retired in 2014 at age 74.

On November 25, 1988, Bistra and I went out to shop for a carpet for my study, acting on a tip from her close friend, Cherie Hoyt. It was the day after Thanksgiving. On the way home we were driving along a stretch of divided highway on Route 16 in Revere, MA, when a car traveling the other way jumped the median strip, hit a small tree on the median and became airborne; it sailed over two lanes, struck a gasoline tanker 6 feet off the ground and crashed down on the passenger side of my car where Bistra was sitting. She died an hour later in Massachusetts General Hospital from a massive traumatic head injury. I suffered a subdural hematoma (cerebral hemorrhage) when my head hit the side window, putting me in the hospital for five days. The driver of the first car was dead at the scene with an alcohol blood level of .28, over 3 times the legal limit.

On the day after the accident the phone rang in my hospital room. It was Joske on the line from Toronto. I don't know how he learned of the catastrophe and Bistra's death. He consoled me and I told him I had no reason to live, that all meaning was gone, that I wished I had died, not Bistra. His reply stays with me even now. "You must go on," he told me. "You can't let this stop you. You simply must go on."

"Why?" I asked. "What for? Tell me one reason!"

"Because she would never forgive you if you didn't," was his answer.

This thought somehow pulled me back from despair and I am forever grateful to Joske for that. It's one among many reasons to be grateful to my teacher, mentor and friend, Joske Agassi.

Joske and Judith became my pillars of strength in my struggle to find a new normal after the calamity. Together with Cherie and other friends, they sustained me through mourning and its rituals, through painful recovery from post traumatic stress disorder (PTSD).

Cherie, especially, helped me through the difficult times like anniversaries and gravesite visits. My close friendship with her grew into love, and in time I faced my next life decision: whether-or-not to cement my bond with Cherie and start a family. Could I, just shy of my 50th birthday, still grieving and emotionally fragile, be a good partner and companion to her and a good father as well? Here it was Judith who helped me decide, when she said as I left her at the curb at Logan Airport, "It would do you a power of good, Charlie [to be a father]".

Judith's encouragement and Cherie's confidence that I would be a good father, and a belief we came to share that bringing a new life into the world would be not only a joy for us but a powerful answer to the tragedy of Bistra's death, gave me the courage I needed.

Cherie got pregnant about a month before I turned 50. Joske and Judith were present when we recited our marriage vows on August 17, 1991. Our cherished son, Samuel Murray Sawyer, came into this world January 11, 1992.

Judith and Joseph Agassi with Charlie and Sam Sawyer (Photo by Cherie Hoyt)

Some Images

Aaron, Tirzah, Judith and Joseph Agassi (Photo by Charlie Sawyer)

N. Bar-Am, S. Gattei (eds.), *Encouraging Openness*, Boston Studies in the Philosophy and History of Science 325, DOI 10.1007/978-3-319-57669-5

Tirzah and Joseph Agassi (Photo by Charlie Sawyer)

Agassi and Jarvie (Photo by Charlie Sawyer)

Agassi and Jarvie (Photo by Charlie Sawyer)

Agassi and Polanyi (Photo by Charlie Sawyer)

Joseph, Tirzah and Judith Agassi with Hennie and Karl Popper

About the Authors

Yakir Aharonov is Professor of Theoretical Physics at Chapman University, as well as Emeritus Professor at Tel-Aviv University. A distinguished professor in the Perimeter Institute. Among his many achievements are major contributions to nonlocal and topological effects in quantum mechanics, to quantum field theory and to the foundations of measurement theory and time in quantum mechanics. The recipient of Wolf Prize and many other international awards.

yakir@post.tau.ac.il

Joan Bailey is retired, having been faculty at New England College in Henniker, NH and administrator, Dean, Sr. Vice President for Academic Affairs and Vice President for Mission at The College of New Rochelle in New Rochelle, NY. Mother of twins, one a surgeon the other a PhD candidate in chemistry, and Joan is a grandmother of a wonderful grandson, Arlo.

jebailey007@gmail.com

Michael Banton is Professor Emeritus of Sociology at the University of Bristol since 1992. President, Royal Anthropological Institute, 1987–1989; Member, UN Committee on the Elimination of Racial Discrimination, 1986–2001 (Chairman 1996–1998). Most recent book: *What We Now Know About Race and Ethnicity* (Berghahn Books, 2015).

michaelbanton58@gmail.com

Nimrod Bar-Am heads the Rhetoric and Philosophy of Communication unit at the Communication Department of Sapir College, Israel. He published *Extensionalism: the Revolution in Logic* (Spinger 2008) and *In Search of a Simple Introduction to Communication* (Springer 2016), as well various peer reviewed papers, most notable of which are "Extensionalism in Context" and "A Review of Thomas S. Kuhn *The Structure of Scientific Revolutions* 50th Anniversary Edition". He was Agassi's

N. Bar-Am, S. Gattei (eds.), *Encouraging Openness*, Boston Studies in the Philosophy and History of Science 325, DOI 10.1007/978-3-319-57669-5

research assistant (1993–1998), and they co-authored "Meaning: from Parmenides to Wittgenstein" and "Popper and the Establishment".
ngbaram@gmail.com

Chemi Ben-Noon is Associate Professor at The Academic Center of Law and Science. Multi-disciplinary researcher in Israeli Law, Political Science, and the Humanities. President of the Arbitration Federation.
benoon@netvision.net.il

William Berkson was a student of Joseph Agassi at the University of Illinois, and did his PhD under Karl Popper at the LSE. He has published on philosophy of science and Jewish ethics, and is also a type designer. He is currently director of the Jewish Institute for Youth and Family.
berkson@mentsh.com

Lawrence A. Boland, FRSC, is a Professor Emeritus at Simon Fraser University. He received a PhD in mathematical economics from the University of Illinois in 1966. He began teaching economics at the University of Wisconsin and was a visiting instructor at Boston University.
boland@sfu.ca

Mario Bunge has earned his doctorate in physics in 1952 at the Universidad Nacional de La Plata, Argentina, and was professor of theoretical physics at his alma mater as well as at the Universidad de Buenos Aires, Temple University, and University of Delaware. He held a named chair at McGill University between 1966 and 2009. He has also been a visiting professor in Australia, Denmark, Germany, Italy, México, USA, Switzerland, and Uruguay. He earned 21 honorary doctorates and four honorary professorships, and is a member of five academies, among them the Royal Society of Canada, the Institut International de Philosophie, the Academia Argentina de Ciencias, and the Academia Argengtina de Ciencias Exactas, Físicas y Natureles. He is a Humboldt and a Guggengheim fellow, and a Prince of Asturias laureate. He authored more than 500 papers and 70 books in physics, philosophy of science, semantics, epistemology, ontology, ethics, political philosophy, and science policy; among them is the nine-volume *Treatise on Basic Philosophy* (1974–1989). He founded the Universidad Obrera Argentina (1938–1943), the journal *Minerva* (1944–1945), the Society for Exact Philosophy (1971-to date), and the Asociación Mexicana de Epistemología (1976–1979).
marucho.bunge@gmail.com

Rafe Champion took an Honors degree in Agricultural Science and Masters degrees in Sociology and the History and Philosophy of Science. Most of his professional life was spent in policy, research and planning in public health and welfare services, with a spell as a technical writer and editor with architects and engineers. He has published on many other topics in including the philosophy of Karl Popper. http://www.the-rathouse.com
rafe.champion@gmail.com

Michael Chiariello is Professor of Philosophy at St. Bonaventure University in New York, former dean of Clare College and director of their Franciscan Heritage Study Abroad Program in Perugia, Italy.
mchiarie@sbu.edu

Daniel B. Cohen has run an agricultural research, development and consulting company called "Maccabee Seed Company" for over thirty years. This focuses on field crops, and in particular oilseeds. He had a dialog over four decades with anthropologist Sidney Mintz on culture, food and agriculture.
daniel.b.cohen@me.com

Eliahu Cohen did his Ph.D. under Yakir Aharonov and Lev Vaidman at Tel-Aviv University. He is currently a postdoctoral research associate at the University of Bristol.
eliahu.cohen@bristol.ac.uk

Robert S. Cohen is Emeritus Professor of Physics and Philosophy at Boston University. He was chair of both Physics and Philosophy Departments as well as Dean of the College of Arts and Sciences.
robertscohen@hotmail.com

Ute Deichmann is Professor and founding director of the Jacques Loeb Centre for the History and Philosophy of the Life Sciences at Ben-Gurion University of the Negev; apl. Prof. University of Köln.
uted@post.bgu.ac.il

Christopher R. Donohue is Historian of the National Human Genome Research Institute at the National Institutes of Health. He is the editor of a forthcoming (December 2017) special issue on the history of the Human Genome Project and of genomic science for the *Journal of the History of Biology*. He is also associate editor of the *Ideology and Politics* Journal.
christopher.donohue@nih.gov

Fred Eidlin (1942–2016) was Professor of Political Science at the University of Guelph from 1973 to 2010. At the time of his death, he was affiliated with the Faculty of Humanitarian Studies at Charles University, Prague.

Avshalom C. Elitzur was a senior lecturer at Bar-Ilan University and a visiting professor at Joseph Fourier University. He is currently a founding member of Iyar, the Israeli Institute for Advanced Research. Elitzur has been conducting a multidisciplinary research, but is mostly known for the Elitzur-Vaidman bomb-tester, as well as other quantum effects and novel measurement techniques.
avshalom@iyar.org.il

Miriam Farhi Rodrig is independent scholar and business woman situated in Tel Aviv. She holds a Ph.D. in Philosophy from Tel Aviv University. She published a few articles, in particular on the critical examination of global problems. She actively participates in international conferences on philosophy, political science and economics. She is also a relentless marathon runner.

miriam.farhi@gmail.com

Maurice A. Finocchiaro is a graduate of M.I.T. and U.C. Berkeley; Distinguished Professor of Philosophy (Emeritus) at the University of Nevada, Las Vegas; recipient of awards from the National Science Foundation, Guggenheim Foundation, American Council of Learned Societies, National Endowment for the Humanities, and International Society for the Study of Argumentation; and author of many books and articles on logic and argumentation theory, the history and philosophy of science, Gramsci, and Galileo.

maurice.finocchiaro@unlv.edu

Lucien Foldes is Emeritus Professor of Economics at the London School of Economics, attached to the Systemic Risk Centre. He has spent most of his academic career at LSE, where he was for a time a colleague of Joseph Agassi. His early work was on business and industrial subjects and economic theory, while later years have been devoted to mathematical work on optimal saving and portfolio problems.

l.foldes@lse.ac.uk

Stefano Gattei graduated in Milan (2003) and earned his Ph.D. at the University of Bristol (2004). He has taught at various Italian universities and has held research positions at Columbia University (2009), Harvard University (2012), the Chemical Heritage Foundation, Philadelphia (2014–2016), and The Huntington Library (2016–2017). He is currently Eleanor Searle Visiting Professor of History of Science and Technology at Caltech. Besides a number of articles in both philosophy and history of science in peer-reviewed journals, he is the author of *Thomas S. Kuhn's 'Linguistic Turn' and the Legacy of Logical Positivism* (Ashgate, 2008), and *Karl Popper's Philosophy of Science: Rationality Without Foundations* (Routledge, 2009). He co-edited, with Joseph Agassi, the fourth volume of Paul K. Feyerabend's collected philosophical papers, *Physics and Philosophy* (Cambridge University Press, 2016). His latest work, *Early Biographiesof Galileo*, is forthcoming with Princeton University Press.

sgattei@caltech.edu

Malachi H. Hacohen (PhD, Columbia) is Bass Fellow and Associate Professor of History and Jewish Studies at Duke University, and Director of the Council for European Studies and of the Religions and Public Life Initiative at the Kenan Institute for Ethics. His research focuses on Jewish European history, and he has published on the Central European Jewish intelligentsia, Cold War liberalism, nation and empire in Austrian history, and cosmopolitanism and Jewish Identity. His *Karl Popper – The Formative Years: Politics and Philosophy in Interwar Vienna* (Cambridge University Press, 2000) won the Herbert Baxter Adams Prize of the AHA and the Victor Adler State Prize. He had Fellowships from the ACLS, Fulbright, Mellon, and Whiting Foundation. He was a fellow at the Center for Advanced Studies in the Behavioral Sciences in Stanford, the National Humanities Center, and the IFK in Vienna. He presently leads an international project on "Empire, Socialism and Jews". His *Jacob & Esau: Jewish European History Between Nation and Empire* is forthcoming with Cambridge University Press.

mhacohen@duke.edu

Sapir Handelman is the co-laureate of the 2010 Peter Becker Award for Peace and Conflict Research. He is an associate professor (senior lecturer) and the director of the Conflict Studies Program at Achva Academic College. Dr. Handelman leads the Minds of Peace initiative and was the first scholar to receive the Lentz fellowship in Peace and Conflict Research three times in a row. Dr. Handelman was a post-doctoral fellow and an associate at the psychology department of Harvard University.

sapir.handelman@gmail.com

Jagdish Hattiangadi is a friend and former colleague of Joseph Agassi in the Department of Philosophy at York University in Toronto, Ontario, Canada. He is currently engaged in preparing for publication the book length typescript that he has described in his article as "forthcoming".

jagdish@yorku.ca

Hilde Hein was a colleague of Joseph Agassi in the Department of Philosophy at Boston University, as well as a neighbor and friend. Her books include *The Exploratorium: The Museum as Laboratory* (1990), *The Museum in Transition: A Philosophical Perspective* (2000), and *Public Art: Thinking Museums Differently* (2006). She is a Resident Scholar at the Brandeis Women's Studies Research Center.

hilde@brandeis.edu

Ian C. Jarvie was a student of Karl Popper when he met Joseph Agassi. They have been colleagues and friends since that time. They share an interest in philosophy of the social sciences and both are ardent film fans. They have collaborated extensively on scholarly projects. Their joint book, *A Critical Rationalist Aesthetics*, was published by Rodopi in 2011.

jarvie@yorku.ca

John Kekes is the author of many books, the latest of which are *The Human Condition* (Oxford, 2010), *The Nature of Philosophical Problems* (Oxford, 2014), *How Should We Live?* (Chicago 2014), and *Human Predicaments* (Chicago, 2016).

jonkekes@nycap.rr.com

Nathaniel Laor MD, PhD (Phil), is Professor of Psychiatry and Chair of the Department of Medical Education, Sackler Faculty of Medicine; Professor at the Department of Philosophy, Lester and Sally Entin Faculty of the Humanities, Tel Aviv University; Clinical Professor, Yale Child Study Center; Director of Cohen-Harris Resilience Center and Chair of the Research and Clinical Committees, Association for Children at Risk (R.A.), Israel. He is the co-author, with Joseph Agassi, of *Diagnosis: Philosophical and Medical Perspectives*, Kluwer, 1990; and the co-editor, with Ian C. Jarvie, of *Critical Rationalism: Essays for Joseph Agassi*, Kluwer, 1995, in 2 vols.

nlaor@me.com

Richard G. Lipsey is Emeritus Professor of Economics at Simon Fraser University, Vancouver B.C., Canada. He is former holder of a chair at the London School of Economics, the new University of Essex England, and Queens University, Ontario Canada. Author of over 150 professional articles, a first year university textbook now in its 13th edition, and co-author of an award-winning book (the Schumpeter Prize) *Economic Transformation: General Propose Technologies and Long Term Economic Growth*.

rlipsey@sfu.ca

Michael R. Matthews is an honorary associate professor in the School of Education at the University of New South Wales. He has degrees in Geology, Psychology, Philosophy, History and Philosophy of Science, and Education. For 25 years he edited the journal *Science & Education*; he has authored and edited numerous books; and published many articles in the fields of philosophy of education and history and philosophy of science.

m.matthews@unsw.edu.au

Abraham Meidan holds a Ph.D. from Tel-Aviv University (1982). He is an independent scholar, published a few books in Philosophy, and since 1983 the CEO and owner of a software vendor (WizSoft) that develops, among other things, data-mining algorithms. He is the co-author, with Joseph Agassi, of *Philosophy from a Skeptical Perspective*, Cambridge University Press, 2008; and *Beg to Differ: The Logic of Disputes and Argumentation*, Springer, 2016.

Abraham@wizsoft.com

Amir Meital completed his Ph.D. in Ancient History at Haifa University. He is the head of "H'evra" program for political activism at Sapir Academic College, Israel. Dr. Meital is also the developer and academic advisor of innovative Bedouin academic program at Sapir College. As an Ancient Greece Scholar Dr. Meitals' main interests are Greek Historiography and Philosophy, especially mutual intellectual influence between scientific and historical thinking in the Archaic and Classical era. In his research Dr. Meital has established a close relation between the structure of the classic Greek Polis and the ways Greek Philosophers explained a variety of natural phenomena.

amirmeital@gmail.com

Margaret Ng B.A.; M.A. (HKU); Ph.D. (Bost); B.A. (Laws)(Cantab) is a barrister-at-law in private practice. She served as a member of the Legislative Council of Hong Kong 1995–1997, 1998–2012 and is also a noted commentator and writer.

margaretng@siroswald.com

Ilkka Niiniluoto finished his Ph. D. in the University of Helsinki in 1974 with a doctoral thesis on scientific theories and inductive logic. He has been Associate Professor of Mathematics in 1973–1977 and Professor of Theoretical Philosophy in 1977–2014. His main works in the philosophy of science include *Truthlikeness* (D. Reidel, 1987) and *Critical Scientific Realism* (Oxford University Press, 1999). He has written also on philosophical logic, epistemology, philosophy of technology, philosophy of culture, and history of philosophy.

ilkka.niiniluoto@helsinki.fi

Zuzana Parusniková is a researcher at the Institute of Philosophy, The Czech Academy of Sciences. Her main research interests are critical rationalism and the early modern philosophy, especially René Descartes and David Hume.

parusnikova@volny.cz

Nicholas Rescher is Distinguished University Professor of Philosophy at be University of Pittsburgh where he has also served as Chairman of the Philosophy Department and a Director of the Center for Philosophy of Science. He was awarded the Alexander von Humboldt Prize for Humanistic Scholarship in 1984, the Belgian Prix Mercier in 2005, and the Aquinas Medal of the American Catholic Philosophical Association in 2007, the Founder's Medal of the Metaphysical Society of America in 2016, and the Helmholtz Medal of the Germany Academy of Sciences (Berlin/Brandenburg) in 2016. In 2011 he received the premier cross of the Order of Merit (Bundesverdienstkreuz Erster Klasse) of the Federal Republic of Germany in recognition of contributions to philosophy and to German-American cooperation in this domain.

rescher@pitt.edu

Sheldon Richmond is an independent scholar, author of *Aesthetic Criteria: Gombrich and the Philosophies of Science of Popper and Polanyi,* and co-editor with Ronald Swartz of Joseph Agassi, *The Hazard Called Education: Essays, Reviews and Dialogues on Education from Forty-Five Years.* Recently retired from working in systems and information technology with the Canadian Federal Government, he publishes critical reviews and essays, and engages in socratic dialogue at international conferences.

askthephilosopher@gmail.com

Raphael Sassower is professor and occasional chair in the Department of Philosophy, University of Colorado, Colorado Springs. His main areas of interest are postmodern technoscience and the philosophy of the social science with special focus on political economy. Among his latest book are *Democratic Problem-Solving: Dialogues in Social Epistemology*, with Justin Cruickshank (2017), *Compromising the Ideals of Science* (2015), and *The Price of Public Intellectuals* (2014).
rsassowe@uccs.edu

Charles Sawyer is an author, photographer, blues musician and retired software engineer. He teaches software engineering at Harvard Extension School in Cambridge, Massachusetts. He lives in Boxford, Massachusetts, with his wife Cherie Hoyt. Their son, Sam, lives in California.
sawyer@fas.harvard.edu

Michael Segre is currently Professor of History of Science at the University of Chieti. His book, including an interview with Agassi, *Accademia e Società* (2004), was the basis for his more recent *Higher Education and the Growth of Knowledge* (2015).
segre@unich.it

Ronald Swartz received a Ph.D. from New York University (1971) from the Department of Philosophy and History of Education. He is currently a Professor Emeritus of Education at Oakland University, Rochester, Michigan. Swartz is the author of *From Socratesto Summerhill and Beyond: Towards a Philosophy of Education for Personal Responsibility*. Swartz also co-authored (with Perkinson, H., and Edgerton, S.) *Knowledge and Fallibilism: Essays on Improving Education*. Swartz co-edited (with Richmond, S.) Joseph Agassi's *The Hazard Called Education: Essays, Reviews, and Dialogues on Education from Forty-Five Years*.
swartz@oakland.edu

Kiichi Tachibana is a Professor of Ethics, Faculty of Education and Human Studies, Akita University, Akita, Japan. He published "Can thc Japanese Learn to Welcome Criticism Openly", in *Karl Popper: A Centenary Assessment*, Ashgate, 1006, Vol. I, ch. 13; "Is Taishō Demokurashī the Same as Taishō Democracy?", in *Japan's Multilayered Democracy*, Lexington Books, 2015, ch. 3. He has also published Japanese translations of Magee's*Popper* and Agassi's *The Continuing Revolution*.
tachiba@ed.akita-u.ac.jp

John R. Wettersten is retired Associate Professor of the Philosophy of Science at Universität Mannheim. Among his more than 100 publications are four books: *Learning from Error* (with William Berkson), 1984; *The Roots of Critical Rationalism*, 1992; *Whewell's Critics: Have they prevented him from doing good?*,

2005; *How Do Institutions Steer Events? An Inquiry into the Limits and Possibilities of Rational Thought and Action*, 2006. Recent articles include: "New Social Tasks for Cognitive Psychology; Or, New Cognitive Tasks for Social Psychology", *The American Journal of Psychology*, 127, 4, 2014, pp. 403–418; "Critical Rationalists Much too Narrow Contribution to *Der Positvismusstreit*: Alternative Political Theories which Conform to Fallibilist, Individualist Methodologies Were Needed then and Are Needed Today", *Journal for Classical Sociology*, 15, 2015, pp. 102–114; "The roles which Otto Selz and Karl Popper played in twentieth century psychology and philosophy of science", *Philosophy of the Social Sciences*, March 2016; "A Way Forward beyond Karl Popper's and Donald T. Campbell's Dead-end Evolutionary Epistemologies", *The American Journal of Psychology*, 2016.
wettersten@t-online.de

Ian S. Winchester is a graduate in physics from the University of Alberta and has a doctorate in philosophy (DPhil) from Oxford. He has taught physics for the University of Alberta, philosophy at Oxford, history and philosophy and higher education at the University of Toronto (OISE) as well as at universities in Sweden and China, and is presently a Professor and former dean at the Werklund School of Education at the University of Calgary. Since 1982 he has been the Editor of the journal *Interchange: A Quarterly Review of Education*, originally the official journal of the Ontario Institute for Studies in Education and since 1998 the Editor of the *Journal of Educational Thought*, the official journal of the Werklund School of Education. He has been delighted that his wife Gunilla and he have been friends of Joseph Agassi and his wife Judith since the early 1980s.
winchest@ucalgary.ca

Jan Woleński is Professor of Philosophy at the University of Information, Technology and Menagement, Rzeszow, Poland. He published 30 books and 700 papers in all branches of philosophy.
jan.wolenski@uj.edu.pl

Chen Yehezkely wrote his doctoral dissertation under the supervision of Joseph Agassi and Nathaniel Laor. He is the author of *Closed Education in the Open Society: Kibbutz Education as a Case Study* (2012).
chenyehe@gmail.com

Index

Note: The names of Joseph Agassi and Karl R. Popper are not indexed.

N. Bar-Am, S. Gattei (eds.), *Encouraging Openness*, Boston Studies in the Philosophy and History of Science 325, DOI 10.1007/978-3-319-57669-5

CPSIA information can be obtained
at www.ICGtesting.com
Printed in the USA
BVOW06*0217110717
489035BV00003B/61/P

9 783319 576688